ALFONSINE ASTRONOMY

STUDIES AND SOURCES

ALFONSINE ASTRONOMY

This collection is created under the auspices of the European Research Council project ALFA: Shaping a European scientific scene, Alfonsine astronomy, CoG 723085.

Alfonsine Astronomy

The Written Record

Edited by
RICHARD L. KREMER
MATTHIEU HUSSON AND
JOSÉ CHABÁS

BREPOLS

D/2022/0095/104
ISBN 978-2-503-59520-7
eISBN 978-2-503-59521-4
DOI 10.1484/M.ALFA-EB.5.124044

Printed in the E.U. on acid-free paper

Table of Contents

Introduction
Richard L. Kremer, Matthieu Husson and José Chabás 7

Part 1
Contexts of Practice and Diffusion as Attested
by Manuscripts and Manuscript Collections

The *Libro de las tablas alfonsíes*:
New Documentary and Material Sources
Laura Fernández Fernández 21

Alfonsine Astronomy and Astrology in Fourteenth-Century Oxford:
The Case of MS Bodleian Library Digby 176
Jean-Patrice Boudet & Laure Miolo 57

Exploring a Late-Fifteenth-Century Astrologer's Toolbox:
British Library Add MS 34603
Richard L. Kremer 107

From Computus Material to Preacher's Toolbox:
Manufacturing a Bat-Book Almanac in the Fifteenth Century
Alexandre Tur 143

Bohemian King Wenceslas IV's Copy of the Alfonsine Tables and Their Place
Within his Astronomical and Astrological Corpus
Eric Ramírez-Weaver 199

Part 2
Authors, Texts and their Receptions in Various Milieus

Editing the Tables of 1322 by John of Lignères
José Chabás, Marie-Madeleine Saby 243

John of Lignères's *Quia ad inveniendum loca planetarum*:
An Edition and Translation
Alena Hadravová, Petr Hadrava 257

New Texts and Tables Attributed to John of Lignères: Context and Analysis
José Chabás 303

Work Cohesion as a Test of Manuscript Transmission: The Case of John of Lignères' *Tabule magne*
Matthieu Husson 317

Retracing the Tradition of John of Genoa's *Opus astronomicum* Through Extant Manuscripts
Laure Miolo 343

All In: Fifteenth-Century Manuscripts Devoted to Giovanni Bianchini's Astronomy
Glen Van Brummelen 381

Postface

From Documents to Data: The Digital Projects of ALFA
Galla Topalian, Matthieu Husson 403

RICHARD L. KREMER, MATTHIEU HUSSON
AND JOSÉ CHABÁS

Introduction

Setting the Context

The ALFA Project is devoted to the study of the history of mathematical astronomy, that is, the art of computing astral positions, as it was practiced in Europe from the end of the thirteenth century to well into the sixteenth century. The discipline was structured around the Alfonsine Tables, a large set of astronomical tables with canons by Isaac ben Sid and Judah ben Moses ha-Cohen. Compiled circa 1271 in Toledo under the patronage of Alfonso X of Castile, this work was heir to Arabic astronomical traditions developed across the Iberian Peninsula during the previous two centuries. Beginning in 1320, the Castilian Alfonsine Tables were recast in Paris, resulting in what we now call the Parisian Alfonsine Tables. These materials circulated widely and fostered astronomical activities throughout Europe. Alfonsine astronomy was shaped around this set of tables and a significant number of new works were produced: texts explaining the use of tables (called canons), texts on astronomical instruments, mathematical and theoretical texts, almanacs, calendars, and ephemerides (lists of eclipses or of daily positions of all the planets). Together, these materials form the corpus of Alfonsine works, of which there are a few hundred, extant in more than 900 manuscript codices and dozens of printed editions. These manuscripts and imprints comprise the written record of Alfonsine astronomy and provide the focus of this volume.

We are not the first, of course, to consider these particular astronomical sources preserved in major European libraries. Earlier generations of bibliographers and cataloguers identified and described many of the Alfonsine manuscripts and the individual works they contain, even if they did not envision the material as an 'Alfonsine corpus'.[1] More recent scholarship has focused on the individual works, usually the tables and canons of

* This work was supported by the European Research Council project ALFA: Shaping a European scientific scene, Alfonsine astronomy, CoG 723085, PI Matthieu Husson. The papers collected here were presented and discussed at an ALFA Conference, held September 2019 in Prague.

1 Cf. J. C. Houzeau and A. Lancaster, *Bibliographie générale de l'astronomie*, 2 vols, new ed. (Brussels: Xavier Havermans, 1964); Wilhelm Schum, *Beschreibendes Verzeichniss der amplonianischen Handschriftensammlung zu Erfurt* (Berlin: Weidmannsche Buchhandlung, 1887); Ernst Zinner, *Verzeichnis der astronomischen Handschriften des deutschen Kulturgebietes* (Munich: Beck, 1925); Lynn Thorndike and Pearl Kibre, *Catalogue of Incipits of Mediaeval Scientific Writings*, rev. and augmented ed. (Cambridge: Mediaeval Academy of America, 1963); Ernst Zinner, *Geschichte und Bibliographie der astronomischen Literatur in Deutschland zur Zeit der Renaissance*, 2d ed. (Stuttgart: Hiersemann, 1964). For the more recent concept of an 'Alfonsine corpus', see José Chabás and Bernard R. Goldstein, *The Alfonsine Tables of Toledo* (Dordrecht: Kluwer, 2003), pp. 248–49.

medieval computational astronomy.[2] It has richly documented the intellectual content of the computational tools by identifying the structural frames of tables (sidereal or tropical longitudes), the underlying numerical parameters and mathematical algorithms, treatment of the motions of the eighth sphere, and the procedural steps required to compute positions with the tables. However, the number of manuscripts considered has remained relatively small, and the relations among works found in given codices and among various codices collected by given actors have not been thoroughly explored.[3] Considering the 'Alfonsine corpus' rather than simply isolated, individual works can give us new insights into the practice of mathematical astronomy in medieval Latin Europe.

In light of this, the ALFA Project and the authors of this volume aim to follow the development of Alfonsine astronomy on the manuscript level. What can the codices tell us about how medieval astronomers actually computed eclipses? What notions of 'efficiency' drove them to rearrange tables to enhance their user friendliness?[4] How did they compare and select among the differing computational tools available to them? Did they ever evaluate their computational results against empirical or philosophical evidence? How did they arrange copies of tabular works and other textual materials in individual codices or as books in their personal libraries? How did the practices of manuscript cultures (and, later, early print cultures) shape the practices of mathematical astronomy? In the contexts of manuscript and print, how did the demands of the university impact the teaching and understanding of mathematical astronomy? By what vehicles did mathematical practices travel to different milieus within Europe? Finally, how did Arabic, Hebrew, and Byzantine Greek materials intermingle with Latin and vernacular manuscripts across the fourteenth and fifteenth centuries? By following the manuscripts, we hope to uncover details about 'how they worked' that are generally not recoverable

2 Examples include G. J. Toomer, 'A Survey of the Toledan Tables', *Osiris* 15 (1968), pp. 5–174; Francis S. Benjamin, Jr., and G. J. Toomer, *Campanus of Novara and Medieval Planetary Theory:* Theorica Planetarum (Madison: University of Wisconsin Press, 1971); Bernard R. Goldstein, *The Astronomical Tables of Levi Ben Gerson* (New Haven: Archon Books, 1974); J. D. North, 'The Alfonsine Tables in England', in *Prismata, Naturwissenschaftsgeschichtliche Studien: Festschrift für Willy Hartner*, ed. Y. Maeyama and W. G. Satzer (Wiesbaden: Steiner, 1977), pp. 269–301; Emmanuel Poulle, *Les Tables alphonsines avec les canons de Jean de Saxe: Édition, traduction et commentaire* (Paris: Éditions du centre national de la recherche scientifique, 1984); Mercè Comes, Roser Puig, and Julio Samsó (eds), *De Astronomia alphonsi regis* (Barcelona: Universidad de Barcelona, 1987); Marie-Madeleine Saby, 'Les Canons de Jean de Lignères sur les tables astronomiques de 1321: Édition critique, traduction et étude', unpublished thesis, Paris, École Nationale des Chartes, 1987; Fritz S. Pedersen, *The Toledan Tables: A Review of the Manuscripts and the Textual Versions with an Edition* (Copenhagen: C.A. Reitzels Forlag, 2002); Beatriz Porres de Mateo, '*Les Tables astronomiques de Jean de Gmunden: Édition et étude comparative*', unpublished PhD thesis, Paris, École Pratique des Hautes Études, 2003; José Chabás and Bernard R. Goldstein, *The Astronomical Tables of Giovanni Bianchini* (Leiden: Brill, 2009); idem, *A Survey of European Astronomical Tables in the Middle Ages* (Leiden: Brill, 2012); Benno van Dalen, *Islamic Astronomical Tables: Mathematical Analysis and Historical Investigation* (Farnham: Ashgate, 2013); José Chabás and Bernard R. Goldstein, *Essays on Medieval Computational Astronomy* (Leiden: Brill, 2015); José Chabás, *Computational Astronomy in the Middle Ages: Sets of Astronomical Tables in Latin* (Madrid: Consejo Superior de Investigaciones Científicas, 2019); Matthieu Husson, Clemency Montelle and Benno van Dalen (eds), *Editing and Analysing Astronomical Tables: Towards a Digital Information System for the History of Astral Sciences* (Turnhout: Brepols, 2022).

3 By way of comparison, a recent critical edition of the Toledan Tables examined more than 250 manuscripts. Pedersen, *Toledan Tables*, pp. 37–43.

4 José Chabás and B.R. Goldstein, 'Computing Planetary Positions: User-Friendliness and the Alfonsine Corpus', *Journal for the History of Astronomy*, 44 (2013), 257–76.

when the surviving historical documents were produced hundreds of years after the initial composition of the texts.

Our choice to follow the manuscripts raises several implications. Firstly, many of the earlier cataloguers were not historians of mathematical astronomy and thus did not always recognize the complexity of materials recorded in the manuscripts. For reasons explored by some of the essays in this volume, much of the Alfonsine corpus has been preserved in what are called composite manuscripts, codices that contain a wide range of diverse material. It is not uncommon to find various sets (or partial sets) of astronomical tables and canons, mathematical texts, texts on instruments, theoretical texts from university curricula, computational notes, texts on medical or meteorological astrology, computus material, and star catalogues mixed together in a single codex. Conversely, rarely do we find only one work in one book. By attending to the manuscript level, the ALFA Project is uncovering new works, new authors, and new relationships among tables, findings that will both deepen and broaden our understanding of Alfonsine astronomy.

Secondly, considering the physical materiality of the manuscripts can yield new information about the contexts in which Alfonsine astronomy was practiced. Marks of ownership can reveal early users, owners, or collectors; scribal colophons, watermarks, and bindings can localize places of production; marginal glosses or computations, underlining, rubrication, and soiled or heavily thumbed folios can suggest how some codices were 'used'. The types of materials bound into a composite manuscript can indicate whether the book was initially intended for a monastery library, a university classroom or professor's study, a courtly patron, or a specialized 'professional' such as a physician, consulting astrologer, calendar maker or mathematician. With such evidence, we can start to track Alfonsine astronomy as a European-wide scientific achievement and a set of practices involving many actors, milieus and sites.

Thirdly, by developing digital tools to manage and interrogate more than 900 manuscript codices carrying hundreds of works and thousands of copies of various astronomical tables, the ALFA Project can explore large-scale trends that have remained hidden to previous generations of scholarship. We do not intend to create a critical edition of the Parisian Alfonsine Tables, extant in more than 170 manuscript witnesses, but we can ask which types of tables were more frequently copied or commonly grouped with others or with texts. We can ask whether particular tables are found mostly in Italy, England, or Central Europe (Cracow, Prague, Vienna). We can track the strategies of early collectors, like William Reed in Oxford or Johannes de Wasia in Paris, who appear to have deliberately assembled copies of Alfonsine material for historical posterity rather than computational or pedagogical use. And once we are able to add machine reading to our repertoire of digital tools, we might explore, at ever finer levels of granularity, the movement of individual tables across the corpus of Alfonsine manuscripts.

The essays in this volume were written before ALFA's digital tools were fully functional. Rather than offering surveys based on 900 manuscripts, these essays present case studies of selected manuscripts or smaller groups of manuscripts. They illustrate the kinds of questions we can ask when conducting a history of Alfonsine astronomy at the manuscript level. More synthetic and integrative studies of the Alfonsine written record will be published during the third and final phase of the ALFA Project, after we have completed a second phase which will examine mathematical practices in the corpus.

Part 1: Alfonsine codices in circulation and collections

The essays of Part 1 examine individual manuscripts containing Alfonsine works. The authors of these essays seek not merely to list the contents, but also to reconstruct the cultural, astronomical, and mathematical worlds in which the manuscripts were initially copied, compiled, used, and collected. In some cases, individual scribes, patrons, owners, and even bookbinders can be identified from physical evidence in the surviving codices. In others, the original books have disappeared, forcing our authors to examine early inventories, library catalogues, or other references to manuscripts once known. Drawing on various types of evidence, the essays of Part 1 seek to contribute to the history of the book as well as to the history of astronomy.

The essays of Part 1 also illustrate a feature of Alfonsine astronomy that sets it apart from most other early traditions of the astral sciences. Surviving Alfonsine codices are usually contemporary, or nearly so, with the composition of the texts they contain. Most surviving sources for early Babylonian, Chinese, Greek, Sanskrit, Persian and Arabic astronomies are preserved as individual tablets or manuscripts created centuries after astronomers had initially authored the materials.[5] The Alfonsine manuscript corpus, therefore, offers historians a level of direct physical evidence about astronomical practice and manuscript culture that is not often available for other early astronomical traditions. The essays in Part 1 seek to interrogate this evidence at the level of the individual codex.

Laura Fernández Fernández reviews what is known about undoubtedly the most highly coveted manuscript of Alfonsine astronomy, the missing copy of the Castilian Alfonsine Tables, composed in the 1270s by two scholars, Isaac ben Sid and Judah ben Moses ha-Cohen, at the court of Alfonso X. Piecing together evidence assembled by earlier historians and from newly discovered sources, Fernández considers how the Castilian canons and tables might have travelled to Paris by the 1320s. Moreover, she discusses early reports of French and Italian translations, and how the Parisian Latin version circulated back to the Hispanic Kingdoms of the fourteenth and fifteenth centuries. She also pinpoints several references, in sixteenth-century and later book inventories and sales catalogues, to illuminated copies of Castilian 'Alfonsine Tables' and wonders whether an illuminated manuscript, produced in Alfonso's scriptorium and now held at the Parisian Bibliothèque de l'Arsenal, may once have contained a copy of the Castilian Alfonsine Tables.

Jean-Patrice Boudet and Laure Miolo examine a codex now at the Bodleian Library. A composite manuscript compiled and bound by William Reed, a fellow at Oxford's Merton

5 For introductions to the preservation of sources in these traditions, see Eleanor Robson, 'Reading the Libraries of Assyria and Babylonia', in *Ancient Libraries*, ed. Jason König, Katerina Oikonomopolou, and Greg Woolf (Cambridge: Cambridge University Press, 2013), pp. 38–56; Mathieu Ossendrijver, *Babylonian Mathematical Astronomy: Procedure Texts* (New York: Springer, 2012); Nathan Sivin, *Granting the Seasons: The Chinese Astronomical Reform of 1280, with a Study of Its Many Dimensions and a Translation of Its Records* (New York: Springer, 2009), pp. 227–47; Olaf Pedersen, *A Survey of the Almagest, with Annotation and New Commentary by Alexander Jones* (Berlin: Springer, 2011), pp. 11–25; Clemency Montelle and Kim Plofker, *Sanskrit Astronomical Tables* (Cham: Springer, 2018); E. S. Kennedy, 'A Survey of Islamic Astronomical Tables', *Transactions of the American Philosophical Society*, N. S. 46 (1956), 123–77; David A. King, Julio Samsó, and Bernard R. Goldstein, 'Astronomical Handbooks and Tables from the Islamic World (750–1900): An Interim Report', *Suhayl* 2 (2001), 9–105; Benno van Dalen, 'A New Survey of Islamic Astronomical Handbooks with Descriptions of More Than 200 Arabic and Persian Zijes' (unpublished manuscript, 2007).

College from 1344–57, this codex reveals a 'community of learning' at this College, since most of its texts were authored by Oxford masters or Merton fellows during the middle third of the fourteenth century.[6] Boudet and Miolo show how astronomer and bibliophile William Reed assembled a large personal library and then donated hundreds of books to several college libraries. In the Bodleian manuscript, they argue, Reed 'consciously gathered' materials concerning the astral sciences written by his contemporaries, and he presented them for use by the next generation of Oxonian scholars. To illustrate some of the practices reflected in Reed's codex, Boudet and Miolo examine in more detail a solar almanac, computed from the Parisian Alfonsine Tables for the years 1341–44; a list of planetary conjunctions and related astrological interpretations; and a precisely computed birth horoscope for a date in 1317. This single codex thus reveals the interests of mid-century Merton scholars in theoretical texts on astrometeorology, the computation of planetary positions and eclipses, and practical astrological prediction.

In his essay, **Richard L. Kremer** investigates another composite manuscript that he labels a 'toolbox', compiled at the end of the fifteenth century by a well-known Swabian astrologer and calendar maker, Marcus Schinnagel. In the 1480s, Schinnagel had produced, for an unknown patron, a large polyptych (1 × 3 metres in size, with multiple wings) compendium of astrological and calendrical texts, astronomical tables, and eclipse predictions. Kremer argues that the toolbox manuscript, now held at the British Library, is closely related to the content of the polyptych. Both sources combine much material copied from astronomical texts printed in Southern Germany during the late fifteenth century. Observing how Schinnagel snatched individual tables from different sources (especially from the 1492 printed edition of the Parisian Alfonsine Tables), rearranging their order in the codex, and dropping in several of his own newly computed tables, Kremer concludes that this astrologer was not interested in collecting well defined 'works' by known authors; rather, he filled his toolbox with miscellaneous tables that he used to cast horoscopes and construct annual astrological calendars.

Alexandre Tur finds another toolbox in a quite different codicological format, a bat-book almanac now held at the Bibliothèque nationale de France. Recently profiled by J. P. Gumbert, who catalogued about sixty known exemplars dating from the thirteenth through sixteenth centuries, these small books contain up to twenty leaves, folded down to a hand-sized package and placed in a sheath designed to be hung from the owner's belt.[7] Tur discovers that the Paris bat-book, dated to 1456 and signed by an otherwise unattested 'frater Paulus de Kignin', contains material from John of Gmunden's widely distributed *Kalendarium*, composed twenty years earlier in Vienna, and was probably revised for liturgical use in a Franciscan community in Northern Italy. Extremely portable, this sanctorale and compendium of official Franciscan liturgical material took Alfonsine astronomy into an unusual context. Tur concludes, however, that as the sole surviving exemplar of John of Gmunden's calendar in this format, the Paris bat-book probably

6 Cf. Constant Mews and John N. Crossley (eds), *Communities of Learning: Networks and the Shaping of Intellectual Identity in Europe, 1100–1500* (Turnhout: Brepols, 2011).

7 J.P. Gumbert, *Bat Books: A Catalogue of Folded Manuscripts Containing Almanacs or Other Texts* (Turnhout: Brepols, 2016).

represents an idiosyncratic prototype for a Franciscan convent and not a serial production of a format that was not preserved.

Eric Ramírez-Weaver considers the cultural and art historical significance of Alfonsine astronomy at the Prague court of Wenceslaus IV, notably reviewing the visual programmes (frontispieces and illuminated initials) in three luxurious manuscripts prepared around 1400. Even as Wenceslaus's political fortunes were declining, his courtiers designed books whose contents would have emphasized links between the King and Alfonso X of Castile, portraying both courts as centres of erudition and astral sophistication. This essay argues that the Alfonsine Tables in Prague served not merely computational but also cultural and political purposes. And they might have offered their embattled patron, the King, a curriculum of cosmic harmony as 'a model for earthly peace'.[8]

By considering 'books' rather than simply 'texts', the authors of Part 1 have opened new vistas to the history of Alfonsine astronomy. It is true that some of its texts were authored by Paris masters or were curated by fellows of Merton College for study by Oxford scholars; but these materials also circulated widely beyond the university milieu. Alfonsine codices were compiled by practicing astrologers seeking patrons, reformatted for liturgical use in monastic houses, decorated and adorned to encourage a beleaguered and inept king. Reports and rumours of the Castilian Alfonsine Tables, initially produced and copied in King Alfonso X's scriptorium, continued into the nineteenth century, even if no physical codices have been found. The Alfonsine Tables, as shown by the authors of Part 1, circulated both as books and as imaginaries.

Part 2: Authors and texts in various milieus

The essays of Part 2 turn from the particular codex to the individual work or author. These contributions ask how particular works have been preserved in surviving manuscript witnesses and how broader manuscript cultures shaped the diffusion, over two centuries, of Alfonsine astronomy across Europe. In some of these essays, the authors show how the manuscript witnesses make it difficult to define boundaries for a given work. Other essays examine how the preparation of critical editions can reveal particular scribal practices. Still others investigate how reputations constructed for given authors affect textual attribution or decisions about what material to bind into a single codex.

Although Part 2 retains a focus on the manuscript level, its chapters consider what the manuscripts can tell us about the identity of 'Alfonsine' works for the historical participants who copied or collected the materials. Additionally, they ask how the agency of authorship was distributed among scribes, compilers, commentators, patrons, 'actual' authors, and the *auctoritas* or attributed intellectual creators. Indeed, by following the manuscripts, the ALFA Project is problematizing the concept of a 'critical edition' for certain Alfonsine works, especially the Parisian Alfonsine Tables. Alfonsine astronomers and their scribes combined different canons with different sets of tables; they rearranged the order of

8 Eric Ramírez-Weaver, 'Bohemian King Wenceslas IV's Copy of the Alfonsine Tables and Their Place Within his Astronomical and Astrological Corpus," in *Alfonsine Astronomy: The Written Record*, ed. by Richard L. Kremer, Matthieu Husson and José Chabás (Turnhout: Brepols, 2022), pp. 199-240 (p. 234).

chapters within canons and the order of tables within sets of tables. Even the appearance of printed editions in 1483 and 1492 did not standardize a single version of the Parisian Alfonsine Tables. Such are the issues raised by the essays of Part 2.

José Chabás and Marie-Madeleine Saby consider the challenge of editing the *Tables of 1322* by John of Lignères, a work that is not found, in its entirety, in any given manuscript witness. As a member of the group of Parisian astronomers who in the 1320s reworked Castilian material into what we now call the Parisian Alfonsine Tables, John of Lignères assembled a set of thirty-two tables (mostly for spherical astronomy, eclipses and limited planetary motions) that illustrate the transition from the Toledan Tables of eleventh-century al-Andalus to Alfonsine astronomy. Chabás and Saby discuss the criteria they developed for selecting, from the more than 30 manuscripts containing related material, five witnesses for their forthcoming critical edition. These include both external (date and milieu of production) and internal (legibility, composition, and layout) factors. Their goal is to document one of the sets of astronomical tables most broadly diffused during the Alfonsine era.

Alena Hadravová and Petr Hadrava offer a first edition of one of John of Lignères' canons, the *Quia ad inveniendum loca planetarum*, that is, instructions for computing planetary longitudes and possibilities for eclipses with the Parisian Alfonsine Tables. Although relatively short (usually filling only a few folios), the *Quia* is quite variable as recorded in the extant manuscripts, a finding that Hadravová and Hadrava explain by suggesting that scribes may have struggled to interpret and formulate these canons, which are among the earliest of Latin Alfonsine astronomy. They describe the ten manuscripts, dating from the mid-fourteenth through the mid-fifteenth centuries, collated for the edition. They also deploy a statistical method of binary correlation of variant readings to create a computer-aided *stemma codicum* of these witnesses. An English translation follows the edited version of John's Latin text.

José Chabás examines several short texts and tables that he recently found in Madrid and Vatican manuscripts. A *Canon supra kalendarium magistri Johannes de Lineriis* and a table of mean syzygy times from 1321–96 is uniquely preserved in a fourteenth-century Madrid manuscript. By comparing these times with those found in other syzygy tables more firmly attributed to John of Lignères and to John of Murs, another early Parisian Alfonsine astronomer, Chabás concludes that the attribution of the Madrid manuscript to John of Lignères can be accepted. In two Vatican manuscripts, however, Chabás found another text and several tables attributed to John of Lignères. Although the tabular material is computed from the Parisian Alfonsine Tables, particularities in the meridians and the twenty-eight-year intervals found in the tables convince Chabás that these tables cannot have been authored by John of Lignères. Rather, the attribution to John of Lignères reveals the authority that this name had achieved a century after his death.

Matthieu Husson considers the authority of John of Lignères from a different angle, asking how another set of tables, firmly attributed to him, was copied into seven manuscripts over a period ranging from the mid-fourteenth to the late-fifteenth century. John's *Tabule magne* are comprised of a canon of eleven sections and eleven individual tables. Coining the phrase 'work cohesion', Husson asks how these twenty-two parts 'stuck together' or were rearranged, abridged, or reconfigured in the manuscript witnesses. He finds that most of the witnesses contain 'procedural gaps', where missing canon sections would prevent users, strictly following the textual instructions, from completing tasks described by subsequent sections. Some canon sections do not relate to any specific table and thus were less 'sticky'. Likewise, the twenty manuscripts bearing the tables often blend in material from John's *Tables of 1322*; canons, however, generally do not blend material from other canons. Nonetheless, Husson concludes that for Alfonsine scribes, the *Tabule magne* did not exhibit strong cohesion and did not circulate as tightly organized sets of working instructions. Instead, the manuscripts reveal the pragmatic variability of scribes as they redacted the astronomical canons and tables they copied.

Laure Miolo introduces John of Genoa, a Parisian astronomer of the 1330s who authored four short works related to eclipse calculations. She offers the first biographical sketch of this little-known figure, defines and dates his works, and identifies more than thirty manuscripts that witness this material. Most copied was his table of lunar and solar velocities and apparent radii of these bodies during eclipses. The accompanying canon is known in fewer manuscripts, suggesting a low level of 'cohesion' (Husson's term) between the tables and canon. John's more theoretical *Canones eclipsium* is known in only seven manuscripts of diverse provenance (England, Italy, Germany, and France). His final work, a didactic calculation of the solar eclipse of 1337, is found in only three manuscripts. Miolo's study thus provides yet another example of Alfonsine astronomers copying and preserving pragmatic computational tools more frequently than they did theoretical or didactic works.

Glen van Brummelen also considers the opus of a single author, in this case the mid-fifteenth-century Italian mathematician, Giovanni Bianchini, some of whose work would be printed in the 1490s. Although dozens of manuscripts are witnesses to Bianchini's five major texts, van Brummelen lists eight codices that exclusively contain his works. Several of these are luxurious, illuminated presentation codices, written by a single professional hand. Others are copied in various hands, sometimes quite casually, in codices overflowing with marginal notes that undoubtedly served as toolboxes for working astronomers/astrologers. These eight manuscripts, all produced in Italy during the 1460s, provide another example of 'cohesion' among texts that van Brummelen explains by suggesting that Bianchini's mathematical innovations required readers to move among his various works. In any case, the works of no other author during the Alfonsine period exhibit as much cohesion as do Bianchini's.

Like those in Part 1, the essays in Part 2 seek to understand individual works, authors, and their *oeuvres* by paying close attention to the manuscript witnesses and the physical evidence they contain. The generally unknown scribes who copied the texts, the patrons or practitioners who selected what to assemble and bind, and the later readers who annotated the folios or preserved the codices, all helped define what we now call the 'Alfonsine corpus'. The diffusion of Alfonsine material across Europe over the course of three centuries was not simply a matter of discrete works being identically copied and carried from one place to another. Instead, the evolution of manuscript cultures and the needs of Alfonsine practitioners shaped and reconfigured the material as it moved through time and space. By following these processes, the essays of this volume begin to question some of the fundamental historiographical notions of 'author', 'work', and 'edition' long deployed by historians of the medieval sciences.

Finally, the essays of this volume begin to illustrate how Alfonsine astronomy developed a life beyond the medieval university and Latin learning. Unlike many popular texts taught at university (e.g. Euclid's *Elements*, Sacrobosco's *Sphaera*, Aristotle's *De caelo*, or, later, Peurbach's *Theoricae novae planetarum*), the works of Alfonsine astronomy did not become 'scholasticized' or surrounded by commentaries and super-commentaries, questions, or disputations. Few Alfonsine texts ever became prescribed within official university curricula. Instead, the manuscript witnesses to Alfonsine works show less structured, more modular and flexible textual traditions, with individual codices tuned to the singular needs of their compilers and structured as composite manuscripts. It is certainly true that, during the first half of the fourteenth century, Paris and Oxford were decisive milieus in which the Parisian Alfonsine Tables and other important Alfonsine works were compiled. Moreover, in the fifteenth century, universities in Cracow and Vienna encouraged significant astronomical activity. Yet the courts of Castile, France (King Charles V), and Bohemia (King Wenceslas) also nourished astronomical activity, as did Holy Roman Emperors Frederick III and Maximilian I in Innsbruck. Religious orders, at times, also provided a context for the collection and compilation of Alfonsine materials. Likewise, noble families, such as the prominent d'Este of Ferrara, patronized important Alfonsine astronomers. Indeed, many 'working professionals', be they physicians, surgeons, astrologers, or calendar makers, required astronomical tables to cast the horoscopes they interpreted for their patrons. Following the Alfonsine manuscripts takes us beyond the confines of the universities.

In a time of crisis and fragmentation for late-medieval European society (plague, famine, depopulation, war, papal contestation, Hussite 'heresies', the 'fall' of Constantinople, etc.), Alfonsine astronomers assembled tools—composite manuscripts bearing modular works—that seem to reflect the fluidity of the social worlds in which they lived. Yet those manuscripts also offered them access to the reassuring, cultural presence of a stable, predictable, and mathematically describable view of the cosmos.

Digital tools and the written record

The last essay in this volume offers a self-reflexive analysis of how the tools of digital humanities have shaped collaborative researches within the ALFA Project. **Gala Topalian and Matthieu Husson** discuss the epistemic and methodological choices required to move from 'documents' (medieval manuscripts) to 'data' (digital artefacts) that can be processed, interrogated, and published for wider access. A digital survey has been constructed to curate information about the 390 discrete Alfonsine works found in 900 manuscripts and dozens of early printed editions. A new digital tool has been developed to store digital representations of astronomical tables and to facilitate their quantitative analysis, description, and critical edition. A text-oriented database based on TEI/XML technology will enable finer grained and more flexible modelling of a select number of manuscript witnesses. Formalizing these three ways of describing sources, Topalian and Husson conclude, has standardized vocabulary, clarified understandings of the objects under analysis, and enriched the research questions being explored by the ALFA team.

Bibliography

Benjamin, Francis S., Jr., and G. J. Toomer, *Campanus of Novara and Medieval Planetary Theory:* Theorica Planetarum (Madison: University of Wisconsin Press, 1971).

Chabás, José, *Computational Astronomy in the Middle Ages: Sets of Astronomical Tables* (Madrid: Consejo Superior de Investigaciones Científicas, 2019).

———, and Bernard R. Goldstein, *The Alfonsine Tables of Toledo* (Dordrecht: Kluwer, 2003).

———, and ———, *The Astronomical Tables of Giovanni Bianchini* (Leiden: Brill, 2009).

———, and ———, *A Survey of European Astronomical Tables in the Middle Ages* (Leiden: Brill, 2012).

———, and ———, 'Computing Planetary Positions: User-Friendliness and the Alfonsine Corpus', *Journal for the History of Astronomy*, 44(2013), 257–76.

———, and ———, *Essays on Medieval Computational Astronomy* (Leiden: Brill, 2015).

Comes, Mercè, Roser Puig, and Julio Samsó (eds), *De Astronomia alphonsi regis* (Barcelona: Universidad de Barcelona, 1987).

Dalen, Benno van, 'A New Survey of Islamic Astronomical Handbooks with Descriptions of More Than 200 Arabic and Persian Zijes' (unpublished manuscript, 2007).

———, *Islamic Astronomical Tables: Mathematical Analysis and Historical Investigation* (Farnham: Ashgate, 2013).

Goldstein, Bernard R, *The Astronomical Tables of Levi Ben Gerson* (New Haven: Archon Books, 1974).

Gumbert, J. P., *Bat Books: A Catalogue of Folded Manuscripts Containing Almanacs or Other Texts* (Turnhout: Brepols, 2016).

Houzeau, J. C. and A. Lancaster, *Bibliographie générale de l'astronomie,* 2 vols, new ed. (Brussels: Xavier Havermans, 1964).

Husson, Matthieu, Clemency Montelle, and Benno van Dalen (eds), *Editing and Analyzing Astronomical Tables: Towards a Digital Information System for the History of Astral Sciences* (Turnhout: Brepols, 2022).

Kennedy, E. S., 'A Survey of Islamic Astronomical Tables', *Transactions of the American Philosophical Society*, N. S. 46 (1956), 123–77.

King, David A., Julio Samsó, and Bernard R. Goldstein, 'Astronomical Handbooks and Tables from the Islamic World (750–1900): An Interim Report', *Suhayl* 2 (2001), 9–105.

Mews, Constant, and John N. Crossley (eds), *Communities of Learning: Networks and the Shaping of Intellectual Identity in Europe, 1100–1500* (Turnhout: Brepols, 2011).

Montelle, Clemency, and Kim Plofker, *Sanskrit Astronomical Tables* (Cham: Springer, 2018).

North, J. D., 'The Alfonsine Tables in England'. In *Prismata, Naturwissenschaftsgeschichtliche Studien: Festschrift für Willy Hartner*, ed. Y. Maeyama and W. G. Satzer (Wiesbaden: Steiner, 1977) pp. 269–301.

Ossendrijver, Mathieu, *Babylonian Mathematical Astronomy: Procedure Texts* (New York: Springer, 2012).

Pedersen, Fritz S, *The Toledan Tables: A Review of the Manuscripts and the Textual Versions with an Edition* (Copenhagen: C. A. Reitzels Forlag, 2002).

Pedersen, Olaf, *A Survey of the Almagest, with Annotation and New Commentary by Alexander Jones* (Berlin: Springer, 2011).

Porres de Mateo, Beatriz, 'Les Tables astronomiques de Jean de Gmunden: Édition et étude comparative', unpublished PhD thesis, Paris: École Pratique des Hautes Études, 2003.

Poulle, Emmanuel, *Les Tables alphonsines avec les canons de Jean de Saxe: Édition, traduction et commentaire* (Paris: Éditions du centre national de la recherche scientifique, 1984).

Robson, Eleanor, 'Reading the Libraries of Assyria and Babylonia', in *Ancient Libraries*, ed. Jason König, Katerina Oikonomopolou, and Greg Woolf (Cambridge: Cambridge University Press, 2013), pp. 38–56.

Saby, Marie-Madeleine, 'Les Canons de Jean de Lignères sur les tables astronomiques de 1321: Édition critique, traduction et étude', unpublished thesis, Paris, École Nationale des Chartes, 1987.

Schum, Wilhelm, *Beschreibendes Verzeichniss der amplonianischen Handschriftensammlung zu Erfurt* (Berlin: Weidmannsche Buchhandlung, 1887).

Sivin, Nathan, *Granting the Seasons: The Chinese Astronomical Reform of 1280, with a Study of Its Many Dimensions and a Translation of Its Records* (New York: Springer, 2009).

Thorndike, Lynn, and Pearl Kibre, *Catalogue of Incipits of Mediaeval Scientific Writings*, rev. and augmented ed. (Cambridge: Mediaeval Academy of America, 1963).

Toomer, G. J, 'A Survey of the Toledan Tables', *Osiris* 15 (1968), 5–174.

Zinner, Ernst, *Verzeichnis der astronomischen Handschriften des deutschen Kulturgebietes* (Munich: Beck, 1925).

———, *Geschichte und Bibliographie der astronomischen Literatur in Deutschland zur Zeit der Renaissance*, 2d ed. (Stuttgart: Hiersemann, 1964).

Part 1

Contexts of Practice and Diffusion as Attested by Manuscripts and Manuscript Collections

LAURA FERNÁNDEZ FERNÁNDEZ

The *Libro de las tablas alfonsíes*: New Documentary and Material Sources

Introduction

The original manuscript of the Alfonsine Tables has unfortunately not been preserved;[1] however, we have precise information about why the tables were compiled and their authorship thanks to MS 3306 at the National Library in Madrid (BNE).[2] The book, a composite paper manuscript made up of several scientific treatises, belonged to the library of Juan Fernández de Velasco (1550–1613), VI Condestable de Castilla,[3] and was

* This work was supported by the ERC project ALFA: Shaping a European scientific scene, Alfonsine astronomy, CoG 723085, PI Matthieu Husson. I am sincerely grateful to José Chabás, Richard L. Kremer, and Julio Samsó for their help, their suggestions, and their stimulating comments to improve this work. I would also like to thank Matthieu Husson and the entire ALFA team, especially Eric Ramirez-Weaver and Alexandre Tur, for their first readings and comments.

1 I use the term 'Alfonsine Tables' only to refer to the astronomical tables created in Toledo by two Jewish astronomers under the commission of Alfonso X, which is the equivalent of the Castilian Alfonsine Tables according to the nomenclature proposed by José Chabás and Bernard R. Goldstein, *The Alfonsine Tables of Toledo* (Dordrecht: Kluwer, 2003).

2 It is a composite and multiple text manuscript, acephalous, which includes an unknown astrological text of Arabic origin partially written in Latin and partially in Castilian (ff. 1r–34v), the canons of the *Libro de las tablas alfonsíes* (ff. 34v–72r), a Castilian translation of John of Saxony's canons that I discuss later (ff. 74r–87v), a fragmentary treatise on the astrolabe (ff. 88r–94v), and a partial copy of several treatises of the Alfonsine *Libro del saber de astrología* (ff. 98r–302v). The texts are from different periods and hands, although the one we are dealing with can be dated to the beginning of the sixteenth century (after 1515). A digital copy is available at: http://bdh-rd.bne.es/viewer.vm?id=0000011059&page=1. Chabás and Goldstein, *The Alfonsine Tables*, pp. 12–15; they also include a transcription of the Castilian canons and a detailed astronomical commentary. Laura Fernández Fernández, *Arte y ciencia en el* scriptorium *de Alfonso X el Sabio* (Seville/El Puerto de Santa María: Universidad de Sevilla/Cátedra Alfonso X, 2013), pp. 207, 269–71, 335.

3 The library, along with Condestable's goods, was inventoried and priced in 1608 (for a partial inventory of the library made *c.* 1600 see Madrid, BNE MS 7840). In 1608, our book was registered as follows: 'Un libro de astrología, escrito de mano, en lengua castellana antigua, cubierto de terciopelo morado tasado en 30 r'; (A book of astrology, written by hand in old Castilian, and covered with purple velvet priced at 30 r). Madrid, Archivo de Protocolos, legajo 24850, ff. 260–521v. José M. Fernández Pomar, 'Manuscritos del VI Condestable de Castilla en la Biblioteca Nacional', *Helmantica*, XVIII (1967), 89–108; Gregorio de Andrés, 'La biblioteca manuscrita del Condestable Juan Fernández de Velasco († 1613)', *Cuadernos bibliográficos*, 40 (1980), 5–22; Fernández Fernández, *Arte y ciencia*, pp. 269-71.

Laura Fernández Fernández • Complutense University, Madrid

Alfonsine Astronomy: The Written Record, ed. by Richard L. Kremer, Matthieu Husson, José Chabás, Turnhout, 2022 (*Alfonsine Astronomy: Studies and Sources*, 1), pp. 21-55
© BREPOLS PUBLISHERS 10.1484/M.ALFA.5.124923

bought by the librarian Juan de Iriarte for the Royal Library in 1736.[4] It is well known that this manuscript includes the sole extant copy of the canons of the *Libro de las tablas alfonsíes* (ff. 34ᵛ–72ʳ).

According to the prologue of the canons, work on the Alfonsine Tables started in 1263, when Judah ben Moses ha-Cohen and Isaac ben Sid began observations in the city of Toledo with the aim of improving on and correcting a precedent work, the Toledan Tables. These were compiled in Toledo about two hundred years previously, by a group of astronomers led by Ṣāʿid al-Andalusī (1029–70), which included Azarquiel (d. 1100). How Judah ben Moses ha-Cohen and Isaac ben Sid were entrusted with this mission and where they made the observations is complicated to ascertain; moreover, what 'observation' meant at this time is also difficult to define, but both authors were familiar with practical astronomy.

Judah ben Moses, the main collaborator of the Alfonsine scientific workshop, was mentioned in the *Lapidario* as 'muy entendudo en el arte de la astronomía' (very knowledgeable in the art of astronomy);[5] undoubtedly he was not merely a translator, but a scientist who knew the practical side of his activity. The first news we have of him is his participation in the Latin translation of Azarquiel's *Book of the Azafeha*, together with the Christian translator, Guillelmus Anglicus, between 1225 and 1231. In this book, he is named 'Iuda filius Mosse Alchoen, professione t[…]ex merito sciencie astronomus dictus'[6] and is quoted as translating under the vigilance of a supervisor, which indicates a certain working structure; the fact that he was assigned to the same task for so many years shows his status as a beginner. Apart from this isolated event concerning which we have no further information, Judah worked at the service of Alfonso X for the rest of his life. From 1243 to 1250, he translated the *Lapidario* with Garcí Pérez, a cleric, and in 1254 he began the *Libro conplido en los iudizios de las estrellas*. In 1256, he translated the *Libro de las figuras de las estrellas fixas* for the first time with the cleric Guillén Arremón Daspa, and following this, on 6 February 1259, he finished the first translation of the *Libro dell alcora* together with Johan Daspa, also a cleric. On 26 February 1259, he completed the *Libro de las cruzes*, working again with Johan Daspa, and in 1276 he made a second revision of the *Libro de las figuras de las estrellas fixas* with another Jewish collaborator named Samuel and two Italian

4 This information can be consulted in the Royal Library's book of accounts, *Libro de asiento de los libros que se compraban para la Biblioteca desde el año 1716 hasta el 1738*, BNE, MS 18841, f. 310ᵛ. Our book was registered then as follows: 'Libro de astrología sin nombre de autor. Itt. cánones de las Tablas Alfonsíes compuestos por el bachiller Francisco de Morales. Libro del espera, del astrolabio, del quadrant etc. mandados escrivir por el rey Dn. Alonso el Sabio'.

5 The *Lapidario* is a collection of four 'lapidaries', concerning the magical properties of stones, books that all can be credited to the Alfonsine milieu. Although the general prologue of the work dates the translation to 1243–50, the royal manuscript today preserved in the Escorial Library, RBME MS h-I-15, was probably made at the beginning of the 1270s. Ana Domínguez Rodríguez, *Astrologia y arte en el "Lapidario" de Alfonso X el Sabio* (Madrid: Edilan, 1984); Fernández Fernández, *Arte y ciencia*, pp. 135–210.

6 The word that begins with 't' has been interpreted in different ways: Millás Vallicrosa suggested *tabulae* or *traductor*, and Hilty suggested *tabib*, whose meaning in Arabic would be akin to doctor or physician. José M. Millás Vallicrosa, *Estudios sobre Azarquiel* (Madrid-Granada: CSIC, 1943–50), pp. 453–54; Gerold Hilty, 'El libro conplido de los iudizios en las estrellas', *Al-Andalus*, 20 (1955), 1–74, (p. 16). Three copies of this manuscript are preserved, two of which cite only Guillelmus Anglicus, MS BnF, lat. 7195 and MS BnF, lat. 16652, and in the other, from the cathedral of Toledo, currently MS BNE 10053, the entire translation is attributed to Judah. This fact has also triggered different hypotheses, including the assignment of the work exclusively to Judah by Norman Roth, 'Jewish translators at the court of Alfonso X', *Thought. A Review of Culture and Ideas*, 60 (1985), 439–55 (p. 442).

collaborators, Joan de Mesina and Joan de Cremona. Some authors have suggested that, in addition to these titles, he was also involved in the redaction of the *Libro de las formas et las ymágenes* and that he was responsible for the translation into Castilian of Ptolemy's *Quadripartitum*, the *Libro de la mágica de las signos* (the famous *Picatrix* in its Latin version), and even the *Cánones de Albateni* in collaboration with Isaac ben Sid. As we can see, Judah ben Moses remained in the King's team of scientific collaborators throughout his life, working for Alfonso X even before he ascended the throne and remaining with him until his death. This loyalty put Judah in the favour of the monarch, who distinguished him as his 'alfaquim e su merced' and made him a beneficiary of Jerez de la Frontera's estate by bequeathing a house to him in 1266.[7]

With regard to Isaac ben Sid, also known as Rabiçag de Toledo, we hardly have any data except for the information provided by the prologues of his works. However, we have enough to picture him as a competent astronomer, well acquainted with the use of observation instruments. He recorded the solar eclipses that took place in 1266 and 1267, and the lunar eclipse of 1263, phenomena probably related to the observation for the Alfonsine Tables.[8] In fact, the prologue of the canons asserts, 'e rectificamos muchos eclipsis de los solares y de los lunares', suggesting they observed many solar and lunar eclipses. He was the translator and/or author of several treatises for the *Libro del saber de astrología*, a vast compendium completed in 1278 (*Libro del astrolabio redondo, Libro del astrolabio llano, Libro de la lámina universal, Libro del quadrante señero, Libro de las armellas, Libro del relogio de la piedra de la sombra, Libro del relogio del agua, Libro del relogio del argento vivo, Libro del palacio de las horas*), translated the *Cánones de Albateni* and the *Tablas de Azarquiel*, and he wrote the *Libro del quadrante con que rectifican* in 1277.[9] In addition to the translation of Arabic scientific texts into Castilian, Rabiçag de Toledo also collected and copied Arabic sources in which the practical side of his work is evident.[10] Rabiçag, referred to by the King on several occasions as 'nuestro sabio' (our wise man), was respectfully referenced

7 For further information and a bibliography of the production of Judah ben Moses, see Fernández, *Arte y ciencia*, pp. 59–63.

8 This information was left by a young student named Isaac Israeli, who in 1310 established an astronomical treatise in which he relates that Rabiçag, whom he quotes as Rabbi Isaac ha-hazzan b. Sîd, made the observations of the eclipses in the city of Toledo under the orders of King Alfonso, observations he recorded in his own handwriting. David Romano, 'Le opere scientifiche di Alfonso X e l'intervento degli ebrei', in *Oriente e Occidente nel Medioevo: filosofia e scienze* (Roma: Academia Nazionale dei Lincei, 1971), pp. 677–711 (pp. 689 and 703); Norman Roth, 'Jewish Collaborators in Alfonso's Scientific Work', in *Emperor of Culture: Alfonso X the Learned of Castile and His Thirteenth-Century Renaissance*, ed. by Robert I. Burns (Philadelphia: University of Pennsylvania Press, 1990), pp. 59–71, (p. 68); Chabás and Goldstein, *The Alfonsine Tables*, p. 141.

9 For further information about the production of Isaac ben Sid, see Fernández Fernández, *Arte y ciencia*, pp. 63–65.

10 Evidence of this activity is MS Or. 152, preserved at the Biblioteca Medicea Laurenziana in Florence. It is a unitary miscellaneous codex written in Arabic. According to the colophons, two of its treatises were copied in Toledo, in 1265 and 1266, and on page 75 there is an interesting note written in Arabic with Hebrew script where Isaac ben Sid talks about the process of copying of one of the treatises and explains how he was able to make a few instruments, proposing better construction. Further information can be found in Donald R. Hill, 'A Treatise on Machines, by Ibn Muādh Abū Abd Allāh al-Jayyānī', *Journal for the History of Arabic Science*, 1 (1977), 33–46; Abdelhamid I. Sabra, 'A Note on Codex Biblioteca Medicea-Laurenziana, Or. 152', *Journal for the History of Arabic Science*, 1 (1977), 276–83; Joan Vernet, 'Un texto árabe de la corte de Alfonso X el Sabio', *Al-Andalus*, 43 (1978), 405–21; Mª Victoria Villuendas, 'A Further Note on a Mechanical Treatise Contained in Codex Medicea Laurenziana Or. 152', *Journal for the History of Arabic Science*, 2 (1978), 395–96; Julio Samsó, *Las ciencias de los antiguos en al-Andalus* (Almería: Fundación Ibn Tufayl, 2011), pp. 249–57.

by other astronomers such as Abraham Zacut and Judah ben Asher II as the author of the Alfonsine Tables.[11]

The prologue of the canons of the Alfonsine Tables specifies that the observations were made between 1263 and 1272; therefore, the final version of the *Libro de las tablas* had to have been made after that period. As a matter of fact, Rabiçag was using the solar parameters of the new tables in 1277 when he was writing the *Libro del quadrante con que rectifican*[12] and the *Libro del astrolabio llano*;[13] hence, the royal manuscript, which was the final part of the project, must have been made during the 1270s, probably during the second half of the decade, contemporaneous with other important scientific works produced in the royal *scriptorium*.[14]

The aim of this paper is to trace, as much as possible, the elaboration of the Alfonsine Tables and their dissemination in other territories, historical periods, and intellectual milieus through documentary and material sources.

1. The making of the Alfonsine Tables

In addition to the BNE MS 3306, the oldest source that documents the origin of the Alfonsine Tables is the *Historia eclesiástica de la imperial ciudad de Toledo*, a work written by the Jesuit Jerónimo Román de la Higuera (1538–1611).[15] This author places a team of scholars at the service of the King working on the *Palacio de Galiana*, Toledo, the royal palace where Alfonso was born. This residence, originally the Palace of Yahya al- Ma'mūn (d. 1075), the Muslim King of Toledo, was later converted into the residence of the Christian sovereigns in the city. Although Román de la Higuera misunderstood important data or directly invented others, this work is a relevant piece in the historiography of the Alfonsine Tables.[16] On the one hand, the author says that the original book of the *Tablas*

11 Chabás and Goldstein, *The Alfonsine Tables*, pp. 20, 139, 226, 236; José Chabás, and Bernard R. Goldstein, *Astronomy in the Iberian Peninsula: Abraham Zacut and the Transition from Manuscript to Print* (Philadelphia: American Philosophical Society, 2000), p. 49.

12 Julio Samsó, 'Alfonso X and Arabic Astronomy', in *De Astronomia Alphonsi Regis*, ed. by Mercè Comes, Roser Puig, and Julio Samsó (Barcelona: Instituto Millás Vallicrosa, 1987), pp. 23–38.

13 José Chabás, 'Were the Alfonsine Tables of Toledo First Used by Their Authors?', *Centaurus*, 45 (2003), 142–50.

14 According to the data provided for the prologues and the manuscripts preserved, the *modus operandi* of the royal *scriptorium* consisted of working with the texts, and probably with the images, in copies conceived as draft material. Once the sources had been translated, revised and the new texts written, the final versions were copied into illuminated manuscripts. Unfortunately, we do not have all the works made in the royal *scriptorium*; a few are preserved in manuscripts commissioned by the King, others are known only from later copies, and others only because they have been cited in other texts. The Alfonsine manuscripts related to scientific matters currently preserved are the *Libro de las cruzes*, MS BNE 9294, the *Libro conplido en los iudizios de las estrellas*, MS BNE 3065, the *Lapidario*, MS RBME h-I-15, the *Libro de las formas et las ymágenes*, RBME MS h-I-16, the *Libro del saber de astrología*, Biblioteca Histórica Marqués de Valdecilla MS 156, the *Compendio tabular*, Bibliothèque de l'Arsenal MS 8322, and the *Libro de astromagia*, BAV MS Reg. lat. 1283pt.A. The entire corpus of Alfonsine scientific manuscripts and their later copies is discussed in Fernández Fernández, *Arte y ciencia*.

15 Madrid, BNE MS 1289, libro 22, capítulo 12, ff. 122–23; a digital copy can be found in: http://bdh-rd.bne.es/viewer.vm?id = 0000014636&page = 1.

16 Part of this text was copied by other authors such as Nicolás Antonio, *Bibliotheca hispana vetus sive Hispanorum* (Rome: Typographia Antonii de Rubeis, 1696), T. I, liber VIII, cap. V, p. 55, and José Rodríguez de Castro, *Biblioteca española: tomo primero que contiene la noticia de los escritores rabinos españoles* (Madrid: Imprenta Real de la Gazeta,

Alfonsíes had previously been kept in Toledo at the library of the Monastery of San Juan de los Reyes,[17] but he does not give any information about the later disposition of the book. Higuera also asserts that the book was held by Juan de Herrera (1530–97), architect of Felipe II (1527–98). This is a significant point to consider, because as we will see later, Juan de Herrera was closely linked to Alfonsine production, specifically to the Alfonsine Tables.[18] And, immediately after mentioning this, Román de la Higuera, without naming his source, provides the prologue of the missing book of the *Libro de las tablas*:

> Los homes dados a la sapiencia cuydaron que si n<o> conmunicauan los sus aueres, e facia<n> que los demas touiessen en ello parte, menguarian sus fechos, e por esso ouieron sabor de facer libros, que non moriesen con ellos, e desta guissa eran de pro, assi a los homes de su tiempo cuemo a los que enpos dellos auían de uenir, e por esso la poca remenbrança e oluidança de lo que con luene tiempo auían adquirido, facían que después de mucho tiempo e después de lueñe afán se perdíe lo ya sauido, e catado que se sauia mucho (BNE MS 1289, ff. 122^{v}–123^{r}).[19]

In the 1860s Rico y Sinobas asserted that Román de la Higuera was here recreating a modern *topos* instead of copying a thirteenth-century text and claimed that he had not been able to find a similar sentence in any Alfonsine scientific manuscript.[20] However, this idea of the responsibility of bequeathing knowledge by copying books fits with the general tone present throughout the Alfonsine cultural project.

1781), II, pp. 643–44, and accepted without doubt by modern bibliography until Moritz Steinschneider, *Die hebräischen Übersetzungen des Mittelalters und die Juden als Dolmetscher* (Berlin: Kommissionsverlag des Bibliographischen Bureaus, 1893), pp. 616–26, drew attention to the contradiction in the data provided by Román de la Higuera. Manuel Rico y Sinobas also analysed this material in his well-known edition, *Los Libros del Saber de Astronomía del Rey D. Alfonso X de Castilla*, 5 vols. (Madrid: Tipografía de Don Eusebio Aguado, 1863–67). Nevertheless, none of the modern scholars cite the complete version of Higuera and thus they miss part of the details provided by him. For this reason, it is worth pursuing a complete analysis of this source.

17 This book collection was compiled with fonds of the old Toledan Franciscan Library and new acquisitions for the new Royal Foundation. The Monastery of San Juan de los Reyes was directly promoted by Queen Isabel 'la Católica' (1451–1504) as a key piece in the religious topography of the reign, and its library became an important intellectual centre. Among its books there were scientific manuscripts; in fact, Román de la Higuera mentions a few of them. This library sadly disappeared during the Spanish War of Independence (1808–14). It was burnt on the night of 19 December 1808, when the French army occupied the monastery, and unfortunately most of its manuscripts disappeared. Among the lost books was the inventory of the library; therefore, it is impossible to check Higuera's information. Antolín Abad Pérez, 'La biblioteca franciscana de Toledo (1284–1808)', *Anales toledanos*, 20 (1984), 9–36, and 'Relación sobre el incendio de San Juan de los Reyes (1808) y vicisitudes posteriores hasta 1864', *Toletum: Boletín de la Real Academia de Bellas Artes y Ciencias Históricas de Toledo*, 4 (1969), 169–88; Fernández Fernández, *Arte y ciencia*, pp. 249–50.

18 Rico y Sinobas believed that the manuscript that Higuera referred to in his *Historia eclesiástica de la imperial ciudad de Toledo* was the *Libro del saber de astrología*, another Alfonsine manuscript, not the *Tables*. In previous works, I agreed but as we discuss later, Juan de Herrera knew the *Alfonsine Tables* and had materials relating to them in his library. This fact suggests that Román de la Higuera was maybe providing accurate information about the manuscript.

19 Because of the punctuation and the use of a few words, it seems that Román de la Higuera's model was not the original text but a later copy. I thank Francisco Bautista for his help with this transcription. A translation would be: 'Men, given knowledge, thought that if they did not communicate and share what they possessed with others, they would be doing wrong, and therefore they were interested in making books which would not die with them and thus would be useful both to the individuals of their time and to those who would come after them; and so the poor memory and the forgetfulness of what they had acquired over the course of time meant that after a long time and a lot of work, what was already known was lost, and it was considered that much was known'.

20 Rico y Sinobas, I (1863), p. LXXVII; V (1867), p. 44.

Furthermore, Higuera was the first author to mention the 'legendary' meeting of scientists in Toledo; according to information he provides, the group was made up of Muslims, Jews, and sages from Gascony (France) and Paris, sent for by the King, and the text specifies that when they finished, they went back to their territories.

There is no information about French people working at the *scriptorium* or the chancellery of Alfonso X; however, the King's interest in having scientists and scholars from other territories at his service is clear.[21] We do know the identity of several Italian collaborators whose presence at the court was linked both to the *scriptorium* and to the chancellery. Among them were Buenaventura de Siena, Joan de Mesina, Joan de Cremona, Aegidio de Tebaldi, and Pietro de Regio. As for the Muslims who may have worked at court, although it is very common to find comments on this subject in the abundant bibliography of this period, the truth is that the only written reference we have is the participation of Master 'Bernardo el Arábigo', the Arab who in 1277 participated in the second translation of the *Libro de la açafeha* together with Abraham, a Jew. He was probably a convert from Islam; in fact, he was cited as '*christiano novo*' in the Repartimiento of Murcia.[22] The King also wanted to recruit the scientist Muhammad b. Ahmad al-Riqūtī al-Mursī from Murcia, although he failed in the attempt; and recently, Julio Samsó has considered the possibility of a connection with scientists from Cairo.[23] In fact, the embassy of Mamluk Sultan Baybars (1223–77), who arrived in Seville in 1260 with numerous gifts for Alfonso X, is well known.[24] In any case, the Islamic context, especially with regard to the sources used, is essential to understanding some of the approaches present in the Alfonsine Tables, as is the role of the Jewish astronomers.[25]

---

21 The Partida II, law 31, describes the creation, organization, and structure of the 'Estudios generales', the equivalent of a university college. The text specifies that the teachers and pupils were 'extraños e de lugares departidos' (foreigners and from different places). Alfonso X, *Las Siete Partidas, Volume 2: Medieval Government: The World of Kings and Warriors*, ed. by Samuel Parsons Scott and Robert I. Burns (Philadelphia: University of Pennsylvania Press, 2001), pp. 527–31. In Seville in 1254 the 'Estudio de latín e arábigo' (Latin and Arabic college) was founded and in 1260 (not in 1254 as many authors say), the King demanded the archbishop and the city council establish some mosques near the *Alcázar* he had given them in the partition of the city, because he planned to dedicate them 'para morada de los físicos que vinieron de allende, y para tenerlos más cerca é que en ellas fagan la su enseñanza á los que les hemos mandado que nos lo enseñen por el su gran saber, ca por eso los hemos ende traido' (for the lodging of the physicians who came from afar, to have them nearer, and for them to teach those of us who have commanded them to teach us through their great knowledge; that is why we have brought them). It is interesting to note that the same year, 1260, an embassy arrived in Seville from Cairo. Diego Ortiz de Zúñiga, *Anales eclesiásticos y seculares de la muy noble y muy leal ciudad de Sevilla* (Madrid: Imprenta Real, 1677), p. 90.

22 Evelyn S. Procter, 'The Scientific Works of the Court of Alfonso X of Castille: The King and his Collaborators', *The Modern Language Review*, 40 (1945), 12–29; Julio Samsó, 'Dos colaboradores científicos musulmanes de Alfonso X', *Llull*, 4 (1981), 171–79; Fernández Fernández, *Arte y ciencia*, pp. 68–72.

23 Julio Samsó, *On Both Sides of the Straits of Gibraltar. Studies in the History of Medieval Astronomy in the Iberian Peninsula and the Maghrib* (Leiden: Brill, 2020), pp. 827–58. I am grateful to Julio Samsó for sending me this chapter before it was published.

24 Pedro Martínez Montávez, 'Relaciones de Alfonso X de Castilla con el sultán mameluco Baybars y sus sucesores', *Al-Andalus*, 27 (1962), 343–76.

25 Apart from the many coincidences between these canons and the Toledan Tables, identified by Chabás and Goldstein, *The Alfonsine Tables*, Samsó has pointed out that chapters 36 and 53 'make sense only in an Islamic context'. Samsó, *On both sides*, p. 835. In a recent paper, Samsó has also pointed out the connection of the Parisian Alfonsine Tables and the Andalusī-Maghribī tradition; see Julio Samsó, 'Ibn Ishāq and the Alfonsine Tables', *Journal for the History of Astronomy*, 50 (2019), 360–65.

It seems that Román de la Higuera does more than simply reproduce historical data (it is not clear from where or whom he obtained the information); he creates or disseminates a legend about that episode. Moreover, it is very likely that the appearance of French astronomers in Higuera's narrative is a result of their role in the process of assimilation of the Alfonsine Tables in Paris rather than their involvement with the actual development of the work. As we will see, the fame of one set of astronomical tables promoted by Alfonso X was quite widespread in the sixteenth century; indeed, Román de la Higuera refers to this book as 'unas tablas tan famosas como todos saben' (such famous tables, as everybody knows).

Beyond the veracity of the information provided by Higuera, what is unquestionable is that his words respond to an intellectual and scientific milieu in which the memory of the Alfonsine Tables played a prominent role. However, this information does not mesh well with the poor material record we have of the Alfonsine Tables and it raises an important question: why were not more copies of that work preserved than the one found in BNE MS 3306? In the following pages, I present new data to further analyse this problem.

2. The dissemination of the Alfonsine Tables and their arrival in Paris

How the Alfonsine Tables began their dissemination outside the Crown of Castile is still not clear, but knowledge of this material in other territories can be documented at least in the first decades of the fourteenth century. It seems that the Castilian set arrived in Italy first,[26] but its arrival in Paris before 1321 was particularly significant for the history of medieval astronomy and for the survival of the Alfonsine work in the coming centuries.[27] We do not know what arrived in Paris, if it was the whole work, canons and tables, or just the tables. John of Murs in his *Expositio Regis Alfonsii circa tabulas ejus, c.* 1321, said he had the tables but not the canons, which, according to Samsó, could be the reason for the need to write new canons. As a matter of fact, Samsó suggests in a recent book that there may have been two different versions of the tables with two different computational

26 In his 1318 work *Theorica planetarum*, Thadeus of Parma provided data that certainly originated from the Castilian Tables. In 1363, Andalò di Negro (1260–1334) mentioned that the Alfonsine Tables were made in 1272; this information was not present in the texts of the Parisian tradition, so it must have found its way to Andalò di Negro through another channel of transmission directly linked with the information provided by the Alfonsine Tables in an original version. Johan L. E. Dreyer, 'On the Original Form of the Alfonsine Tables', *Monthly Notices of the Royal Astronomical Society*, 80 (1920), 243–62 (p. 252); Richard I. Harper, 'Prophatius Judaeus and the Medieval Astronomical Tables', *Isis*, 62/1 (1971), 61–68; John D. North, 'The Alfonsine Tables in England' in *Prismata: Festschrift für Willy Hartner*, ed. by Y. Maeyama and W. G. Salzer (Wiesbaden: Franz Steiner, 1977), pp. 269–301, (p. 270); Olaf Pedersen, 'The 'Theorica Planetarum' and its Progeny', in *Filosofia, scienza e astrologia nel trecento europeo*, ed. by Graziella Federici Vescovini and Francesco Barocelli (Padua: Il poligrafo, 1992), pp. 53–78; Chabás and Goldstein, *The Alfonsine Tables*, p. 248; Chabás, *Computational Astronomy*, p. 238; C. Philipp E. Nothaft, 'Critical Analysis of the Alfonsine Tables in the Fourteenth Century: The Parisian *Expositio tabularum Alfonsii* of 1347', *Journal of History of Astronomy*, 46 (2015), 76–99 (p. 82).

27 According to Chabás and Goldstein, in *The Alfonsine Tables*, pp. 246–47, the French physician and astrologer Geoffrey of Meaux (*c.* 1310–*c.* 1348) in 1320–21 was the first documented source in Paris to refer to the Alfonsine material (in BnF, MS lat. 7281, f. 160^{v}, and Biblioteca Capitular de Toledo, MS 99–5, f. 14^{r}). See Nothaft, 'Critical Analysis', p. 89.

systems, an initial version with computed sidereal positions whose content would have been reflected in the Castilian canons preserved in the BNE MS 3306, and a second recension that gave tropical positions and would correspond to the material later adapted in the Parisian tables. This second version would represent the evolution towards tropical astronomy due to the influence of al-Battānī's tables and their assimilation by Jewish astronomers since the twelfth century.[28] Samsó's suggestion about the existence of two versions of the tables would fit perfectly with the *modus operandi* of the *scriptorium* of Alfonso X. All the outstanding works of the cultural project promoted by the monarch were subject to constant revision and thus came in several versions as new sources and materials became known and incorporated; so this hypothesis would make sense in this framework. According to this argument, the translation of the al-Battānī canons in the Alfonsine circle would have inspired this new version of the tables made by Jehuda and Rabiçag.[29]

Regardless of whether there could have been two versions of the tables, it is important to consider the linguistic channel used for their transmission, irrespective of the content, to other territories. We do not know if the Alfonsine Tables, initially written in Castilian (old Spanish), were translated into Latin during the rule of Alfonso X,[30] or if the Latin translation was done in Paris or elsewhere, or even if it actually existed. Unfortunately there is no linguistic study concerning the canons of the versions of the tables written in Latin that might help us to identify traces of Castilian vocabulary or other lexical peculiarities that shed light on this issue. What it is clear is that if the work received in Paris at the beginning of the fourteenth century was written in Castilian, the scholars involved in the assimilation process must have been proficient in this language, perhaps not enough so as to translate but sufficiently to comprehend the content and to write a Latin version that included the updates obtained in Paris. This detail could help us to understand why Parisian astronomers wrote new canons, preserving the memory of King Alfonso as the main promoter of the work and consequently keeping the Alfonsine era (31 May 1252) and radix position given for the meridian of Toledo, but erasing the references of the Jewish authors, an element that they were not interested in maintaining. Had the tables arrived in Paris without canons, which is another possibility to be taken into account, the astronomers involved in their reception and revision would have had full freedom to write new canons without the information concerning the Jewish authors but keeping the essential data of the tables from Castille.

---

28 Samsó, *On Both Sides*, pp. 827–58.

29 Alfonso X, *Los cánones de Albateni*, ed. by George Bossong (Tübingen: Max Niemeyer, 1978).

30 Julio Samsó considers that a Latin translation of the Castilian text could have been produced under the policy of the *fecho del Imperio*, i.e. the imperial candidacy of Alfonso X, as had happened with other scientific works, such as the Latin translation of the *Cuadripartito* and the *Libro conplido en los iudizios de las estrellas* (there is also a Latin translation of the *Libro de la mágica de los signos*, the famous *Picatrix*, but we cannot know if the translation was commissioned by the King or was done later). Julio Samsó, 'La ciencia española en la época de Alfonso el Sabio' in *Alfonso X, Toledo* (Madrid: Ministerio de Cultura, Dirección General de Bellas Artes y Archivos,1984), pp. 89–102 and *On Both Sides*, p. 858. If this were so, the Latin version of the Alfonsine Tables would have been written between 1272, when the observations were completed, and 1275, when the project of the *fecho del Imperio* formally ended, thus practically at the same time as the Castilian version was written. On the other hand, this chronology fits with their presence within the reign of the Italian collaborators, Pietro de Reggio and Egidio de Tebaldis, who produced the aforementioned Latin translations.

In any case, what is indisputable is that in Paris important astronomers assimilated and adapted the tradition of the Alfonsine Tables, whatever their content, with great success, producing new material would circulate to other European territories (Italy, England, Germany, and Poland) and generate the so-called Parisian Alfonsine Tables.[31] Among these were John of Murs, John of Vimond, John of Lignères, John of Genoa, and John of Saxony. The latter made new canons in 1327 for this Parisian version, canons that were selected by Erhard Ratdolt, the Venetian editor, who in 1483 first printed a version of the Parisian Alfonsine Tables.[32]

As we have seen, we do not know precisely when or how the Alfonsine Tables arrived in Paris, but it is interesting to point out that the Crown of Castile had strong ties to the French monarchy and that diplomatic embassies frequently travelled to the neighbouring court.[33] The exchange of books between both kingdoms can be ascertained thanks to Alfonso X's will.[34] In this document, two manuscripts gifted by the King of France, Louis IX (1214–70), to his cousin Alfonso are explicitly mentioned: a moralized Bible, the *Biblia de Toledo* or *Biblia de San Luis*,[35] and a copy of Vincent of Beauvais's (*c*. 1190 - *c*. 1267) *Speculum Historiale*.[36] If Alfonso received books from his French homonym, it seems plausible that he also gifted books to his relatives.[37] Unfortunately, there is no documentary trace to confirm this possibility, although as we shall see later, there must have been an exchange of books between the courts at some point. Concerning the documented embassies between these kingdoms, it may be noteworthy that on 20 May 1280, three members of the court travelled to Aix-en-Provence to meet Charles of Salerno (1254–1309) to solicit his mediation between Alfonso X and Philippe III (1245–85) concerning the problem of

31 Chabás and Goldstein, *The Alfonsine Tables*, pp. 243–306. Chabás, *Computational Astronomy*, pp. 238–39.

32 For printed editions of the Parisian Alfonsine Tables see 'The Legacy of Alfonsine Astronomy' in Chabás and Goldstein, *The Alfonsine Tables*, pp. 243–306; Laura Fernández Fernández, 'Las Tablas Astrónomicas de Alfonso X el Sabio. Los ejemplares del Museo Naval de Madrid', *Anales de Historia del Arte*, 15 (2005), 29–50; Fernández Fernández, *Arte y ciencia*, pp. 339–41.

33 According to the documentary data provided by George Daumet, *Mémoire sur les relations de la France et de la Castille de 1255 à 1320* (Paris: Fontemoing et C^le^, 1913) and thanks to the references mentioned by other authors such as Jerónimo Zurita (1512–80), we can document at least twelve Castilian embassies in France between 1266 and 1305 with different aims and purposes. I thank Oscar García Villaroel for sharing this information with me.

34 Manuel González Jiménez, *Diplomatario Andaluz* (Seville/El Monte: Caja de Huelva y Sevilla, 1991), pp. 559–60.

35 Toledo, Tesoro de la Catedral, three volumes without shelfmarks; New York, Morgan Library MS M. 240.

36 '[…] e los cuatro libros que llaman Espejo istorial que mandó fazer el Rey Luis de Francia […] E mandamos otrosi que las dos Biblias et tres libros de letra gruesa, cobiertas de plata, e la otra en tres libros historiada, que nos dio el rey Luis de Francia'. González Jiménez, pp. 559–60. ([…] and the four books called Historical Mirror that King Louis of France ordered to be made […] And we also send the two Bibles and three books of thick lettering, covered with silver, and the other in three books with images, that King Louis of France gave us).

37 The French translation of the *Escala de Mahoma* under the commission of the King could also be evidence of this exchange. The *Mi'rāŷ* was translated into three different languages. First into Castilian as the *Escala de Mahoma* (now lost), by a Jew named Abraham; then into Latin as *Liber de mundo et coelo* (BnF MS lat. 6064 and BAV MS Vat. lat. 4072); and into French as *Livre de l'echelle* (Bodleian Library, MS Laud. Misc. 537). Both translations were produced by Buenaventura de Siena, an Italian member of the Castilian chancellery and scribe of King Alfonso. In the prologue of the French translation, Buenaventura acknowledges in 1264 that he is not an expert in French and that the text has errors, and he asks French people not blame him for he considers it better to have with mistakes than not at all. Consequently, this translation was done with the desire that it would be read by a French audience. And it is also significant to point out that the Latin and French translations introduce the King as '*Romanorum rex*', '*Rois des Romeins*', not just the King of Castile, therefore signifying an international projection. Fernández Fernández, *Arte y ciencia*, pp. 56–58.

succession of the throne of Castile.[38] The negotiation was a success and they went on to Paris with Charles.[39] The group included Pelayo Pérez, Archdeacon of Astorga, Bellus de Arculis [*Bellum de Archulis*], 'miles et porterius' of the King's chamber, and Pietro de Reggio, 'Magister Petrus de Regio domini Alfonsi regis Castelle protonotarius', member of the chancellery and one of the collaborators in the royal *scriptorium*, specifically in the scientific team.[40]

As members of those royal embassies, important bishops also travelled to Paris. We should not forget the scholarly relationships between episcopal centres as a potential channel of transmission for the Alfonsine Tables. In 1286 and in 1292, the Archbishop of Toledo, Gonzalo Pérez Gudiel (*c.* 1238–99), a pertinent scholar who also collaborated with the intellectual project of Alfonso X, travelled to France in order to negotiate peace with the neighbouring kingdom on behalf of Sancho IV (1258–95).[41] I would also like to underline the embassies organized in 1296 and 1306, the former on behalf of Sancho IV and the latter for Fernando IV (1285–1312). On both occasions, Nicolás, doctor and advisor to the royal house from the time of Alfonso X, 'medicum et consiliarium', travelled.[42] This official, trained in Montpellier and apparently more famous for his political skills than his scientific efforts, boasted at court of his knowledge of astronomy and music; the idea of Nicolás flattering the French monarch with a copy of the Alfonsine Tables might well fit his personality.[43]

Along with the movements of members of the Castilian court into French territory, we should also consider the presence of emissaries from the neighbouring kingdom in the Peninsula. In 1276, Alfonso spent a long period in the city of Vitoria because of his illness. There, the monarch met with Count Robert of Artois (1250–1310), who, on behalf of King Philip III (1245–85), signed a treaty on 7 November 1276.[44]

As we can see, several participants played important roles in both territories; one of these movements could have been used as channel of transmission of books and new ideas.

---

38 The unexpected death of the heir of Castile, Fernando de la Cerda (1255–75), while fighting the Merinids, induced the problem of succession. The legal regulations of the reign established that the line of succession should be that of Alfonso de la Cerda, son of Fernando (and nephew of Philippe III, King of France), but the Castilian nobility generally agreed that the new heir should be Alfonso X's second son, Sancho, who was officially proclaimed the heir in 1278. This action triggered a series of events that involved French diplomacy and ended with Sancho's rebellion against his own father. See H. Salvador Martínez, *Alfonso X, the Learned: A Biography* (Leiden/Boston: Brill, 2010), pp. 257–91; for the embassies see Daumet, pp. 167–69; Procter, p. 26; Antonio Ballesteros Beretta, *Alfonso X el Sabio* (Barcelona: El Albir, 1984), p. 924.

39 'The embassy was successful, and its envoys accompanied Charles from Aix-en-Provence to Paris where their presence was noted and their names duly reported to Edward I [King of England, 1239–1307] by his agent, Maurice de Craon'. Procter, p. 26.

40 Pietro di Reggio [Petro de Regio], in collaboration with Egidio de Tebaldi [Aegidius de Tebaldis, Aegidio de Tebaldi], produced one of the Latin translations of the *Libro conplido en los iudizios de las estrellas* (see note 59).

41 Gonzalo Pérez Gudiel was Bishop of Cuenca, Burgos, and, finally, Toledo. He was also a member of the Royal Chancellery. In 1273 (when he was Archdeacon of Toledo) and in 1280 (as part of the Curial court in Viterbo), Pérez Gudiel commissioned an inventory of his possessions, including his books — a rich collection with numerous scientific manuscripts. The inventories can be consulted in Ramón Gonzálvez Ruiz, *Hombres y libros de Toledo* (Madrid: Fundación Ramón Areces, 1997), pp. 461–549; Francisco J. Hernández and Peter Linehan, *The Mozarabic Cardinal. The Life and Times of Gonzalo Pérez Gudiel* (Florence: Sismel-Edizioni del Galluzo, 2004), pp. 476–505.

42 Daumet, pp. 120, 122, 132, 133, 214, 218–24, 227–28.

43 Juan Torres Fontes, 'Un médico alfonsí: Maestre Nicolás', *Revista Murgetana*, 6 (1954), 9–16.

44 Martínez, p. 382.

With regard to the circulation of books and scholars, it is worth quoting the words of a Parisian astronomer, recently identified as Geoffrey of Meaux,[45] who claimed to have seen a Castilian version of the *Book of the Fixed Stars* taken from the King's bookshelves and also a celestial globe made for Alfonso himself.[46] As Chabás and Goldstein pointed out,[47] it is unclear whether this book was seen, in Castile or in Paris, although Nothaft guesses that the book was seen in Paris.[48] The source is not sufficiently explicit to confirm this claim, but it shows the transmission of Alfonsine works from Castile (provably Seville) to Paris.

In Paris, before 1373, King Charles V of France (1338–80) commissioned a French translation of the Alfonsine Tables. The inventory of the Royal Library in the Louvre, written by Gilles Mallet in 1373, introduces this French copy, which, according to the document, was very well written, in two columns, and *enluminees d'or*, meaning it was a rich, illuminated manuscript with golden elements. The book had to have played an important role in the royal collection, not just for its luxury appearance, but because it was signed by the monarch himself:

> 595. Les Tables Alphons, roy de Castelle, translatees en francois du commandement du roy Charles le quint, et sont en un cayer de parchemin sanx aiz, royees par dessus de vert et de jaune, tres bien escriptes de lettre de forme, à deux coulombes et enluminees d'or. Comm.: *dessus dictes*[49]. Fin: *table du moyen*. Et sont signées au dos dudit derrenier foillet CHARLES.[50]

Unfortunately, the French copy is lost (according to the inventories, it disappeared after 1413); therefore it is impossible to know its content precisely. Nor do we know who produced the French translation. The manuscript seems incomplete, as the inventory records just one quire without binding, but its features are those of a finished book, not a work in progress. We could think that the quire contained just the tables, but Gilles Mallet specifies the kind of writing as *lettre de forme* and the *mise en page* being in two columns; clearly he is describing a folio with text, not with tables, presumably the canons. The question is: which canons? The information provided is not enough to specify whether the manuscript contained the Castilian or the Parisian version. Nevertheless, when in the

45 Nothaft, pp. 88–89.

46 'Vidi namque librum stellarum fixarum scriptum in Hispanico continentem radices stellarum fixarum eodem modo cum tabulis. Qui liber extractus fuit de armario Regis Alfontii, sicut dixit mihi ille qui extrahi eum procuravit. Vidi etiam stellas fixas situatas isto modo in spera solida facta pro ipsomet Alfontio'. A book on the fixed stars written in Spanish that contains the radices of the fixed stars in the same manner as the tables. This book was taken from King Alfonso's book cabinet as I was told by the person who administered its removal. I have also seen the fixed stars placed in this manner on a solid sphere made for Alfonso himself. Nothaft, p. 94. As Nothaft notes, this passage was previously cited by North, pp. 289–90, who misreads *sicut dixit* as *Servus* (?) *dixit*.

47 Chabás and Goldstein, *The Alfonsine Tables*, pp. 247–48.

48 'Since our author makes no allusion whatsoever to having visited the Iberian Peninsula, it would seem that a copy of the *Libro* — together with a celestial globe based on it — had been removed from King Alfonso's private library and transported to France at some point between the 1270s and 1347'. Nothaft, p. 82.

49 In the inventory of 1411 (MS. français 2700, f. 109v), it is specified that this text begins in the II folio.

50 Léopold Delisle, *Recherches sur la librairie de Charles V*, 3 vols (Paris: Champion, 1907), II, p. 100. 'The Alfonsine Tables, King of Castile, translated into French by order of Charles V, and they are in a parchment quire without binding, decorated with green and yellow bands, beautifully written and illuminated with gold. Beginning: *dessus dictes*. End: *table du moyen*. And signed on the back of the last page. Charles'.

inventory a book of astronomical tables is linked to a recognizable author, such as John of Lignères (item 599),[51] the copyist indicates it. Had King Charles V commissioned a French translation of the tables of Alfonso X, it is logical to think that the text used as model would be that made in Castile or at least a version that was considered relevant and hence supported by a qualified author, but as we can see, the inventory only specifies 'Les Tables Alphons, roy de Castelle'. This small detail leaves the door open to speculation that this French copy had been translated from the version that came from Castile rather than the Parisian Alfonsine Tables. Beyond these disquisitions on the possible content of this French copy, this record is also particularly interesting since it mentions the materiality of the manuscript and helps us to consider how the Alfonsine royal manuscript could have been made.

The Royal Library in the Louvre also housed several copies of the tables written in Latin, but the descriptions do not clarify if these items refer to the Castilian or the Parisian set. One of them, item 592, again had the King's signature at the end of the book, so it also had to be considered an important piece in the royal collection.[52] In addition to these items directly related to the Alfonsine Tables, it is important to note that in the inventories of the Royal Library, written in 1411, 1413, and 1424, there are references to two other Alfonsine manuscripts. The first is recorded as follows:

> 714. Un livre d'astronomie, qui semble estre de Arte notoria, escript en espagnol, de lettre en forme, a deux coulombes, très parfaitemente bien figuré, et de bonnes couleurs de'enluminure de Boulougne, et contient en tout cinq cayers, dont le premier commence, au II foillet en rouge lettre: *estas son las figuras*, et ou derrenier *ocio aneiello de mercurio*. Couvert d'une pele de parchemin.[53]

According to its material description, its illumination, and the *incipit* and *explicit*, the manuscript must be the *Libro de astromagia*, kept at the Vatican Library (BAV MS Reg. lat. 1283pt.A, ff. 1r–36v). This manuscript also left the Royal Library, as it was not recorded in the inventory dated in 1423 and it started its own periplus until reaching Rome.[54] The

---

51 'Un livre appellée le Canon maistre Jehan de Linières sur les tables de Alphonse, roy de Castille, en latin et en parchemin, et n'y a autres choses que tables, et commence ou II fo.: *tabula notarum annis*'. Delisle, II, p. 101. (A book called the Canon of Master John of Lignères on the Tables of Alfonso, King of Castile, in Latin, and on parchment, and there is nothing more than tables and starts on the second folio: *tabula notarum annis*).

52 592. 'Les Tables Alphons, en un livre couvert de parchemin. Comm.: tabula veri motus. Fin: tabula proportionis. Et sont signées en la fin CHARLES. – 4s.- / 593. Les Tables Alphons et alia, couvertes de parchemin, escriptes de lettre de forme, en parchemin. Comm.: medius cursus solis. Fin: tabule differentie ascensionum. – Modici valoris. 2 s.' / '594. Tabule Alphonsi, en cayers, couvertes de parchemin et escriptes en parchemin. Comm.: tabula medii motus. Fin: prima secunda tercia. – Nihil'. Delisle, II, pp. 99–100.

53 Delisle, II, p. 117. (A book of astronomy, which seems to be of Arte notoria, written in Spanish, well written, two-column, very well figured, and with good colours, Boulougne illumination, and contains five quires, and the first begins in the second folio with red letters: 'estas son las figuras' and the last 'ocio aneiello de mercurio'. Covered by a parchment.) The previous item, 713, is also a manuscript of 'Ars notoria', written in Spanish, which was also registered in the first inventory redacted in 1373.

54 For further information on the *Libro de astromagia*, see A. Warburg, *La rinascità del paganesimo antico. Contributi alla storia della cultura* (Florence: La nuova Italia 1966); David Pingree, 'Between the "Ghāya" and the "Picatrix"', *Journal of the Warburg and Courtauld Institutes*, 44 (1981), 27–56; A. D'Agostino, *Astromagia: ms. Reg. lat. 1283* (Naples: Liguori 1992); Alejandro García Avilés, 'Two Astromagical Manuscripts of Alfonso X', *Journal of the Warburg and Courtauld Institutes*, 59 (1996), 14–23; Fernández Fernández, *Arte y ciencia*, pp. 289–319.

other was a French copy of the *Libro de las formas et las ymágenes*,[55] another partially preserved Alfonsine manuscript, with just the prologue and the index, currently at the Escorial (RBME MS h-I-16):

> 616. Trente neuf cayers en papier du livre des formes, figures et ymages qui sont ès cieux, translatez d'espagnol en françois par Pierre Leraut, jadiz maistre des pors et passaige en la senechaucie de Beaucaire, du commandement de Monseigneur le duc de Berry, dont le premier cayer commence *Au nom du père et du filz*. Et sont touz yceulz cayers liez en une couverture de parchemin. – Non prisié. Nihil.[56]

According to the inventory, the translation into French was commissioned by John, Duke of Berry (1340–1416), and was produced by Pierre Leraut, identified as Pierre Lesant, seneschal at the service of the King of France, *maître des ports de la sénéchaussée de Beaucaire-Nîmes*, who certainly knew the Spanish language well.[57] If the duke ordered this translation, it is logical to think that he had in his possession a manuscript with the Spanish version, perhaps the one produced in the Alfonsine *scriptorium*, or at least a close copy. Unfortunately, apart from the general prologue and the index kept at the Escorial, there is no trace of this book nor of the French copy in current collections; therefore there are multiple gaps in the information that we have.[58]

In addition to these recognizable titles, the inventories of the Louvre Royal Library also include two Latin copies of the *Liber Razielis*,[59] as well as three exemplars of Haly Abenragel's *Liber in judiciis astrorum*.[60] Both works were also translated in the Alfonsine *scriptorium*,[61] which shows that Alfonso X's books circulated beyond the borders of his

55 Alfonso X, *Lapidario and Libro de las formas & las ymagenes*, ed. by Roderic C. Diman and Lynn W. Winget (Madison: Hispanic Seminary of Medieval Studies, 1980); García Avilés, 'Two Astromagical'; Fernández Fernández, *Arte y ciencia*, pp. 281–88.

56 Delisle, II, p. 103. (Thirty-nine quires on paper of the book of the forms, figures, and images that are in the heavens, translated from Spanish into French by Pierre Leraut, master of ports of the seneschal of Beaucaire, by order of the Duke of Berry, and the first quire begins *Au nom du père et du filz*. And all the quires are gathered in a parchment cover. It is not common.)

57 The word 'Leraut' seems to be a misunderstanding of Delisle. García Avilés, 'Two Astromagical', p. 18; cf. G. Dupont-Ferrier, *Gallia regia ou état des officiers roy aus des baillages et des senéchaussées de 1328 à 1515*, 6 vols (Paris: Imprimierie nationale, 1942–61), I, pp. 302–3; C. Lagomarsini, 'Le cas du compilateur compilé', *Journal of the International Arthurian Society*, 3 (2015), 55–71 (p. 64).

58 As García Avilés pointed out, the *Libro de las formas et las ymágenes* did not appear, or at least with this title, in the inventory of the library of the Duke of Berry made in 1416, but in the document it is said that there were two Spanish manuscripts of magic in the hands of M. Arnoul Belin, treasurer of Sainte Chapelle de Bourges, 'Deux gros livres de magique escripts en espagnol, l'un couvert de pel rouge et l'autre d'une blanche pel sans ais, lesquels M. Arnoul Belin a eu comme l'an dit'. Alfred Hiver de Beauvoir, *La Librairie de Jean duc de Berry au Château de Mehun-sur-Yevre* (Paris: Auguste Aubry, 1860), item 81, p. 43; García Avilés, 'Two Astromagical', p. 18.

59 699 and 700. Delisle, II, p. 115.

60 671, 675 and 721. Delisle, II, pp. 111, 119.

61 The Castilian translation of the *Liber Razielis* is lost, but thanks to the Latin copy (BAV MS Reg. lat. 1300), we know it was produced under the commission of King Alfonso X by 'magister Johannes clericus', probably Johan Daspa. The *Libro conplido de los iudizios en las estrellas* was translated into Latin twice, and both translations were commissioned by the King during different periods. The first one was produced by *Alvarus*, identified as Álvaro de Oviedo, and is preserved just in two manuscripts (BAV MS Palat. lat. 1370, ff. 65ʳ–77ʳ, and RBME MS J-II-17). The second was produced by Aegidio de Tebaldi and Pietro de Reggio, and in this case, there are multiple copies preserved. This version rapidly spread and was used for new translations into French, English, German, Dutch, and Hebrew. For further information and a bibliography, see Alfonso X, *El libro conplido en los iudizios de las estrellas*, ed. by Gerold

kingdom and highlights the great interest that French monarchs, especially Charles V, had in the scientific production commissioned by their Castilian homonyms.[62]

3. The reception of the Alfonsine Tables and their Parisian version in the Hispanic kingdoms during the fourteenth and fifteenth centuries

3.1. The Crown of Castile

Curiously, in Castile after the death of King Alfonso X and for almost two centuries after, the influence of his astronomical tables, both the original and the later Parisian version, was practically non-existent. Between 1284 and the second half of the fifteenth century, there are hardly any documentary traces or evidence of their use.[63]

There is a mention of the Alfonsine Tables in a treatise on astrology, dated 1438 (BNE MS RES/2) and attributed to Enrique de Villena (*c.* 1384–1428).[64] The author explains how the movements of the Sun and the Moon are 'concordados et examinados en las tablas alfonsias del anden del çielo estrellado'.[65] Villena, a nobleman related to the royal houses of Castile and Aragon, devoted his life to study.[66] Although he died in Castile in the service of King Juan II (1405–54), his nephew, Villena spent a long period at the court of King Martì I (1356–1410) in Barcelona, and had a close relationship with some members of the Hebraic community in Barcelona and Zaragoza. When Villena died, King Juan II

Hilty (Madrid: Real Academia Espanola, 1954); Gerold Hilty, 'El libro conplido en los iudizios de las estrellas', *Al-Andalus*, 20 (1955), 1–74; Fernández Fernández, *Arte y ciencia*, pp. 107–12, 320–22. In the Louvre Royal Inventory, there are also three examples of the French translation of this work: 669, 672, and 673, Delisle, II, p. 111 (maybe also 670, draft material).

62 Jean-Patrice Boudet studied the connection between the two monarchs and how they used Salomon as a model of a learned King. J.-P. Boudet, 'Le modèle du roi sage aux XIIIe et XIVe siècles: Salomon, Alphonse X et Charles V', *Revue historique*, 647 (2008), 545–66.

63 With regard to the presence of the exemplars of the Parisian Alfonsine Tables in Spanish institutions, apart from the BNE MS 4238 linked to the Crown of Aragon we discuss later, Chabás and Goldstein, *The Alfonsine Tables*, pp. 292–93, discuss BNE MS 7856, a fourteenth-century manuscript containing the tables and canons of John of Saxony explicitly mentioning *Jacobus de P[e]rusia etc.* [the final sign is the abbreviation of *et caetera*, not *scripsit* as some authors read], and following Millás Vallicrosa (1942, 227–30), BNE MS 10002, an early fifteenth-century manuscript, probably of Italian origin. BNE MS 7856 comes from the Royal Library in Madrid, so it is not easy to specify a previous owner without an in-depth study of the different book collections gathered in the palace; BNE MS 10002 belonged to Cardinal Francisco Javier de Zelada who lived and created his library in Rome. After his death, his manuscripts were donated to the library of Toledo Cathedral; this is why the book is currently in the BNE, so it cannot be considered as a piece of information to document the knowledge of the Parisian Alfonsine Tables in the Iberian Peninsula during the late Middle Ages.

64 José M. Millás Vallicrosa, 'El "Libro de Astrología" de Don Enrique de Villena', *Revista de Filología Española*, 27 (1943), 512–42 (p. 23); Enrique de Villena, *Tratado de astrología atribuido a Enrique de Villena*, ed. by Pedro M. Cátedra and Julio Samsó (Madrid-Barcelona: Editorial Labor-Río Tinto Minera, 1981), p. 157.

65 BNE MS RES/2, f. 19vb.

66 Emilio Cotarelo y Mori, *Don Enrique de Villena. Su vida y obras* (Madrid: Estudio Tipográfico Sucesores de Rivadeneyra, 1896). Elena Gascón Vera, 'Nuevo retrato histórico de Enrique de Villena', *Boletín de la Real Academia de la Historia*, 175 (1978), 107–43; Pedro M. Cátedra, 'Para la biografía de don Enrique de Villena' *Estudi General*, 1 (1981), 29–33 and 'Algunas obras perdidas de Enrique de Villena, con consideraciones sobre su obra y su Biblioteca', *El Crotalon: Anuario de Filologia Española*, 2 (1985), 53–75; Derek C. Carr and Pedro M. Cátedra, 'Datos para la biografía de Enrique de Villena', *La Coronica*, 11 (1983), 293–99.

ordered the expurgating of his library. This action was justified by religious intentions, but in reality it was an organized political strategy in the court to find a scapegoat in a period of great difficulty.[67] The act was carried out by Bishop Lope de Barrientos (1382–1469), who decided to burn fifty scientific books, which, according to his criteria, dealt with necromancy and occult sciences. Barrientos kept the rest of the book collection. Whether a copy of the Alfonsine Tables was among those books is something we cannot confirm, but it seems plausible since Villena explicitly mentions them in his work.

Apart from this short note, we do not have, or have not yet found, other references to the Alfonsine Tables in scientific works produced in Castile. In fact, information about astronomical and astrological practices in this territory during the fourteenth century is not abundant and the promotion of this kind of work by the court disappeared. This could be related to the position of Christian orthodoxy against Muslim and Judaic communities; the latter groups were closely related to the knowledge and practice of natural sciences, especially astrology. Nevertheless, there are enough references to conclude that astrological practices remained alive in Castile at that time,[68] but people involved in these practices probably used other tables or tools to make horoscopes.

An 'Alfonsine tradition' in Castile was not recovered until Nicolaus Polonius occupied the chair of astronomy, *cathedra astrologia*, at the University of Salamanca (*c.* 1460). This scholar, probably from Poland as his name suggests, brought with him a version of the Parisian Alfonsine Tables called the *Tabulae resolutae*, initially developed in Cracow for university students and later adapted to other Central European cities. In Salamanca, Polonius composed new tables for the meridian of this city.[69]

The Polonius tables and canons are kept in a fifteenth-century manuscript of Castilian origin, now in Oxford (Bodleian Library MS Canon Misc. 27, ff. 122^{v}–129^{r})[70] and in a

67 After his death, Villena was accused of practising necromancy and magic, and although he was defended by important authors, most of the nobility and common people judged his excessive liking for study and science as inappropriate and gladly accepted this accusation. For further information, see Elena Gascón Vera, 'La quema de los libros de don Enrique de Villena: una maniobra política y antisemítica', *Bulletin of Hispanic Studies*, Oct 1 (1979), 317–24; Derek C. Carr, 'Arabic and Hebrew Auctoritates in the Works of Enrique de Villena' in *From Arabye to Engelond. Medieval Studies in Honour of Mahmoud Manzalaoui*, ed. by A. E. Christa Canitz and G. R. Wieland (Ottawa: University of Ottawa Press, 1999), pp. 39–60.

68 In addition to the mention of astrological practices in medieval literature and Castilian chronicles, the copies of the *Libro conplido* confirm the interest on this matter during the fourteenth and fifteenth centuries. Luis M. Vicente García, *Estrellas y astrólogos en la literatura medieval española* (Madrid: Laberinto, 2006); Fernández Fernández, *Arte y ciencia*, pp. 129–34.

69 Guy Beaujouan, *Manuscrits scientifiques médiévaux de l'Université de Salamanque et de ses 'Colegios mayores'* (Bordeaux: Yéret, 1962), p. 16; José Chabás and Beatriz Porres de Mateo, 'Los cánones de las "Tabulae Resolutae" para Salamanca: origen y transmisión', *Cronos: cuadernos valencianos de historia de la medicina y de la ciencia*, 1 (1998), 51–83. José Chabás, 'The Diffusion of the Alfonsine Tables: The Case of the *Tabulae resolutae*', *Perspectives on Science*, 10 (2002), 168–78; José Chabás, 'The University of Salamanca and Renaissance of Astronomy during the Second Half of the 15th Century', in *Universities and Science in the Early Modern Period*, ed. by Morchedai Feingold and Víctor Navarro Brotóns (Dordrecht: Springer, 2006), pp. 29–36 (p. 31). Chabás, *Computational Astronomy*, pp. 311–20.

70 For reproductions of several folios, see: https://digital.bodleian.ox.ac.uk/inquire/p/f. 5659bcc-a7ab-4129-bd37-00c397d37f68. In addition to the tables and canons written by Polonius, this manuscript includes the canons and tables of Jacob ben David Bonjorn, some tables associated with Ibn al-Kammād, the canons of John of Lignères and his Tables of 1322. Chabás and Goldstein, *The Alfonsine Tables*, p. 293. The Bodleian Library digital catalogue repeats the incorrect date (fourteenth century, third quarter) published by Jonathan J. G. Alexander and Otto Pächt, *Illuminated Manuscripts in the Bodleian Library* (Oxford: Clarendon Press, 1966), p. 67, instead of the correct date, *c.* 1461.

fifteenth-century manuscript in Lisbon (Torre do Tombo, Manuscritos da Livraria, tt-msliv n.º 2115); its canons are also found in a sixteenth-century manuscript in Cambridge, (Trinity College, MS O.3.13, ff. 74v–80v). The manuscript in Oxford is especially interesting, because, in addition to Polonius's text, it has a set of beautiful drawings of the zodiacal constellations (with the stars in red and captions written in Castilian ff. 152r–157v). The captions provide material that links certain stars to the lunar mansions or clarifies aspects of the figure. The set of zodiacal constellations is located at the end of the book, after the Polonius tables. Besides their beauty, these illustrations are a very interesting part of the history of the Alfonsine Tables because the latter allow us to consider whether or not they had a star catalogue.

In 1986, Kunitzsch wrote an interesting paper on the star catalogue appended to the Alfonsine Tables, or rather the Parisian Alfonsine Tables.[71] The author asserted that 'like every complete *zij*, or work of astronomical tables, the Alfonsine Tables [he was talking about the Parisian version], also contain a star catalogue' that was included in many manuscript copies. He further explained that 'no collaborators of Alfonso's were involved in the establishment of this catalogue', which was the same as that in the translation of Gerard of Cremona's *Almagest*, but adopted to the value of precession used in the star catalogue found in Alfonso's *Libro del saber de astrología* (Ptolemy +17°8').[72]

Today, with a much greater number of examples analysed, we know that most manuscript copies of the Parisian Alfonsine Tables do not contain a star catalogue and where they do, they are not all the same.[73] Why do only certain manuscripts contain the star catalogue? Does this detail have anything to do with the content and dissemination of the Alfonsine Tables produced in Castile? Did the Castilian Tables have a star catalogue or was it included in the early Parisian version? As usual, there are no conclusive answers; but if the coordinates in some tables are the same as in other Alfonsine materials, it is logical to assume that the Alfonsine Tables had a star catalogue from the beginning. Indeed, if we analyse the content of the Castilian canons in MS 3306, a table with the fixed stars is expressly mentioned in Chapters 39 and 41.[74]

What might the table of fixed stars in the Alfonsine Tables have looked like? Did it include iconic apparatus? Ptolemy's star catalogue in the Latin translation of Gerard of Cremona was also disseminated in a group of manuscripts called *Ṣūfī Latinus*, but in this case the longitudes of the stars are the same as al-Ṣūfī's value (Ptolemy +12°42'). These star

71 Paul Kunitzsch, 'The Star Catalogue Commonly Appended to the Alfonsine Tables', *Journal for the History of Astronomy*, 17/2 (1986), 89–98 (p. 89).

72 Kunitzsch, 'The Star Catalogue', p. 90. The same argument appears in Paul Kunitzsch, 'Star Catalogues and Star Tables in Mediaeval Oriental and European Astronomy', *Indian Journal of History of Science*, 21/2 (1986), 113–22 (p. 117).

73 I owe thanks to Richard L. Kremer for this information. The star catalogues preserved in manuscripts and printed editions have variations in their data and even in the names of the stars. For example, the catalogue in the 1483 *editio princeps* derives from Prosdocimo's catalogue (José Chabás, 'From Toledo to Venice: the Alfonsine Tables of Prosdocimo de' Beldomandi of Padua (1424)', *Journal for the History of Astronomy*, 38 (2007), 269–81) and the star catalogue published in 1492 has twenty-nine stars with different names, which, according to Kunitzsch, were compiled from other sources and added to the star catalogue by the circle of John of Gmunden (*c.* 1380–1442) in Vienna (Kunitzsch, 'The Star Catalogue', pp. 92–93). Different kinds of star catalogues can be found in José Chabás and Bernard R. Goldstein, *A Survey of European Astronomical Tables in the Late Middle Ages* (Leiden: Brill, 2012), pp. 185–99.

74 Chabás and Goldstein, *The Alfonsine Tables*, pp. 78, 84.

catalogues are illustrated with images of the constellations derived from the visual tradition of al-Ṣūfī's *Book of the Fixed Stars*.[75] With regard to the constellations of Oxford Bodleian MS Canon Misc. 27, although the style of these drawings follows the fifteenth-century trend, it is worth noting that the iconography of the constellations is the same as in the Alfonsine manuscripts. Like the *Ṣūfī latinus* corpus, the Alfonsine iconography of constellations stems from the figurative tradition derived from al-Ṣūfī's *Book of the Fixed Stars* but from a different brand of dissemination with specific features.[76] Moreover, the information presented by the Castilian captions in MS Canon Misc. 27 fits with the data provided by the *Libro de las figuras de las estrellas fixas*, the first treatise of the aforementioned *Libro del saber de astrología*. These particularities seem to point to a Castilian source for this material. What was the formal model for this manuscript? Could the painter be using a lost copy of the Alfonsine Tables alongside a star catalogue with figures? Certainly, if we think of most scientific manuscripts made in the Alfonsine *scriptorium*, the idea of the Alfonsine Tables as an illuminated manuscript with an iconic repertoire makes perfect sense. These questions once again are difficult to resolve, but it is clear that there is a direct connection between the images in the Oxford manuscript and Alfonsine materials.[77]

In addition to the knowledge brought to, and developed by, Polonius in Salamanca, the Parisian Alfonsine Tables were also known in Castile in the final decades of the fifteenth century and throughout the sixteenth century. Proof of this is the anonymous Castilian translation of John of Saxony's canons preserved in a paper manuscript (dated *c.* 1490) in the Escorial (RBME MS T-III-29, ff. 120^r–150^r)[78] and the sixteenth-century Castilian translation made by the 'bachiller Franciso de Morales, clérigo presbítero' preserved in BNE MS 3306, folios. 74^r–87^v. Both translations were made from the text recorded in the *editio princeps* of 1483.[79] Who commissioned the first translation into Spanish, where was it made, and by whom are questions yet to be resolved. Clarifying these points would contribute to a better overall view of this sequence of dissemination. The second translation is also linked to numerous interrogations. The type of writing is known as *italic*, whose

75 Kunitzsch, 'Star Catalogues', p. 117. For the 'Sufi Latinus' manuscripts, see Marie-Thérèse Gousset, 'Le Liber de locis stellarum fixarum d'Al-Sûfi, ms. 1036 de la Bibliothèque de l'Arsenal à Paris: une réattribution', *Arte medievale*, 2 (1984), 93–107; Paul Kunitzsch, 'Ṣūfī Latinus', *Zeitschrift der Deutschen Morgenländischen Gesellschaft*, 115 (1965), 65–74; Paul Kunitzsch, 'The Astronomer Abū 'l-Husayn al-Ṣūfī and his Book on the Constellations', *Zeitschrift für Geschichte der Arabisch-Islamischen Wissenschaften*, 3 (1986), 56–81.

76 Laura Fernández Fernández, 'Arab Stars in the Castilian Sky: Al-Ṣūfī's *Book of fixed stars* Amongst the Manuscripts of Alfonso X', in *The Stars in the Classical and Medieval Traditions*, ed. by Alena Hadravová, Petr Hadrava, and Kristen Lippincott (Prague: Scriptorium, 2019), pp. 93–114.

77 This relationship between the figures in MS Canon Misc. 27 and the Islamic imprint in the Alfonsine manuscripts was pointed out by Gousset, p. 94 and Kunitzsch, 'The Astronomer', p. 81. On the other hand, Alejandro García Avilés, 'Arte y astrología en Salamanca a finales del siglo XV', *Anuario del Departamento de Historia y Teoría del Arte*, 6 (1994), 39–60 (p. 45) suggested that the visual source of this manuscript might be an Islamic celestial globe rather than a manuscript.

78 This manuscript in its old signature, *olim*. iii.Q.26, was cited by Rodríguez de Castro, p. 645 and Rico y Sinobas, V, pp. 24–25. The codex is a composite miscellaneous book with a Latin text of Gasparo Contarini (1483–1542), dated in Venice in 1527 (ff. 1^r–119^v). The book, presently in El Escorial, belonged to the library of the Conde Duque de Olivares (1587–1645).

79 The first text was edited by José Martínez Gázquez, who also published a few fragments of the second one. John of Saxony, *Las Tablas de los movimientos de los cuerpos çelestiales del illuxtrisimo Rey Don Alonso de Castilla: seguidas de su Additio: traducción castellana anónima de los cánones de Juan de Sajonia*, ed. by José Martínez Gázquez (Murcia: Universidad de Murcia, 1989).

use, first extended in the Aragon area, spread through Castile at the end of the sixteenth century; therefore, we could consider that this copy was made during the latter decades of that century.[80] Neither Rico y Sinobas nor Martínez Gázquez, the scholars who studied this translation, provided any information about Francisco de Morales.[81] As mentioned previously, the text is part of BNE MS 3306, a composite manuscript that had belonged to Juan Fernández de Velasco, whose library was well stocked with scientific books;[82] therefore, it is logical to think that the translation was made within his circle of influence.

The question of the disappearance of the Alfonsine Tables for almost two centuries is always present in studies of history of science on the Iberian Peninsula and the answer probably has more to do with political than with scientific aspects. Science is not neutral, it is subordinated to the political and social framework in which it develops; in this case, astronomical tables were intimately linked to the monarch. In addition to their scientific purpose, the Alfonsine Tables were conceived as an important piece in the construction of the King's legacy. The epoch of their calculation was the year 1252, when Alfonso was proclaimed king, and in the prologue of the canons, the *era alfonsí* was established, equivalent to the eras of other historically renowned characters such as Alexander the Great or Caesar. One of the possible reasons for their 'disappearance' from the Castilian scientific milieu could be directly related to the desire to tarnish the memory of the King. This *damnatio memoriae* was used as part of a political strategy constructed within the framework of the court, firstly to justify the revolt of his son, Sancho, and later to vindicate the need to halt his lineage and support the arrival of the Trastámara House in the government.[83] In this political movement, the King, previously distinguished for his wisdom, was transformed into an astrologer or even a sorcerer interested in occult sciences. Furthermore, the main aspect of this legend, constructed to slander Alfonso X, was his excessive promotion and practice of science and consequently his arrogance was so intense, according to his enemies, that he had even questioned the work of God.[84]

---

80 The type of writing is known as *italic*, whose use first extended in the Aragon area and spread through Castile at the end of the sixteenth century; therefore, we could consider that this copy was made during the latter decades of that century. Rico y Sinobas dated it ambiguously: V, pp. 42–43 (sixteenth century), I, p. LVI (seventeenth century); Martínez Gázquez, p. 16, also dated it to the sixteenth century.

81 A *bachiller*, scholar, and priest, called Francisco de Morales Cabrera (1564–1614) oversaw the Latin Chair, and from 1591 also the Greek Chair, at the University of Salamanca. Enrique Esperabe de Arteaga, *Historia pragmática e interna de la Universidad de Salamanca y los Reyes* (Salamanca: Imprenta de Francisco Núñez, 1914), p. 600. Perhaps this scholar was the one who was entrusted with the translation.

82 It is not strange that Juan Fernández de Velasco's son, Bernardino Fernández de Velasco y Tovar (1609–52), commissioned Francisco García Ventanas, a mathematician at his service, with the last edition of the Parisian Alfonsine Tables, the only one made in Spain. This new edition was printed in Madrid in 1641. *Tabulae Alphonsinae Perpetuae Motuum Coelestium denuo restitutae et illustratae à Francisco García Ventanas Mathematico* (Madrid: Officina Regia, 1641).

83 The Trastámara lineage began its rule in Castile in 1369 with Enrique II (1334–79) after the murder of his half-brother King Pedro I (1334–69).

84 This episode is known in Alfonsine historiography as the 'blasphemy of King Alfonso'. For further information, see Jerry R. Craddock, 'Dynasty in Dispute: Alfonso X el Sabio and the Succession to the Throne of Castile and Leon in History and Legend', *Viator*, 17 (1986), 214–19; Bernard R. Goldstein, 'The Blasphemy of Alfonso X: History or Myth', in *Revolution and Continuity: Essays in the History and Philosophy of early Modern Science*, ed. by Peter Barker and Roger Ariew (Washington, DC: Catholic University of America Press, 1991), pp. 143–53; Leonardo Funes, 'La blasfemia del Rey Sabio: itinerario narrativo de una leyenda (primera parte)', *Incipit*, 13 (1993), 51–70, 'La blasfemia del Rey Sabio: itinerario narrativo de una leyenda (segunda parte)', *Incipit*, 14 (1994), 69–101, and 'La leyenda de

It is remarkable that when, in the rest of Europe, Alfonso X was respected and indissolubly linked to astronomical science thanks to his tables, in the Castilian historiographical sources, the monarch became a negative figure. However, despite the pernicious vision of Alfonso X orchestrated by a faction of the court, the ideal of the Learned King was defended by other authors such as don Juan Manuel (1282–1348). In the fifteenth century, a few Castilian chronicles tried to recover the intellectual reputation of the monarch, pointing to his *tablas de astrología* as one of his main achievements.[85] In the sixteenth century, important scholars claimed the cultural production promoted by the Learned King, and as we have seen, the fame of his tables continued to grow. The laudatory words about the Alfonsine Tables that Jerónimo Román de la Higuera incorporated in his aforementioned book, *Historia eclesiástica de la imperial ciudad de Toledo*, should be understood within this framework of recognition of Alfonso X. Nevertheless, the negative view of the King was kept alive as a result of the new versions of the legend disseminated by Juan de Mariana (1536–1624) and Jerónimo Zurita (1512–80), and it remained active until the nineteenth century.[86]

3.2 The Crown of Aragon

On the other hand, in the Crown of Aragon scientific activity promoted by the court continued during the late Middle Ages; in fact, King Pere 'el Ceremoniós' (1319–87) commissioned his own tables.[87] The Royal Library had to offer a significant number of scientific books in order to support the work of the astronomers and physicians patronized first by King Pere, and later by his son Joan I 'el Caçador' (1350–96). Proof of this is in the letter that *el Ceremoniós* wrote to his archivist on 24 October 1359 to allow Dalmau Ses Planes, an astronomer from Perpignan in his service, to consult all the astrological books in

---

la blasfemia del Rey Sabio: revisión de su itinerario narrativo', e-Spania [Online], online since 1 October 2016, accessed 22 March 2020. URL: http://journals.openedition.org/e-spania/25873; George Martin, 'Alphonse X maudit son fils', *Atalaya*, 5 (1994), 151–78; Amaia Arizaleta, 'De la soberbia del rey: dos formas breves en la construcción historiográfica', in *Tipología de las formas narrativas breves románicas medievales*, III, ed. by Juan M. Cacho Blecua and Mª Jesús Lacarra (Zaragoza/Granada: Universidad de Zaragoza/Universidad de Granada, 2004), pp. 79–110; Isabel de Barros Días, 'A blasfémia do Rei Sabio: os antecedentes da lenda', in *Estudios de literatura medieval. 25 años de la AHLM*, ed. by Antonia Martínez Pérez and Ana Luisa Baquero Escudero (Murcia: Universidad de Murcia, 2012), pp. 189–96; Isabel de Barros Días, 'La blasfemia del Rey Sabio: vicisitudes de una leyenda (nuevas hipótesis respecto a la datación y la posición relativa del texto portugués)', *Anuario de Estudios Medievales*, 45/2 (2015), 733–52.

85 Fernando Pérez de Guzmán (*c.* 1377–1460) evoked the influence that Alfonso X had over the kingdom even after his death in his work, *Loores a los claros varones de España*: 'Non le valiendo ciencia, / Franqueza, esfuerzo e potencia / De que tanto fué dotado. / Vive por caballería / Este rey, pero es muerto. / Aunque duerme está despierto / Por Tablas de Astrologia. / Ordena, rige e guia / Con Leyes nuestras memorias / Deléitanos con Estorias; / Orna con Filosofia'. (Not worth the science, openness, effort, and power, which were his gifts. He lives by chivalry, this king though he is dead, though he sleeps he is awake, by tables of astrology, he commands, rules, and guides, with laws, our memories. He delights us with stories, he decorates with philosophy). Marcelino Menéndez y Pelayo, *Antología de poetas líricos castellanos. La poesía en la Edad Media* (Madrid: CSIC, 1944), T. 4, p. 259. For an interesting analysis of Alfonso X in some fifteenth-century chronicles, see Jean-Pierre Jardin, 'La figure du roi Alphonse X chez quelques chroniqueurs du XVe siècle', *Cahiers de linguistique hispanique médiévale*, 20 (1995), 75–96.

86 Antonio Rivera García, 'La leyenda sobre la blasfemia de Alfonso X: un episodio de la conflictiva relación entre especulación teórica y razón de estado', *eHumanista* 31 (2015), 426–51.

87 José M. Millás Vallicrosa, *Las Tablas astronómicas del Rey Pedro el Ceremonioso* (Madrid/Barcelona: CSIC, 1962).

the royal archive with the aim of creating his own work.[88] Thanks to such documentation, we know that in 1361 the monarch also commanded Dalmau Ses Planes and Pere Gilbert to construct a celestial sphere for the Royal Library, 'cambra on estan los libres' (room where the books are), with the figures of the constellations and planets, 'signes, steles, planets e alters figures diverses' (signs, stars, planets, and other figures). Depicted by Ramón Torrent, a painter from Barcelona, this curious and very luxurious artefact was made of gold, silver, and tin, was depicted in blue 'atzur d Acra' (ultramarine blue) and exhibited in the Royal Palace.[89]

In this scientific atmosphere, the tradition of the Alfonsine Tables was kept alive, at least in a documentary sense. In 1373, Ses Planes specifically mentions them in the prologue of the work *Taules i almanac*.[90] Additionally, in the inventory of goods preserved in the Royal Palace in Barcelona [Palau Reial Major] recorded after the death of King Martí 'l'Humà' (1356–1410), there are two important items related to them.[91] The first mentions a book of the tables, probably without the canons, apparently produced in Mallorca[92] and written in Catalan:

> 24. Yĕhudà ben Mošé ha-Kohen, Iṣḥāq ben Sid, Taules alfonsines, Traductor: Anònim. Text: Item, un altre libre appellat Taules alfonsines, en català, scrit en pergamins ab posts de fust cubert de cuyro vermell emprenptades ab senyal reyal a cada part, ab dos tancadors de fil groch e vermell. Lo qual comença: 'Taules de les diffarèncias', e faneix en vermelló: 'in Maiorica'.[93]

---

88 Antonio Rubió y Lluch, *Documents per l'historia de la cultura catalana mig-eval* (Barcelona: Institut d'Estudis Catalans, 1908–21), I, 190: CXCI. The place, as Rubió clarifies, was Cervera, not Girona. The same day, the King wrote to Ses Planes to command him to work with Pere Gilbert, 'magistrum in artibus et in astrorum scienciam peritissimum'. José Chabás, *L'Astronomia de Jacob Ben David Bonjorn* (Barcelona: Institut d'Estudis Catalans, 1992), p. 16. The document specifies the prohibition of consulting two books of Ali ben Ragel, one of which was probably the *Judiciis Astrologie*, the Latin translation of the *Libro conplido en los iudizios de las estrellas* commissioned by Alfonso X. According to Vernet and Romano, the book was translated into Catalan during the reign of King Pere 'el Ceremoniós', and King Joan, on 11 October 1386 (one year before his coronation), commissioned a new copy. Unfortunately, the Catalan copy is also lost. Joan Vernet and David Romano, *Bartomeu de Tresbéns. Tractat d'Astrologia* (Barcelona: Biblioteca Catalana d'Obres Antigues, 1957–58), p. 202; Fernández Fernández, *Arte y ciencia*, pp. 132–33.

89 Rubió y Lluch, *Documents*, I, pp. 199–200: CCI, CCII, and 1921, II, pp. 139–42: CXLI. Curtains for protecting the globe are also mentioned in the document.

90 This work must be the result of the observations made by Ses Planes and Gilbert between 1360 and 1366, but it was written by Ses Planes in 1373. Lynn Thorndike, 'Introduction and Canon by Dalmatius to Tables of Barcelona for the Years 1361–1433', *Isis* 26 (1937), 310–20; Chabás, *L'Astronomia*, pp. 16–17.

91 'Inventari parcial de la biblioteca reial elaborat, a la mort de Martí I 'l'Humà' (1396–1410), per Pau Nicolau, secretari del rei, per autoritat de Francesc Fonolleda, en benefici de la reina vídua, Margarida de Prades.' Barcelona - Arxiu de la Corona d'Aragó - Cancelleria reial, Registres – reg. 2326. This is a partial inventory published for the first time by Jaume Massó i Torrents, 'Inventari dels béns mobles del rey Martí d'Aragó', *Revue Hispanique*, 12/42 (1905), 413–590. A selection with the scientific items can be consulted in Chabás, *L'Astronomia*, pp. 28–37, and online via Sciencia.cat: *La ciència en la cultura catalana a l'Edat Mitjana i el Renaixement* (https://www.sciencia.cat/db/cercador.htm?doc=19).

92 In 1381, King Joan wrote to the governor of Mallorca to request astrological treatises and the Jewish astronomer Vidal Afrahim was working there, so the city must have been an established location for copying scientific manuscripts. Michael A. Ryan, *A Kingdom of Stargazers: Astrology and Authority in the Late Medieval Crown of Aragon* (Ithaca: Cornell University Press, 2011), p. 114.

93 (Another book called Taules alfonsines, in Catalan, written on parchment, with board binding covered with red leather, engraved with the royal arms, with two yellow and red thread fastenings. Which begins 'Taules de les diffarèncias', and ends, in red, 'in Maiorica'.)

This parchment codex had to be a relevant copy, or at least, it had a special significance in the Royal Library since its binding included the royal heraldry, a peculiarity that not all the books showed. In addition to this manuscript, another Catalan copy is mentioned, on paper:

> 238. Yěhudà ben Mošé ha-Kohen, Iṣḥāq ben Sid, Taules alfonsines, Traductor: Anònim. Text: 'Ítem, un altre libre appellat Taules alfonsines, en romanç, scrit en paper ab posts de paper engrutades e cubert de cuyro vermell, ab dos tancadors de bagua. Lo qual comença: 'Per ço és atrobat', e faneix: 'havets aüdes altres'.[94]

Therefore, the Alfonsine Tables were also translated into Catalan and they were copied on several occasions. Although this inventory was made after King Martí's death, the astronomical books had been commissioned or acquired by King Pere, and particularly by Joan I. The latter became Jean de Berry's brother-in-law following Joan's marriage to Yolande de Bar, so he could have been a point of connection between both territories and cultural milieus.[95]

Unfortunately, none of the books of the tables mentioned in this inventory have been preserved. We must not forget that this list of books corresponds to part of the goods given to the widowed Queen Margarita de Prades (*c.* 1387–1429) and that they were probably sold, meaning they left the royal archive, and were sadly either lost or simply have not yet been identified.[96]

The Parisian Alfonsine Tables were also known in the Crown of Aragon and they were even adapted to the meridian of Morella (BNE MS 4238), although the number of copies of the Parisian Alfonsine Tables on the Peninsula is minimal compared to those preserved in other parts of Europe.[97]

94 (Another book called Taules alfonsines, in romance, written on paper, with paper covers and covered with red leather, with two clasps. Which begins 'Per ço és atrobat' and ends 'havets aüdes altres'.)

95 If we examine the documentation, this connection is evident. For example, on 11 October 1388, in Zaragoza, King Joan (1350–96) requested that Jean de Berry give the Parisian astrologer Guillem Lunell permission to travel from France to Aragon. Rubió y Lluch, *Documents*, I, 354: CCCXCVI; Ryan, pp. 118–19. García Avilés, 'Two Astromagical', pp. 20–21, suggests that the *Libro de las formas et las ymágenes* and the *Libro de astromagia* could have arrived in France thanks to the close ties between Joan I and the Duke of Berry, and even Charles VI of France. If so, we should first pinpoint the transmission of the book from Castile to Aragon.

96 Jaume Riera i Sans, 'La bibliothèque du roi Martin', in *Association Internationale de Bibliophilie. XXII Congrès. Actes et Communications, Barcelona, 2001* (Barcelona: Associació de Bibliòfils de Barcelona, 2005), pp. 105–17 (p. 117). There is also a seventeenth-century copy made by Fray Manuel Ribera (1652–1736) among the books from the Convento de la Merced, in Barcelona, which was used by Félix Torres Amat, *Memorias para ayudar a formar un diccionario crítico de los escritores catalanes y dar alguna idea de la antigua y moderna literatura en Cataluña* (Barcelona: Imprenta de J. Verdaguer, 1836), p. 715. In this list, it is said that both books were in the library of this convent: 'TAULES Alfonsines: en catalá, scrit en pergamins. Comensa: Taula de las diffarencias, é faneix: in Maioricae. Bib. De D. Martin, Merced. Taules alfonsines: en romans, en paper. Comensa: Perço es trobat, é faneix: havets agudes altres. Idem. n. 237'.

97 BNE MS 4238 is a composite, multiple-text manuscript dated to the end of the fourteenth and the early-fifteenth centuries. The manuscript contains the canons by John of Saxony and the Parisian Alfonsine Tables adapted to Morella. See José Chabás, 'Astronomía alfonsí en Morella a finales del siglo XIV', *Cronos*, 3 (2000), 381–91; Chabás and Goldstein, *The Alfonsine Tables*, p. 292.

3.3 The Kingdom of Navarre

To complete the picture of the Hispanic Christian kingdoms, we turn to Navarre. In the courtly documentation of the fourteenth century some astrologers appeared, denoted as 'sol', 'estorlogo', or 'astrologuo'. All came from other territories, Foix, Avignon, Burgundy, Germany (*maestre Reymar*), and apparently they went to the court to cover specific demands. There are also references to two Jews as 'maestre de astrolabio' (master of astrolabe). In 1416, Johan de Sancto Archangelo received fifty-eight sueldos for the creation of an 'estrument de astrologia' (astrology instrument),[98] a sophisticated and precious artefact made of silver with 'figuras de estreillas' (figures of the stars) commissioned by King Carlos III (1361–1425). In 1430 García Arnaldo de Suescun, *estorlogo*, is related to another artefact called an *estrelario*.[99] As we can see, astronomical and astrological practices were present in Navarre and scholars and texts circulated in this territory; but there is no trace of the Alfonsine Tables or the Parisian version in this documentation. It is worth mentioning that the only medieval fragment of astronomical tables preserved in the Castilian language, in this case a translation of the Toledan Tables, is an unclassified bifolio in the Archivo General de Navarra. It apparently dates to the fourteenth century and, after falling into disuse, served as a binding for the cover of a book by 'los calcateros y sastres' (the guild of shoemakers and tailors) of Pamplona. Apparently the Toledan Tables were used throughout the fourteenth century. The fact that the fragment has been found in a book on guilds outside the scientific context of court or university suggests that those tables were widely disseminated, even though no further copies have been found.[100]

4. The documental recovery of the Alfonsine Tables in Spain during the sixteenth century

Despite the silence of the Castilian sources until the fifteenth century, the inventories of sixteenth-century libraries provide us with valuable information to document new stages in the history of the Alfonsine Tables and their impact. The tables did not disappear from Castile; rather they had been 'disconnected' but re-emerged in the sixteenth century, at least from a documental point of view. This is evidenced not only by their presence in several inventories but also by the copy of the Castilian canons preserved in BNE MS 3306, with which I started this chapter.

One of the earliest pieces of information we have, here quoted for the first time in relation to the history of the Alfonsine Tables, is the inventory of Juan de Guzmán, III

98 AGN, Comptos. Registros, 1ª Serie, nº 344, f. 103; Fernando Serrano Larráyoz, 'Astrólogos y astrología al servicio de la monarquía navarra durante la Baja Edad Media (1350–1446)', *Anuario de Estudios Medievales*, 39 (2009), 539–53, p. 548. Iohannes de Sancto Archangelo was the author of one set of astronomical tables 'Tabule Johannis Archangeli ad inveniendum facillime vera loca omnium planetarum', Bodleian Library MS lat. Misc. d. 88, ff. 60–84, and he is also mentioned by Simon de Phares as the author of the treatise 'Equatorium planetarum facilis compositionis', BnF MS lat. 7443, ff. 243ᵛ–246, and named in f. 229ʳ. Jean-Patrice Boudet, *Lire dans le ciel. La bibliothèque de Simon de Phares, astrologue du xve siècle* (Brussels: Centre d'études des manuscrits, 1994), pp. 146, 151.

99 Serrano Larráyoz, 'Astrólogos y astrología', p. 548.

100 Serrano Larráyoz, 'Astrólogos y astrología', p. 549; José Chabás, 'The Toledan Tables in Castilian: Excerpts of the Planetary Equations', *Suhayl*, 11 (2012), 179–88.

duke of Medina Sidonia (1466–1507).[101] This nobleman died in 1507 in Seville, probably of the plague. The inventory of his belongings provides important data about his library of circa 230 books, a significant part of which were devoted to scientific topics, especially medicine and astronomy. It is likely that the scientific collection came from the library of his father, don Enrique Pérez de Guzmán y Meneses (*c.* 1434–92), who lived in his residence in Sanlúcar de Barrameda (Cádiz) between 1478 and 1492.[102]

In addition to the scientific books, the duke owned a few instruments, 'un estrolatio que tiene quatro ruedas de laton y otro triabulo de laton e otros antojos de laton metidos en su caxa' (one astrolabe with four brass circles, and one brass triangle [a quadrant?], and a brass eyeglass in its box), so he clearly was interested in scientific matters.[103]

Among these books, there are four related to astronomical tables,[104] two of which concerned the Alfonsine Tables: item 75 is a paper book, 'otro libro, de las tablas alfonsyes', and item 140, on parchment, 'otro libro en pergamino, que son las Tablas Alfonsyes'. We know that both were manuscripts, as the few printed books in the inventory are identified by expressions like *libro de molde* or *libro de forma*.[105] Furthermore, both books were written in Castilian, as evidenced by their titles; had they been Latin texts, the copyist would have specified it in the inventory as is done with all the Latin books registered. The fact that there is no Latin copy of the tables in this library or in any contemporary library we know, either handwritten or printed, I think is evidence enough to consider that these entries refer to the Castilian version of the tables and not to the Parisian version. Furthermore, the presence of these copies of the Alfonsine Tables in this library is proof of the knowledge of the work and its circulation, at least among a select group of people.[106]

The Guzmán lineage was very close to the royal house; in fact, one of its members, Alonso Pérez de Guzmán (1256–1309), known as *el Bueno*, was a direct collaborator, first

101 Granada, Archivo de la Real Chancillería de Granada, ARCHGR/01RACH//CAJA 764, pieza 13. The inventory was published in Miguel A. Ladero Quesada and Mª Concepción Quintanilla Raso, 'Bibliotecas de la alta nobleza castellana en el siglo XV', in *Colloque International de la Casa de Velázquez. Livre et lecture en Espagne et en France sous l'Ancien Regime* (Paris: Casa de Velázquez, 1981), pp. 47–62 (pp. 57–59); Miguel A. Ladero Quesada, *Guzmán. La casa ducal de Medina Sidonia en Sevilla y su reino. 1282–1521* (Madrid: Editorial Dykinson, 2015).

102 For Guzmán lineage, see Ladero Quesada, *Guzmán*.

103 Among the astronomical and astrological manuscripts, we find the 'espera mundi con sus glosas', two 'Tolomeo', 'Abumasar de magis comuncionibus', several treatises of astrology and alchemy, one book described as 'se dize conpendio de las estrellas', and another 'que dize encima de Aben Ragel'; the latter could be the *Libro conplido en los iudizios de las estrellas*, also commissioned by Alfonso X. Another interesting book is the *arte notoria*, with the royal heraldry, lions, and castles in its tables (it is not clear if the heraldry was in the binding or if the book had tables with these elements), 'Otro libro de arte notoria, que tiene castillos e leones en las tablas'. We also find the *Estoria de España*, the *General Estoria*, and the III *Partida*, all of them from Alfonso X's intellectual productions.

104 Item 150, 'que es glosa de las tablas toledanas' and item 158, 'otro libro enquadernado en pergamino, que dize encima Tablas del Blanco'. These 'Tablas del Blanco' could very well be the tables of Giovanni Bianchini, first printed in 1495. Bianchini was called *Blanchinus* in Latin, which was probably translated into Castilian as *Blanco*. I thank José Chabás for this information.

105 There are just five printed books in the whole inventory: 'Otras oras yluminadas de mano, y la escritura de **molde**, de pargamino, con una cerradura de plata', 'un libro de poesía de **molde**, guarnecido de terciopelo negro', 'unas oras de **molde** viejas', 'otro libro de **molde** de rezar, con un cuero colorado' f. 10^v, 'otro libro de Ovidio, de **molde**', and 'otro libro de **forma**, de noviceas', f. 26^r.

106 I wonder if the existent copy of the Castilian canons, in the hands of Juan Fernández de Velasco (BNE MS 3306), was made from one of the books mentioned in the library of the third duke of Medina Sidonia, since these lineages were linked through Juan Fernández de Velasco's mother, Ana Pérez de Guzmán y Aragón, daughter of the sixth duke of Medina Sidonia.

of Alfonso X, and later of his son, Sancho IV, and his lineage remained linked to the royal house throughout the fourteenth and fifteenth centuries. As we can see, there was a direct connection between this family and the court and therefore a member of the Guzmán family may have known and acquired a copy of the Alfonsine Tables. What happened to these books is not known; unfortunately, none of our exemplars of interest are recorded in the catalogues of current public institutions.

To this new reference we must add the valuable information provided by Gonzalo Argote de Molina (1548–96), an influential character on the Spanish intellectual scene during the second half of the sixteenth century. Gonzalo Argote de Molina had a remarkable library in Seville containing important manuscripts, including several directly related to Alfonso X. We know the list of books that Argote had in his studio in Seville thanks to one document kept in the BNE MS 18554/23, folios 1–2v: 'Libros de mano nunca impressos tocantes a Historia de Espana que se veen en Sevilla en el estudio de Goncalo Argote de Molina'.[107] Among them is an item directly related to our topic: 'Las Tablas Alfonsies originales que mando escreuir el Rey don Alonso el sabio escriptas en pergamino yluminadas', in fact, according to the description, the book could be the royal manuscript of the tables that Alfonso X commissioned.[108]

Another inventory from Argote's studio in the Colombina Library (Biblioteca Capitular Colombina MS 57-3-16, ff. 144–7) specifies even more information: 'Las tablas alfonsinas originales que mandó escribir el rey don Alonso el Sabio, escritas en pergamino, iluminados con adornos de oro'.[109] Both documents indicate that the manuscript was the original commissioned by Alfonso X, written on parchment and illuminated with golden ornaments, the same features found on the French copy in the Louvre Library.

How the Alfonsine Tables arrived in Gonzalo Argote de Molina's studio is an enigma. He was one of the most remarkable scholars at that time and had received permission from King Felipe II to consult any archive or library he needed for his work; therefore, he had access to nobiliary collections, as well as municipal and religious archives.[110] The book of the Alfonsine Tables in his studio could be the one previously owned by don Juan Pérez de Guzmán. As a matter of fact, Argote de Molina had a close relationship with this family, documented the Guzmán lineage, and wrote the book *Linage y sucesión de la*

107 A digital copy is available at: http://bdh-rd.bne.es/viewer.vm?id = 0000250259&page = 1. In addition to the *Tablas Alfonsíes*, Gonzalo Argote de Molina had in his studio these Alfonsine manuscripts: 'El uso del Astrolabio escripto en pergamino por mandado del Rey don Alonso el Sabio. Libro de philosophia escripto por el moro Azbrani hecha por mandado del Rey don Alonso el Sabio. Repartimiento original antiguo de la ciudad de Seuilla y su tierra hecho por el Rey don Alonso el Sabio'. See Laura Fernández Fernández, 'El MS. 8322 de la Bibliothèque de l'Arsenal y su relación con las tablas alfonsíes. Hipótesis de trabajo', *Alcanate*, 7 (2010/11), 235–68, (pp. 248–49) and *Arte y ciencia*, pp. 334, 341.

108 (The original Alfonsine Tables that King Alfonso the Learned ordered to be written, on parchment and illuminated.)

109 (The original Alfonsine Tables that King Alfonso the Learned ordered to be written, on parchment and illuminated with gold ornament.)

110 On 3 July 1576, the monarch signed a royal charter, *real cédula*, granting him permission to consult any document of interest to his work, which allowed Argote access to all kinds of books and documents. Celestino López Martínez, *Algunos documentos para la biografía de Argote de Molina* (Seville: Imprenta y Librería de Eulogio de las Heras, 1921), p. 96.

casa de Guzmán y de la de Ponce de León. Following Argote's death in 1596, the fate of his manuscripts and the exemplars he had borrowed is unclear.[111]

A last reference to this record is a sale catalogue published in 1804 by a bookstore in Madrid, the *Librería de Claros*, located in Arenal street. The catalogue states: 'Catálogo de manuscritos especiales de España anteriores al año 1600 que logró juntar en la mayor parte un curioso andaluz',[112] and the Alfonsine Tables, handwritten and illuminated, are mentioned in it: 'Tablas astronómicas, alfonsinas, iluminadas'.[113]

This list has been studied by several scholars and raises important questions concerning the whereabouts of these manuscripts, since we have no information about their sale or acquisition by institutions such as the National Library or the Royal History Academy.[114] Not to mention that the sale of such important manuscripts would not have gone unnoticed in Madrid at that time. I wonder whether the bookstore only possessed (and was offering for sale) the inventory itself, not the books. Aside from this 1804 reference, we have no further information on the subject, leaving the manuscript of the Alfonsine Tables lost to the subsequent historiography.

To finish our documentary and material history of the *Libro de las tablas alfonsíes*, I would like to consider the library of Juan de Herrera, the aforementioned architect of King Felipe II, as it provides new and important data on the dissemination of the astronomical tables related to Alfonso X.

In addition to his role as main architect of the court, Herrera distinguished himself for his intellectual work and collected a considerable number of books. We are aware of his library thanks to the four inventories of his belongings he made during his life.[115] To create this library, he acquired books from different places, even asking for specific titles outside of the kingdom. This is the case of his request to Cristobal de Salazar, secretary of the Spanish Embassy in Venice. In 1584, Herrera asked him for several scientific books, including 'Las Tablas del rey D. Alonso en vulgar italiano'.[116] The solicited copy was certainly

---

111 The library and activity of this character have been studied by Inora Pepe Sarno, 'La biblioteca di Argote de Molina. Tentativo de catalogo della sezione manoscritti', in *Studi di letteratura spagnola* (Rome: Società Filologica Romana, 1967), pp. 165–262; Gregorio de Andrés, 'Códices del Escorial procedentes de Gonzalo Argote de Molina, con la edición de dos inventarios de sus manuscritos', *Cuadernos para investigación de la filología hispánica*, 10 (1988), 7–38.

112 'Catalogue of special manuscripts of Spain prior to 1600, united in their majority by a curious Andalusian man'. Although Inoria Pepe Sarno questioned the identity of this *curioso andaluz*, Gregorio de Andrés asserted without doubt that the character was Argote de Molina himself. Pepe Sarno, p. 182; Andrés, 'Códices', p. 10.

113 (Astronomical tables, Alfonsine, illuminated.)

114 Andrés, 'Códices', pp. 10–11, pp. 17–31; Pepe Sarno, pp. 177–83.

115 The inventories were published by Agustín Ruiz de Arcaute, *Juan de Herrera. Arquitecto de Felipe II* (Madrid: Instituto Juan de Herrera, 1997 [Madrid: Espasa Calpe, 1936.]), pp. 150–71; Luis Cervera Vera, *Los cuatro testamentos otorgados por Juan de Hererra* (Santander: Fundación Juan de Herrera, 1997); Francisco J. Sánchez Cantón, *La librería de Juan de Herrera* (Madrid: CSIC, 1941).

116 The books Herrera asked for were mainly for the library of the Royal Mathematical Academy founded by Felipe II in 1582, a project in which Herrera was involved from the beginning. In fact, Herrera was the first director of the institution until his death in 1597. For the complete list of books that Herrera asked of Salazar see Eugenio Llaguno y Amirola, *Noticias de los arquitectos y arquitectura de España desde su restauración* (Madrid: Imprenta Real, 1829), Tomo II, pp. 360–62; Ruiz de Arcaute, pp. 99–100. Regarding the Royal Mathematical Academy, see Pedro García-Barreno, 'The Madrid Mathematical Academy of Phillip II', *Bollettino di Storia delle Scienze Matematiche*, XX (2000), 87–188.

sent to Herrera as it appears as 'Las tablas de el Rey don alonso en ytaliano manoescritas sin figuras'[117] in the inventory of his library made several years after the request.[118]

As Chabás and Goldstein explained,[119] the Parisian Alfonsine Tables arrived in Italy in the first half of the fourteenth century, but as we have seen so did the original Alfonsine Tables, even before their arrival in Paris. According to the astronomer and mathematician Prosdocimo de' Beldomandi (*c.* 1370–1428), Jacopo Dondi of Padua (1290–1359) had composed new tables for this city based on the Parisian Alfonsine Tables, an endeavour that would be completed by Prosdocimo himself.[120] The tables and canons of Prosdocimo are known in several Latin manuscripts; Chabás provides details of fifteen copies, one of which (Biblioteca Nazionale Centrale di Firenze MS Conv. Sopp. I.III.23), has the canons translated into Italian. Could this Italian copy of the Prodoscimo tables have anything to do with Juan de Herrera's request? Was there another version of the Alfonsine Tables or the Parisian Alfonsine Tables written in Italian? As far as I know, nobody has considered whether the Alfonsine Tables, or their Parisian version, could have been translated into Italian. Beyond the Italian copy of the Prosdocimo tables, it seems that no other exemplars are preserved. However, thanks to the inventory of Juan de Herrera's library, we can confirm that a version of the tables written in Italian was known in Venice and that a manuscript with that content was sent to Madrid.[121]

In addition to this Italian copy, two other items interest us in Herrera's library inventory: a folder (*cartapazio*) with material on the Alfonsine astronomical tables written on parchment, 'Un cartapazio de cosas tocantes a las tablas de el rrey don alonso escrito en pergamino',[122] and a Latin copy of the *Libro de las armellas* and the rest of the astronomical instruments (I suppose those explained in the *Libro del saber de astrología*), 'Libro manoescripto del rrey don alonso sobre las armyllas y todos los instrumentos astronómicos en latín'.[123]

The connection between Juan de Herrera and the Alfonsine scientific material is not strange. As noted above, he was entrusted with the mission of illustrating the copy of the *Libro del saber de astrología* produced in 1562[124] and, according to Jerónimo Román de la Higuera in his *Historia eclesiástica de la imperial ciudad de Toledo*, the original manuscript of the Alfonsine Tables was once in the possession of the architect. We cannot know

117 (The tables of King Alfonso, in Italian, handwritten without figures). It may be that the specification 'sin figuras' is proof of the existence of copies with figures, and therefore with illustrations.

118 Ruiz de Arcaute, p. 161.

119 Chabás and Goldstein, *The Alfonsine Tables*, p. 292.

120 Chabás, 'From Toledo to Venice', p. 270.

121 Just out of curiosity, because I have not been able to find more information, there is a mention of one manuscript of the *Tabule Alphonsine* in the library of San Antonio, in Venice, recorded by Giacomo Filipo Tomasini in his work *Bibliothecae Venetae manuscriptae publicae et privatae* (Udine: Typis Nicolai Schiratti, 1650), p. 1. The manuscript is linked to Iohaneis de Rubeis (https://data.cerl.org/thesaurus/cnp00297682).

122 Ruiz de Arcaute, p. 160. (A folder of things concerning the tables of King don Alonso written on parchment).

123 Ruiz de Arcaute, p. 157. (Manuscript book of King don Alonso about the *armyllas* and all their astronomical instruments in Latin).

124 In 1505, the *Libro del saber* was bought by the Cardenal Cisneros (1436–1517) from King Fernando 'el Católico' (1452–1516) for the library of the university college he founded in Alcalá de Henares, the Colegio de San Ildefonso, the origin of the Complutense University. There, this manuscript was copied on several occasions, one of them in 1562 for Carlos, Prince of Asturias (1545–68), son of Felipe II, and Juan de Herrera was responsible for making its figures (RBME MS h-I-1). Fernández Fernández, *Arte y ciencia*, p. 248.

whether Higuera was right; but it is clear that Herrera knew and owned material related to these tables.

5. Material traces of the royal manuscript: a proposal

A significant number of manuscripts commissioned by King Alfonso remained linked to the royal treasury; but many others, especially those related to scientific issues, left the court and were acquired by other prominent personalities in the political and cultural sphere of the time and were preserved in their circles.[125] The royal manuscript of the Alfonsine Tables could have been sold or gifted and thus have assumed a history outside the court. However, did it even disappear at all? Is there no material trace left of the royal manuscript? Regarding this question, I would like to draw attention to one of the preserved Alfonsine codices, the compendium now in Paris, at the Bibliòtheque de l'Arsenal MS 8322.[126] This illuminated manuscript commissioned by Alfonso X is a miscellaney whose content is entirely devoted to astronomical tables: the *Canones de Albateni, Canones de Azarquiel, Tablas de Azarquiel,* and *Libro del cuadrante señero,* a book related to tabular content written by Isaac ben Sid in 1277, who was, as we have seen, one of the authors of the Alfonsine Tables. There is no doubt that this book was produced in Alfonso's scriptorium. Its textual and material features date it to the end of 1270s or early 1280s. Despite containing different texts, the manuscript was conceived as a compilation with a unitary character, both in terms of its material dimensions and theoretical approach. However, the manuscript is not complete; it lacks a first section with the general prologue, the *intitulatio,* and the usual table of contents. In fact, the codex was manipulated during the fifteenth century, as we can see on its first folio, completed on paper, not parchment, and written in *gotica rotunda,* not *gotica textualis* like the rest of the manuscript, that replaced the missing original folio. Furthermore, a close codicological examination reveals that at the beginning of the codex, thirteen queries have disappeared, approximately one hundred folios, as well as other queries inside the manuscript. These circumstances made me ask whether this codex might be the surviving part of a compilation, and whether part of those lost queries contained a copy of the missing Alfonsine Tables. During the second half of the 1270s, the *scriptorium* was entrusted to produce these kinds of works, compendia of all the scientific material developed during previous decades; therefore, mine is a plausible hypothesis. Who manipulated the manuscript and removed the first part and when they did so is again an enigma, but the codex was in the hands of a Venetian nobleman called Jacobus Contarenus, apparently a member of the Contarini family, which was well-known for its political and diplomatic activity and also for its intellectual position. According to an inscription on folio 23^r of the manuscript, Contarenus gifted the book to King Manuel I of Portugal (1469–1521) in 1496. Sometime after that, the manuscript returned to Spain, to

125 This happened with the *Libro de las cruzes, Libro conplido en los iudizios de las estrellas, Lapidario, Libro del saber de astrología, Libro de las formas et las ymágenes* (except for the prologue and index), *Libro de astromagia,* and the *Compendio* in Arsenal library. Further information about this topic can be found in Fernández Fernández, *Arte y ciencia.*

126 A digital copy available at: https://gallica.bnf.fr/ark:/12148/btv1b71003376

Seville, and it was housed in the library of Diego Ortiz de Zúñiga (1636–80) who gifted it to Juan Lucas Cortés (1624–1701), thanks to whom it would finish its periplus, after many vicissitudes, in the Bibliothèque de l'Arsenal.[127]

As I suggested a decade ago, the royal manuscript was manipulated and transformed throughout the course of the fifteenth century. It travelled from Castile to Venice, into the hands of Contarini, wherever he may have been, then to Lisbon, returning years later to Seville, and concluding its journey in Paris after a period in Lyon.

Having assumed the possibility that the *Libro de las tablas alfonsies* was the missing part of MS 8322, I began to ask what this book would look like. According to the documentary references analysed so far and to the features of the other scientific manuscripts commissioned by Alfonso X, it must have been an illuminated manuscript. The canons and tables of al-Battānī in MS 8322 undoubtedly offer us a direct element of comparison; and the prologues in other scientific manuscripts, like the *Lapidario* or the *Libro de las formas et de las ymágenes*, as well as the rest of the Alfonsine production, suggest that our book had an opening image portraying the King and his collaborators. We could also speculate on the presence of a constellation set, a topic directly connected to the subject and often used in other scientific manuscripts promoted by Alfonso. This is a possibility that certainly seems plausible if we recall the set of constellations present in MS Canon Misc. 27 analyzed above.

Furthermore, we can know with certainty who wrote at least the numerical tables in the manuscript. This is because Diego Ortiz de Zúñiga provided news of a document in the archive of the Cathedral of Seville, unfortunately undated, in which the name of one of the King's scribes, Suer Meléndez, is specified, 'que le faze las tablas e numerانças de los sus libros' (who makes the tables and numbers of his books).[128]

6. Conclusion

The history of the astronomical tables commissioned by Alfonso X, their dissemination, and their reinterpretation, is an important episode in the history of science. It is a history marked by the constant movement of books, scholars, and ideas, tracing multiple routes of knowledge and intellectual development over the centuries.

As we have seen, the *Libro de las tablas alfonsíes* was more than a scientific success. These tables were part of the history of Alfonso X for a long time, and despite being relegated and apparently disappearing from the scientific scene, they remained in the collective memory of the Kingdom of Castile through chronicles and other documentation. It is not surprising that some scholars in Spain during the sixteenth and seventeenth centuries were interested in this work and what it meant. Furthermore, these scholars, as those in the rest of Europe, were fully aware of the relevance of medieval manuscripts with regard to access to knowledge of earlier times, as well as their historical significance, and their beauty; thus, they made a significant effort to locate, identify, copy, and collect these sources. Proof of this lies in the numerous medieval books preserved in Renaissance libraries.

127 For further information, see Fernández Fernández, 'El MS. 8322'.

128 Ortiz de Zúñiga, p. 90.

For the time being, the exemplars of the Alfonsine Tables mentioned in the inventories of Juan Pérez de Guzmán, Gonzalo Argote de Molina, and Juan de Herrera's libraries have not appeared; but there are many archives and libraries yet to be explored. I trust that we will be able to find new references in the future to guide our search.

Manuscript sources

Cambridge, Trinity College, O.3.13
El Escorial, Real Biblioteca del Monasterio de El Escorial, h-I-1
El Escorial, Real Biblioteca del Monasterio de El Escorial, h-I-15
El Escorial, Real Biblioteca del Monasterio de El Escorial, h-I-16
El Escorial, Real Biblioteca del Monasterio de El Escorial, J-II-17
El Escorial, Real Biblioteca del Monasterio de El Escorial, T-III-29
Florence, Biblioteca Medicea Laurenziana, Or. 152
Florence, Biblioteca Nazionale Centrale, Conv. Sopp. I.III.23
Lisbon, Torre do Tombo, Manuscritos da Livraria, tt-msliv n.º 2115
Madrid, Biblioteca Histórica Marqués de Valdecilla, 156
Madrid, Biblioteca Nacional de España, 1289
Madrid, Biblioteca Nacional de España, 3065
Madrid, Biblioteca Nacional de España, 3306
Madrid, Biblioteca Nacional de España, 4238
Madrid, Biblioteca Nacional de España, 7840
Madrid, Biblioteca Nacional de España, 7856
Madrid, Biblioteca Nacional de España, 9294
Madrid, Biblioteca Nacional de España, 10002
Madrid, Biblioteca Nacional de España, 10053
Madrid, Biblioteca Nacional de España, 18841
Madrid, Biblioteca Nacional de España, 18554/23
Madrid, Biblioteca Nacional de España, RES/2
New York, Morgan Library, M.240
Oxford, Bodleian Library, Can. Misc. 27
Oxford, Bodleian Library, lat. d. 88
Paris, Bibliothèque nationale de France, lat. 6064
Paris, Bibliothèque nationale de France, lat. 7195
Paris, Bibliothèque nationale de France, lat. 7281
Paris, Bibliothèque nationale de France, lat. 7443
Paris, Bibliothèque nationale de France, lat. 16652
Paris, Bibliothèque de l'Arsenal, 8322
Seville, Biblioteca Capitular Colombina, 57–3-16
Toledo, Biblioteca Capitular de Toledo, 99–5
Vatican, Biblioteca Apostolica Vaticana, Pal. lat. 1370
Vatican, Biblioteca Apostolica Vaticana, Reg. lat. 1283pt.A
Vatican, Biblioteca Apostolica Vaticana, Reg. lat. 1300
Vatican, Biblioteca Apostolica Vaticana, Vat. lat. 4072

Bibliography

Abad Pérez, Antolín, 'Relación sobre el incendio de San Juan de los Reyes (1808) y vicisitudes posteriores hasta 1864', *Toletum: Boletín de la Real Academia de Bellas Artes y Ciencias Históricas de Toledo*, 4 (1969), 169–88.

———, 'La biblioteca franciscana de Toledo (1284–1808)', *Anales toledanos*, 20 (1984), 9–36.

Alexander, Jonathan J. G., and Otto Pächt, *Illuminated Manuscripts in the Bodleian Library* (Oxford: Clarendon Press, 1966).

Andrés, Gregorio de, 'La biblioteca manuscrita del Condestable Juan Fernández de Velasco († 1613)', *Cuadernos bibliográficos*, 40 (1980), 5–22.

———, 'Códices del Escorial procedentes de Gonzalo Argote de Molina, con la edición de dos inventarios de sus manuscritos', *Cuadernos para investigación de la filología hispánica*, 10 (1988), 7–38.

Antonio, Nicolás, *Bibliotheca hispana vetus sive Hispanorum* (Rome: Typographia Antonii de Rubeis, 1696).

Alfonso X, *El libro conplido en los iudizios de las estrellas*, ed. by Gerold Hilty (Madrid: Real Academia Espanola, 1954).

———, *Lapidario and Libro de las formas & las ymagenes*, ed. by Roderic C. Diman and Lynn W. Winget (Madison: Hispanic Seminary of Medieval Studies, 1980).

———, *Las Siete Partidas, Volume 2: Medieval Government: The World of Kings and Warriors*, ed. by Samuel Parsons Scott and Robert I. Burns (Philadelphia: University of Pennsylvania Press, 2001).

———, *Los cánones de Al-bateni*, ed. by George Bossong (Tübingen: Max Niemeyer, 1978).

Arizaleta, Amaia, 'De la soberbia del rey: dos formas breves en la construcción historiográfica', in *Tipología de las formas narrativas breves románicas medievales*, III, ed. by Juan M. Cacho Blecua and Mª Jesús Lacarra (Zaragoza/Granada: Universidad de Zaragoza/Universidad de Granada, 2004), pp. 79–110.

Ballesteros Beretta, Antonio, *Alfonso X el Sabio* (Barcelona: El Albir, 1984).

de Barros Días, Isabel, 'A blasfémia do Rei Sabio: os antecedentes da lenda' in *Estudios de literatura medieval. 25 años de la AHLM*, ed. by Antonia Martínez Pérez and Ana Luisa Baquero Escudero (Murcia: Universidad de Murcia, 2012), pp. 189–96.

———, 'La blasfemia del Rey Sabio: vicisitudes de una leyenda (nuevas hipótesis respecto a la datación y la posición relativa del texto portugués)', *Anuario de Estudios Medievales*, 45 (2015), 733–52.

Beaujouan, Guy, *Manuscrits scientifiques médiévaux de l'Université de Salamanque et de ses 'Colegios mayores'* (Bordeaux: Yéret, 1962).

Boudet, Jean-Patrice, 'Le modèle du roi sage aux XIIIe et XIVe siècles: Salomon, Alphonse X et Charles V', *Revue historique*, 647 (2008), 545–66.

———, *Lire dans le ciel. La bibliothèque de Simon de Phares, astrologue du XVe siècle* (Bruxelles: Centre d'études des manuscrits, 1994).

Carr, Derek C., 'Arabic and Hebrew Auctoritates in the Works of Enrique de Villena', in *From Arabye to Engelond. Medieval Studies in Honour of Mahmoud Manzalaoui*, ed. by A. E. Christa Canitz and G. R. Wieland (Ottawa: University of Ottawa Press, 1999), pp. 39–60.

———, and Pedro M. Cátedra, 'Datos para la biografía de Enrique de Villena', *La Coronica*, 11 (1983), 293–99.

Cátedra, Pedro M., 'Para la biografía de don Enrique de Villena', *Estudi General*, 1 (1981), 29–33.

———, 'Algunas obras perdidas de Enrique de Villena con consideraciones sobre su obra y su Biblioteca', *El Crotalon: Anuario de Filología Española*, 2 (1985), 53–75.

Cervera Vera, Luis, *Los cuatro testamentos otorgados por Juan de Hererra* (Santander: Fundación Juan de Herrera, 1997).

Chabás, José, *L'Astronomia de Jacob Ben David Bonjorn* (Barcelona: Institut d'Estudis Catalans, 1992).

———, 'Astronomía alfonsí en Morella a finales del siglo XIV', *Cronos*, 3 (2000), 381–91.

———, 'The Diffusion of the Alfonsine Tables: The Case of the *Tabulae resolutae*', *Perspectives on Science*, 10 (2002), 168–78.

———, 'Were the Alfonsine Tables of Toledo First Used by Their Authors?', *Centaurus*, 45, 2003, 142–50.

———, 'The University of Salamanca and the Renaissance of Astronomy during the Second Half of the 15th Century', in *Universities and Science in the Early Modern Period*, ed. by Morchedai Feingold and Víctor Navarro Brotóns (Dordrecht: Springer, 2006), pp. 29–36.

———, 'From Toledo to Venice: the Alfonsine Tables of Prosdocimo de' Beldomandi of Padua (1424)', *Journal for the History of Astronomy*, 38 (2007), 269–81.

———, 'The Toledan Tables in Castilian: Excerpts of the Planetary Equations', *Suhayl*, 11 (2012), 179–88.

———, and Bernard R. Goldstein, *Astronomy in the Iberian Peninsula: Abraham Zacut and the Transition from Manuscript to Print* (Philadelphia: American Philosophical Society, 2000).

———, and———, *The Alfonsine Tables of Toledo* (Dordrecht: Kluwer, 2003).

———, and ———, *A Survey of European Astronomical Tables in the Late Middle Ages* (Leiden: Brill, 2012).

———, and Beatriz Porres de Mateo, 'Los cánones de las Tabulae Resolutae para Salamanca: origen y transmisión', *Cronos: cuadernos valencianos de historia de la medicina y de la ciencia*, 1 (1998), 51–83.

Cotarelo y Mori, Emilio, *Don Enrique de Villena. Su vida y obras* (Madrid: Estudio Tipográfico Sucesores de Rivadeneyra, 1896).

Craddock, Jerry R., 'Dynasty in Dispute: Alfonso X el Sabio and the Succession to the Throne of Castile and Leon in History and Legend', *Viator*, 17 (1986), 214–19.

D'Agostino, Antonio, *Astromagia: ms. Reg. lat. 1283*, Naples: Liguori, 1992.

Daumet, George, *Mémoire sur les relations de la France et de la Castille de 1255 à 1320* (Paris: Fontemoing et C^le^, 1913).

Delisle, Léopold, *Recherches sur la librairie de Charles V*, 3 vols (Paris: Champion, 1907).

Domínguez Rodríguez, Ana, *Astrologia y arte en el "Lapidario" de Alfonso X el Sabio* (Madrid: Edilan, 1984).

Dreyer, J. L. E., 'On the Original Form of the Alfonsine Tables', *Monthly Notices of the Royal Astronomical Society*, 80 (1920), 243–62.

Esperabe de Arteaga, Enrique, *Historia pragmática e interna de la Universidad de Salamanca y los Reyes* (Salamanca: Imprenta de Francisco Núñez, 1914).

Fernández Fernández, Laura, 'Las Tablas astrónomicas de Alfonso X el Sabio: Los ejemplares del Museo Naval de Madrid', *Anales de Historia del Arte*, 15 (2005), 29–50.

———, 'El MS. 8322 de la Bibliothèque de l'Arsenal y su relación con las tablas alfonsíes. Hipótesis de trabajo', *Alcanate*, 7 (2010/11), 235–68.

———, *Arte y ciencia en el* scriptorium *de Alfonso X el Sabio* (Seville/El Puerto de Santa María: Universidad de Sevilla/Cátedra Alfonso X, 2013).

———, 'Arab stars in the Castilian sky: Al-Sūfī's *Book of fixed stars* amongst the manuscripts of Alfonso X', in *The Stars in the Classical and Medieval Traditions*, ed. by Alena Hadravová, Petr Hadrava, and Kristen Lippincott (Prague: Scriptorium, 2019), pp. 93–114.

Fernández Pomar, José M., 'Manuscritos del VI Condestable de Castilla en la Biblioteca Nacional', *Helmantica*, 18 (1967), 89–108.

Funes, Leonardo, 'La blasfemia del Rey Sabio: itinerario narrativo de una leyenda (primera parte)', *Incipit*, 13 (1993), 51–70.

———, 'La blasfemia del Rey Sabio: itinerario narrativo de una leyenda (segunda parte)', *Incipit*, 14 (1994), 69–101.

———, 'La leyenda de la blasfemia del Rey Sabio: revisión de su itinerario narrativo', e-Spania [Online], online since 1 October 2016, accessed 22 March 2020. URL: http://journals.openedition.org/e-spania/25873.

García Avilés, Alejandro, 'Arte y astrología en Salamanca a finales del siglo XV', *Anuario del Departamento de Historia y Teoría del Arte*, 6 (1994), 39–60.

———, 'Two astromagical manuscripts of Alfonso X', *Journal of the Warburg and Courtauld Institutes*, 59 (1996), 14–23.

García-Barreno, Pedro, 'The Madrid Mathematical Academy of Phillip II', *Bollettino di Storia delle Scienze Matematiche*, 20 (2000), 87–188.

García Ventanas, Francisco, *Tabulae Alphonsinae perpetuae motuum coelestium denuo restitutae et illustratae à Francisco García Ventanas mathematico* (Madrid: Officina Regia, 1641).

Gascón Vera, Elena, 'La quema de los libros de don Enrique de Villena: una maniobra política y antisemítica', *Bulletin of Hispanic Studies*, 1 October 1979, pp. 317–24.

———, 'Nuevo retrato histórico de Enrique de Villena', *Boletín de la Real Academia de la Historia*, 175 (1978), 107–43.

Goldstein, Bernard R., 'The Blasphemy of Alfonso X: History or Myth', in *Revolution and Continuity: Essays in the History and Philosophy of early Modern Science*, ed. by Peter Barker and Roger Ariew (Washington, DC: Catholic University of America Press, 1991), pp. 143–53.

González Jiménez, Manuel, *Diplomatario Andaluz* (Seville/El Monte: Caja de Huelva y Seville, 1991).

Gonzálvez Ruiz, Ramón, *Hombres y libros de Toledo* (Madrid: Fundación Ramón Areces, 1997).

Gousset, Marie-Thérèse, 'Le Liber de locis stellarum fixarum d'Al-Sûfi, ms. 1036 de la Bibliothèque de l'Arsenal à Paris: une réattribution', *Arte medievale*, 2 (1984), 93–107.

Harper, Richard I., 'Prophatius Judaeus and the Medieval Astronomical Tables', *Isis*, 62 (1971), 61–68.

Hernández, Francisco J., and Peter Linehan, *The Mozarabic Cardinal. The Life and Times of Gonzalo Pérez Gudiel* (Florence: Sismel-Edizioni del Galluzo, 2004).

Hill, Donald R., 'A Treatise on Machines, by Ibn Muādh Abū Abd Allāh al-Jayyānī', *Journal for the History of Arabic Science*, 1 (1977), 33–46.

Hilty, Gerold, 'El libro conplido de los iudizios en las estrellas', *Al-Andalus*, 20 (1955), 1–74.

Hiver de Beauvoir, Alfred, *La Librairie de Jean duc de Berry au Château de Mehun-sur-Yevre* (Paris: Auguste Aubry, 1860).

Jardin, Jean-Pierre, 'La figure du roi Alphonse X chez quelques chroniqueurs du XVe siècle', *Cahiers de linguistique hispanique médiévale*, 20 (1995), 75–96.

John of Saxony, *Las Tablas de los movimientos de los cuerpos çelestiales del illuxtrisimo Rey Don Alonso de Castilla: seguidas de su Additio: traducción castellana anónima de los cánones de Juan de Sajonia*, ed. by José Martínez Gázquez (Murcia: Universidad de Murcia, 1989).

Kunitzsch, Paul, 'Ṣūfī Latinus', *Zeitschrift der Deutschen Morgenländischen Gesellschaft*, 115 (1965), 65–74.

———, 'The Astronomer Abū'l-Husayn al-Sūfī and his Book on the Constellations', *Zeitschrift für Geschichte der Arabisch-Islamischen Wissenschaften*, 3 (1986), 56–81.

———, 'Star Catalogues and Star Tables in Mediaeval Oriental and European Astronomy', *Indian Journal of History of Science*, 21 (1986), 113–22.

———, 'The Star Catalogue Commonly Appended to the Alfonsine Tables', *Journal for the History of Astronomy*, 17 (1986), 89–98.

Ladero Quesada, Miguel A., *Guzmán. La casa ducal de Medina Sidonia en Sevilla y su reino. 1282–1521* (Madrid: Editorial Dykinson, 2015).

———, and Mª Concepción Quintanilla Raso, 'Bibliotecas de la alta nobleza castellana en el siglo XV', in *Colloque International de la Casa de Velázquez. Livre et lecture en Espagne et en France sous l'Ancien Regime* (Paris: Casa de Velázquez, 1981), pp. 47–62.

Lagomarsini, Claudio, 'Le cas du compilateur compilé: une œuvre inconnue de Rusticien de Pise et la réception de *Guiron le Courtois*', *Journal of the International Arthurian Society*, 3 (2015), 55–71.

Llaguno y Amirola, Eugenio, *Noticias de los arquitectos y arquitectura de España desde su restauración* (Madrid: Imprenta Real, 1829).

López Martínez, Celestino, *Algunos documentos para la biografía de Argote de Molina* (Seville: Imprenta y Librería de Eulogio de las Heras, 1921).

Martin, George, 'Alphonse X maudit son fils', *Atalaya*, 5 (1994), 151–78.

Martínez Montávez, Pedro, 'Relaciones de Alfonso X de Castilla con el sultán mameluco Baybars y sus sucesores', *Al-Andalus*, 27 (1962), 343–76.

Massó i Torrents, Jaume, 'Inventari dels béns mobles del rey Martí d'Aragó', *Revue Hispanique*, 12 (1905), 413–590.

Menéndez y Pelayo, Marcelino, *Antología de poetas líricos castellanos. La poesía en la Edad Media* (Madrid: CSIC, 1944).

Millás Vallicrosa, José M., 'El "Libro de Astrología" de Don Enrique de Villena', *Revista de Filología Española*, 27 (1943), 512–42.

———, *Estudios sobre Azarquiel* (Madrid/Granada: CSIC, 1943–50).

———, *Las Tablas astronómicas del Rey Pedro el Ceremonioso* (Madrid/Barcelona: CSIC, 1962).

North, John D., 'The Alfonsine Tables in England', in *Prismata: Festschrift für Willy Hartner*, ed. by Y. Maeyama and W. G. Salzer (Wiesbaden: Franz Steiner, 1977), pp. 269–301.

Nothaft C., Philipp E., 'Critical Analysis of the Alfonsine Tables in the Fourteenth Century: the Parisian *Expositio tabularum Alfonsii* of 1347', *Journal of History of Astronomy*, 46 (2015), 76–99.

Ortiz de Zúñiga, Diego, *Anales eclesiásticos y seculares de la muy noble y muy leal ciudad de Sevilla* (Madrid: Imprenta Real, 1677).

Pedersen, Olaf, 'The "Theorica Planetarum" and its Progeny', in *Filosofia, scienza e astrologia nel trecento europeo*, ed. by Graziella Federici Vescovini and Francesco Barocelli (Padua: Il poligrafo, 1992), pp. 53–78.

Pepe Sarno, Inora, 'La biblioteca di Argote de Molina. Tentativo de catalogo della sezione manoscritti', in *Studi di letteratura Spagnola* (Rome: Società Filologica Romana, 1967), pp. 165–262.

Pingree, David, 'Between the "Ghāya" and the "Picatrix"', *Journal of the Warburg and Courtauld Institutes*, 44 (1981), 27–56.

Procter, Evelyn S., 'The Scientific Works of the Court of Alfonso X of Castille: The King and his Collaborators', *Modern Language Review*, 40 (1945), 12–29.

Rico y Sinobas, Manuel, *Los Libros del Saber de Astronomía del Rey D. Alfonso X de Castilla*, 5 vols (Madrid: Tipografía de Don Eusebio Aguado, 1863–67).

Riera i Sans, Jaume, 'La bibliothèque du roi Martin', in *Association Internationale de Bibliophilie. XXII Congrès. Actes et Communications, Barcelona, 2001* (Barcelona: Associació de Bibliòfils de Barcelona, 2005), pp. 105–17.

Rivera García, Antonio, 'La leyenda sobre la blasfemia de Alfonso X: un episodio de la conflictiva relación entre especulación teórica y razón de estado', *eHumanista*, 31 (2015), 426–51.

Rodríguez de Castro, José, *Biblioteca española: tomo primero que contiene la noticia de los escritores rabinos españoles* (Madrid: Imprenta Real de la Gazeta, 1781).

Romano, David, 'Le opere scientifiche di Alfonso X e l'intervento degli ebrei', in *Oriente e Occidente nel Medioevo: filosofia e scienze* (Rome: Academia Nazionale dei Lincei, 1971), pp. 677–711.

Roth, Norman, 'Jewish Translators at the Court of Alfonso X', *Thought. A Review of Culture and Ideas*, 60 (1985), 439–55.

———, 'Jewish Collaborators in Alfonso's Scientific Work', in *Emperor of Culture: Alfonso X the Learned of Castile and His Thirteenth-Century Renaissance*, ed. by Robert I. Burns (Philadelphia: University of Pennsylvania Press, 1990), pp. 59–71.

Rubió y Lluch, Antonio, *Documents per l'historia de la cultura catalana mig-eval*, 2 vols (Barcelona: Institut d'Estudis Catalans, 1908–21).

Ruiz de Arcaute, Agustín, *Juan de Herrera. Arquitecto de Felipe II* [1936] (Madrid: Instituto Juan de Herrera, 1997).

Ryan, Michael A., *A Kingdom of Stargazers: Astrology and Authority in the Late Medieval Crown of Aragon* (Ithaca: Cornell University Press, 2011).

Sabra, Abdelhamid I., 'A note on Codex Biblioteca Medicea-Laurenziana, Or. 152', *Journal for the History of Arabic Science*, 1 (1977), 276–83.

Salvador, Martínez, H., *Alfonso X, the Learned: A Biography* (Leiden: Brill, 2010).

Samsó, Julio, 'Dos colaboradores científicos musulmanes de Alfonso X', *Llull*, 4 (1981), 171–79.

———, 'La ciencia española en la época de Alfonso el Sabio', in *Alfonso X, Toledo* (Madrid: Ministerio de Cultura, Dirección General de Bellas Artes y Archivos, 1984), pp. 89–102.

———, 'Alfonso X and Arabic Astronomy', in *De Astronomia Alphonsi Regis*, ed. by Mercè Comes, Roser Puig, and Julio Samsó (Barcelona: Instituto Millás Vallicrosa, 1987), pp. 23–38.

———, *Las ciencias de los antiguos en al-Andalus* (Almería: Fundación Ibn Tufayl, 2011).

———, 'Ibn Ishāq and the Alfonsine Tables', *Journal for the History of Astronomy*, 50 (2019), 360–65.

———, *On Both Sides of the Straits of Gibraltar: Studies in the History of Medieval Astronomy in the Iberian Peninsula and the Maghrib* (Leiden: Brill, 2020).

Sánchez Cantón, Francisco J., *La librería de Juan de Herrera* (Madrid: CSIC, 1941).

Serrano Larráyoz, Fernando, 'Astrólogos y astrología al servicio de la monarquía navarra durante la Baja Edad Media (1350–1446)', *Anuario de estudios medievales*, 39 (2009), 539–53.

Steinschneider, Moritz, *Die hebräischen Übersetzungen des Mittelalters und die Juden als Dolmetscher* (Berlin: Kommissionsverlag des Bibliographischen Bureaus, 1893).

Thorndike, Lynn, 'Introduction and Canon by Dalmatius to Tables of Barcelona for the Years 1361–1433', *Isis* 26 (1937), 310–20.

Tomasini, Giacomo F., *Bibliotecae Venetae manuscriptae publicae et privatae* (Udine: Typis Nicolai Schiratti, 1650).

Torres Amat, Félix, *Memorias para ayudar a formar un diccionario crítico de los escritores catalanes y dar alguna idea de la antigua y moderna literatura en Cataluña* (Barcelona: Imprenta de J. Verdaguer, 1836).

Torres Fontes, Juan, 'Un médico alfonsí: Maestre Nicolás', *Revista Murgetana*', 6 (1954), 9–16.

Tratado de astrología atribuido a Enrique de Villena, ed. by Pedro M. Cátedra and Julio Samsó (Madrid/Barcelona: Editorial Labor-Río Tinto Minera, 1981).

Vernet, Juan, 'Un texto árabe de la corte de Alfonso X el Sabio', *Al-Andalus*, 43 (1978), 405–21.

———, and David Romano, *Bartomeu de Tresbéns. Tractat d'astrologia* (Barcelona: Biblioteca Catalana d'Obres Antigues, 1957–58).

Vicente García, Luis M., *Estrellas y astrólogos en la literatura medieval Española* (Madrid: Laberinto, 2006).

Villuendas, Mª Victoria, 'A further note on a mechanical treatise contained in Codex Medicea Laurenziana Or. 152', *Journal for the History of Arabic Science*, 2 (1978), 395–96.

Warburg, Aby M., *La rinascita del paganesimo antico. Contributi alla storia della cultura* (Florence: La nuova Italia 1966).

JEAN-PATRICE BOUDET & LAURE MIOLO

Alfonsine Astronomy and Astrology in Fourteenth-Century Oxford: The Case of MS Bodleian Library Digby 176

Introduction

The manuscript, Oxford, Bodleian, Digby 176, is a key witness for better understanding the astronomical and astrological practices and innovations of a group of practitioners trained in Oxford in the mid-fourteenth century. This group of scholars sharing a similar background and interest in the 'science of the stars' was closely linked to Merton College. The institution, founded in 1264 by Walter of Merton (d. 1277), constitutes the first *collegium* in Oxford.[1] Allying the religious model of a community ruled by formal statutes and the University's ideal of corporation, it gathered bachelors in arts close to their *Inceptio* and masters of arts studying theology. The *esprit de corps* of such an autonomous community, living together, studying together, and supporting each other, fostered collaborations and the development of intellectual circles. Although the 'house of the scholars of Merton' (*domus scolarium de Mertone*) was founded with the idea of training high-level ecclesiastics who

* This project received funding from the European Research Council (ERC) under the European Union's Horizon 2020 research and innovation programme (grant agreement No 723085). Our thanks go to José Chabás, Matthieu Husson, and Richard L. Kremer for their judicious remarks on preliminary versions of this article.

1 A primitive foundation established by Walter of Merton already existed in 1262 in Surrey. The founder allocated his manors of Malden, Farleigh, and Chessington to the Augustinian priory of Merton in order to support twelve *scolares* as well as his kin, who were intending to study in Oxford. In 1264, when the first statutes were written (under the patronage of the Bishop of Winchester and the Augustinian priory of Merton), the main part of the community was in Oxford, and some young students remained in Surrey. The final installation of the whole community in Oxford took place in 1274, when new statutes were composed. On the history of Merton College: Percy S. Allen and Heathcote W. Garrod, *Merton Muniments* (Oxford: Clarendon Press, 1928); J. Roger L. Highfield, *The Early Rolls of Merton College, Oxford, with an Appendix of Thirteenth-Century Oxford Charters* (Oxford: Clarendon Press, 1964); J. Roger L. Highfield and Geoffrey H. Martin, *A History of Merton College* (Oxford: Oxford University Press, 1997).

Jean-Patrice Boudet • University of Orléans

Laure Miolo • University of Cambridge

Alfonsine Astronomy: The Written Record, ed. by Richard L. Kremer, Matthieu Husson, José Chabás, Turnhout, 2022 (*Alfonsine Astronomy: Studies and Sources*, 1), pp. 57-105
© BREPOLS PUBLISHERS 10.1484/M.ALFA.5.124924

would serve the Church and the Kingdom on the model of their founder, this institution would host, in the first half of the fourteenth century, scholars who went on to produce some important mathematical works.[2]

The scientific activities, which took place at Merton College less than a century after its founding, are often eclipsed by the so-called calculators. This circle, represented by such figures as Thomas Bradwardine, Richard Swineshead, and John Dumbleton, focused on questions related to natural philosophy, which were solved with mathematical and logical methods.[3] However the emergence of this class of calculators involved in mathematical physics has often eclipsed another group of scholars who were well trained in astronomy and astrology and responsible for several innovations in this field. Members of each circle were not strangers to one another, as some of them were contemporary fellows of the college. The two groups are not opposed; yet they do represent different interests and practices. The self-regulated community of Merton College within the University of Oxford likely played an important role in the emergence of these intellectual groups. The sense of belonging to a common community of scholars and a common place to live and to study, including the mutual support they gave to each other, created a collective emulation around specific areas of knowledge. However the influence of certain individuals in the construction of such groups cannot be neglected. The circle of astronomers and astrologers greatly benefitted from key figures, such as Simon Bredon (d. 1372) and William Reed (d. 1385), who played the role of patrons, providing subsidies, books, and without a doubt scientific expertise.[4] The codex Oxford, Bodleian Library, Digby 176 reveals the activities and intellectual exchanges of this group of individuals. It also allows us to better understand the role played by William Reed in this circle.

Digby 176 is a parchment manuscript of 119 folios. In the nineteenth century, this codex had been highlighted for its exceptional list of meteorological observations dating from 1338 to 1344. Such observations were made and recorded by William Merle, who is often associated with Merton College because of his friendship with Simon Bredon and who was a rector of Driby in Lincolnshire from 1331 to 1347.[5] He is also the author of two astro-meteorological works, the *De pronosticatione aeris* (1340) and the *De futura temperie*

2 Walter of Merton entered into King Henry III's service in 1236 and was also supported by the Bishop of Durham, Nicholas Farnham (d. 1257). After years acting as a royal clerk in the chancellor's household, he became chancellor from 1261 to 1265 and by July of 1274 had secured the bishopric of Rochester. Cf. Geoffrey H. Martin, 'Merton, Walter of (*c.* 1205–77)', in the *Oxford Dictionary of National Biography*, 2004 https://www-oxforddnb-com.ezp.lib.cam.ac.uk/view/10.1093/ref:odnb/9780198614128.001.0001/odnb-9780198614128-e-18612.

3 James A. Weisheipl, 'Ockham and the Mertonians', in J. I. Catto, ed., *The History of the University of Oxford. I: The Early Oxford Schools* (Oxford: Oxford University Press, 1984), pp. 607–58; Edith D. Sylla, *The Oxford Calculators and the Mathematics of Motion, 1320–1350: Physics and Measurements by Latitudes* (New York: Garland, 1991).

4 See, for example, their wills: Frederick M. Powicke, *The Medieval Books of Merton College* (Oxford: Clarendon Press, 1931), pp. 82–91; for a more recent edition of William Reed's will, see Rodney M. Thomson, *The University and College Libraries of Oxford* (London: The British Library, 2015).

5 See *Consideraciones temperiei pro 7 annis per Magistrum Willelmum Merle: The Earliest Known Journal of the Weather, 1337–1344*, facsimile with English translation ed. by G. J. Symons (London: E. Stanford, 1891); Lynn Thorndike, *A History of Magic and Experimental Science*, 7 vols (New York: Columbia University Press, 1923–58), III (1934), pp. 141–42.

aeris pronosticanda, contained in Digby 176, which consists of instructions for elaborating astrological weather forecasts.[6]

Although the manuscript was well studied by Lynn Thorndike, it was not considered by John North in his important article on the introduction of the Alfonsine Tables in England.[7] His student, Hilary Carey, came back to this manuscript for her study on astrology and astrologers at the English court and universities in the late Middle Ages.[8]

However, despite Hilary Carey's important contribution to the history of this manuscript, new insights into its content, the history of its composition, as well as the milieu to which it belonged are required. This manuscript indeed reflects the influence of Alfonsine astronomy in England. It also preserved the richest collection of predictions on the planetary conjunctions of the fourteenth century (for the conjunctions of 1325, 1345, 1357, 1365, and 1367). This codex also addresses the problem of the complementarity between astronomy and astrology in practice, thanks to what it reveals of the close collaboration between John Aschenden and William Reed.

By focusing on Digby 176, we aim to retrace scholarly collaborations in a specific milieu, through intellectual and material exchanges. Careful study of the manuscript and its content help us to demonstrate that some scientific innovations depended on personal links and quick transmissions. Within this framework, Merton College is thus a very good example of a 'community of learning' and one of scientific knowledge.[9] This also suggests that fourteenth-century Merton College was a catalyst for innovations and practices in the field of the *scientia stellarum*. The works contained in this manuscript and the links we are able to reconstruct among some of these authors highlight the complementarity of astrological and astronomical practices.

The first part of this paper considers the history of the manuscript from its conception to its integration into the larger collection and the benefaction plan of William Reed. The relationships and collaborations of several scholars are highlighted with an analysis of certain texts contained in this manuscript and a palaeographical study of specific sections. The second part explores the reception of Alfonsine astronomy in England in detail, particularly through the example of William Reed's *Almanak Solis*. Furthermore, we demonstrate why this codex can be considered as one of the richest collections of predictions on planetary conjunctions. To do so, we focus on, among others, John Aschenden's treatises. Part two concludes with a case study of a nativity likely made for an individual linked to the Merton circle of astronomers and astrologers.

---

6 The *De pronosticatione aeris* or *Opusculum de mutatione aeris* is in Oxford, Bodleian Library Digby 147, fols 125^r–138^v. This treatise is based on Greek and Arabic sources. The *De futura temperie aeris pronosticanda* can be read in Digby 176 (ff. 3^r–4^r) and in Oxford, Bodleian Library Digby 97, ff. 128^v–129^r.

7 J. D. North, 'The Alfonsine Tables in England', in *ΠΡΙΣΜΑΤΑ: Naturwissenschaftsgeschichtliche Studien; Festschrift für Willy Hartner*, ed. by Y. Maeyama and W. G. Saltzer (Wiesbaden: F. Steiner, 1977), pp. 269–301. North has studied parts of this manuscript in *Horoscopes and History* (London: The Warburg Institute, 1986), pp. 136–39, but in a very partial way.

8 Hilary M. Carey, *Courting Disaster: Astrology at the English Court and Universities in the late Middle Ages* (London: MacMillan, 1992), pp. 63–77, 82, 85–89, 181–95.

9 See Constant Mews and John N. Crossley (eds), *Communities of Learning: Networks and the Shaping of Intellectual Identity in Europe, 1100–1500* (Turnhout: Brepols, 2011), especially Mew's article, 'Communities of Learning and the Dream of Synthesis: The Schools and Colleges of Thirteenth-Century Paris', pp. 109–35.

1. A manuscript in the Oxonian and Mertonian milieu

1.1. A composite manuscript

We begin by considering the structure of the manuscript and its composition. MS Digby 176 consists of nineteen different codicological units, originally belonging to different fourteenth-century manuscripts (see Appendix 2). This is a collection of astronomical and astrological works, including several treatises composed by former Oxford masters and Merton College fellows, such as William Merle (who was not a fellow of Merton College, but who was at Oxford from 1335-*c.* 38),[10] Simon Bredon (fellow from 1330–41), John Aschenden (fellow from 1337-*c.* 55), Reginald Lambourne (fellow from 1353–63) and William Reed (fellow from 1344–57), owner and compiler of this manuscript.[11] The authors of the other treatises are almost all contemporaries of William Reed: Geoffrey of Meaux, John of Murs, Levi ben Gerson, John of Saxony, and Walter Elveden.

All these parts were bound together by William Reed, a former fellow of Merton College, of which he was sub-Warden (vice-rector) at the end of his stay in Oxford. He was also a Doctor of Theology, Archdeacon of Rochester, and Provost of the collegiate Church of Wingham (Kent). He ended his career serving as the Bishop of Chichester from 1368 until his death. His best known work is an adaption of the Alfonsine Tables for the Oxford meridian and their canons, which he composed around 1340 and which mainly rely on John of Lignères' Tables of 1322 and *Tabule magne* dating to 1325.[12] William Reed is said to be the author of a 'solar almanac' or a set of tables recording the Sun's position over the years 1341–44 (in degrees, minutes, and seconds for each day, at the longitude of Oxford), perhaps composed in collaboration with Simon Bredon.[13] William Reed also provided computations for John Aschenden's prognostications, as is mentioned in the table of contents of Digby 176.[14] These calculations are based on his own adaptation of the Alfonsine tables. According to the content of the manuscript, William Reed was closely linked to a dynamic group of Merton scholars specialized in the field of 'quadrivium', and more particularly in the science of the stars. We review these relationships in more detail

10 William Merle was associated with Merton scholars; he borrowed Simon Bredon's medical book and was likely also connected to John Aschenden, as the latter borrowed some information from William for his *Summa iudicialis de accidentibus mundi.* See Keith Snedegar, 'The Works and Days of Simon Bredon. A Fourteenth-Century Astronomer and Physician', in *Between Demonstration and Imagination: Essays in the History of Science and Philosophy, presented to John D. North*, ed. by Lodi Nauta and Arjo Vanderjagt (Leiden: Brill, 1999), pp. 285–309.

11 See, in particular, James A. Weisheipl, 'Repertorium Mertonense', *Mediaeval Studies*, 31 (1969), 174–224; Rodney Thomson, 'William Reed, Bishop of Chichester (d. 1385) – Bibliophile?', in *The Study of Medieval Manuscripts of England: Festschrift in Honor of Richard W. Pfaff*, ed. by George H. Brown and Linda E. Voigts (Tempe-Turnhout: Arizona for Medieval and Renaissance Studies-Brepols, 2010), pp. 281–93.

12 Richard Harper, 'The Astronomical Tables of William Rede', *Isis*, 66 (1975), 369–78; North, 'The Alfonsine Tables in England', pp. 274 and 300; José Chabás, *Computational Astronomy in the Middle Ages. Sets of Astronomical Tables in Latin* (Madrid: Consejo Superior de Investigaciones Científicas, 2019), Chapter 10.

13 The association between both scholars is also highlighted by Snedegar, 'The Works and Days of Simon Bredon', pp. 290, 307; on the *Almanak Solis*, see *infra.*

14 According to the table of contents, William Reed made the calculations for the conjunctions of 1345 and 1365. Digby 176, f. 1^{v}: '[…] coniunctionum 3um planetarum superiorum anno Domini 1345 precedentium quas magister W. Reed calculavit et magister Johannes Aschenden pronosticavit'. 'Pronosticatio coniunctionis magne Saturni et Jovis anno Christi 1365 quam magister Willelmus Red calculavit et Johannes Asshenden pronosticavit'.

below, but we can already say that Simon Bredon, John Aschenden, William Merle, and Reginald Lambourne were part of this group. It seems that Digby 176 was part of William Reed's plan to keep a record of his fellows' writings and to collect information on planetary conjunctions, especially the triple conjunction of 1345.

1.2. MS Digby 176: A bibliophile's book

A piece of evidence provides a more precise timeframe for the final assembly of this manuscript. Reginald Lambourne's weather prediction for the years 1368–74 allows us to deduce that the manuscript was compiled by William Reed sometime after 1368 and obviously before his death in 1385.[15] This prediction, from a letter written by Lambourne while he was a monk of Eynsham abbey, was addressed to William Reed upon his elevation to the bishopric of Chichester in 1368.[16] The compilation of all the parts now preserved in Digby 176 thus occurred after William Reed had already left Merton College. While serving as Bishop of Chichester, he likely finalized a book acquisition plan, which he had begun at least two decades previously. Digby 176 was obviously part of this project. Although William Reed assembled this codex for his personal use, his attempt to create a testimony of the scientific activities of his former fellows and himself, as well as his contemporaries' productions, reveals that he did not compile this manuscript solely for personal use. The final destiny of this volume, as with many others from his library, was for it to be used by future generations of scholars, and also for his own salvation.[17]

William Reed's endeavour to collect results from a goal of providing Oxford students and scholars with access to books. This is supported by his *inter vivos* donations to several Oxonian colleges in 1374 and by his will written in 1382. Merton was not the only college to which he donated books, since his project was larger than supplying books to his former institution. His great vision was to take advantage of his substantial network in order to acquire a large collection of books. Therefore, the various phases of book acquisition strongly relied on bequests or gifts he received.[18] As we shall see shortly, since the donors are explicitly mentioned, Digby 176 is, *per se*, an example of these different steps of procurement spread over the years. The fact that William Reed's main aim was to provide books for Oxonian students' needs and studies is legible in the thematic classification

15 Digby 176, fols 40^r^–41^v^. On Reginald Lambourne, see Carey, *Courting Disaster*, pp. 69–72.

16 Digby 176 is the only witness of Reginald Lambourne's work. Both letters contained in the manuscript (fols 50^{r}–53^{v} and fols 40^{r}–41^{v}) were written in Eynsham abbey, where he was a monk from 1363.

17 Gifts of manuscripts had different and co-existing motivations: gratitude to the College, a vow to help future generations of scholars, remembrance, and salvation. Concerning the last two dimensions, the donation of books, and more particularly of a whole private library, can be compared with the foundation of anniversary masses in commemoration to the donor. The College, contrary to the University, held the responsibility of the individual commemoration of its fellows, which it fulfilled via its obituaries, which is also the case with the ex-libris formulae. Cf. Nathalie Gorochov, 'La mémoire des morts dans l'université de Paris au XIIIe siècle', in Hanno Brand, Pierre Monnet, Martial Staub (eds), *Memoria, communitas, civitas: Mémoires et consciences urbaines en Occident à la fin du Moyen Âge* (Ostfildern: J. Thorbecke, 2003), pp. 117–29; on the ex-libris as a medium of memory (through a case study of the ex-libris of the Sorbonne), Gilbert Fournier, 'Les Conditions d'une réussite: le livre et la *memoria* au collège de Sorbonne (XIIIe-XVe siècle)', in *Scriptoria e biblioteche nel basso medioevo (secoli XII–XV), atti del LI Convegno storico internazionale, Todi, 12–15 ottobre 2014* (Spoleto: Fondazione Centro Italiano di Studi sull'Alto Medioevo, 2015), pp. 475–502.

18 For an overview of the provenance of some of William Reed's manuscripts, see Thomson, 'William Reed, Bishop of Chichester', pp. 289–91.

of the two indentures of 1374.[19] Both documents are related to his lifetime donations to Merton and Exeter Colleges, and show the care with which he correlated his books to the prerequisite of the University's curricula.[20] Although many of the manuscripts gifted and bequeathed by him were related to the educational standards, he did not seem to rely entirely on the University's curriculum. Some of his manuscripts contain rare or high-level texts, which are far from being required readings. They do, however, correspond to William Reed's own interests. In the latter case, it is not the educational utility of the texts that is of utmost importance for him, but their intellectual interest. Digby 176 can be considered to belong to this last category of books; it is not a student handbook, but an expert compendium of important astrological and astronomical material. However, both handbooks and specialised books were indistinguishably donated or bequeathed by the Bishop of Chichester.

Like William Reed's book acquisition project, the dispersal of his private library was carefully planned. The first milestone was achieved in October 1374 with the aforementioned indentures (*indenturae*) made in London in favour of Merton and Exeter Colleges. He donated twenty-five books to Exeter College — an institution welcoming students in the arts from the diocese of Exeter where William Reed had been born — and twenty pounds sterling for the restoration of the *libraria communis*.[21] To Merton College, he gave one hundred books, as well as one-hundred pounds sterling for repairs of the *libraria communis*.[22] In both indentures, a recurrent recommendation appears: the books had to be chained in the library for the common use by members of the college (*ad usum communem sociorum ibidem studentium*). That same year he presumably made a new donation to Queen's College, since the account rolls show evidence of the journey of two fellows to London to receive books and money from the Bishop of Chichester.[23]

As suggested by both indentures, William Reed's plan was based not only on book donations but also, more pragmatically, on the maintenance of the library buildings. Moreover, his recommendation to chain books derives from his desire to provide the whole community access to his books. He was more heavily involved in his former house, Merton College, to which he gave most generously. In this respect, he was committed to

19 See the edition of both sources, in Thomson, *The University and College Libraries of Oxford*, pp. 666–73 and 866–74.

20 These classifications lead Rodney Thomson to refer to 'the educational vision' of William Reed. The contents of both indentures correspond to the ambition of each college: Exeter College, for example, was reserved for students in the arts faculty only. See Thomson, 'William Reed', p. 292; Thomson, *The University and College Libraries of Oxford*, pp. 666–73 and 866–74.

21 Thomson, *The University and College Libraries of Oxford*, p. 666: '[...] quod dictus procurator accepit et habuit in die confeccionis presencium ex dono domini venerabilis patris viginti libras sterlingorum ad reparacionem librarie communis eiusdem aule. Item habuit et recepit viginti quinque libros in dicta libraria ad usum communem sociorum ibidem studencium cathenandos [...]'.

22 Thomson, *The University and College Libraries of Oxford*, p. 856: '[...] quod dicti procuratores acceperunt et habuerunt in die confectionis presencium ex dono dicti venerabilis patris centum libras sterlingorum ad reparacionem librarie communis eiusdem domus; item habuerunt et receperunt centum libros in dicta libraria ad usum communem sociorum ibidem studencium cathenandos [...]'. The donation also includes ten astronomical instruments: the list describes these instruments in parallel with the ninety-nine books (and not one hundred).

23 Regarding this donation to Queen's College, see Thomson, 'William Reed', p. 285.

the construction of a new building for the Merton College Library, which began a year after he offered a significant donation in 1375 and was completed in 1380.[24]

Although Digby 176 was not included in this first gift, it was part of his larger plan. The absence of the manuscript in the 1374 donation may be due to the fact that it had not yet been assembled at that date. A second milestone occurred in 1382 with William Reed's will, which was executed at his death in 1385.[25] To his relative, Richard Pestour (fellow of Exeter and Merton Colleges) and additional kin studying in Oxford he bequeathed one hundred books related to diverse faculties, some of which exceeded university curricula. After Pestour's death, these books would be kept by the masters of Merton and Exeter Colleges.[26] Several other bequests for Oxonian institutions are mentioned in the will. They received books and a certain amount of money; Merton College received one hundred books, one hundred pounds of gold for the repairs to the library, and one hundred pounds to be kept in a chest for the use of the *socii*. He also gave one hundred books and twenty pounds of gold to the recent foundation of New College, twenty books and twenty pounds of gold to Exeter College, ten books each to Balliol, Oriel, and Queen's Colleges, and eighteen books to citizens and churches of Oxford.[27]

Between 1374 and 1382, William Reed gave and bequeathed a total of about five hundred and thirty volumes. Although he probably did not own all these books at the same time — as he gave some of them away in 1374 —, he had gathered one of the most important private libraries of England at that time.[28] However not all these volumes, according to their *ex-libris* or *ex-dono*, were commissioned by the Bishop of Chichester; he acquired them from individuals he knew in Kent, London, Oxford, and Chichester. Most of the books were related to the Oxford curricula, but he also had more peculiar scientific manuscripts which were not prescribed by university regulations. William Reed must be

24 William Reed went to the Dominicans' convent of London with John Bloxham, Warden of Merton College, and William Humberville, chief mason, in order to explore the library of the Preachers. It is also noteworthy that John Bloxham was one of the witnesses to the 1374 donation to Merton College. On the new library building and the involvement of the Bishop of Chichester in its construction, see Heathcote William Garrod, 'The Library Regulations of A Medieval College', *The Library*, 3 (1927), 312–35; Garrod, 'An Indenture between William Reed, Bishop of Chichester, and John Bloxham and Henry Stapilton, Fellows of Merton College, Oxford, London 22 October 1373', ed. by J. Roger L. Highfield, *Bodleian Library Record*, 10 (1982), 9–19; Highfield and Martin, *A History of Merton College*, pp. 89–91.

25 Cf. the edition of William Reed's will in Powicke, *The Medieval Books of Merton College*, pp. 87–91.

26 Powicke, *The Medieval Books of Merton College*, p. 87: '[…] Item lego Magistro Ricardo Pestour et scolaribus de genere meo Oxonie in studio litterarum proficere volentibus C libros diverse facultatis eisdem intitulatos ipsis, iuxta discretionem custodis et magistri domorum de Merton et Stapelton in Oxonia post mortem dicti magistri Ricardi provide liberandos […]'.

27 The Merton library in question is likely the new one, the *reparatio* ([…] *C libras in auro ad reparationem librarie ibidem*) is perhaps related to the furnishing of the new building. Furthermore, Arundel College in Kent is also mentioned following the bequests made to the Oxford Colleges. It received, *inter alia*, thirteen books. See Powicke, *The Medieval Books of Merton College*, pp. 87–88; Thomson, 'William Reed', p. 284.

28 As a comparison and since the number of books in the renowned private collections of Richard de Bury (d. 1345) and of Simon Langham (d. 1376) cannot be determined, two of the largest private English libraries were those of John Ergome (d. 1385) and Adam Easton (d. 1397). The former was an Augustinian friar in York who owned around 300 volumes, which he bequeathed to his convent; the latter was a Benedictine monk who would go on to become a cardinal in Avignon. He possessed 228 volumes that he gave to his former monastery, the Norwich Cathedral priory. Cf. Jenny Stratford and Teresa Webber, 'Bishops and Kings: Private Book Collections in Medieval England', in *The Cambridge History of Libraries in Britain and Ireland*, ed. by Elisabeth Leedham-Green and Teresa Webber (Cambridge: Cambridge University Press, 2006), pp. 178–217.

considered as a real bibliophile, aware of scholars' needs but also eager to document the scientific activities of his time.

MS Digby 176 was part of his bequest to Richard Pestour, mentioned in his 1382 will. The manuscript appears in another indenture, dated 1400, recording the delivery, by John Bampton and Richard Pestour, of forty-five books from William Reed's library to the Rector of Exeter College and the Warden of Merton College.[29] This document can be considered as the execution of William's will. It represents almost half of the provision of this donation of one-hundred books. Therefore, this source only concerns part of the delivery of this donation, and it seems likely that another act was written for the execution of the rest of the provision. The indenture implies that the collection of books was divided between Merton and Exeter Colleges, in order to be managed and preserved by each of them. The text preceding the description of the forty-five volumes is to be linked to Exeter College, as it develops a series of prescriptions related to the care of the books to be kept in Exeter and to their use, with an explicit hierarchy including Richard Pestour, William Reed's kin, and the other *socii* of the College. The list numbers the forty-five volumes, summarizes their content, and gives the first words of the second folio as key words. The rest of the legacy of the Bishop of Chichester, which should have comprised fifty-five volumes, was probably devoted to Merton College. This suggests that another list, no longer extant, was established contemporarily for this institution, modeled on the 1400 Exeter College list.

Although MS Digby 176 is not described among the forty-five volumes for Exeter College, internal evidence suggests that it was part of William Reed's will to his kinsman Richard Pestour. In fact, folio 2^{r} displays an *ex-dono*, which clearly demonstrates that the manuscript was part of this will:[30]

> Liber scolarium de genere venerabilis patris domini Willelmi Reed, episcopi Cicestrie, Oxon[ie] successive studentium, ex dono venerabilis patris predicti, per custodem et rectorem domorum de Merton et Stapelton in Oxon[ia], vel per eorum librarios, eisdem scolaribus juxta facultates et merita ipsorum cuiusque, ad tempus sub caucione juratoria provide liberandus.[31]

The fact that the manuscript was part of the delivery to Merton College is attested by the mention of 'Merton', likely written contemporarily to the transfer of the books, in the upper margin of folio 1^{v}, above the table of contents.

MS Digby 176 was still in the college in 1483. Indeed, the volume is described in the indenture established to list the books to be assigned to the new Warden, Richard

---

29 On this indenture, see Thomson, 'William Reed', pp. 285–86; Thomson, *The University and College Libraries of Oxford*, pp. 670–80.

30 This *ex-dono* was not mentioned by Rodney Thomson. He refers to a similar *ex-dono* transcribed in an Exeter College manuscript during the seventeenth century (but no longer extant), and says that this kind of 'inscription does not appear in any surviving book'. Thomson, 'William Reed', p. 287 n. 22. However, Digby 176 does display such an *ex-dono* and is perhaps one of the rare examples to testify this will.

31 'Book of the scholars from the kin of the venerable father master William Reed, Bishop of Chichester, and of the future students of Oxford. Gift from the venerable father aforementioned, to the care and administration of the houses of Merton and Stapelton in Oxford, or to their librarians, and should be temporarily delivered against caution to the same students of the local faculties and the deserving ones'.

Fitzjames (d. 1522).[32] The Warden of Merton College had a separate library, for which he received books when he was elected. It is likely that the head of the college selected part of this book collection. It is not thus surprising that Richard Fitzjames asked for MS Digby 176, as he patronised astrologers and commissioned the zodiac archway which is still visible today and connects the Warden's house to the Hall of the College.[33] The manuscript can be identified in the indenture, as follows: '9. Item liber de astronomia cum multis aliis contentis, 2 fol. *Notandum quod dies*'.[34] The first words of the second folio correspond to the incipit of folio 4^{r} in William Merle's *Regule ad futuram aeris temperiem prenosticandam*.

After its sojourn at Merton College, the manuscript passed to a certain John Collston, whose signature is legible on folios 2^{v} and 87^{r} and which could be dated either to the very end of the fifteenth century or the early sixteenth century. The mathematician Thomas Allen (d. 1632) acquired the manuscript according to his library's shelf mark '5' on folio 3^{r}. In 1632, the manuscript joined the main part of Allen's collection in Sir Kenelm Digby's library, who donated his books to the Bodleian Library in 1634. Digby's signature and motto are also legible on folio 3^{r}.[35]

Therefore, Digby 176 was part of William Reed's greater project of providing his books to Oxonian scholars. However, this codex was also conceived as a testimony of contemporaries' scientific activities in the field of the science of the stars. As we shall argue, William Reed consciously gathered the records of these practices from several individuals. The different texts, notes, and quires provide a clear picture of the milieu and collaborations of the Bishop of Chichester, who maintained close links with other Mertonians.

1.3. The origin of the manuscript: Within a Mertonian context

The first flyleaf of the manuscript provides us with important information about how this collection was gathered. The verso of the first flyleaf displays a detailed *ex-libris* and a table of contents, both written by Walter Robert, William Reed's clerk (see Fig. 1):

> Liber magistri Willelmi Reed, episcopi Cicestrie, cuius partem habuit ex dono reverendi domini sui magistri Nicholai de Sandwyco, partem emit de executoribus reverendi patris domini Thome de Bradewardina, archiepiscopi Cantuarie, partem emit de executoribus magistri Ricardi Camsale, partem ipse magister Willelmus scripsit et partem scribi

32 Cf. Powicke, *The Medieval Books of Merton College*, pp. 167–68, n° 537. The indenture is edited in Thomson, *The University and College Libraries of Oxford*, pp. 951–58. Richard Fitzjames was trained in Oxford and was a fellow of Merton College from *c.* 1468–83. He was the Warden of Merton College from 1483 to 1507, Henry VII's almoner in 1495, Bishop of Rochester in 1497, Bishop of Chichester in 1504, and eventually Bishop of London from 1506 until his death. Cf. S. Thompson, 'Fitzjames, Richard (d. 1522), Bishop of London', in *Oxford Dictionary of National Biography*, (Oxford: 2012), [accessed: 03.02.2020] <https://www-oxforddnb-com.ezp.lib.cam.ac.uk/view/10.1093/ref:odnb/9780198614128.001.0001/odnb-9780198614128-e-9612>.

33 See: Hilary M. Carey, 'Henry VII's Book of Astrology and the Tudor Renaissance', *Renaissance Quaterly*, 65 (2012), 661–710.

34 Thomson, *The University and College Libraries of Oxford*, p. 953.

35 Digby 176, f. 3^{r}: 'Vindica te tibi. Kenelme Digby'.

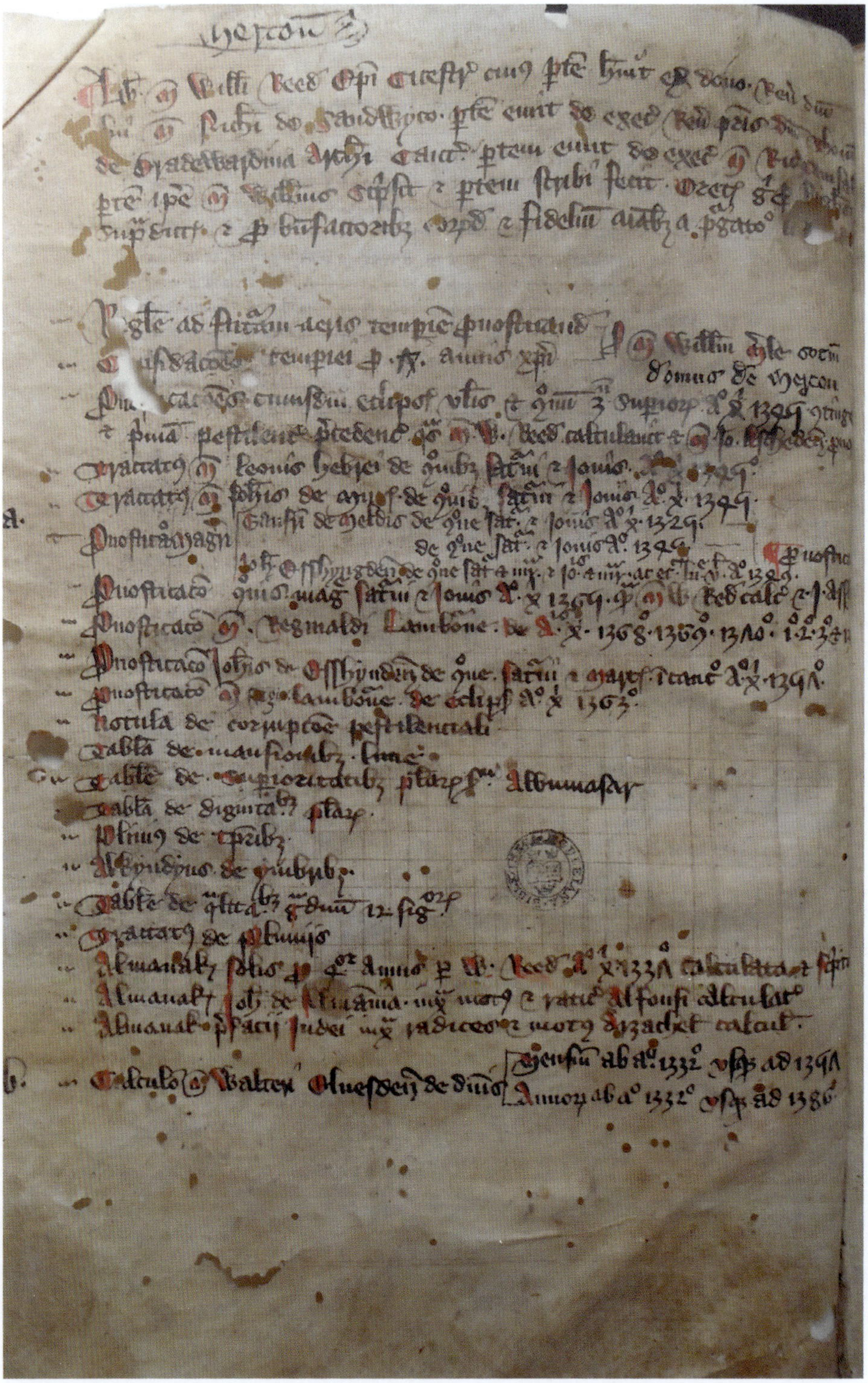

Figure 1. MS Oxford, Bodleian Library, Digby 176, folio 1[r]: *ex-libris* and table of contents. Authors' photograph.

fecit. Oretis igitur pro singulis supradictis, et pro benefactoribus eorundem et fidelium animabus a purgatorio liberandis.[36]

The *ex-libris* specifies that William acquired part of the manuscript from Nicolas of Sandwich's donation, part from the executor of Thomas Bradwardine, part from Richard Campsale, part that William Reed himself wrote, and part that had been copied by scribes. It is followed by a request for prayers for the persons mentioned and for their salvation.

It is difficult to know which part belonged to the aforementioned former owners as there is no internal evidence such as owner marks. However, one can say with confidence that all the individuals mentioned are linked in one way or another to William Reed.

Firstly, Nicholas of Sandwich (fl. 1347) became William's patron when he was orphaned.[37] He was a wealthy clerk who owned vast estates in Kent and spent his life in Oxford. He raised William, and he supplied him with books and money while he was in Oxford. He probably helped William to obtain his positions in Kent: initially as Archdeacon of Rochester, and later as Provost of Wingham. The detailed *ex-libris* of William's extant manuscripts allow us to say that he had at least fourteen books that were given to him by Nicholas, and that he purchased at least thirty books with Nicholas's money.[38] As Nicholas was still alive in 1347, we can assume that his donation (now preserved in Digby 176) likely occurred when William Reed was in Oxford between 1337 and 1357. However, it is difficult to allot a specific quire to a gift from Nicholas of Sandwich.

The second individual mentioned in the inscription is the well-known mathematician, theologian, and former fellow of Merton College (between 1323 and 1335), Thomas Bradwardine.[39] He was much older than William, and it is unlikely that they met at Merton. However, it is specified that William Reed purchased a part of the codex from the executors of Thomas Bradwardine's will, who is mentioned as the Archbishop of Canterbury. His election to the See of Canterbury took place in 1348, a year before his death from the black plague. It is likely that Thomas was already deceased when William acquired the manuscript. Thus, this acquisition may well have occurred after 1349, perhaps soon after Thomas's death.[40]

---

36 'Book of master William Reed, Bishop of Chichester, whose one part was acquired from the gift of the venerable master Nicholas of Sandwich, a part purchased from the executors of the will of the venerable father master Thomas Bradwardine, Archbishop of Canterbury, a part purchased from the executors of master Richard Campsale, a part written by master William, and a part commissioned. Pray for the aforementioned individuals and to their benefactors, and for freeing the souls of the faithful from the purgatory'.

37 He is twice mentioned — in laudatory terms — in William Reed's will, who founded anniversary masses for his late patron. See one of these notes: 'Item volo quod omnia superius per me data et legata anime Magistri Nicholai de Sandewyco qui me a puericia usque ad provectam etatem educavit sicut et mee cedant in premium perpetuum et salutem et quia plura superius recitavi legata que ante factionem istius testamentu fuerint soluta et liberata et expedita [...]', Powicke, *The Medieval Books of Merton College*, p. 90.

38 See Powicke, *The Medieval Books of Merton College*, pp. 29–30; Alfred B. Emden, *A Biographical Register of the University of Oxford to A. D. 1500* (Oxford: Clarendon Press, 1957–1959), pp. 1639–40 (henceforth *BRUO*). Thomson, 'William Reed', p. 282.

39 On his life see *BRUO*, pp. 244–46; for an updated bibliography, see Edith D. Sylla, 'Bradwardine, Thomas', in *Complete Dictionary of Scientific Biography*, (New York: 2008), pp. 377–80, [accessed: 03.02.2020] <https://link-gale-com.ezp.lib.cam.ac.uk/apps/doc/CX2830905519/GVRL?u = cambuni&sid = GVRL&xid = 61d8fe01>

40 It should also be noted that William Reed had a copy of Thomas Bradwardine's theological treatise *De causa Dei*, which he commissioned while he was a fellow of Merton College with funds given by Nicholas of Sandwich (New College, MS 134). It was included in manuscripts given by William Reed to New College.

Even more interesting is the reference to Master Richard Camsale, or Campsale. Two different Richard Campsale can be considered; one was a fellow of Merton College (1306–22) and author of *Questions* on Aristotle's *Physics* who died around 1330, the other was the Rector of Saint-Martin's Church, Canterbury who acted as the executor of Simon Bredon's will and also bequeathed several of his books to him.[41] Consequently, the Richard Campsale mentioned in Digby 176 is likely the later, since his link to William Reed is Simon Bredon. Indeed, the latter's 1368 describes in great detail the books and all the other belongings he left to him.[42] One of the originalities of this will lies in the short description of the content of each book, as well as in the itemization of the bequests and beneficiaries. Simon Bredon bequeathed seventy-eight volumes to individuals and Oxford Colleges. Among the institutions, Merton College was the biggest beneficiary, with twenty-eight books donated.[43]

From Simon, Richard Campsale received the following: twenty shillings, a silver cup without cover, a quire containing Profatius's *Almanach*, a treatise on sines, Albumasar's *Ysagoga minor*, John Mauduith's trigonometric tables of 1310 (John Mauduith was a fellow of Merton: 1309–13),[44] a book of medical astrology addressed to Robert of Beaumont, second Earl of Leicester (d. 1168), which begins *Que in gloriosissimis*, Pseudo-Albert the Great's *De herbis, lapidibus et animalibus*, *Experimenta* concerning snake flesh, the *Centiloquium* ascribed to Bethen, canons on the Toledan Tables, Albumazar's *Introductorius maius* in addition to a chronicle and Simon Bredon's own prayerbook, and, finally, Simon's best robe.[45]

---

41 Cf. Edward A. Synan, 'Richard of Campsall, an English Theologian of the Fourteenth Century', *Mediaeval Studies*, 14 (1952), 1–8.

42 Powicke, *The Medieval Books of Merton College*, pp. 82–86; Thomson, *The University and College Libraries of Oxford*, pp. 1266–69 (only the books are mentioned).

43 Oriel College received four books, Balliol College three, Exeter College two, University College one, and Queen's College one. It is noteworthy that Simon Bredon also bequeathed a copy of the Bible to the Augustinian priory of Merton in Surrey (conditional upon the payment of a fee to his executors), which was linked to the eponym college: 'Item Bibliam meam lego Prioratui de Merton sub condicione quod solvant executoribus meis [...]'. Powicke, *The Medieval Books of Merton College*, p. 85.

44 John Mauduith or Maudith (d. 1343) was a native of Worcester diocese and a fellow of Merton College from 1309 until 1319. He belonged to the college's first generation of astronomers and mathematicians. He received several ecclesiastical benefices from 1319 to 1343 and was part of the entourage of the Bishop of Durham, Richard of Bury. He likely finished his life as Dean of Auckland in the Diocese of Durham. John Mauduith is known for his trigonometrical tables and canons composed in 1310, which are mainly based on the Toledan Tables, and for a list of eleven stars (1316) derived from the treatise *Novus quadrans* by Profatius Judeus. Mauduith also practiced and taught medicine. Cf. *BRUO*, pp. 1243–44; J. D. North, 'Medieval Star Catalogues and the Movement of the Eighth Sphere', *Archives internationales d'histoire des sciences*, 20 (1967), 71–83; North, *Richard of Wallingford: An Edition of his Writings*, 3 vols (Oxford: Clarendon Press, 1976), I, pp. 3–19, 192–97, II, pp. 155–58 and III, pp. 157–58.

45 Powicke, *The Medieval Books of Merton College*, pp. 83–85: '[...] Item lego domino Ricardo de Camsale rectori ecclesie Sancti Martini Cantuarie viginti solidos. [...] Item lego domino Ricardo de Camsale unum ciphum argenteum qui non habet cooperculum. [...] Item quaternum meum in quo continentur *Almanac* Prefacii [*sic*], et opus de sinibus, et *Minus introductorium* Albumasar, et *Tabule Mauduit*, et opus liber qui incipit *Que in glorio[si]ssimis*, et Albertus *De herbis, lapidibus et animalibus*, et *Experimenta de corrio serpentis* et *Centilogium Betem*, et *Canones Azarchelis*, istum quaternum cum omnibus contentis in eo et Albumazer in *Maiori introductorio* lego domino Ricardo de Camsale, rectori Sancti Martini Cantuarie. Item eidem lego cronica Martini. [...] et iurnale meum de laudibus et horis lego domino Ricardo Camsale [...] Item lego domino Ricardo Camsale meliorem robam meam'. For these different treatises see: *Almanach Dantis Aligherii sive Profhacii Judaei Montispessulani*, ed. by J. Boffito and C. Melzi d'Eril (Florence: Olschki, 1908); José Chabás and Bernard R. Goldstein, 'The *Almanac* of Jacob ben Makhir', in *Editing and Analysing Numerical Tables: Towards a Digital Information System for the History of Astral Sciences*, ed. Mathieu Husson, Clemency Montelle and Benno van Dalen (Turnhout: Brepols, 2022), pp. 53-78; Abū Ma'shar, *The Abbreviation of*

The book descriptions provided in Simon Bredon's will are short, and do not detail all the content of the volumes; it is therefore difficult to link one of these items to Digby 176. However, there is one common mention between Digby 176 and the will. The almanac of Profatius Judeus (Jacob ben Makhir Ibn Tibbon) is listed among the treatises linked to the donation to Richard Campsale. A copy of this work was also present in Digby 176, as mentioned in the table of contents (f. 1^r). It was originally located just after John of Saxony's almanac (ff. 73^r–86^r).[46] Profatius's treatise is unfortunately no longer extant and it thus remains difficult to assert that the copy mentioned in Simon Bredon's will was the one copied in Digby 176. However, as we discuss shortly, it is interesting to note that Simon Bredon likely owned folios 71^r–86^r, corresponding to a section dedicated entirely to almanacs.

Although we cannot determine which part William Reed acquired from Nicholas of Sandwich or from the executors of Thomas Bradwardine's will, it is possible to identify the parts acquired from Richard Campsale. Since Simon Bredon's handwriting has a few peculiarities, we are able to determine which parts he copied.

Simon Bredon was a physician, theologian, astronomer and fellow of Merton College from 1330 to 1341.[47] He completed his regency in arts in 1340 and undertook studies in theology and medicine which he likely completed in 1348.[48] He was also involved in the day-to-day life of the College, as was William Reed. His name can be seen in the records of the annual meetings (*scrutinia*) of 1338 and 1339, during which he requested access to the common library of the college for arts students and a distribution of books (*electio*) to the fellows.[49] He served as a procurator of the college at the end of the 1330s and was notably in charge of administrative visits to Merton's estates. Simon Bredon was also strongly committed to the University's administration. In 1333, he pleaded for the Archdeacon of

---

the Introduction to Astrology, together with the Medieval Latin Translation of Adelard of Bath, eds Charles Burnett, Keiji Yamamoto, Michio Yano (Leiden: Brill, 1994); Burnett, 'Astrology for the Doctor in a Work Addressed to Robert, Earl of Leicester', in *De l'homme, de la nature et du monde. Mélanges d'histoire des sciences médiévales offerts à Danielle Jacquart* (Geneva: Droz, 2019), pp. 179–96; Isabelle Draelants, *Le Liber de virtutibus herbarum, lapidum et animalium (Liber aggregationis). Un texte à succès attribué à Albert le Grand* (Florence: Sismel-Edizione dell Galluzzo, 2007). For the *Experimenta de corrio serpentis*, an opuscule on the twelve marvellous virtues of a snakeskin, preserved in quite a number of Latin manuscripts from the thirteenth century onwards, see John William S. Johnson, 'Les *Experimenta duodecim Johannis Paulini*', *Bulletin de la Société française d'histoire de la médecine*, 12 (1913), 257–67. The *Centiloquium* ascribed to Bethen was published in 1493 with Ptolemy's *Quadripartitum* by Bonetus Locatellus in Venice, ff. 119ra-120ra. See also F.S. Pedersen, *The Toledan Tables, A Review of the Manuscripts and the Textual Versions with an Edition* (Copenhagen: C. A. Reitzels Forlag, 2002); Abū Maʿshar al-Balkhi, (Albumasar), *Kitāb al-mudkhal al-kabīr, Liber Introductorii Maioris ad Scientiam Iudiciorum Astrorum*, ed. Richard Lemay (Naples: Istituto universitario Orientale, 1995–1996), vols IV–VII, for the Latin translations of John of Seville and Hermann of Carinthia.

46 See the transcription of the list below. It reads *Almanak Prefacii Judei iuxta radices et motus Arzachel calculat[ur]*.

47 See Charles H. Talbot, 'Simon Bredon (*c*. 1300–1372): Physician, Mathematician and Astronomer', *British Journal for the History of Science*, 1 (1962), 19–30; Snedegar, 'The Works and Days of Simon Bredon'.

48 The strict interdiction of medical studies for the *socii* of Merton College, enforced by the Archbishop of Canterbury John Peckham during his visitation of 1284, does not seem to have had any incidence since the physician and author of the *Rosa anglica medicine*, John Gaddesden (d. 1348) was a fellow of Merton College between 1305–7. On John Peckham's injunctions, see *Registrum epistolarum Fratris Johannis Peckham Archiepiscopi Cantuariensis*, ed. Charles T. Martin (London: Longman, 1885), III, pp. 811–18; see also Martin and Highfield, *A History of Merton College*, pp. 50–51.

49 Cf. *Merton Muniments*, n° xiii–xv. On the *electio* system, see Neil Ripley Ker, 'The Books of Philosophy Distributed at Merton College in 1372 and 1375', in *Books, Collectors and Libraries: Studies in the Medieval Heritage* ed. Andrew Watson (London: Hambledon Press, 1985), pp. 331–78.

Oxford's jurisdiction over the University at the Papal Curia in Avignon. He became the proctor of the University at the very end of the 1330s. From 1348, he held various ecclesiastical benefices and likely practiced medicine in addition to his appointments within the church. In the 1350s, he thus served as a physician to the Earl of Arundel and the wife of the King of Scotland, who was in exile in Hertford.[50] In the 1360s, he greatly benefitted from the protection of the Archbishop of Canterbury, Simon Islip, who replaced Thomas Bradwardine within the see from 1349 to 1366.[51]

By 1355, the Earl of Arundel, his former patron, had secured a position at Chichester for Simon Bredon,[52] who was likely appointed immediately after the Archbishop of Canterbury's death in 1366. In any case, Simon was Canon of Chichester in 1368, when he established his will. The year 1368 also marked the election of William Reed as Bishop of Chichester. Simon Bredon, William Reed, and William Heytesbury (d. 1372–73), another Mertonian and also a canon of Chichester, gathered there.[53] However, even before Chichester, in the early 1360s the three men were appointed to the collegiate church of Wingham in Kent.[54] Links between William Reed and Simon Bredon are also attested by the latter's bequest of a small astrolabe to the Bishop of Chichester.[55] We cannot exclude the possibility that both scholars collaborated around 1340, when William Reed was elaborating his adaption of the Alfonsine Tables to the Oxford meridian and his *Almanak Solis* for the years 1341–44, and Simon Bredon his commentary on the *Almagest*.[56] This demonstrates the close relationships maintained between these Mertonians well after their Oxford years.

Simon Bredon maintained strong links with other Mertonians, such as the astrologer John Aschenden. He even resigned his position of Procurator of Merton College for Aschenden (fellow from 1337 until the 1350s).[57] Aschenden was also greatly involved in the daily life of his college and followed the same academic path as Bredon, studying theology

---

50 Snedegar, 'The Works and Days of Simon Bredon', pp. 301–3.

51 On Simon Islip, Archbishop of Canterbury, see *BRUO*, pp. 1002–6. Thanks to Simon Islip, Simon Bredon was appointed to two rectories in Kent and to the canonry of Wingham; cf. Snedegar, 'The Works and Days of Simon Bredon', p. 301.

52 Although Simon Bredon became Archdeacon of Chichester in 1354, he had resigned this prebendary for another one by 1356.

53 William Heytesbury was one of the executors of Simon Bredon's will. See Powicke, *The Medieval Books of Merton College*, pp. 82–86; *BRUO*, pp. 927–28.

54 William Reed was Provost of Wingham between 1363 and 1368, Simon Bredon was canon there in the early 1360s, and William Heytesbury held prebendaries in Wingham and Ickham in Kent at the same time. Simon Bredon bequeathed two books of canon law (the *Decretals* and the *Decretum*) to William Heytesbury: 'Item lego librum *Decretalium* cum tabula edita super eum, and librum *Decretorum* cum tabula Martini super eum magistro Willelmo de Heghterbury'. He also received several items of furniture (such as a bed with paintings of parrots and cockerels) and clothing. Powicke, *The Medieval Books of Merton College*, pp. 83–84.

55 Simon Bredon's large astrolabe was bequeathed to Merton College: 'Item astrolabium maius lego Aule de Merton et astrolabium minus lego magistro Willelmo Reed'. Powicke, *The Medieval Books of Merton College*, p. 84.

56 The commentary on the *Almagest* Bk III is related to solar motion and the *Almanak Solis*, whose purpose is to provide true solar positions for every day of the year from 1341 to 1344. Cf. Henry Zepeda, '*The Medieval Latin Transmission of the Menelaus Theorem*', unpublished PhD Dissertation, University of Oklahoma, 2013, pp. 282–301 and 637–86; Zepeda, *The First Latin Treatise on Ptolemy's Astronomy: The Almagesti minor (c. 1200)* (Turnhout: Brepols, 2018), pp. 95–98; David Juste, 'Simon Bredon, *Commentary on the Almagest*' (update: 29.10.2019), *Ptolemaeus Arabus et Latinus. Works*, URL = http://ptolemaeus.badw.de/work/76.

57 Simon Bredon called him *dilectus magister*. Cf. Snedegar, 'The Works and Days of Simon Bredon', p. 301.

and medicine.[58] In his prediction for the conjunctions of 1357 and 1365, he explicitly refers to Simon Bredon, who is said to have equated the motion of the eighth sphere.[59] Keith Snedegar has suggested that John Aschenden also consulted Simon Bredon's manuscript of the *Quadripartitum* (Oxford, Bodleian Digby 179), from which he took material for his *Summa iudicialis de accidentibus mundi* (1346–48).[60] We cannot dismiss the notion that Simon Bredon had granted access to his collection of books to his fellows and friends. Intellectual exchanges and collaborations within this circle of scholars were not uncommon. It should be noted that John Aschenden, in his *Summa iudicialis*, also borrowed material from William Merle *Considerationes temperiei pro 7 annis* (only preserved in Digby 176) for the part dedicated to weather forecasting. As we discuss below, the astrologer also collaborated with William Reed for some of his predictions. Furthermore, we know that William Merle and Simon Bredon knew each other. In fact, the latter specified in his will that he had loaned William Merle a volume containing a comment on the *Viaticum* translated by Constantine the African with the *De modo medendi* of Gerardus.[61] More particularly, it is specified that he bequeathed this volume to Roger of Aswardby, who was Master of University College Oxford by 1350, whereas William Merle had not yet returned this book.[62] It seems that Simon Bredon was older than his aforementioned colleagues, since he became a fellow of Merton several years before the others. During their Oxford years, Simon Bredon likely acted as a mentor to them. Once they became alumni of Merton College, Simon Bredon and William Reed, having reached high positions, became something of patrons for their former fellows.

Simon Bredon's works appear to have focused mainly on the mathematical sciences, especially astronomy and arithmetic. During his years at Merton College, he wrote a commentary on the *Almagest* (on some demonstrations of Books I–III) arranged in propositions and proofs (quite similar in its organization to the *Almagesti minor*) and some glosses to three different translations of the *Quadripartitum*.

58 See Keith Snedegar, 'Ashenden, John', *Oxford Dictionary of National Biography*, 2004 (online). He was a Northumbrian who became a fellow of Merton College (1337–55), *magister* from 1340 onwards, and 'a purely academic astrologer'. 'He was probably deceased by 1368, as he is not mentioned in the will of his friend Simon Bredon, which was drawn up in that year'.

59 Digby 176, f. 45r: '[…] Ista patent secundum magistrum Simonem de Bredon qui circa annum Christi 1340 equavit motum 8ve spere cum macima diligencia […]'. This sentence is also discussed in C. Philip E. Nothaft, 'Criticism of Trepidation Models and Advocacy of Uniform Precession in Medieval Latin Astronomy', *Archives for History of Exact Science*, 71 (2017), 211–44 (p. 232).

60 Cf. Keith Snedegar, 'John Ashenden and the *Scientia Astrorum Mertonensis*', unpublished PhD thesis, University of Oxford, 1988. For Digby 179, see David Juste, 'MS Oxford, Bodleian Library, Digby 179' (updated: 11.12.2017), *Ptolemaeus Arabus et Latinus. Manuscripts*, URL = http://ptolemaeus.badw.de/ms/232.

61 'Item Gerardum super *Viaticum* cum Constantino, seu Gerardo, *De modo medendi* in eodem volumine assigno magistro Rogero de Aswardby rogans ut ipse tradat illum librum illi cuius erat et a quo magister Willelmus Merle illum ex accomodato habebat, quod si dominum dicti libri non noverit, det dictum librum alicui medico eo indigenti, pro anima domini dicti libri'. Powicke, *The Medieval Books of Merton College*, pp. 84–85.

62 It is noteworthy that Simon Bredon also bequeathed a book to University College, as well as a missal to Roger Aswardby. Powicke, *The Medieval Books of Merton College*, p. 84. On Roger Aswardby, see Robin Darwall-Smith, *A History of University College, Oxford* (Oxford: Oxford University Press, 2008), pp. 33–35.

1.3.1. Simon Bredon's quires

A palaeographical analysis of the codicological units of MS Digby 176 may help to determine which parts belonged to Simon Bredon and were given later to William Reed by Richard Camsale.

Since he annotated and copied several of his manuscripts, Simon Bredon's handwriting is well known. It is a typical fourteenth-century Anglicana (English *cursiva antiquior*) script, including more cursive and condensed features when he annotates, which become neater and more spaced out when he copies a main text (more formal, closer to a *libraria* script) (Fig. 2).[63] The ascenders of *l, h, b,* and *k* are looped and large at the top. The left side of the loop stroke is bold, whereas the right side is thinner. Simon Bredon's *ductus* includes: 1) the *d* is typical of the Anglicana script with a loop, including a bold and curved diagonal stroke; its shape is quite circular but has a tendency to lean to the left. Whether it is included in a more cursive script or not, Simon Bredon emphasized the internal stroke by exerting a pressure on the nib; 2) two types of *r* are used, the sharp v-shaped *r* with marked right and left shoulders and a tail descending beneath the baseline, and the more cursive, round *r*, which is also used when it follows the letter *o*; 3) the *s* is perhaps Bredon's most distinctive letter, appearing in three forms; most frequently, he traces a one-stroke round *s*, with an upper lobe extended to the right. This typical documentary form is employed at the beginning or at the end of a word. From time to time, but less frequently, he uses the 8-shaped *s* when it is the final letter of a word. And most of the time in the middle of a word, Simon Bredon employs the straight *s*; 4) the *g* is a typical 8-shaped letter. This handwriting is a fine example of a neat Anglicana script; it can be compared with a professional English scribe's handwriting. Indeed it can be sometimes difficult, especially in its more cursive form, to distinguish from other standardized English *cursiva antiquior* script.

Among his extant astronomical manuscripts, Simon Bredon copied Ptolemy's *Quadripartitum* (Oxford, Bodleian Library, Digby 179; cf. Fig. 2). This is a unique manuscript bringing together the three translations of the *Quadripartitum* by Plato of Tivoli, Egidius de Tebaldis, and William of Moerbeke. Simon Bredon is the scribe of the main text and also responsible for the marginal glosses.[64] He also owned and copied a part of the composite manuscript, Oxford, Bodleian Digby 168 (ff. 1–146). The part that belonged to Simon Bredon is composed of several booklets gathered by him. He copied folios 1–107 and 131–35,

63 On the English *cursiva antiquior*, see Albert Derolez, *The Paleography of Gothic Manuscript Books: From the Twelfth to the Early Sixteenth Century* (Cambridge: Cambridge University Press, 2003), pp. 134–41.

64 For Simon Bredon's handwriting, see also Oxford, Bodleian Library Digby 178 and London, BL Harley 625; for his autograph manuscripts, see Andrew G. Watson, 'A Merton College Manuscript Reconstructed: Harley 625; Digby 178, ff. 1–14, 88–115; Cotton Tiberius B. IX, ff. 1–4, 225–35', *Bodleian Library Record*, 9 (1976), 207–17. For Digby 179, see *supra*, note 60 and Gudrun Vuillemin-Diem and Carlos Steel, *Ptolemy's Tetrabiblos in the Translation of William of Moerbeke. Claudii Ptolemaei liber iudicialium* (Leuven: Leuven University Press, 2015), pp. 3–6. It seems to us that Simon Bredon is responsible not only for the glosses in Digby 179 but also is the scribe of the main text. Although the script is neater than the glosses and the font is larger, the main text presents all the characteristics of Simon Bredon's handwriting. He used two different pens; the main text is written with a thicker pen, whereas the glosses are copied with a thin pen that emphasizes the broad and thin strokes.

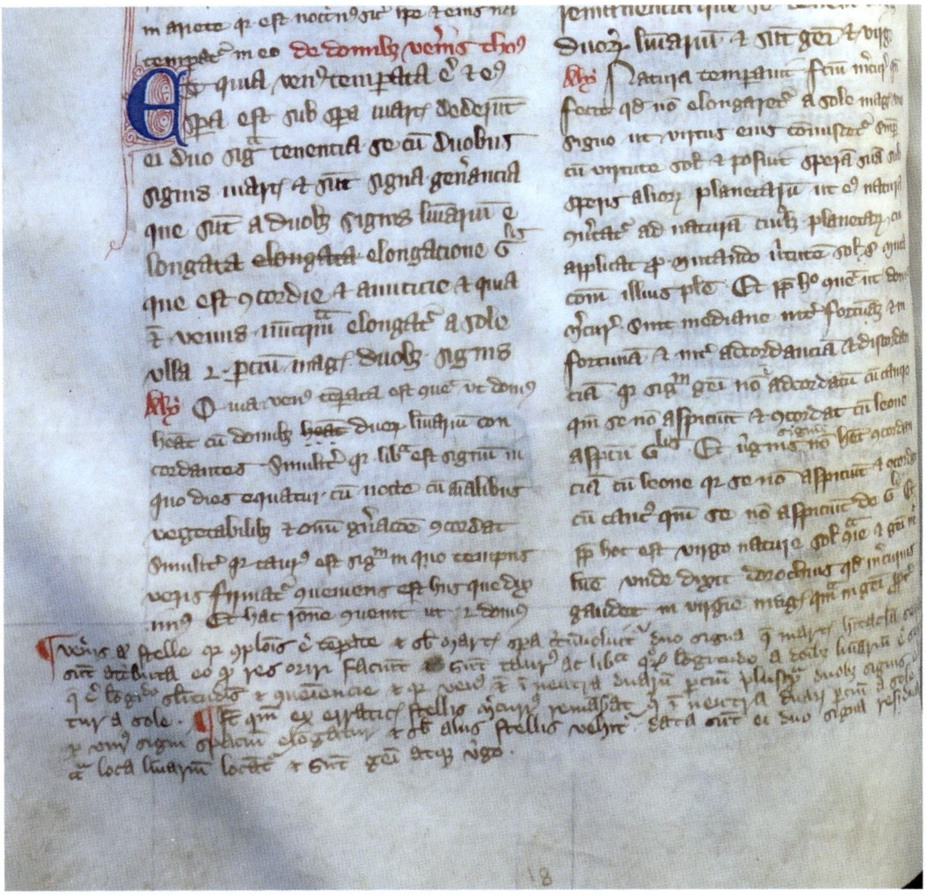

Figure 2. MS Oxford, Bodleian Library, Digby 179, folio 34^{v}: Simon Bredon's handwriting. Authors' photograph.

and likely acquired the different quires contained in folios 108–46.[65] It is noteworthy that MS Digby 168 contains Parisian Alfonsine materials such as John of Saxony's canons of 1327, followed by excerpts from the Alfonsine Tables and John of Lignères' canon *Quia ad inveniendum loca planetarum.*[66] This manuscript, displaying a section of Simon Bredon's

65 MS Digby 168, ff. 108–15, were copied in Italy, and he may have acquired them during his stay in Avignon in 1333; ff. 116–30 were written in England either at the very end of the thirteenth century or during the early fourteenth century; ff. 131–35 were composed in England in the fourteenth century; ff. 136^{v}–38^{r} (England); ff. 139^{r}–44^{v} (France?), and ff. 145^{r}–46^{r} were copied in France. We cannot neglect the notion that he acquired this French section on his way to Avignon.

66 Digby 168: John of Saxony, *Tempus est mensura motus* (ff. 131ra–35va); Alfonsine Tables (excerpts ff. 139^{r}–44^{v}, 146^{v}); John of Lignères, *Quia ad inveniendum loca planetarum* (ff. 145rb–46rb). On the canons *Quia ad inveniendum loca planetarum*, see the critical edition of Petr Hadrava and Alena Hadravova in this volume. For John of Saxony's canons, see Emmanuel Poulle, *Les Tables alphonsines avec les canons de Jean de Saxe* (Paris: CNRS éditions, 1984); for the tables, see the *editio princeps*: *Tabule astronomice illustrissimi Alfontii regis Castelle* (Venice: Ratdolt, 1483).

commentary on the *Almagest* (ff. 21^{r}–39^{r}), was likely assembled by him in the 1340s. It is thus one of the earliest witnesses to the reception of Alfonsine Parisian works in England.

Two pieces of evidence link Simon Bredon and William Reed and could suggest that some parts acquired by Reed from Richard Campsale actually came from Bredon. Firstly, MS Digby 176 displays the 1341–1344 solar almanac based on the Alfonsine Tables adapted by William Reed.[67] The mention of the *Almanak Solis* in the table of contents written by Walter Robert reads, 'Almanak Solis pro 4^{or} annis per Willelmum Reed anno Christi 1337 calculata et scripta'. The attribution to William Reed is explicit here. The tables giving daily positions of the Sun for the years 1341–44 can be found before the anonymous astrometeorological treatise, *Sapientes Indii de pluviis*, and John of Saxony's *Almanach* for the years 1349–80.[68] However, the *Almanak Solis* is part of an independent binion displaying the tables and some annotations on folios 71^{v} and 72^{v} (in the lower margin). Another witness of this *Almanak Solis* can be found in MS Digby 178 (ff. 11^{r}–13^{r}), which belonged to Simon Bredon. Part of this manuscript, copied by Simon Bredon (ff. 1–14 and 88–115), originated from a larger manuscript bequeathed by him to Merton College.[69] This codex, reconstructed by Andrew Watson, consists of London, British Library Harley MS 625; Digby 178 (ff. 1–14 and 88–115) and London, British Library Cotton Tiberius B IX, (ff. 1–14 and 225–35).[70] Simon Bredon copied the *Almanak Solis* in Digby 178 and wrote, in the lower margin of folios 11^{r}–13^{r}, his arithmetical treatise dedicated to the square numbers: *Conclusiones quinque de numero quadrato*. This part is also signed by him on folio 13^{r}: 'has conclusiones recom[m]endo ego Simon de Bredone volenti circa quadraturam circuli laborare'. Following the *Almanak Solis* he also copied a table of lunar latitude which does not appear in MS Digby 176.

Digby 178 and Digby 176 are the earliest witnesses of this *Almanak Solis*.[71] There are few variants in the entries and the two copies seem to have been written contemporarily. The scribe's handwriting in Digby 176 is notably similar to that in Simon Bredon's copy. Since the *ductus* is the same, it appears that the *Almanak Solis* in Digby 176 was copied by Simon.[72] However, he is not responsible for the annotations in the lower margins of folios 71^{v} and 72^{v}. This evidence again raises the question of the attribution of this set of tables, which were credited to William Reed or to Simon Bredon. The only strong evidence of William Reed's authorship lies on the table of contents of Digby 176. Furthermore, the table of solar declinations attributed to Simon Bredon, and often linked to the *Almanak*

67 See our edition of the *Almanak Solis* in Appendix 1. On William Reed's adaption of the Alfonsine tables for the Oxford meridian, see North, 'The Alfonsine Tables in England', pp. 274–79. The *Almanak Solis* is situated in Digby 176, ff. 71^{r}–72^{v}. For other witnesses of this almanac, see below.

68 On both texts, see Stuart Jenks, 'Astrometeorology in the Middle Ages', *Isis*, 74 (1983), 185–210; José Chabás and Bernard R. Goldstein, 'The Master and the Disciple: the Almanac of John of Lignères and the Ephemerides of John of Saxony', *Journal for the History of Astronomy*, 50 (2019), 82–96.

69 Digby 178 is now a composite manuscript, of which ff. 15^{r}–87^{v} were owned and copied by the physician and astronomer Lewis of Caerleon in the 1480s. On Lewis of Caerleon, see Pearl Kibre, 'Lewis of Caerleon, Doctor of Medicine, Astronomer, and Mathematician (d. 1494)', *Isis*, 43 (1952), 100–08.

70 This manuscript was once owned by John Dee. See Watson, 'A Merton College Manuscript Reconstructed', pp. 207–17.

71 Digby 57 (England, early fifteenth century), shows the *Almanak Solis* (ff. 103^{r}–5^{r}); however it seems to have been newly adapted since the values are only given in degrees and minutes, and there is a discrepancy of one minute for all of the entries.

72 Thus, Carey's assertion that the *Almanak Solis* in our manuscript is an autograph of William Reed must be rejected. Carey, *Courting Disaster*, p. 65.

Solis, is only found in two fifteenth-century manuscripts.[73] Thus, it is difficult to attribute this table to Simon Bredon on this basis. Furthermore, the link between the table and the *Almanak Solis* does not seem relevant, as the values differ. In any case, the fact that Simon Bredon owned the *Almanak Solis* and copied both examples again underlines the link between the two men.

Although the quite standardized English *cursiva antiquior* of Simon Bredon remains difficult to differentiate from contemporary scripts, it seems that he is the author of the annotation in the lower margin of the excerpts of John of Saxony's *Almanach*; and he could also be responsible for the copying of this fragment.[74] We cannot dismiss the notion that he is also the scribe of the short note related to Ptolemy's *Quadripartitum* on folio 87^{r}.[75] In that case, he could have been the owner of the quaternion containing the horoscopes of an individual born in 1317.

1.3.2. Walter Robert and William Reed

Walter Robert was William Reed's secretary. He was notably in charge of writing his master's *ex-dono*, the table of contents, and the destination of the manuscripts.[76] In MS Digby 176, folio 1^{r}, he copied the *ex libris* and the table of contents (Fig. 1):[77]

> Regule ad futuram aeris temperiem pronosticandam per magistrum Willelmum Merle, Considerationes temperiei pro 7 annis Christi [socium domus de Merton[78]].
> Pronosticationes cuiusdam eclipsis visibilis et coniunctionum 3um planetarum superiorumanno Domini 1345, contingentium et primam pestilenciam precedentium quas magister W. Reed calculavit et magister Johannes Aschenden pronosticavit.
> Tractatus magistri Leonis Hebrei de coniunctionibus Saturni et Iovis anno Christi 1345.
> Tractatus magistri Johannis de Muris de coniunctionibus Saturni et Iovis anno Christi 1345.
> a. Pronosticatio magistri Gaufridi de Meldis - de coniunctione Saturni et Jovis anno Christi 1325.
> - de coniunctione Saturni et Iovis anno Christi 1345.
> Johannis de Esshyndem de coniunctione Saturni et Martis et Jovis et Martis ac eclipsi Lune visibili anno 1349.

73 London, BL Egerton 889, an early fifteenth-century manuscript belonging to the Cantabrigian astronomer, John Holbroke (d. 1437), preserves this *Tabula declinatio Solis* (f. 18^{v}) with the attribution to Simon Bredon. On this MS, see David Juste, 'MS London, British Library, Egerton 889' (update: 21.06.2018), *Ptolemaeus Arabus et Latinus. Manuscripts*, URL = http://ptolemaeus.badw.de/ms/50. Oxford, Ashmole 191, f. 77^{r}, shows another table of solar declination, attributed to Simon Bredon, interpolated with zodiac sign positions.

74 Digby 176, ff. 73^{r}–86^{r} (the annotation is located on f. 73^{r}). If we compare the tables copied by Simon Bredon in Digby 178, ff. 4^{v}–8^{v}, it seems that the excerpts of the *Almanach* of John of Saxony have been copied by a similar script.

75 Digby 176, f. 87^{r}: 'Ptholomeus libro primo [written above the line '2°'] capitulo 3°, trigonus Arietis, Leonis et Sagitarii pertinet ad angulum inter septentrionem et occidens … - … De ventis hec maxime intellige'. See Ptolemy, *Quadripartitum*, 1493, II, 3, f. 32ra (Gilles of Thebaldis's translation).

76 Cf. *BRUO*, p. 1579; Thomson, 'William Reed', p. 289. Walter Robert wrote in Merton College MS 168, f. iiiv: 'Orare etiam dignemini pro Waltero Roberti scriba et notarius dicti venerabilis patris, qui suprascriptos titulos et titulos aliorum 99 librorum per dictum patrem […]'.

77 The *ex-libris* and the table of contents are written in the same script, which is Walter Robert's handwriting. We do not think that the *ex-libris* is in William Reed's hand, as suggested by Thomson. 'William Reed', p. 289.

78 'socium domus de Merton' was added later by another hand.

Pronosticatio coniunctionis magne Saturni et Jovis anno Christi 1365, quam magister Willelmus Red calculavit et Johannes Asshenden pronosticavit.
Pronosticatio magistri Reginaldi Lambourne de annis Christi 1368°, 1369°, 1370°, 1°, 2°, 3°, et 1374°.
Pronosticatio Johannes de Esshynden de coniunctione Saturni et Martis in Cancro anno Christi 1357°.
Pronosticatio magistri Reginaldi Lambourne de eclipsi anno Christi 1363°.
Notula de corruptione pestilenciali.
Tabula de mansionibus Lune.
Tabule de superioritatibus planetarum secundum Albumasar.
Tabula de dignitatibus planetarum.
Plinius de temporibus.
Alkyndus de ymbribus.
Tabule de qualitatibus graduum 12 signorum.
Tractatus de pluviis.
Almanak Solis pro 4^{or} annis per Willelmum Reed anno Christi 1337 calculata et scripta.
Almanak Johannis de Almannia iuxta motus et radices Alfonsi calculatur.
Almanak Prefacii Judei iuxta radices et motus Arzachel calculat[ur].[79]
b. Calculatio magistri Walteri Elvesden
— de dominis mensium ab anno 1332° usque ad 1357;
— de dominis annorum ab anno 1332° usque ad 1386.

It is noteworthy that in the table of contents Walter Robert inserts two letters, 'a' and 'b', in the margin facing, respectively, Geoffrey of Meaux's works and Walter of Elvenden's sets of tables. The letters indicate that, despite their order in the table of contents, both works are in fact part of the same codicological unit; indeed, Walter of Elvenden's tables are copied in the same quire as Geoffrey of Meaux's writings (see infra fols 19^{r}–29^{r})[80]. This itemization was likely more convenient for the cataloguing of the works. However, this also suggests that the manuscript was not assembled as it is preserved today. We have of course already mentioned that the *Almanach* of Profatius is no longer extant. In addition, Walter Elveden's tables are not located at the end of the manuscript (ff. 19^{r}–21^{v} and 22^{v}–23^{r}). The manuscript now ends with six quires dedicated to eclipse tables and *figure celi* from 1341 to 1346 (based on the Oxford meridian, except the year 1342 based on the London meridian), which are not mentioned in the table of contents, including a canon for using the eclipse tables of 1341; it is specified that they are 'super longitudinem et latitudinem Oxonie et secundum radices et motus Alfonsi in 9ª spera' (f. 99^{v}).[81]

79 The relevance of this description should be noted because Profatius's *Almanac* is indeed based on the Toledan Tables. See G. J. Toomer, 'Prophatius Judaeus and the Toledan Tables', *Isis*, 64 (1973), 351–55; Chabás and Goldstein, 'The *Almanac* of Jacob ben Makhir'.

80 Digby 176, ff. 19^{r}–21^{v}; 22^{v}–23^{r}: Walter Elveden's tables of the lords of the months from 1332–57 and the lords of the years from 1332–86. See North, *Horoscopes and History*, pp. 137–38; Carey, *Courting Disaster*, pp. 66–69. The presence of the astrological tables drawn by Walter Elveden and their successful reception in Oxford in the fourteenth century were likely due to the links between the Cantabrigian astrologer and John Aschenden. In fact, the latter clearly refers to Elveden's computations on the partial lunar eclipse of 31 July 1357, see *infra* n. 112.

81 These tables are likely based on William Reed's Oxford Tables.

Walter Robert was also William Reed's personal scribe, as he specifies in Merton MS 168, 'scriba et notarius dicti venerabilis patris'. His handwriting is close to an English *cursiva antiquior libraria* or *textualis*, including strong and sharp chancery script features. The round *d* is often surmounted by a straight stroke above the loop; the right descenders of the *h* and *m* are extended below the baseline and curved towards the left. Walter Robert can be identified as the scribe of Geoffrey of Meaux's astrological judgment on the conjunctions of 1325 and 1345 and on the causes of the Great Plague (ff. 25^r–29^r), as well as Reginald Lambourne's epistle addressed to John of London (27 February 1364) (ff. 50^r–53^v). Although Walter Robert likely acted as a scribe for William Reed, the Bishop of Chichester is said in the *ex libris* to have copied a part of the manuscript. It is likely that he copied the small bifolio inserted between John Aschenden's and Levi ben Gerson's predictions on the conjunction of 1345.[82] This part is entitled, 'Tota ista calculatio est facta pro meridie Oxonie et mense Martii anni 1345' and echoes the table of contents.[83] Although it is mentioned as a calculation, this bifolio actually contains four *figure celi* related to the conjunction: 1) on the conjunction of Jupiter and March; 2) on the conjunction of Saturn and March; 3) on the conjunction of Saturn and Jupiter; and 4) on the Lunar eclipse of the 18 March 1345.

Digby 176 is one of the rare witnesses providing explicit details on its own history and also on the close links among some contemporary astronomers and astrologers. It offers an overview of what could have been the cooperation of these scholars sharing a common background and continuing to collaborate after their years at Oxford. Several strong pieces of evidence make it possible to reconstruct the making of a codex as conceived by its first owner, William Reed, as a key witness of the astronomical and astrological activities of his time. However, Digby 176 is also important because it shows the key role played by William Reed in this circle of scholars. He was likely a friend and a patron for some of them. For some time, John Aschenden and Simon Bredon, among others, were part of the entourage of the Bishop of Chichester. The integration of this manuscript in the larger and philanthropic view that William Reed had of his private collection allowed this important anthology, eventually, to come back to Merton College.

2. Practicing the science of the stars according to MS Digby 176

2.1. A codex reflecting the influence of Alfonsine astronomy in England

The seminal article of John D. North, 'The Alfonsine Tables in England', is still our principal guide in this regard. However, North did not use MS Digby 176 in his paper, because it preserves no copy of the Alfonsine Tables themselves, no astronomical treatises of John of Murs, John of Lignères, and so on, nor the Oxford tables of 1348, and not even William Reed's adaptation of the Alfonsine Tables with his canons.[84] In fact, in MS Digby 176 we

82 The bifolio is located in ff. 13^v–14^r.

83 Digby 176, f. 1^v: 'Pronosticationes cuiusdam eclipsis visibilis et coniunctionum 3um planetarum superiorum anno Domini 1345, contingentium et primam pestilenciam precedentium quas magister W. Reed calculavit et magister Johannes Aschenden pronosticavit'.

84 North, 'The Alfonsine Tables in England', item n° 41, p. 300. We have seen Bodleian Library Ashmole 191, ff. 59^r–61^v for the Reed's canons, ff. 62^r–74^r for his tables.

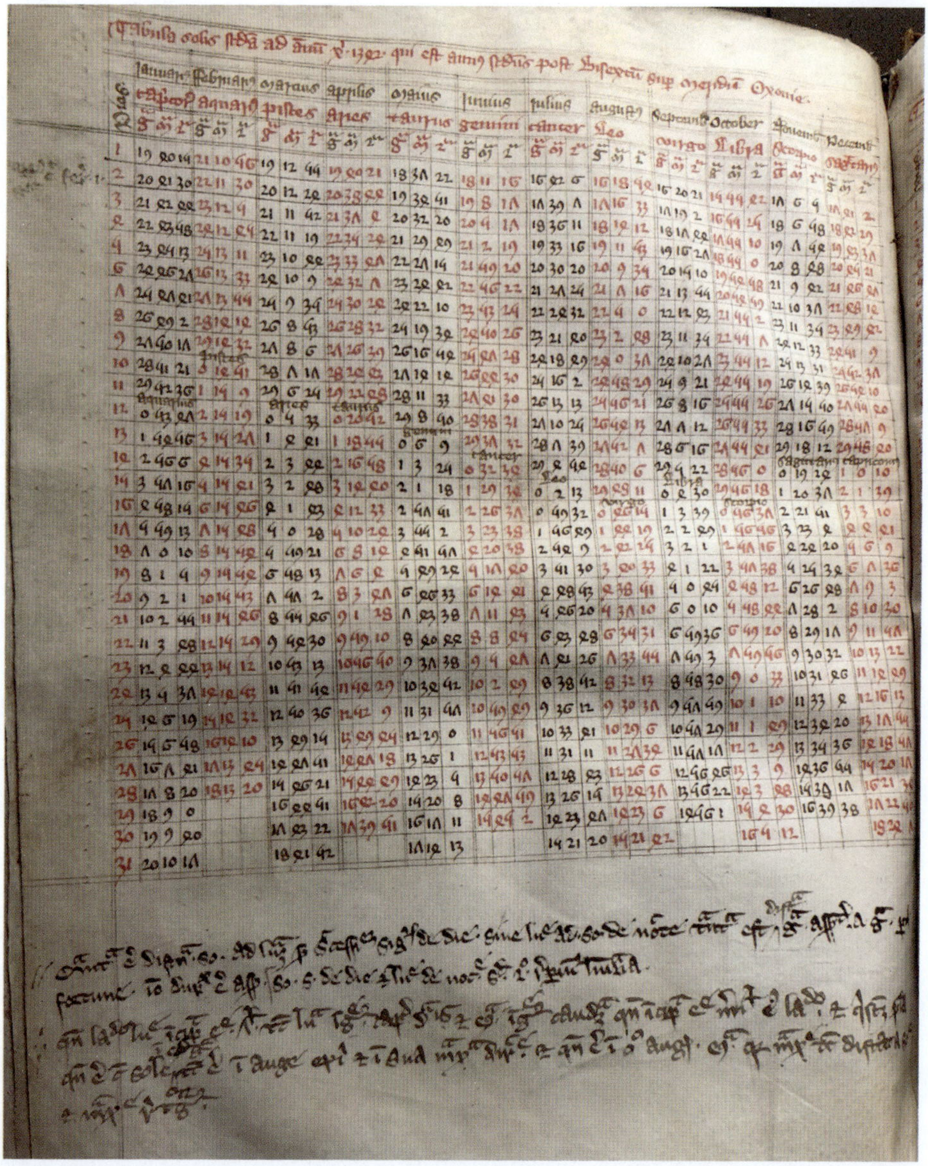

Figure 3. MS Oxford, Bodleian Library, Digby 176, folio 71^{v}: The *Almanack Solis* for 1341–44 (here, 1342). Authors' photograph.

have at least two interesting 'almanacs' based on the Parisian Alfonsine Tables, directly or indirectly, via William Reed's tables for the first one.

The first, on folios 71^{r}–72^{v}, is the *Almanak Solis* for 1341–44,[85] ascribed to William Reed and dated 1337 in the table of contents of the volume. The title given on folio 71^{r} is: 'Tabula

85 See our edition of the *Almanak Solis*, below.

Solis prima ad annum Christi 1341 qui est annus primus post bisextum super meridiem Oxonie'. In fact, it is not a planetary almanac but rather a list of true positions of the Sun during a cycle of four years, indicated in degrees, minutes, and seconds of each zodiacal sign, which is unprecedented in early Alfonsine astronomy.[86] The position of the Sun on 1 January 1341 is thus 19;55,8° Capricorn. For 1 February, it is 21;25,36° Aquarius, etc.[87] The true movement of the Sun in one day is 1;01,15° on 1 January, 1;00,37° on 1 February, etc. The title given on folio 71^{v} is 'Tabula Solis secunda ad annum Christi 1342 qui est annus secundus post bisextum super meridiem Oxonie' (Fig. 3), and the titles on the next two pages correspond to the years 1343 and 1344, the third and fourth years after 1340, which was a leap year. The year 1340 is indeed the epoch of William Reed's mean movement tables,[88] which means that the *Almanak Solis* may have been composed as early as 1337, as its secretary Walter Robert says in the table of contents of the volume. However, it may seem strange that such a document, giving the true locations of the Sun for 1341–44, would be drawn up as early as 1337 and one wonders whether this date might not be an error to be on Walter Robert's part. Yet, he was very conscientious in this table of contents, and the year 1337 is the beginning of a cycle of four years before the cycle 1341–44. So, it appears that William Reed's secretary can be trusted here, and it seems logical to assume that William's tables were composed before the *Almanak Solis*, in 1337 at the latest, but this is not certain.

The second 'allmanac' belonging to Alfonsine astronomy in this manuscript is found in folios 74^{r}–86^{r}, with some planetary ephemerides for 1349–80 which are, in fact, a part of John of Saxony's *Almanach* for 1336–80.[89] The years 1336–48 are therefore missing from this section of the Oxford manuscript, which is incomplete and disorganized. But the presence of the *Almanach* of John of Saxony is interesting, especially since in this *Almanach* the true positions of the Sun are only directly indicated for the years 1336–39.[90] One may, therefore, wonder whether William Reed's *Almanak Solis* was perhaps intended to complement that of John of Saxony for the true locations of the Sun in 1341–44, by reaching an additional precision, down to the second. A fifteenth-century anonymous note within William Reed's tables also alludes to an almanac giving the positions of all the planets, which could be that of John of Saxony.[91]

---

86 However, it did occur in astronomy based on the Toledan Tables; and John of Murs already gave daily solar positions, to seconds, for the single year 1321. See José Chabás and Bernard R. Goldstein 2012, *A Survey of European Astronomical Tables in the Late Middle Ages* (Leiden: Brill, 2012), p. 83. In fact, the presentation of the *Almanak Solis* is inspired by that of Profatius's *Almanach perpetuum*, where the tabulation of the Sun is given in a four-year cycle, beginning 1 March 1301. See *Almanach Dantis Aligherii*, pp. 38–45; Chabás and Goldstein, 'The *Almanac* of Jacob ben Makhir', section 6.

87 The Alfonsine positions, according to *Astromodels* (Lars Gislén), are 19;55,7° Capricorn for 1 January and 21;25,36° Aquarius for 1 February, taking into account a time difference of sixteen minutes between Toledo and Oxford, as indicated in William Reed's tables. See Harper, 'The Astronomical Tables of William Rede', p. 373; North, 'The Alfonsine Tables in England', p. 275. The positions of the *Almanak Solis*, presumably found via William Reed's tables, nearly exactly match those found using the Alfonsine Tables.

88 Ashmole 191, f. 62^{r}. See Harper, 'The Astronomical Tables of William Rede', pp. 377–78, who compares the positions given by the *Almanak Solis*, William Reed's tables, and John of Lignières' tables. See also Chabás. *Computational Astronomy in the Middle Ages*, pp. 207–8.

89 See Chabás and Goldstein, 'The Master and the Disciple', pp. 88–96. We have seen MS Erfurt, Universitats- und Forschungsbibliothek, Amplon. F 386, ff. 62ra-63ra (canons), 63^{v}–107^{v} (tables). The tables for 1349–1380 have been copied on ff. 72^{r}–107^{v} of the Erfurt codex.

90 MS Erfurt, F 386, ff. 64^{r}–65^{v}.

91 See Ashmole 191, f. 64^{v}: 'Nota hic tabulas equacionis Solis et Lune et ceterorum planetarum, ut possis habere verum motum omnium planetarum sicut fit almanach'.

At the end of the codex, folios 100r–119r, some tables of conjunctions and oppositions of the Sun and the Moon in Oxford for the years 1341 and 1343–46 have been copied, as well as in London for the year 1342. These true conjunction and opposition tables are preceded by a canon, folio 99v, and could be compared to William Reed's mean conjunction and opposition tables *in annis expansis*, which are based on the radix of 1341.[92]

The MS Digby 176 is therefore a significant witness to the establishment of Alfonsine astronomy in England in the 1330s. However, the influence of the Alfonsine science of the stars in this codex may also be seen from an astrological point of view.

2.2. The richest preserved collection of predictions on planetary conjunctions of the fourteenth century

From this point of view, MS Digby 176 has already been partially studied by several scholars, but it deserves a more complete description.[93] On folios 25r–26r, the treatise of Geoffrey of Meaux on the conjunction of 1325 has been copied, with the incipit 'Cunctis quorum interest astronomie scire nova Galfridus de Meldis hoc quod in presenti cedula continetur…' According to Geoffrey of Meaux, the conjunction of Saturn and Jupiter will occur '… circa finem maii anno Domini m° ccc° xxv°'. The text begins in the future, but at the top of the second page retrospectively deals with the conjunction in the past tense, referring to or quoting Albumasar's *Great Introduction*, Pseudo-Aristotle's *Liber de proprietatibus elementorum* and *Liber de regimine principum* (i.e. the *Secretum secretorum*), and Pseudo-Ptolemy's *Centiloquium* (the famous *verba* 5 and 8).[94] There is no *figura celi* of this conjunction of 1325 and no more astronomical details here. In folio 26r, a single line is left blank and the ensuing (ff. 26r–29r) sections analyse the astrological cause of the Black Death of 1348. Geoffrey of Meaux looks back to the conjunctions of 1325 and 1345 to do so, referring to Ptolemy's *Quadripartitum* but also to Roger Bacon, which is more original from a French point of view but not surprising from an English one because the dissemination of Roger Bacon's works in Britain was great, contrary to what was observed on the continent.[95] Furthermore, although he had trained at the University of Paris and was one of the physicians of the French court in the 1320s, Geoffrey of Meaux seems to have resided for some time in Oxford.[96] He refers of course to

---

92 Ashmole 191, ff. 64v–65r.

93 Cf. Thorndike, *A History of Magic*, vol. III, Chapter XIX and Appendix 19 on Geoffrey of Meaux, Chapter XX on John of Murs and the conjunction of 1345, Chapter XXI on 'John of Eschenden: specialist in conjunctions' and Appendix 20 on MSS of the astrological writings of John of Eschenden. See also J. D. North, 'Astrology and the Fortune of Churches', *Centaurus*, 24 (1980), 181–211; Carey, *Courting Disaster*, especially pp. 65–78 and 182–95.

94 See Jean-Patrice Boudet, 'Ptolémée dans l'Occident médiéval: roi, savant et philosophe', *Micrologus*, 21 (2013), 193–217 (particularly p. 208); Boudet, 'The Medieval Latin Versions of Pseudo-Ptolemy's *Centiloquium*: A Survey', in *Ptolemy's Science of the Stars in the Middle Ages*, ed. David Juste and others (Turnhout: Brepols, 2020), pp. 283–304.

95 Digby 176, f. 26r: 'Frater Rogerus Bacon in tractatu suo post locorum significationem dicit sic: Singula puncta terre sunt centra diversarum orizontium ad que coni diversarum pyramidum virtutum celestium veniunt'. See *The 'Opus Majus' of Roger Bacon*, 3 vols, ed. by John Henry Bridges (Oxford: Clarendon Press, 1897–1900; repr. Cambridge: Cambridge University Press, 2010), vol. I, pp. 138–39, 250 and 380.

96 On Geoffrey's career and works, in addition to Thorndike, *A History of Magic*, see Ernest Wickersheimer, *Dictionnaire biographique des médecins en France au Moyen Âge*, new ed. (Geneva: Droz, 1979), vol. I, p. 180, and *Supplément* by Danielle Jacquart, p. 83; Lynn Thorndike, *Latin Treatises on Comets, 1238–1368 A.D.* (Chicago: Chicago University Press, 1950), pp. 208–25; Joël Plassard, *Projets de réforme du calendrier à Paris au début du XIVe siècle. Textes édités et commentés*, unpublished thesis, Paris, École des chartes, 1975 (cf. *Positions des thèses…* [1975], 176–81); Danielle Jacquart, 'Médecine

the doctrine of great conjunctions, but claims to rely on Ptolemy; he cites the convergence of the conjunction of the three superior planets in Aquarius with a universal eclipse of the Moon on 18 March 1345 at Oxford longitude as a cause of the Great Plague.[97] This reference to the Oxford meridian could indicate that Geoffrey had left the French court after the advent of Philippe VI of Valois in 1328 and that he stayed in Oxford in the 1340s. It therefore seems likely that he composed this treatise in Oxford, unless the scribe, himself active in Oxford, was responsible for what could be a geographical confusion. In any event, the date of the eclipse given by Geoffrey is not congruent with John Aschenden's estimation,[98] their astronomical substratum is not the same (non-Alfonsine for Geoffrey, Alfonsine for John),[99] and the duration of the eclipse is different for the two astronomers: 3;29,54h according to Geoffrey of Meaux;[100] 3;43h according to John Aschenden.[101]

---

et astrologie à Paris dans la première moitié du XIVe siècle', in *Filosofia, scienza e astrologia nel Trecento europeo*, ed. G. Federici Vescovini and Francesco Barocelli (Padova: Il Poligrafo, 1992), pp. 121–34; Pierre Charon, 'Note sur Geoffroy de Meaux, médecin et astrologue du XIVe siècle', *Bulletin de la société archéologique de Meaux et sa région*, 7 (2010), 101–14; C. Philipp E. Nothaft, 'Critical Analysis of the Alfonsine Tables in the Fourteenth Century: The Parisian *Expositio Tabularum Alfonsii of 1347*', *Journal for the History of Astronomy*, 46 (2015), 76–99; Nothaft, 'Glorious Science or 'Dead Dog'? Jean de Jandun and the Quarell over Astrology in Fourteenth-Century Paris', *Vivarium*, 57 (2019), 51–101.

97 Digby 176, f. 26^v: '…et est omnibus astrologis quod anno Domini 1345, incipiendo annum a ianuarii fuerit eclipsis Lune universalis cum magna mora 18 die marcii una hora post ortum Lune ad longitudinem Oxonie. Et similiter tempore fuerunt duo superiores gradu per gradum coniuncti in Aquario et Mars cum eis in eodem signo infra lumen Iovis'. Geoffrey again refers to the longitude of Oxford when he goes back further to the conjunction of 1325 and its consequences on the mortality of English tycoons: Digby 176, f. 27^r: '…cuius exemplum in astrologi videre possint per coniunctionem trium superiorum que fuit cum eclipsi Solis anno Domini 1325 ad longitudinem Oxonie. Fuit locus Solis et Lune tempore eclipsis in domo septima et Ariete, et coniunctio trium superiorum in domo octava que est mortem, in prima facie signi Geminorum cum stella que vocatur Aldebaran et cum constellatione Capitis Algol, que stelle sunt de natura Martis, significantes maiores quia de prima et de secunda magnitudine et mortem eorum in Anglia quia coniunctio eorum fuit in domo octava, una cum Sole qui ecclipsibatur in Ariete qui maiores significant quid accidit post illam coniunctionem maiorem duorum superiorum cum presencia Martis in magnatibus'.

98 See below. For John Aschenden and William Reed, the middle of the total eclipse of the Moon occurred on 18 March 1345 at 9h 40m p. m., very close to the time given by NASA, 9h 44m (Greenwich). See https://eclipse.gsfc.nasa.gov/LEcat5/LE1301-1400.html (cat. 08079).

99 In his *Kalendarium*, Geoffrey of Meaux showed his opposition to the Alfonsine Tables as soon as they appeared in Paris around 1320: 'Et sciant omnes ad presens devenerit quod ego Gaufridus de Meldis non accepi radices Alfonsi propter quasdam rationes insolubiles quas proposui coram omnibus dum legebam, sed accepi radicas ab antiquis sapientibus approbatas secundum radices Azerchelis positas in tabulis tholetanis'. (fourteenth-century MS CP 340 of the IRHT, Paris, belonging to a private collection, f. 1^v, examined by J.-P. Boudet and C. Vulliez in 2015; see also Paris, BnF, MS lat. 7281, f. 160^v). Translation: 'And let it be known from now on that I, Geoffrey of Meaux, did not accept the radices of Alfonso because of certain insoluble problems, which I put forward in everyone's presence when I lectured [on this subject]; but instead I accepted the radices approved by the ancient sages, in accordance with the radices of Azarquiel laid down in the Toledan Tables'.

100 Digby 176, f. 27^v: 'Nunc restat dicere quanto tempore durabit effectus per illam constellacionem que 18 die martii anno Domini 1345. Notum est omnibus astrologis quod hore equales obscuritatis illius eclipsis lunaris cum qua fuit coniunctio trium superiorum anno Domini supradicto fuerunt tres hore et fuerunt viginti novem minuta et quinquaginta quatuor secunde…' According to Lars Gislén's program *Astromodels*, computing the positions of the Toledan Tables, the true opposition between the Sun and the Moon would have taken place on March 18, 1345 at 9h 7m p. m., i.e. at 9h 23m p. m. in Oxford. The Sun was then at 5;6° Aries, the Moon at 5;6° Libra, Saturn at 15;20° Aquarius, Jupiter at 18;29° Aquarius and Mars at 28;35° Aquarius. Bodleian Library Ashmole 1471, which preserves a copy from the end of the fourteenth century of Geoffrey of Meaux's treatise on the conjunction of 1345 (ff. 102^r–104^v), includes after this text four horoscopes borrowed from John Aschenden's treatise, the first of which concerns this lunar eclipse and includes positions (rounded up to the next degree) that had been computed with the Alfonsine Tables. The other three horoscopes have the same source and concern the three conjunctions of the superior planets.

101 Digby 176, f. 14^r. NASA computes the duration of the partial eclipse as 3;40h and the duration of totality as 1;28h.

On folios 9^{r}–16^{r}, the prediction for the 1345 conjunction by John Aschenden, based on William Reed's calculations, has been copied.[102] On folio 9^{r}, the rubric (crossed out) is: 'Incipiunt pronosticationes de eclipse universali Lune et de coniunctione trium planetarum superiorum que apparaverunt anno Domini 1345° in marcio. Et complete fuerunt iste pronosticationes 20° die predicti mensis marcii anni Christi supradicti'. The incipit of the text is 'Significatio eclipsis Lune universalis iuxta sententiam Ptholomei et Haly 2° *Quadripartiti* secundum quod misceantur significationes aliarum coniunctionum trium superiorum planetarum si fuit ista eclipsis in marcio anno Domini 1345° complete de marcio 18 diebus, 9 horis et 40 minutis'. John Aschenden relies here on Hali Abenrudian's commentary on Book II, Chapter 4, of Ptolemy's *Quadripartitum*, which emphasizes the importance of eclipses at the time of the conjunctions of Saturn and Jupiter.[103] Then he relies on other authorities, namely Messahalla, Albumasar, and Hali Abenragel. On folio 11^{v}, in the middle of the astrological square, the coordinates of the eclipse are given as follows: 'Ffigura eclipsis Lune universalis et fuit 19^{a} die [incompleto] marcii et 22 gra. Libre',[104] but below the square of the initial coordinates, we see: 'Ffigura eclipsis Lune in martis, anno Christi 1345, post 18 diem, 9 horas, 40 minutas'. On folio 12^{r}, the *figura celi* of the conjunction of Saturn and Jupiter can be found: 'Ffigura coniunctionis magne Saturni et Jovis que fuit 21 die marcii et 19 gradu Aquarii'. The ascendant is 15° Taurus, the conjunction in the tenth house at 19° Aquarius. In the upper margin: 'Ffigura coniunctionis magne Saturni et Jovis que fuit in marcio anno Christi 1345° imperfecto 20 diebus, 18 horis et 47 minutis, et fuit predicta coniunctio magne 19 gradu Aquarii'.[105] On folios 13^{v}–14^{r}, we see the horoscopes of the conjunctions Jupiter-Mars and Saturn-Mars, and another square for the eclipse of the Moon. On folio 14^{r}, the time and duration of the eclipse are indicated again: 'Tempus medie eclipsis: 18d, 9h, 40m. Tempus durationis eclipsis: 3h 43m'. On folio 15^{v}, the horoscope of the conjunction Saturn-Mars, 3 March, 18h 6m p. m. at 17° Aquarius is shown.[106] On folio 16^{r}, we see the horoscope of the conjunction Jupiter-Mars, at 15° Aquarius in the third house, although no date is given. The original text ends with a very general remark: 'Ex predictis patet quod effectus proveniens ex eclipisi et ex istis 3[107] coniunctionibus erit caristia[108] et guerra et multe infirmitates et tempestates in temporibus et locis predictis'. Eight lines and two marginal notes have then been added in a more cursive way to explain, *a posteriori*, the appearance of the Black Death: 'Magna mortalitas et magna corruptio aeris, et alia mala que predixi …'[109] Of course, no astrologer was capable of predicting a disaster of this

102 See the table of contents of f. 1^{v}: 'Pronosticationes cuiusdam eclipsis universalis et coniunctionum trium superiorum anno Christi 1345 contingentium et primam pestilentiam precedentium quas Magister W. Reed calculavit et Magister Iohannes Aschenden pronosticavit'. For a summary of this prediction, see Thorndike, *A History of Magic*, III, pp. 326–28.

103 Ptolemy, *Quadripartitum*, 1493, I, 4, ff. 37rb-37vb.

104 According to *Astromodels*, the Alfonsine Tables give a true longitude of 6;39° Libra at 9h 29m p. m. in Oxford on 18 March. There is therefore a large discrepancy here between the Alfonsine model and the positions of the luminaries provided by William Reed.

105 According to *Astromodels*, the Alfonsine Tables give a true longitude of Saturn and Jupiter as 18;42,50° Aquarius at 17h 30m in Oxford. At 18h 47m, Saturn is computed to be at 18;43,6° and Jupiter at 18;43,23°.

106 *Astromodels*: Saturn 16;57,36°; Mars: 16;58,31°.

107 '3' is an interlinear addition.

108 'breviter' is crossed out.

109 See the translation of this addition by Thorndike, *A History of Magic*, III, p. 328.

magnitude, despite John Aschenden's claim in December 1348 at the end of his *Summa iudicialis de accidentibus mundi*, also entitled *Summa anglicana*[110].

From a comparative perspective, two other texts on the conjunction of 1345 were copied after the one written by John Aschenden: Levi ben Gerson's treatise and another tract ascribed to John of Murs. On folios 16^{r}–17^{v}, the 'Tractatus magistri Leonis Hebrei de coniunctione Saturni et Jovis anno Christi 1345' is the only astrological prediction of the famous Jewish scholar to be preserved.[111] On folios 17^{v}–18^{r}, another prognostication is announced in the running title: 'Incipit tractatus Johannis de Muris de coniunctione Saturni et Jovis anno Christi 1345°'. However, the incipit is 'Tres principles ex militia superiori …' and the text shows that it is in fact Firmin de Beauval's opuscule, which is totally different from John of Murs's prediction.[112]

Two others treatises on planetary conjunctions written by John Aschenden have been copied in our manuscript: on folios 30^{r}–33^{r} his astrological judgment concerning the Saturn-Mars conjunction of 1349,[113] and on folios 42^{r}–49^{v} and 34^{r}–40^{r} his treatise on the conjunctions of 1357 and 1365.[114] Here, the second text is interesting for us, because it is possible to compare it with the letter from John of Murs to Pope Clement VI on the same conjunctions, probably written *c.* 1346–47.[115] Like John of Murs, John Aschenden considered the two Saturn-Mars conjunctions of 1357 and the Saturn-Jupiter conjunction of 1365 in the same treatise, finished 10 March 1357. It is noteworthy that the English astrologer did not entirely rely on William Reed's calculations, since a passage at the end of his judgement on the conjunctions of 1357 and 1365 clearly mentions his reliance on the computations of Walter Elveden. In fact, the duration and magnitude of the partial Lunar eclipse of 31 July 1357, and its disastrous effects, are said to have derived from the calculations of the Cantabrigian master.[116]

---

110 Thorndike, *A History of Magic*, III, p. 331. See Johannes Eschuid [*sic*], *Summa astrologiæ iudicialis*, II, 12, 2, (Venice: Johannes L. Santritter 1489), f. 312; according to ISTC ie00109000, this edition is preserved in more than 200 copies. In this apologetic chapter, built as a technical *exemplum*, John Aschenden reproduces his horoscopes of the triple conjunction and the lunar eclipse of 1345, ff. 311vb-312rb.

111 Cf. the edition of Bernard R. Goldstein and David Pingree, 'Levi ben Gerson's *Prognostication for the Conjunction of 1345*', *Transactions of the American Philosophical Society*, 80/6 (1990), 1–60. This text is preserved in Hebrew and in three Latin manuscripts: Bodlleian Library Ashmole 393, ff. 81^{r}–81^{v}; Paris, BnF lat. 7378A, ff. 62va-63ra; Troyes, Médiathèque 62, ff. 107ra-107va (included in Pierre de Ceffons's question, see below).

112 See the rather poor edition given by Hubert Pruckner, *Studien zu den astrologischen Schriften des Heinrich von Langenstein* (Leipzig: Teubner, 1933), pp. 220–21. See also Christopher Schabel and Fritz S. Pedersen, 'Miraculous, Natural, or Jewish Conspiracy?; Pierre de Ceffons, Question on the Black Death, with Astrological Predictions by Gersonides and John of Murs/Firmin de Beauval', *Recherches de Théologie et Philosophie médiévales*, 81 (2014), 137–79, with a new edition of the predictions of Gersonides and Firmin on the conjunction of 1345, inserted in the *questio* of Pierre de Ceffons (pp. 170–79). Firmin's little treatise is also attributed to John of Murs by Pierre de Ceffons, but this claim does not seem to have a solid foundation. An edition of John of Murs's real prediction on the conjunction of 1345 has been given by Jean-Patrice Boudet in 'Jean des Murs, Astrologer', *Erudition and the Republic of Letters*, 4 (2019), 123–45.

113 Cf. Thorndike, *A History of Magic*, vol. III, pp. 334–37 and 720.

114 Thorndike, *A History of Magic*, vol. III, pp. 338–45 and 720–21.

115 See Jean-Patrice Boudet, 'La papauté d'Avignon et l'astrologie', in *Fin du monde et signes des temps. Visionnaires et prophètes en France méridionale (fin xiiie - début xve siècle), Cahiers de Fanjeaux*, 27 (1992), 257–93 (especially pp. 268–79 and 281–84), and the new version of this paper to be published in Id. *Astrologie et politique entre Moyen Âge et Renaissance* (Florence: Sismel-Edizione dell Galluzzo, 2020), text III.

116 Digby 176, ff. 49^{r}-v: 'Considera si eclissetur [*sic*] Sol aut Luna in anno in quo fuerit coniunctio magna, per hoc enim poteris scire omnia accidentia que in illo anno et isto anno Christi 1357° in primo die mensis Augusti iuxta calculationem Galteri Elwinden, incipiendo diem a meridie diei precedentis completi 6 horis et 46 minutis erit eclipsis Lune partialis. Et durabit item eclipsis per 3 horas et 6 minuta et erunt 9 puncta de corpore Lune eclissatur

On folio 42ʳ, the rubric and the incipit are: 'Incipit tractatus Johannis de Eschyndene de significatione coniunctionis Saturni et Martis in Cancro que erit illo anno Christi 1357° in 8° die junii et de significatione coniunctionis magne Saturni et Jovis que erit anno Christi 1365 in 30° die octobris./ Sicut dicit Philosophus [*sic, for* Ptholomeus] in *Centilogio* propositione 50ª: Non obliviscaris…' This apparent confusion between Aristotle and Ptolemy is not entirely accidental, especially since it is subsequently repeated in this treatise: Ptolemy, like Aristotle, is indeed sometimes considered as the Philosopher *par excellence* in texts relating to the science of the stars.[117] However, this prediction is preserved in at least six other codices, including MSS Oxford, Bodleian Library, Ashmole 393 and 392, which rectify the error by speaking of 'Ptholomeus' and not 'Philosophus'.[118] There are also some discrepancies in the text, as well as between Digby 176 and other manuscripts, regarding the time of the occurrence of each of the two planetary conjunctions. Although further examination is required, overall, Aschenden's treatise does not seem to directly answer John of Murs's letter to Pope Clement VI (d. 1352). Moreover, it is not only a text intended, more modestly, for the common usefulness of students in astronomy, especially to his former colleagues at 'Merton Hall', as the author claims, but also for European astrologers and rulers.

In his prologue, John Aschenden refers to Pseudo-Ptolemy's *Centiloquium* and its commentary by 'Haly', to Messahalla, Albumasar, Abraham Ibn Ezra, and Pseudo-John of Seville' *Quadriparitum*. He then gives the coordinates of the two conjunctions and the general motivations for the composition of his treatise:

> Cum ergo hoc anno Christi 1357° in mense junii completis de eodem mense 7 diebus, 22 horis et 30 minutis, incipiendo die a meridie diei precedentis prout faciunt astronomi, erit coniunctio Saturni et Martis in signo Cancri,[119] quod signum est signum mundi prout dicunt astronomie eo quod propinquissimum est nobis; et anno Christi 1365° in mense octobris completis de eodem mense 29 diebus, 14 horis et 22 minutis, erit coniunctio Saturni et Jovis in signo Scorpionis et in nova triplicitate que significabit valde magnas mutationes in mundo et accidencia grandia et terribilia secundum

[*sic*] et 44ª minuta, et est Sol in ista eclipsis in 15° gradibus Leonis et Luna in 15° gradibus Aquarii. Saturnus est in 27° gradibus Cancri, Jupiter [in] 29° Aquarii, Mars in 24 gradu Leonis, Venus in 2° gradu Leonis, Mercurius in 26° gradibus Leonis. In principio istius eclipsis est Luna in ascendente et Sol in 7ª domo, hoc ergo super fortitudinem effetus eiusdem eclipsis. Et quia locus in medio eclipsis est prope gradum occidentis, est significatio istius eclipsis in moribus ac in morum variationibus et in senibus et in mortuis'. 'Galterius Elwinden' must be identified with Walter Elveden. See Keith Snedegar, 'Elveden, Walter', in the *Oxford Dictionary of National Biography*, 2004, online. Elveden had obtained a Master of Arts degree and a common law doctorate by 1350 from the University of Cambridge. His competences in the astronomical-astrological field have been questioned by Carey, *Courting Disaster*, pp. 67–69 and *supra*, note 80, but his calculations regarding the time and duration of this partial eclipse seem to be highly accurate. According to Walter Elveden's calculation, the mid-time of the partial lunar eclipse occurred at 6;46h, quite close to NASA's value of 6;36h. Walter's duration for this partial eclipse, 3;6h, exactly matches NASA's. See https://eclipse.gsfc.nasa.gov/5MCLEmap/1301–1400/LE1357–07-31P.gif (cat. 08108).

117 Boudet, 'Ptolémée dans l'Occident médiéval'.

118 Cf. Thorndike, *A History of Magic*, vol. III, pp. 720–21. See also Lynn Thorndike and Pearl Kibre, *A Catalogue of Incipits of Mediaeval Scientific Writings in Latin*, rev. and augmented ed. (Cambridge: The Medieval Academy of America, 1963), col. 1489.

119 According to *Astromodels*, the Alfonsine Tables would have shown a true conjunction of Saturn and Mars at 20;57,15° Cancer at 21;48h in Toledo, so at 22;4h in Oxford.

> omnes astronomos loquentes de hac materia. Et ideo proposui a diu scripsisse de significationibus istarum coniunctionum …[120]

John Aschenden then goes into his interpretation of the 1357 conjunction of the two evil planets, Saturn and Mars, in Cancer — a harmful phenomenon that occurs every thirty years —[121] by insisting, as John of Murs had done, on the fact that Cancer constitutes the detriment (*detrimentum*) of Saturn and the fall (*casus*) of Mars. The horoscope of the conjunction appears on folio 42^{v} (cf. Fig. 4), with 12° Virgo as ascendant, Mars and Saturn in the eleventh house at 21° Cancer, and this reference in the central text box: Vera latitudo septentrionalis Saturni 5 m^{a}, Martis 18 [m^{a}].[122] The place particularly affected by the conjunction is located in the northern part of the seventh climate, where the Kingdoms of France and England are situated. However, the continuation of the prediction announces a radiant future for England, and, on the contrary, a disastrous one for the Kingdom of France which would undergo a mutation. With Mercury, the planet signifying the Kingdom of France, being in the tenth house at 7° Gemini in the horoscope and undergoing because of the Sun (at 25° Gemini, which means that Mercury will be *combustus et infortunatus per Solem*) 'a transfer of light' (*translatio luminis*)[123] to the benefit of the Moon, which means England, and the Moon being located in the sixth house at 4° Pisces, in conjunction with Jupiter, beneficial planet, and in trine aspect with Saturn and Mars, John Aschenden manages to argue that this mutation of the crown of France will benefit the King of England.[124]

120 'Since then this year 1357, during the month of June, after seven days, twenty-two hours, and thirty minutes of the same month, starting the day at noon of the day preceding, as astronomers do, will be a conjunction of Saturn and Mars in the sign of Cancer, Cancer being the sign of the world closest to us as say astronomers. And the year 1365, during the month of October, after twenty-nine days, fourteen hours, and twenty-two seconds of the same month, will be a conjunction of Saturn and Jupiter in the sign of Scorpio and a new triplicity, that will signify very large changes in the world and great accidents and terrible things, according to the astronomers talking about this matter. And, therefore, I have long offered to write about the meanings of these types of conjunctions …'

121 See Albumasar's *De magnis coniunctionibus*, I, I, 14 and II, VIII, 3 in *Abū Ma'šar on Historical Astrology: The Book of Religions and Dynasties (On the Great Conjunctions)*, 2 vols, ed. Keijo Yamamoto and Charles Burnett (Leiden: Brill, 2000), vol. II, pp. 9 and 80–81.

122 Albumasar takes into account the conjunctions of the planets not only in longitude but also in latitude: Abū Ma'šar al-Balkhī [Albumasar], *Liber introductorii maioris ad scientiam iudiciorum astrorum*, ed. by Richard Lemay, vol. V, *Texte latin de Jean de Séville avec la révision de Gérard de Crémone*, Tractatus VII, diff. 5, pp. 295–96.

123 Albumasar, *Liber introductorii maioris*, vol. V, Tractatus VII, diff. 5, p. 299, for the double meaning of *translatio* in an astrological context. Cf. also the Arabic original text and the English translation in *The Great Introduction to Astrology by Abū Ma'šar*, 2 vols, ed. Keijo Yamamoto and Charles Burnett (Leiden: Brill, 2019), vol. I, pp. 776–77. However, *translatio luminis* seems to come more precisely from Zahel's *Introductorium* in Ptolemy's *Quadripartitum*, 1493, f. 123va: 'Translatio luminis a planeta in planetam est ut separetur planeta levior ab alio ponderosiori et iungatur alteri: tunc quasi coniungit eos et defert naturam primi ad alterum cui iungitur'. See *Works of Sahl and Māshā'allāh*, translated by Benjamin N. Dykes (Minneapolis: Cazimi Press, 2008), p. 17: 'The transfer of light from a planet to a planet is if a lighter planet is being separated from another, heavier one, and it would be joined to another: then it practically conjoins them and bears the nature of the first to the other (to whom he is being joined)'. Here, it is the Moon that takes advantage of the separation of the Sun from Mercury to recover its power, joining Jupiter.

124 Digby 176, ff. 46^{v}–47^{r}: 'Et non tamen reddet Sol lumen Mercurii et regnum Francie ad locum coniunctionis Saturni et Martis utrum etiam ipsi coniunctioni Saturni et Martis eo quod principaliter causa mutationis regni Francie erit hec coniunctio Saturni et Martis. Et quia nec Saturnus nec Mars erit principalis dominus istius coniunctionis sed Luna vendicabit sibi principale dominam istius coniunctionis, ut prius probatum est, Saturnus ergo et Mars transferrent lumen Mercurii receptum a Sole et redditum eis ad Lunam et per consequens transferrent regnum Francie ad Lunam, signacionem Anglie, quam aspicient aspectu amicabili, scilicet de trino aspectu. Et sic transferrent regnum Francie

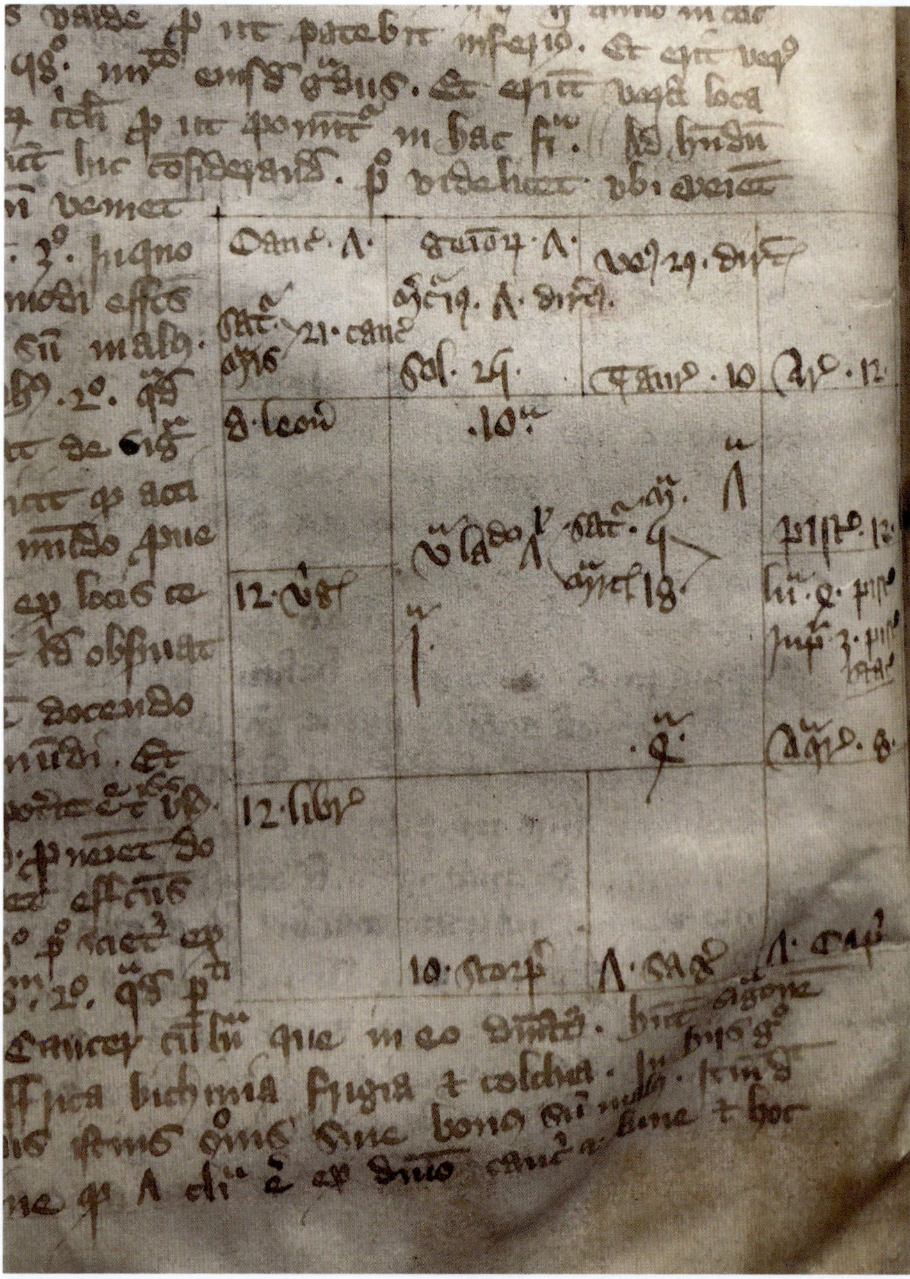

Figure 4. MS Oxford, Bodleian Library, Digby 176, folio 42^{v}: horoscope of the planetary conjunction Saturn-Mars of 1357. Authors' photograph.

ad regem Anglie et ad Anglos et omnem sublimacionem et exaltacionem atque laudem eius regni. Et confirmatur hoc quod hec translatio luminis et translatio regni Francie erit ad Lunam quia cum hoc quod Luna erit principalis domina et disponitrix istius coniunctionis, ipsa etiam erit multum fortunata per eius coniunctionem cum Jove…'

The two astrologers, John of Murs and John Aschenden, agreed in their estimations that this conjunction of Mars and Saturn in 1357 would be disastrous for France, but they were in complete disagreement concerning the *significatores* of this conjunction and, therefore, on the political consequences to be expected. For the French astrologer, for whom Jupiter means the Kingdom of France and Saturn means the Kingdom of England, it is above all the location of Jupiter, in a retrograde position in Pisces at the time of the conjunction, that is worrying because it could indicate a flight of the King of France during an upcoming battle. The identification of Mercury as a planet signifying France and the Moon as a luminary identifiying England led Aschenden on a completely different and much more subtle path of interpretation, allowing him to conclude that the situation was even more favourable for the King of England, destined, according to the wishes of Edward III, to inherit the throne of France. Even if his approach is astrologically well argued, one wonders, with Thorndike,[125] whether the defeat of Jean le Bon at Poitiers in 1356 and his subsequent imprisonment may have helped to guide his judgment, which was completed in March 1357.

Aschenden's text on the conjunction of 1365 begins on folio 34^{r}. Like John of Murs, he observed that this conjunction of Jupiter and Saturn would take place in Scorpio and mark a change of triplicity between the air and the water signs, signifying events of great religious and political importance.[126] The horoscope of the conjunction is copied at the bottom of the page (Fig. 5) and, under the main text, the chronology of the three conjunctions of the upper planets in 1365 (Fig. 6) is presented:[127]

Vera coniunctio	Martis et Jovis in	4to die augusti	anno Christi 1365 apud Oxonie
		20mo gradu Libri	
	Martis et Saturni in	19no die augusti	
		30 gradu Libri	
	Jovis et Saturni in	30mo die octobris	
		8vo gradu Scorpionis	

According to John Aschenden's opinion, which is based not only on the authorities of Albumasar and Messahalla but also on the *De revolutionibus annorum mundi* of Abraham Ibn Ezra and Pseudo-Ovid's *De vetula*, since Venus and Mercury mean, respectively, the faiths of the Saracens and the Christians, the future balance of power between the two religions

125 Thorndike, *A History of Magic*, vol. III, p. 340.

126 Cf. Albumasar's *De magnis coniunctionibus*, I, I, 13 in *Abū Maʿšar on Historial Astrology*, vol. II, pp. 8–9.

127 According to *Astromodels*, the Alfonsine Tables indicate a true conjunction of Mars and Jupiter on 4 August at 19;46,44° Libra (at 0;11h in Toledo; at 0;27h in Oxford), a true conjunction of Mars and Saturn on 18 August at 29;15, 25° Libra (at 4h in Toledo; 4;16h in Oxford), and a true conjunction of Saturn and Jupiter on 29 October at 7;16,46° Scorpio (at 14;20h in Toledo; 14;36h in Oxford).

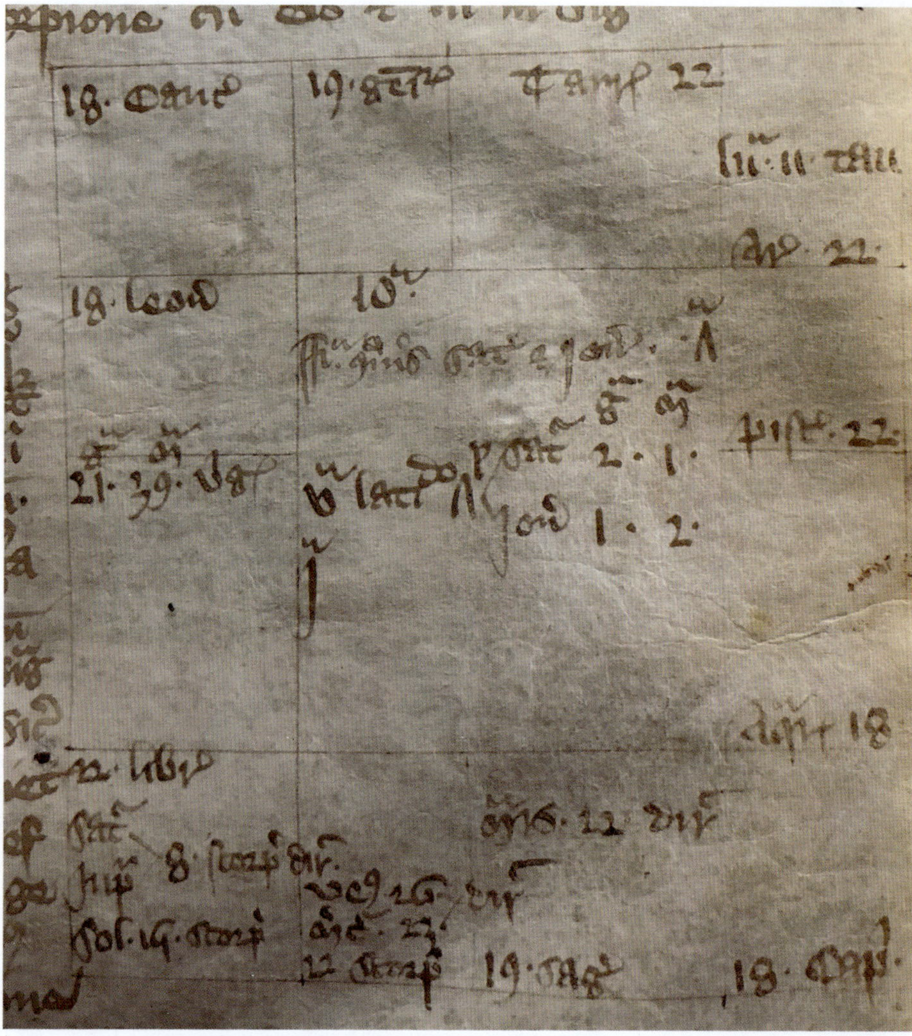

Figure 5. MS Oxford, Bodleian Library, Digby 176, folio 34^{r}: horoscope of the planetary conjunction Saturn-Jupiter of 1365. Authors' photograph.

can be partially deduced from the location of these two planets on the horoscope of the Saturn-Jupiter conjunction. Mercury and Venus are precisely in quasi-conjunction, in the third house of heaven, that of religion (concurrently with the ninth house), and in the twenty-third and twenty-sixth degrees of Scorpio, the sign in which Venus is in *detrimentum*, whereas the ascendant is at 21° 39' Virgo, in the house of Mercury, meaning in one of the two signs of the zodiac where Mercury sees its qualities best emphasized. Mercury, the planet signifying Christianity, is therefore the master of the ascendant (*dominus ascendentis*) of the conjunction of Saturn and Jupiter of 1365 and the fact that Mercury is itself in quasi-conjunction with the planet of Islam, Venus (itself in an unfortunate position), implies the future destruction of Islam by Christianity. However, two complementary remarks in this section allow us to clearly distinguish John Aschenden's judgment from that of John

Figure 6. MS Oxford, Bodleian Library, Digby 176, folio 34ʳ, lower margin: chronology of the three conjunctions of the upper planets in 1365. Authors' photograph.

of Murs. First, the possibility of a *diluvium particulare* in regions most concerned by the conjunction is not very compatible with the *generale passagium ultra mare* advocated by the French astrologer. Second, the domination of Mercury at the time of the conjunction may bring troubles within the Church and be particularly harmful to the Papacy. John Aschenden's view is therefore both pro-English and far less enthusiastic than that of John of Murs about the desirability of a crusade.

Digby 176 is thus a particularly remarkable example of a practitioner's manuscript of world astrology relating to planetary conjunctions. But it also reserved a significant place for the astrology of nativities.

2.3. A good example of astrological practice probably concerning the nativity of a member of the Oxford Mertonian circle

In accordance with Ptolemy's teaching in the *Quadripartitum*,[128] the case in question is based first of all on the horoscope of the Sun-Moon conjunction on 4 December 1317 (f. 90ᵛ), preceding the birth of an individual *aput Combe* (in Berkshire or in Oxfordshire).[129] The birth horoscope of the person concerned can be found on the following page (f. 91ʳ; see Fig. 7):

> Figura nativitatis coniunctionalis et nocturne aput Combe, cuius longitude ab occidente est 13 gradus et 30 mª et eius latitude est 51 gradus et 30 minuta, anno Christi 1317 imperfecto post meridiem, 10 diei decembris per 14 horas equales et 30 minuta unius hore secundum propinquam estimationem laicorum hora gallicantus. Et hec est disposition celi et planetarum ad idem tempus:[130]
> Sol: 28ᵍ Sagitarii, velocis motus, in 2ª quarta ecentrici. [*Add.*: Pars fortune: 9 Leonis.]
> Luna: 12 Piscium, velocis motus, in 2ª quarta epicicli et in 2ª 4ª ecentrici, meridionalis ab ecliptica, occidentalis a Sole sive vespertina.

128 Ptolemy, *Quadripartitum*, 1493, III, 2, ff. 55va-56va.

129 North, in *Horoscopes and History*, p. 138, tried to identify the Combe in question and thought that 'we should probably favour the hamlet south-west of Newbury in Berkshire', but it could more probably be identified with Combe in Oxfordshire. Indeed, the horoscope is computed for 16 minutes east of Toledo, i.e., for the usual meridian of Oxford (thanks to Richard L. Kremer for this remark).

130 'Horoscope of a conjunctional and nocturnal birth at Combe, its longitude from the west is thirteen degrees and thirty minutes, and its latitude is fifty-one degrees and thirty minutes, the year 1317 being imperfect, in the afternoon of the tenth day of December, by fourteen equal hours and thirty minutes of an hour according to the nearest estimation of the lays for the time from cockcrow. And this is the position of heaven and the planets at the same time'.

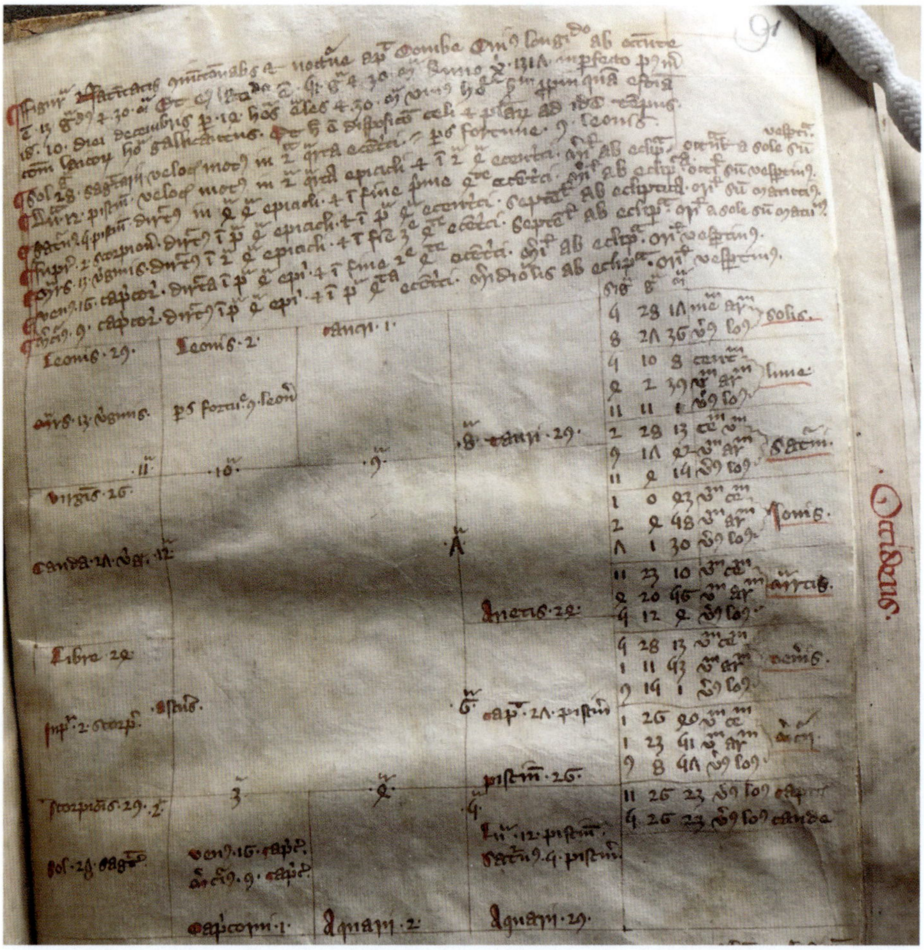

Figure 7. MS Oxford, Bodleian Library, Digby 176, folio 91r: horoscope of an individual born *aput Combe*, 10 December 1317. Authors' photograph.

Saturnus: 5 Piscium, directus, in 4^a 4^a epicicli et in fine prime 4te ecentrici, meridionalis ab ecliptica, occidentalis sive vespertinus.
Jupiter: 2 Scorpionis, directus, in prima 4^a epicicli et in prima 4^a ecentrici, septentrionalis ab ecliptica, orientalis sive matutinus.
Mars: 13 Virginis, directus, in 2^a 4^a epicicli et in finem 3^e 4te ecentrici, septentrionalis ab ecliptica, orientalis a Sole sive matutinus.
Venus: 16 Capricornis, directa in prima 4^a epicicli et in fine 2^e 4te ecentrici, meridionalis ab ecliptica, orientalis, vespertinus.
Mercurius: 9 Capricornis, directus, in prima 4^a epicicli et in prima 4ta ecentrici, meridionalis ab ecliptica, orientalis, vespertinus.

It is interesting to compare these positions, as reported at the top of the page relative to the geometrical models of Ptolemy and as mentioned in the *figura celi*, especially to the right of it, with values computed by *Astromodels*, i.e., based on the Alfonsine Tables for the date in question:

			Digby 176, f. 91r		Astromodels		
Sig^a	g^a	m^a			Sig^a	g^a	m^a
5	28	17	medium argumentum	Solis	5	28	18
8	27	36	verus locus —		8	27	36
5	10	8	centrum	Lune	5	10	8
4	2	39	verum argumentum —		4	2	39
11	11	1	verus locus —		11	11	1
2	28	13	centrum verum	Saturni	2	28	38
9	17	42	verum argumentum —		9	17	42
11	4	15	verus locus —		11	4	15
1	0	43	verum centrum	Jovis	1	1	6
2	4	58	verum argumentum —		2	4	59
7	1	30	verus locus —		7	1	29
11	23	10	verum centrum	Martis	11	23	30
4	20	56	verum argumentum —		4	21	1
5	12	4	verus locus —		5	11	55
5	28	13	verum centrum	Veneris	5	28	13
1	11	53	verum argumentum —		1	11	53
9	15	1	verus locus —		9	15	11
1	26	40	verum centrum	Mercurii	1	26	40
1	23	51	verum argumentum —		1	23	51
9	8	57	verus locus —		9	8	56
11	26	32	verus locus	Capitis	11	26	21
5	26	23	verus locus	Caude	5	26	21

We do not know exactly when these calculations were made (*c.* 1330–40), but the comparison with *Astromodels* is conclusive and attests to the fact that we are dealing with a highly skilled user of the Alfonsine Tables, fully worthy of the reputation of a Merton College 'calculator'. At the bottom of the page, a note refers to the fourth part of Alcabitius's *Liber introductorius* to determine that the Moon is the planet *yleg* (i.e. *dator vite*), Jupiter the planet *alkocoden* (i.e. *dator annorum*), and Mercury the planet *almubtaz* (i.e. the planet dominating the nativity after the *yleg* and the *alkocoden*), with these three planets being decisive in calculating the life expectancy of the person in question.[131] Since this part of

131 On the combined notions of *hyleg* (Arabic *haylāǧ*, Latin *dator vite*) and *alkocoden* (Arabic *al-kadkhudāh*, Latin *dator annorum*) in a birth horoscope and the calculations of direction to which these aphetic points can be subjected, see Al-Qabīsī (Alcabitius), *The Introduction of Astrology. Editions of the Arabic and Latin Texts and an English Translation*, eds Charles Burnett and others (London: The Warburg Institute, 2004), IV, 4–6, pp. 110–15 and 319–25, and Paul Kunitzsch, *Mittelalterliche astronomisch-astrologische Glossare mit arabischen Fachausdrücken* (Munich: Bayerische

the manuscript seems to have been owned by Simon Bredon, it could have been done, if not by Simon himself (he was born *c.* 1310), at least a member of his entourage, perhaps another Mertonian.

Conclusion

MS Oxford, Bodleian Library Digby 176 raises the problem of the complementary practices between Alfonsine astronomy and astrology in a very interesting way, with some of the predictions for planetary conjunctions and eclipses (for 1345, 1357, and 1365) having been made by John Aschenden on the basis of William Reed's and Walter Elveden's calculations. This codex shows that Merton College became, in the fourteenth century, a breeding ground for scholars specialized in the sciences of *quadrivium*, collaborating throughout the rest of their lives as Merton *alumni*, particularly in the field of astronomy-astrology. The individual born in 1317 near Combe was probably one of them.

Examination of this codex confirms the early use of the Parisian Alfonsine Tables in England, a practice that clearly played a role in the improvement of prediction techniques, notably in the refinement of the astrology of planetary conjunctions, championed by John Aschenden. We are faced with a highly sophisticated, cutting-edge scientific milieu, where innovation was fostered by a rapid flow of information and technical data within what can be called a 'community of knowledge' (Constant Mews) or a 'textual community' (Brian Stock)[132] that was also a spiritual community (one thinks of the various legacies that represented the foundation of the constitution of the manuscript) that William Reed seems to have led already when he was a *socius* of the College between 1344 and 1357.

Although Merton College had become a nursery for calculators, the practice of astrology remained politically and religiously correct as would have befitted such an institution founded by an English theologian. It was, therefore, the astrology of revolutions and conjunctions that was favoured there more than that of nativities, to say nothing of questions and elections that are more easily assimilated to divination. It was within these limits that John Aschenden situated his activity at the end of his *Summa iudicialis de accidentibus mundi*:

> Non te intromittas de scientiis illicitis cuiusmodi sunt nigromantia, magica, geomantia, nec de questionibus astrologiæ nec de electionibus sit tibi magna cura. Questiones enim et electiones, ut dicit Haly, sunt res debiles nec rem naturalem sequuntur. Et Ptolemeus etiam in 2° [libri] *Quadripartiti*, capitulo primo, eas expresse reprobat. De istis scientiis illicitis dimittas quæ ponunt in fatalem necessitatem. Planetæ enim non

Akademie der Wissenschaften, 1977), pp. 31–32, 35–37, and 49–50. *Almubtaz* (from Arabic *al-mubtazz*, 'who reigns', 'who dominates') is, in a birth horoscope, a dominant planet, because it enjoys the greatest number of essential and accidental dignities. Cf. Al-Qabīsī (Alcabitius), *The Introduction of Astrology*, IV, 7, p. 325: 'Almubtaz vero qui preest nativitati signifat esse nati post hilesg et quodchodeuh'. The note also refers to proposition 38 of a *Centiloquium*, but it is not Pseudo-Ptolemy's *Centiloquium*, *Centiloquium Hermetis*, or *Centiloquium Bethen*.

132 Brian Stock, *The Implications of Literacy: Written Language and Models of Interpretation in the Eleventh and Twelfth Centuries* (Princeton: Princeton University Press, 1983). This concept has, in particular, been taken up and exploited by Alain Boureau in *L'Empire du livre. Pour une histoire du savoir scolastique (1200–1380)* (Paris: Les Belles Lettres, 2007), Chapter II.

necessitant sed disponunt: iudicia enim astrorum sunt medium inter necessarium et possibile, sicut dicit Ptolemeus in *Centiloquio*, propositione prima, nullam ergo necessitatem agunt stelle et fatum etiam theologi et physici et quasi omnes sane mentis reprobant et condemnant.[133]

133 Johannes Eschuid [*sic*], *Summa astrologiæ iudicialis de accidentibus mundi*, II, 12, 3, f. 306rb [*sic* for 314rb]: 'Do not involve yourself in illicit sciences like nigromancy, magic, geomancy, and do not care a lot about questions of astrology or elections. For questions and elections, as Haly says, are weak things that do not follow natural data. So Ptolemy also, in the first chapter of the second book of *Quadripartitum*, explicitly rejects it. Renounce these illicit sciences that are based on a fatal necessity. In fact, planets do not necessarily require but predispose: [astrological] judgments are intermediate between necessity and possibility, as says Ptolemy in the first proposition of the *Centiloquium*, therefore, the stars operate without any necessity. This is why theologians, physicians, and almost all sane people reject and condemn fatalism'. In fact, Ptolemy does not explicitly speak of interrogations and elections, but it is Hali Abenrudian, in his commentary to Chapter 1 of Book II of the *Quadripartitum* (Venice: 1493, f. 29vb), who considered these aspects of astrology to be of lesser value than revolutions and nativities: '[…] questiones tamen et electiones sunt res debiles satis, nec rem naturalem consequuntur'. John Aschenden goes on to cite the first proposal of Pseudo-Ptolemy's *Centiloquium*: Ptholomeus, *Quadripartitum*, f. 107ra: 'Astrologus autem non debet dicere rem specialiter sed universaliter, ut qui eminus videt aliquam rem. Sic enim facit qui considerat rem secundum materiam suam et non venit ad eius certam cognitionem. Per materiam habemus de re cognitionem dubiam, per formam vero certam. Et hec iudicia que trado tibi sunt media inter necessarium et possibile'.

Appendix 1: Edition of the *Almanak Solis*

Note concerning the apparatus:

For a better understanding we have added an abscissa line with numbers and an ordinate line with letters. The location of each variant is given with an abscissa and ordinate coordinates, i.e., **x,y: D1:** variant, **D2:** variant.

D1: Oxford, Bodleian Library, Digby 176
D2: Oxford, Bodleian Library, Digby 178

B and **C**: **D1** shows the zodiac line followed by the line of the months; **D2** shows the line of the months before the zodiac line. For the following tables, **D1** and **D2** both present the line of the months followed by the zodiac line.

1	2	3	4	5	6	7	8	9	10	11	12	13	14	15	16	17	18	19	20	21	22	23	24	25	26	27	28	29	30	31	32	33	34	35	36	37	
Tabula solis prima ad annum Christi 1341 qui est annus primus post bisextum super meridiem Oxonie																																					**A**
Dies	**Januarius**			**Februarius**			**Martius**			**Aprilis**			**Maius**			**Junius**			**Julius**			**Augustus**			**September**			**October**			**November**			**December**			**B**
	Capricornus			**Aquarius**			**Pisces**			**Aries**			**Taurus**			**Gemini**			**Cancer**			**Leo**			**Virgo**			**Libra**			**Scorpio**			**Sagittarius**			**C**
	Grad	Min	Sec	Grad	Min	Sec	Grad	Min	Sec	Grad	Min	Sec	Grad	Min	Sec	Grad	Min	Sec	Grad	Min	Sec	Grad	Min	Sec	Grad	Min	Sec	Grad	Min	Sec	Grad	Min	Sec	Grad	Min	Sec	**D**
1	19	55	8	21	25	36	19	27	21	19	54	30	18	51	17	18	25	4	16	55	54	16	32	50	16	34	36	16	10	9	17	21	8	17	55	57	**E**
2	20	56	23	22	26	13	20	26	50	20	52	52	19	48	46	19	22	6	17	52	58	17	30	30	17	33	16	17	9	53	18	21	46	18	57	22	**F**
3	21	57	37	23	26	46	21	26	18	21	51	12	20	46	15	20	19	6	18	50	0	18	28	9	18	31	58	18	9	39	19	22	36	19	58	50	**G**
4	22	58	51	24	27	20	22	25	44	22	49	36	21	43	43	21	16	8	19	47	5	19	25	50	19	30	42	19	9	31	20	23	35	21	0	16	**H**
5	24	0	5	25	27	53	23	25	9	23	47	56	22	41	9	22	13	9	20	44	9	20	23	32	20	29	26	20	9	30	21	24	30	22	1	43	**I**
6	25	1	20	26	28	13	24	24	34	24	46	14	23	38	37	23	10	12	21	41	15	21	21	14	21	28	9	21	9	32	22	25	26	23	3	10	**J**
7	26	2	33	27	28	36	25	23	58	25	44	30	24	36	4	24	7	13	22	38	21	22	18	59	22	27	0	22	9	35	23	26	25	24	4	37	**K**
8	27	3	53	28	28	53	26	23	15	26	42	37	25	33	25	25	4	14	23	35	31	23	16	47	23	25	53	23	9	41	24	27	22	25	6	5	**L**
9	28	5	10	29	29	11	27	22	27	27	40	43	26	30	46	26	1	17	24	32	43	24	14	37	24	24	45	24	9	46	25	28	21	26	7	34	**M**
10	29	6	15	o Pisces	29	29	28	21	37	28	38	47	27	28	6	26	58	18	25	29	52	25	12	29	25	23	38	25	9	52	26	29	30	27	9	6	**N**
11	o Aquarius	7	28	1	29	48	29	20	46	29	36	52	28	25	27	27	55	18	26	27	4	26	10	21	26	22	34	26	10	0	27	30	43	28	10	40	**O**
12	1	8	39	2	29	55	o Aries	19	54	o Taurus	34	56	29	22	43	28	52	20	27	24	16	27	8	3	27	21	30	27	10	7	28	31	51	29	12	6	**P**
13	2	9	48	3	30	3	1	18	59	1	32	58	o Gemini	20	1	29	49	20	28	21	30	28	6	9	28	20	34	28	10	20	29	33	3	o Capricornus	13	36	**Q**
14	3	10	57	4	30	10	2	18	2	2	30	50	1	17	20	o Cancer	46	21	29	18	47	29	4	13	29	19	43	29	10	38	o Sagittarius	34	17	1	15	6	**R**
15	4	12	3	5	30	18	3	16	57	3	28	42	2	14	19	1	43	23	o Leo	16	6	o Virgo	2	17	o Libra	18	51	o Scorpio	10	57	1	35	20	2	16	36	**S**
16	5	13	3	6	30	21	4	15	52	4	26	34	3	11	41	2	40	25	1	13	25	1	0	21	1	17	59	1	11	14	2	36	43	3	18	6	**T**
17	6	14	1	7	30	27	5	14	45	5	24	25	4	8	53	3	37	26	2	10	42	1	58	26	2	17	11	2	11	33	3	37	56	4	19	37	**U**
18	7	14	59	8	30	31	6	13	37	6	22	15	5	5	55	4	34	27	3	8	2	2	56	32	3	16	24	3	11	55	4	39	13	5	21	4	**V**
19	8	15	53	9	30	32	7	12	29	7	20	4	6	3	15	5	31	28	4	5	23	3	54	43	4	15	45	4	12	20	5	40	27	6	22	32	**W**
20	9	16	48	10	30	25	8	11	17	8	17	46	7	0	11	6	28	30	5	2	48	4	53	1	5	15	10	5	12	54	6	41	41	7	24	0	**X**
21	10	17	43	11	30	15	9	10	1	9	15	26	7	57	28	7	25	32	6	0	15	5	51	21	6	14	35	6	13	28	7	42	55	8	25	26	**Y**
22	11	18	36	12	29	59	10	8	44	10	13	8	8	54	33	8	22	33	6	57	43	6	49	40	7	14	0	7	14	3	8	43	59	9	26	52	**Z**
23	12	19	31	13	29	43	11	7	27	11	10	48	9	51	37	9	19	35	7	55	11	7	48	4	8	13	28	8	14	39	9	45	24	10	28	18	**AA**
24	13	20	22	14	29	22	12	6	8	12	8	28	10	48	42	10	16	37	8	52	38	8	46	23	9	12	55	9	15	16	10	46	40	11	29	43	**BB**
25	14	21	3	15	29	0	13	4	48	13	6	7	11	45	46	11	13	37	9	50	7	9	44	48	10	12	19	10	15	53	11	47	57	12	31	8	**CC**
26	15	21	44	16	28	36	14	3	27	14	3	41	12	42	49	12	10	39	10	47	36	10	43	18	11	12	1	11	16	32	12	49	13	13	32	30	**DD**
27	16	22	25	17	28	12	15	2	2	15	1	14	13	39	50	13	7	42	11	45	7	11	41	47	12	11	36	12	17	13	13	50	29	14	33	51	**EE**
28	17	23	4	18	27	27	16	0	31	15	58	45	14	36	53	14	4	45	12	42	38	12	40	18	13	11	12	13	17	53	14	51	49	15	35	11	**FF**
29	18	23	43				16	59	3	16	56	16	15	33	56	15	1	47	13	40	9	13	38	50	14	10	50	14	18	33	15	53	11	16	36	27	**GG**
30	19	25	24				17	57	33	17	53	47	16	30	59	15	58	47	14	37	42	14	37	19	15	10	27	15	19	18	16	54	33	17	37	43	**HH**
31	20	25	0				18	56	2				17	28	2				15	35	15	15	35	56				16	19	58				18	38	53	**II**

3,HH: D1: 25; **D2:** 24 | **29,H: D1:** 219; **D2:** 19

1	2	3	4	5	6	7	8	9	10	11	12	13	14	15	16	17	18	19	20	21	22	23	24	25	26	27	28	29	30	31	32	33	34	35	36	37	
Tabula solis secunda ad annum Christi 1342 qui est annus secundus post bisextum super meridiem Oxonie																																					A
Dies	Januarius			Februarius			Martius			Aprilis			Maius			Junius			Julius			Augustus			September			October			November			December			B
	Capricornus			Aquarius			Pisces			Aries			Taurus			Gemini			Cancer			Leo			Virgo			Libra			Scorpio			Sagittarius			C
	Grad	Min	Sec	Grad	Min	Sec	Grad	Min	Sec	Grad	Min	Sec	Grad	Min	Sec	Grad	Min	Sec	Grad	Min	Sec	Grad	Min	Sec	Grad	Min	Sec	Grad	Min	Sec	Grad	Min	Sec	Grad	Min	Sec	D
1	19	40	15	21	10	56	19	12	55	19	40	21	18	37	22	18	11	16	16	42	6	16	18	54	16	20	21	15	55	42	17	6	5	17	41	2	E
2	20	41	30	22	11	30	20	12	24	20	38	44	19	34	51	19	8	17	17	39	7	17	16	33	17	19	2	16	55	25	18	6	58	18	42	29	F
3	21	42	44	23	12	5	21	11	52	21	37	4	20	32	20	20	5	17	18	36	11	18	14	12	18	17	44	17	55	10	19	7	54	19	43	37	G
4	22	43	58	24	12	45	22	11	19	22	35	24	21	29	49	21	2	19	19	33	16	19	11	53	19	16	27	18	55	0	20	8	48	20	45	21	H
5	23	45	13	25	13	11	23	10	44	23	33	47	22	27	15	21	59	20	20	30	20	20	9	35	20	15	10	19	54	58	21	9	42	21	46	47	I
6	24	46	27	26	13	33	24	10	9	24	32	7	23	24	42	22	56	22	21	27	25	21	7	16	21	13	55	20	54	59	22	10	37	22	48	14	J
7	25	47	41	27	13	55	25	9	35	25	30	24	24	22	10	23	53	25	22	24	32	22	5	0	22	12	43	21	55	2	23	11	35	23	49	42	K
8	26	49	2	28	14	14	26	8	53	26	28	32	25	19	34	24	50	26	23	21	40	23	2	48	23	11	35	22	55	7	24	12	33	24	51	9	L
9	27	50	17	29	14	32	27	8	6	27	26	39	26	16	54	25	47	28	24	18	49	24	0	37	24	10	27	23	55	12	25	13	31	25	52	37	M
10	28	51	21	0 Pisces	14	51	28	7	17	28	24	43	27	14	14	26	44	30	25	16	2	24	58	29	25	9	21	24	55	19	26	14	39	26	54	10	N
11	29	52	36	1	15	9	29	6	25	29	22	48	28	11	33	27	41	30	26	13	13	25	56	21	26	8	16	25	55	26	27	15	50	27	55	40	O
12	0 Aquarius	53	47	2	15	19	0 Aries	5	33	0 Taurus	20	52	29	8	50	28	38	31	27	10	25	26	54	13	27	7	12	26	55	33	28	16	59	28	57	9	P
13	1	54	56	3	15	27	1	4	41	1	18	55	0 Gemini	6	9	29	37	32	28	7	39	27	52	7	28	6	16	27	55	41	29	18	12	29	58	40	Q
14	2	56	6	4	15	35	2	3	44	2	16	58	1	3	25	0 Cancer	32	34	29	4	54	28	50	6	29	5	22	28	56	0	0 Sagittarius	19	24	1 Capricornius	0	10	R
15	3	57	16	5	15	41	3	2	48	3	14	40	2	1	18	1	29	34	0 Leo	2	13	29	48	11	0 Libra	4	30	29	56	18	1	20	37	2	1	39	S
16	4	58	15	6	15	46	4	1	43	4	12	33	2	57	51	2	26	37	0	59	32	0 Virgo	46	15	1	3	39	0 Scorpio	56	37	2	21	51	3	3	10	T
17	5	59	13	7	15	48	5	0	28	5	10	24	3	55	2	3	23	38	1	56	49	1	44	19	2	2	49	1	56	56	3	23	4	4	4	41	U
18	7	0	10	8	15	54	5	59	21	6	8	14	4	51	57	4	20	38	2	54	9	2	42	25	3	2	1	2	57	16	4	24	20	5	6	9	V
19	8	1	5	9	15	54	6	58	13	7	6	4	5	49	24	5	17	40	3	51	30	3	40	33	4	1	22	3	57	38	5	25	34	6	7	36	W
20	9	2	1	10	15	53	7	57	2	8	3	47	6	46	33	6	14	41	4	48	53	4	38	51	5	0	45	4	58	12	6	26	48	7	9	3	X
21	10	2	55	11	15	46	8	55	46	9	1	28	7	43	38	7	11	43	5	46	20	5	37	10	6	0	10	5	58	44	7	28	2	8	10	30	Y
22	11	3	48	12	15	29	9	54	30	9	59	10	8	40	44	8	8	45	6	43	48	6	35	31	6	59	36	6	59	20	8	29	17	9	11	57	Z
23	12	4	44	13	15	12	10	53	13	10	56	50	9	37	38	9	5	47	7	41	26	7	33	55	7	59	3	7	59	56	9	30	32	10	13	22	AA
24	13	5	37	14	14	53	11	51	54	11	54	29	10	34	52	10	2	49	8	38	52	8	32	13	8	58	30	9	0	33	10	31	46	11	14	49	BB
25	14	6	19	15	14	32	12	50	36	12	52	9	11	31	57	10	59	49	9	36	12	9	30	37	9	57	59	10	1	10	11	33	4	12	16	13	CC
26	15	6	58	16	14	10	13	49	15	13	49	45	12	29	0	11	56	51	10	33	41	10	29	6	10	57	29	11	1	49	12	34	20	13	17	55	DD
27	16	7	41	17	13	45	14	47	51	14	47	18	13	26	1	12	53	53	11	31	11	11	27	34	11	57	17	12	2	29	13	35	36	14	18	57	EE
28	17	8	20	18	13	20	15	46	21	15	44	49	14	23	5	13	50	57	12	28	43	12	26	6	12	56	46	13	3	9	14	36	55	15	20	17	FF
29	18	9	0				16	44	51	16	42	20	15	20	8	14	47	59	13	26	15	13	24	37	13	56	22	14	3	48	15	37	17	16	21	34	GG
30	19	9	40				17	43	22	17	39	51	16	17	11	15	45	2	14	23	47	14	23	6	14	56	1	15	4	30	16	39	38	17	22	50	HH
31	20	10	17				18	41	52				17	14	13				15	21	20	15	21	42				16	5	12				18	24	7	II

7,K: D1: 55; **D2:** 45 | **28,CC: D1:** 59; **D2:** 56 | **33,GG: D1:** 38 was corrected to 37; **D2:** 37 | **37,DD: D1:** 55; **D2:** 35

1	2	3	4	5	6	7	8	9	10	11	12	13	14	15	16	17	18	19	20	21	22	23	24	25	26	27	28	29	30	31	32	33	34	35	36	37	
Tabula solis tercia ad annum Christi 1343 qui est annus tercius post bisextum super meridiem Oxonie																																					A
	Januarius			Februarius			Martius			Aprilis			Maius			Junius			Julius			Augustus			September			October			November			December			B
Dies	Capricornus			Aquarius			Pisces			Aries			Taurus			Gemini			Cancer			Leo			Virgo			Libra			Scorpio			Sagittarius			C
	Grad	Min	Sec	Grad	Min	Sec	Grad	Min	Sec	Grad	Min	Sec	Grad	Min	Sec	Grad	Min	Sec	Grad	Min	Sec	Grad	Min	Sec	Grad	Min	Sec	Grad	Min	Sec	Grad	Min	Sec	Grad	Min	Sec	D
1	19	25	23	20	56	13	18	58	27	19	26	10	18	23	26	17	57	27	16	28	17	16	4	53	16	6	8	15	40	40	16	51	22	17	26	13	E
2	20	26	37	21	56	46	19	57	59	20	24	37	19	20	55	18	54	29	17	25	18	17	2	33	17	4	49	16	40	21	17	52	17	18	27	37	F
3	21	27	51	22	57	22	20	57	27	21	22	55	20	18	25	19	51	29	18	22	22	18	0	14	18	3	29	17	40	4	18	53	10	19	29	3	G
4	22	29	6	23	57	55	21	56	54	22	21	18	21	15	53	20	48	30	19	19	26	18	57	53	19	2	12	18	39	48	19	54	4	20	30	29	H
5	23	30	20	24	58	28	22	56	19	23	19	39	22	13	19	21	45	32	20	16	31	19	55	34	20	0	56	19	39	35	20	54	59	21	31	56	I
6	24	31	34	25	58	56	23	55	45	24	17	58	23	10	48	22	42	34	21	13	36	20	53	16	20	59	40	20	39	33	21	55	54	22	33	26	J
7	25	32	49	26	59	15	24	55	9	25	16	16	24	8	14	23	39	36	22	10	42	21	50	57	21	58	27	21	39	33	22	56	52	23	34	53	K
8	26	34	4	27	59	35	25	54	31	26	14	27	25	5	40	24	36	37	23	7	49	22	48	46	22	57	19	22	39	34	23	57	50	24	36	19	L
9	27	35	17	28	59	55	26	53	45	27	12	34	26	3	1	25	33	40	24	4	59	23	46	36	23	56	11	23	39	41	24	58	49	25	37	48	M
10	28	36	31	0 Pisces	0	12	27	52	56	28	10	38	27	0	21	26	30	40	25	2	10	24	44	26	24	55	4	24	39	46	25	59	54	26	39	18	N
11	29	37	43	1	0	30	28	52	4	29	8	43	27	57	40	27	27	41	25	59	22	25	42	18	25	53	59	25	39	51	27	1	3	27	40	48	O
12	0 Aquarius	38	55	2	0	44	29	51	13	0 Taurus	6	47	28	54	58	28	24	43	26	56	35	26	40	11	26	52	55	26	39	59	28	2	13	28	42	18	P
13	1	40	6	3	0	52	0 Aries	50	20	1	4	50	29	52	14	29	21	44	27	53	47	27	38	3	27	51	55	27	40	6	29	3	23	29	43	51	Q
14	2	41	15	4	0	59	1	49	26	2	2	47	0 Gemini	49	33	0 Cancer	18	44	28	51	1	28	36	3	28	51	2	28	40	14	0 Sagittarius	4	36	0 Capricornus	45	21	R
15	3	42	26	5	1	6	2	48	23	3	0	38	1	46	47	1	15	46	29	48	20	29	34	8	29	50	10	29	40	32	1	5	52	1	46	51	S
16	4	43	36	6	1	11	3	47	18	3	58	32	2	43	59	2	12	48	0 Leo	45	39	0 Virgo	32	13	0 Libra	49	18	0 Scorpio	40	51	2	7	5	2	48	21	T
17	5	44	22	7	1	16	4	46	11	4	56	23	3	41	11	3	9	49	1	42	56	1	30	16	1	48	26	1	41	8	3	8	17	3	49	50	U
18	6	45	22	8	1	20	5	45	5	5	54	14	4	38	23	4	6	50	2	40	16	2	28	23	2	47	40	2	41	28	4	9	32	4	51	19	V
19	7	46	18	9	1	21	6	43	57	6	52	3	5	35	33	5	3	51	3	37	36	3	26	29	3	46	57	3	41	48	5	10	45	5	52	46	W
20	8	47	13	10	1	21	7	42	47	7	49	48	6	32	43	6	0	52	4	34	58	4	24	45	4	46	20	4	42	8	6	12	0	6	54	15	X
21	9	48	7	11	1	14	8	41	32	8	47	30	7	29	49	6	57	55	5	32	26	5	23	4	5	45	46	5	42	40	7	13	14	7	55	42	Y
22	10	49	1	12	1	0	9	40	15	9	45	11	8	26	55	7	54	57	6	29	53	6	21	23	6	45	11	6	43	13	8	14	30	8	57	11	Z
23	11	49	56	13	0	43	10	38	58	10	42	51	9	23	59	8	51	58	7	27	20	7	19	45	7	44	37	7	43	48	9	15	45	9	58	34	AA
24	12	50	49	14	0	24	11	37	41	11	40	32	10	21	4	9	49	0	8	24	47	8	18	7	8	44	5	8	44	25	10	17	1	10	59	59	BB
25	13	51	35	15	0	4	12	36	22	12	38	11	11	18	8	10	46	0	9	22	16	9	16	28	9	43	34	9	45	1	11	18	15	12	1	25	CC
26	14	52	16	15	59	41	13	35	1	13	35	48	12	15	11	11	43	2	10	19	45	10	14	56	10	43	4	10	45	38	12	19	31	13	2	45	DD
27	15	52	56	16	59	17	14	33	39	14	33	30	13	12	15	12	40	5	11	17	15	11	13	26	11	42	43	11	46	17	13	20	47	14	4	8	EE
28	16	53	37	17	58	53	15	32	9	15	30	52	14	9	16	13	37	8	12	14	46	12	11	54	12	41	50	12	48	28	14	22	5	15	5	31	FF
29	17	54	16				16	30	4	16	28	20	15	6	19	14	34	10	13	12	17	13	10	25	13	41	26	13	49	7	15	23	28	16	6	47	GG
30	18	54	56				17	29	12	17	25	55	16	3	23	15	31	14	14	9	49	14	8	56	14	41	3	14	49	50	16	24	51	17	8	3	HH
31	19	55	34				18	27	41				17	0	25				15	7	21	15	7	29				15	50	32				18	9	19	II

4,U: D1: 22; **D2:** 24 | **10,GG: D1:** after 4 another number has been erased; **D2:** 46 | **37,J: D1:** 26; **D2:** 36

1	2	3	4	5	6	7	8	9	10	11	12	13	14	15	16	17	18	19	20	21	22	23	24	25	26	27	28	29	30	31	32	33	34	35	36	37	
Tabula solis quarta ad annum Christi 1344 qui est annus bisextilis super meridiem Oxonie																																					A
	Januarius			Februarius			Martius			Aprilis			Maius			Junius			Julius			Augustus			September			October			November			December			B
Dies	Capricornus			Aquarius			Pisces			Aries			Taurus			Gemini			Cancer			Leo			Virgo			Libra			Scorpio			Sagittarius			C
	Grad	Min	Sec	Grad	Min	Sec	Grad	Min	Sec	Grad	Min	Sec	Grad	Min	Sec	Grad	Min	Sec	Grad	Min	Sec	Grad	Min	Sec	Grad	Min	Sec	Grad	Min	Sec	Grad	Min	Sec	Grad	Min	Sec	D
1	19	10	29	20	41	27	19	43	34	20	10	26	19	7	0	18	40	38	17	11	30	16	48	36	16	50	36	16	26	24	17	37	25	18	12	38	E
2	20	11	44	21	42	4	20	43	1	21	8	47	20	4	29	19	37	41	18	8	33	17	46	15	17	49	16	17	26	8	18	38	19	19	14	3	F
3	21	12	59	21	42	39	21	42	28	22	7	9	21	1	56	20	34	43	19	5	37	18	43	54	18	47	59	18	25	54	19	39	13	20	15	23	G
4	22	14	13	23	43	13	22	41	54	23	5	30	21	59	25	21	31	44	20	2	42	19	41	36	19	46	42	19	25	51	20	40	7	21	16	56	H
5	23	15	27	24	43	48	23	41	20	24	3	49	22	56	52	22	28	45	20	59	46	20	39	17	20	45	26	20	25	49	21	41	2	22	18	23	I
6	24	16	42	25	44	17	24	40	44	25	2	9	23	54	19	23	25	47	21	56	52	21	36	58	21	44	12	21	25	50	22	41	59	23	19	49	J
7	25	17	56	26	44	37	25	40	8	26	0	22	24	51	47	24	22	49	22	53	59	22	34	46	22	43	3	22	25	55	23	42	56	24	21	17	K
8	26	19	11	27	44	57	26	39	23	26	58	28	25	49	8	25	19	49	23	51	7	23	32	35	23	41	55	23	26	0	24	43	54	25	22	47	L
9	27	20	25	28	45	16	27	38	34	27	56	33	26	46	27	26	16	50	24	48	19	24	30	25	24	40	48	24	26	6	25	44	58	26	24	16	M
10	28	21	38	29	45	34	28	37	44	28	54	39	27	43	47	27	13	53	25	45	31	25	28	18	25	39	42	25	26	13	26	46	8	27	25	47	N
11	29	22	51	0 Pisces	45	52	29	36	52	29	52	42	28	41	4	28	10	55	26	42	43	26	26	9	26	38	37	26	26	21	27	47	17	28	27	16	O
12	0 Aquarius	24	4	1	46	9	0 Aries	36	0	0 Taurus	50	46	29	38	23	29	7	55	27	39	55	27	24	1	27	37	36	27	26	28	28	48	27	29	28	46	P
13	1	25	14	2	46	16	1	35	8	1	48	46	0 Gemini	35	42	0 Cancer	4	56	28	37	9	28	21	59	28	36	42	28	26	44	29	49	40	0 Capricornus	30	17	Q
14	2	26	17	3	46	24	2	34	5	2	46	38	1	32	56	1	1	57	29	34	27	29	20	3	29	35	49	29	27	2	0 Sagittarius	50	53	1	31	46	R
15	3	27	35	4	46	31	3	33	20	3	44	30	2	30	8	1	58	57	0 Leo	31	44	0 Virgo	18	8	0 Libra	34	58	0 Scorpio	27	19	1	52	6	2	33	18	S
16	4	28	37	5	46	36	4	31	55	4	42	23	3	27	21	2	55	59	1	29	4	1	16	13	1	34	6	1	27	39	2	53	22	3	34	49	T
17	5	29	35	6	46	42	5	30	48	5	40	13	4	24	32	3	53	1	2	26	23	2	14	17	2	33	17	2	27	59	3	54	35	4	36	17	U
18	6	30	34	7	46	47	6	29	40	6	38	1	5	21	42	4	50	3	3	23	42	3	12	24	3	32	32	3	28	19	4	55	48	5	37	44	V
19	7	31	30	8	46	47	7	28	32	7	35	50	6	18	53	5	47	5	4	21	4	4	10	37	4	31	55	4	28	48	5	57	3	6	39	11	W
20	8	32	25	9	46	48	8	27	17	8	33	31	7	15	59	6	44	6	5	18	31	5	8	55	5	31	19	5	29	21	6	58	17	7	40	38	X
21	9	33	20	10	46	44	9	26	0	9	31	12	8	13	5	7	41	8	6	15	57	6	7	14	6	30	45	6	29	55	7	59	31	8	42	5	Y
22	10	34	15	11	46	29	10	24	45	10	28	54	9	10	10	8	38	9	7	13	26	7	5	37	7	30	11	7	30	30	9	0	46	9	43	34	Z
23	11	35	9	12	46	14	11	23	27	11	26	33	10	7	14	9	35	10	8	10	52	8	3	58	8	29	39	8	31	7	10	2	3	10	44	59	AA
24	12	36	3	13	45	56	12	22	8	12	24	13	11	4	19	10	32	11	9	8	20	9	2	19	9	29	5	9	31	43	11	3	18	11	46	23	BB
25	13	36	48	14	45	36	13	20	49	13	21	52	12	1	21	11	29	14	10	5	51	10	0	44	10	28	39	10	32	23	12	4	34	12	47	46	CC
26	14	37	31	15	45	13	14	19	26	14	19	23	12	58	24	12	26	16	11	3	36	10	59	15	11	28	14	11	33	2	13	5	50	13	49	8	DD
27	15	38	13	16	44	50	15	17	58	15	16	56	13	55	27	13	23	18	12	0	50	11	57	44	12	27	49	12	33	40	14	7	6	14	50	29	EE
28	16	38	53	17	44	26	16	16	29	16	14	28	14	52	31	14	20	22	12	58	21	12	56	15	13	27	26	13	34	21	15	8	28	15	51	48	FF
29	17	39	33	18	44	0	17	15	0	17	11	59	15	49	33	15	17	24	13	55	53	13	54	46	14	27	3	14	35	2	16	9	45	16	53	6	GG
30	18	40	13				18	13	30	18	9	30	16	46	36	16	14	28	14	53	24	14	53	17	15	26	42	15	35	43	17	11	11	17	54	21	HH
31	19	40	51				19	12	0				17	43	39				15	50	57	15	51	46				16	36	30				18	55	37	II

3,Q: D1: 25; **D2:** 24

Appendix 2: Codicological diagram

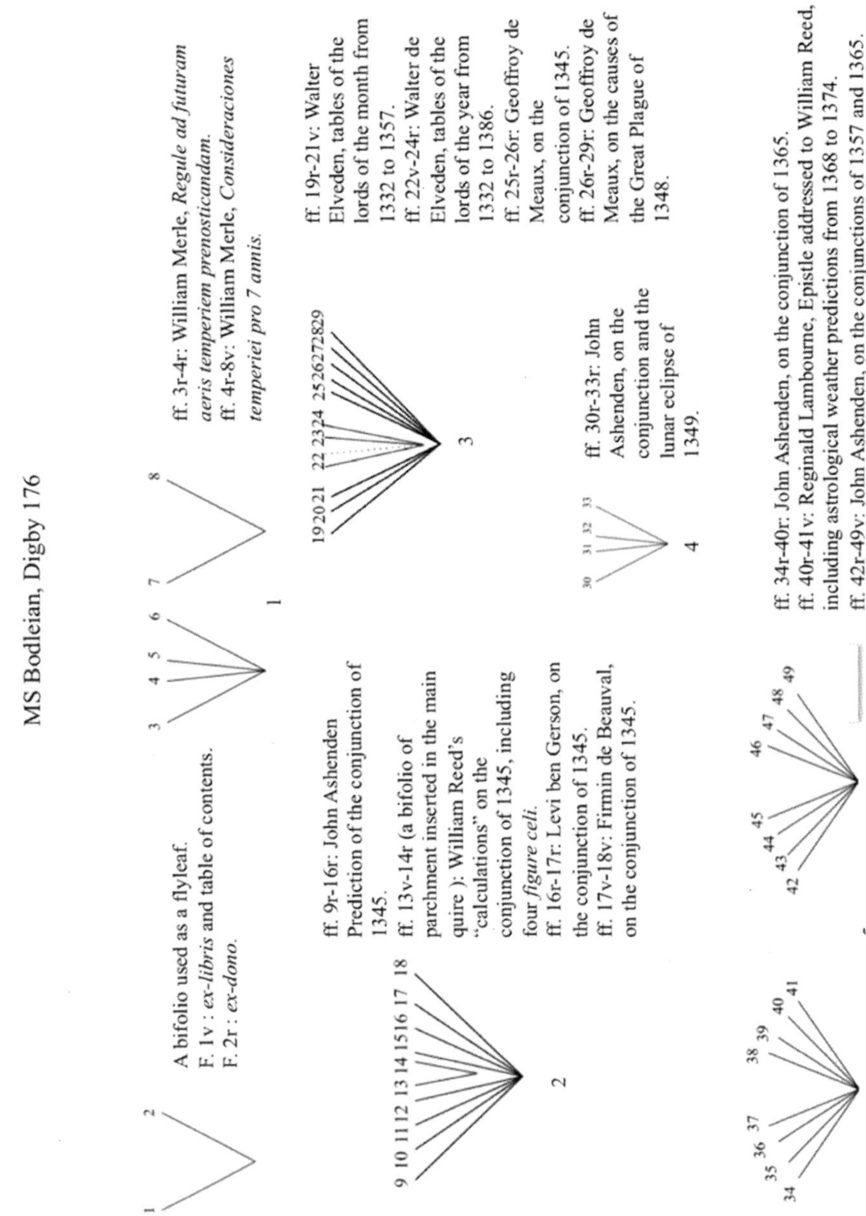

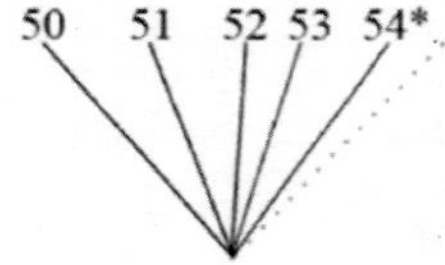

6

ff. 50r-53v: Reginald Lambourne, Epistle to John of London (27 February 1364), including the astrological significance on the Lunar eclipses of March and September 1363 .

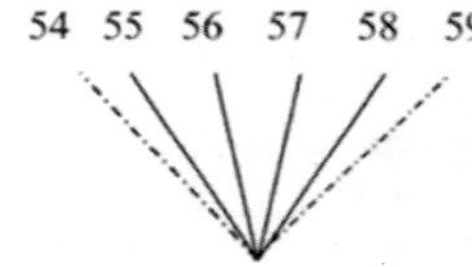

7

ff. 54r-v: Anonymous, *Notula de corruptione pestilanciali.*
ff. 55-59r: Various tables, *Tabula mansionum Lune, Tabula de superioritatibus planetarum, Tabula de dignitatibus planetarum.*

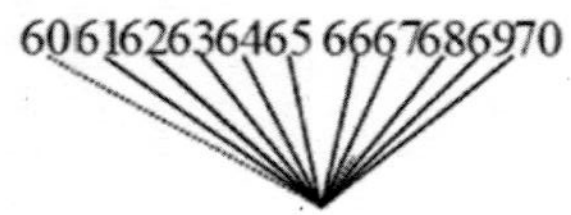

8

f. 60r: *Nota de cognitione ventorum.*
f. 60v: Pliny, excerpts from the *Historia naturalis*: *De temporibus.*
f. 61r: al-Kindī, excerpts from the *De mutatione aeris.*
ff. 67ra-68rb: Anonymous *De pluviis,* inc. *Si sol fuerit in emisperio.*
ff. 68r-69v: Pliny, excerpts from the *Historia naturalis* (XVIII).
f. 69v: Excerpts from Jafar, *Liber imbrium.*
Ff. 69v-70r: Anonymous, *Sapientes indii de pluviis.*

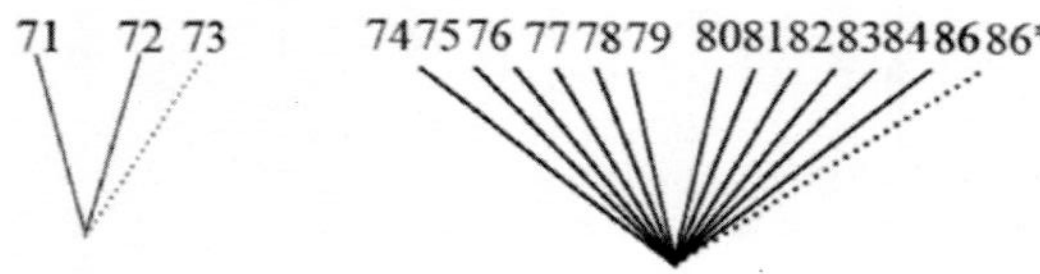

9

ff. 71r-72v: William Reed, *Almanak Solis* (1341-1344).
ff. 73r-86r: John of Saxony, excerpts from the *Almanac* (1349-1380).

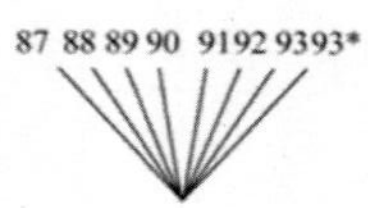

10

ff. 87v-88v: various additions of astrological notes : astrological medicine and astrological weather forecasting.
ff. 89v-90r: computistical table for the year 1333.
ff. 90v-91r: A nativity of an unnamed individual born in Combe (today in the Oxfordshire) dated to 1317.
ff. 91v-93r: computistical tables.

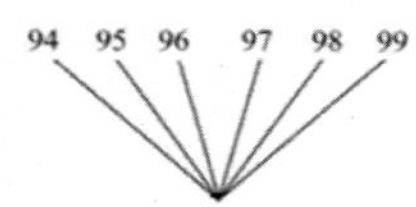

11

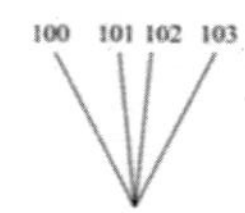

12

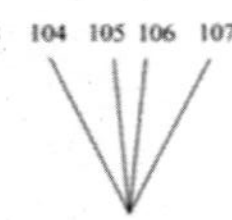

13

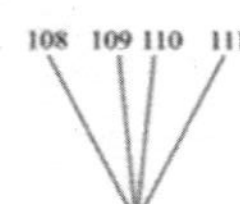

14

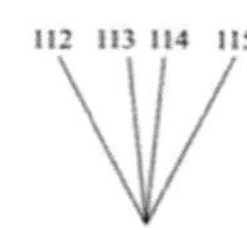

15

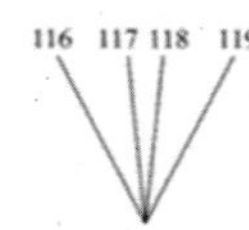

16

ff. 94r-119r: Anonymous, tables of conjunctions and oppositions of the sun and the moon, related to *figure eclipsis* for the year 1341 and the years 1343-1346. They are mainly based on the Oxford meridian, except for the year 1342, which is based on the London meridian.

Manuscript sources

Erfurt, Universitätsbibliothek, CA 2° 386
London, British Library, Egerton 889
Oxford, Bodleian Library, Ashmole 191
Oxford, Bodleian Library, Ashmole 393
Oxford, Bodleian Library, Ashmole 1471
Oxford, Bodleian Library, Digby 57
Oxford, Bodleian Library, Digby 97
Oxford, Bodleian Library, Digby 147
Oxford, Bodleian Library, Digby 168
Oxford, Bodleian Library, Digby 176
Oxford, Bodleian Library, Digby 178
Oxford, Bodleian Library, Digby 179
Oxford, Merton College, 168
Paris, Bibliothèque nationale de France, lat. 7281
Paris, Bibliothèque nationale de France, lat. 7378A
Troyes, Médiathèque, 62

Bibliography

Allen, Percy Stafford, and Heathcote William Garrod, *Merton Muniments* (Oxford: Clarendon Press, 1928).

Almanach Dantis Aligherii sive Profhacii Judaei Montispessulani, ed. by J. Boffito and C. Melzi d'Eril (Florence: Olschki, 1908).

Abū Maʿshar (Albumasar), *The Abbreviation of the Introduction to Astrology, together with the Medieval Latin Translation of Adelard of Bath*, ed. Charles Burnett, Keiji Yamamoto, Michio Yano (Leiden: Brill, 1994).

———, *Kitāb al-mudkhal al-kabīr, Liber Introductorii Maioris ad Scientiam Iudiciorum Astrorum*, ed. Richard Lemay (Naples: Instituto universitario Orientale, 1995–96).

———, *The Great Introduction to Astrology by Abū Maʿšar*, ed. Keiji Yamamoto and Charles Burnett (Leiden: Brill, 2019).

———, *Abū Maʿšar on Historial Astrology: The Book of Religions and Dynasties (On the Great Conjunctions)*, ed. Kieji Yamamoto and Charles Burnett (Leiden: Brill, 2000).

Al-Qabīsī (Alcabitius), *The Introduction of Astrology. Editions of the Arabic and Latin Texts and an English Translation*, ed. Charles Burnett, Keiji Yamamoto, Michio Yano (London: Warburg Institute, 2004).

Boudet, Jean-Patrice, 'La papauté d'Avignon et l'astrologie', in *Fin du monde et signes des temps. Visionnaires et prophètes en France méridionale (fin* XIIIe *- début* XVe *siècle), Cahiers de Fanjeaux*, 27 (1992), 257–93.

———, 'Ptolémée dans l'Occident médiéval: roi, savant et philosophe', *The Medieval Legends of Philosophers and Scholars, Micrologus*, XXI (Florence: Edizioni del Galluzzo 2013), pp. 193–217.

———, 'Jean des Murs, Astrologer', *Erudition and the Republic of Letters*, 4 (2019), 123–45.

———, 'The Latin Medieval Versions of Pseudo-Ptolemy's *Centiloquium*', in *Ptolemy's Science of the Stars in the Middle Ages*, ed. David Juste and others (Turnhout: Brepols, 2020), pp. 283–304.

———, *Astrologie et politique entre Moyen Âge et Renaissance* ((Florence: Sismel-Edizione dell Galluzzo, 2020).

Boureau, Alain, *L'Empire du livre. Pour une histoire du savoir scolastique (1200–1380)* (Paris: Les Belles Lettres, 2007).

Burnett, Charles, 'Astrology for the Doctor in a Work addressed to Robert, Earl of Leicester', in *De l'homme, de la nature et du monde. Mélanges d'histoire des sciences médiévales offerts à Danielle Jacquart* (Geneva: Droz, 2019), pp. 179–96.

Carey, Hilary M., *Courting Disaster: Astrology at the English Court and Universities in the late Middle Ages* (London: MacMillan, 1992).

———, 'Henry VII's Book of Astrology and the Tudor Renaissance', *Renaissance Quaterly*, 65 (2012), 661–710.

Chabás, José, *Computational Astronomy in the Middle Ages. Sets of Astronomical Tables in Latin* (Madrid: Consejo Superior de Investigaciones Científicas, 2019).

———, and Bernard R. Goldstein, 'The Master and the Disciple: the Almanac of John of Lignères and the Ephemerides of John of Saxony', *Journal for the History of Astronomy*, 50 (2019), 82–96.

———, and ———, *A Survey of European Astronomical Tables in the Late Middle Ages* (Leiden: Brill, 2012).

———, and ———, 'The *Almanac* of Jacob ben Makhir', in *Editing and Analysing Astronomical Tables: Towards a Digital Information System for the History of Astral Sciences*, ed. Matthieu Husson, Clemency Montelle and Benno van Dalen (Turnhout: Brepols, 2022), pp. 53-78..

Charon, Pierre, 'Note sur Geoffroy de Meaux, médecin et astrologue du XIVe siècle', *Bulletin de la société archéologique de Meaux et sa région*, 7 (2010), 101–14.

Consideraciones temperiei pro 7 annis per Magistrum Willelmum Merle: The Earliest Known Journal of the Weather, 1337–1344, ed. by G. J. Symons (London: E. Stanford, 1891).

Darwall-Smith, Robin, *A History of University College, Oxford* (Oxford: Oxford University Press, 2008).

Derolez, Albert, *The Paleography of Gothic Manuscript Books: From the Twelfth to the Early Sixteenth Century* (Cambridge: Cambridge University Press, 2003).

Draelants, Isabelle, *Le Liber de virtutibus herbarum, lapidum et animalium (Liber aggregationis). Un texte à succès attribué à Albert le Grand* (Florence: Sismel-Edizione dell Galluzzo, 2007).

Emden, Alfred B., *A Biographical Register of the University of Oxford to A. D. 1500*, 3 vols (Oxford: Clarendon Press, 1957–1959).

Eschuid [*sic*], Johannes, *Summa astrologiæ iudicialis de accidentibus mundi* (Venice: Johannes Lucillus Santritter, 1489).

Fournier, Gilbert, 'Les conditions d'une réussite: le livre et la *memoria* au collège de Sorbonne (XIIIe-XVe siècle)', in *Scriptoria e biblioteche nel basso medioevo (secoli XII–XV), atti del LI Convegno storico internazionale, Todi, 12–15 ottobre 2014* (Spoleto: Fondazione Centro Italiano di Studi sull'Alto Medioevo, 2015), pp. 475–502.

Garrod, Heathcote William, 'The Library Regulations of A Medieval College', *The Library*, 3 (1927), 312–35.

———, 'An Indenture between William Reed, Bishop of Chichester, and John Bloxham and Henry Stapilton, Fellows of Merton College, Oxford, London 22 October 1373', ed. J. Roger L. Highfield, *Bodleian Library Record*, 10 (1982), 9–19.

Goldstein, Bernard R., and David Pingree, 'Levi ben Gerson's Prognostication for the Conjunction of 1345', *Transactions of the American Philosophical Society*, 80 (1990), 1–60.

Gorochov, Nathalie, 'La Mémoire des morts dans l'université de Paris au XIIIe siècle', in *Memoria, communitas, civitas: Mémoires et consciences urbaines en Occident à la fin du Moyen Âge*, ed. Hanno Brand, Pierre Monnet, Martial Staub (Ostfildern: J. Thorbecke, 2003), pp. 117–29.

Harper, Richard, 'The Astronomical Tables of William Rede', *Isis*, 66 (1975), 369–78.

Highfield, J. Roger L., *The Early Rolls of Merton College, Oxford, with an Appendix of Thirteenth-Century Oxford Charters* (Oxford: Clarendon Press, 1964).

———, and Geoffrey H Martin, *A History of Merton College* (Oxford: Oxford University Press, 1997).

Jacquart, Danielle 'Médecine et astrologie à Paris dans la première moitié du XIVe siècle', in *Filosofia, scienza e astrologia nel Trecento europeo*, ed. by G. Federici Vescovini, Francesco Barocelli (Padova: Il Poligrafo, 1992), pp. 121–34

Jenks, Stuart, 'Astrometeorology in the Middle Ages', *Isis*, 74 (1983), 185–210.

Johnson, John William S., 'Les *Experimenta duodecim Johannis Paulini*', *Bulletin de la Société française d'histoire de la médecine*, 12 (1913), 257–67.

Juste, David, 'Simon Bredon, *Commentary on the Almagest*' (updated: 29.10.2019), *Ptolemaeus Arabus et Latinus. Works*, URL: http://ptolemaeus.badw.de/work/76.

———, 'MS Oxford, Bodleian Library, Digby 179' (updated: 11.12.2017), *Ptolemaeus Arabus et Latinus. Manuscripts*, URL: http://ptolemaeus.badw.de/ms/232.

———, 'MS London, British Library, Egerton 889' (updated: 21.06.2018), *Ptolemaeus Arabus et Latinus. Manuscripts*, URL: http://ptolemaeus.badw.de/ms/50.

Ker, Neil Ripley, 'The Books of Philosophy Distributed at Merton College in 1372 and 1375', in *Books, Collectors and Libraries: Studies in the Medieval Heritage*, ed. Andrew Watson (London: Hambledon Press, 1985), pp. 331–78.

Kibre, Pearl, 'Lewis of Caerleon, Doctor of Medicine, Astronomer, and Mathematician (d. 1494)', *Isis*, 43 (1952), 100–08.

Kunitzsch, Paul, *Mittelalterliche astronomisch-astrologische Glossare mit arabischen Fachausdrücken* (Munich: Bayerische Akademie der Wissenschaften, 1977).

Martin, Geoffrey H., 'Merton, Walter of (*c.* 1205–77), Administrator, Bishop of Rochester, and Founder of Merton College, Oxford', in the on-line *Oxford Dictionary of National Biography*, Oxford University Press, 2004 [accessed: 30.01.2020].

Mews, Constant, and John N. Crossley (eds), *Communities of Learning: Networks and the Shaping of Intellectual Identity in Europe, 1100–1500* (Turnhout: Brepols, 2011).

———, 'Communities of Learning and the Dream of Synthesis: The Schools and Colleges of Thirteenth-Century Paris', in Mews and John N. Crossley (eds), *Communities of Learning: Networks and the Shaping of Intellectual Identity in Europe, 1100–1500* (Turnhout: Brepols, 2011), pp. 109–35.

North, J. D, *Horoscopes and History* (London: The Warburg Institute, 1986).

———, 'The Alfonsine Tables in England', in *ΠΡΙΣΜΑΤΑ: Naturwissenschaftsgeschichtliche Studien; Festschrift für Willy Hartner*, ed. Y. Maeyama and W. G. Saltzer (Wiesbaden: F. Steiner, 1977), pp. 269–301.

———, 'Medieval Star Catalogues and the Movement of the Eighth Sphere', *Archives internationales d'histoire des sciences*, 20 (1967), 71–83.

———, *Richard of Wallingford: An Edition of his writings*, 3 vols (Oxford: Clarendon Press, 1976).

———, 'Astrology and the Fortune of Churches', *Centaurus*, 24 (1980), 181–211.

Nothaft, C. Philipp E, 'Critical Analysis of the Alfonsine Tables in the Fourteenth Century: The Parisian *Expositio Tabularum Alfonsii of 1347*', *Journal for the History of Astronomy*, 46 (2015), 76–99.

———, 'Criticism of Trepidation Models and Advocacy of Uniform Precession in Medieval Latin Astronomy', *Archives for History of Exact Science*, 71 (2017), 211–44.

———, 'Glorious Science or 'Dead Dog'? Jean de Jandun and the Quarrel over Astrology in Fourteenth-Century Paris', *Vivarium*, 57 (2019), 51–101.

The Opus Majus of Roger Bacon, 3 vols, ed. John Henry Bridges (Oxford: Clarendon Press, 1897–1900; repr. Cambridge: Cambridge University Press, 2010).

Pedersen, Fritz S, *The Toledan Tables: A Review of the Manuscripts and the Textual Versions with an Edition*, 4 vols (Copenhagen: C.A. Reitzels Forlag, 2002).

Plassard, Joël, *Projets de réforme du calendrier à Paris au début du XIVe siècle: Textes édités et commentés* (unpublished thesis, Paris, École des chartes, 1975).

Poulle, Emmanuel, *Les Tables alphonsines avec les canons de Jean de Saxe* (Paris: CNRS éditions, 1984).

Powicke, Frederick M., *The Medieval Books of Merton College* (Oxford: Clarendon Press, 1931).

Pruckner, Hubert, *Studien zu den astrologischen Schriften des Heinrich von Langenstein* (Leipzig: Teubner, 1933).

Ptolemy, *Quadripartitum, Centiloquium* (Venice: Bonetus Locatellus, 1493).

Registrum epistolarum Fratris Johannis Peckham Archiepiscopi Cantuariensis, 3 vols, ed. Charles T. Martin (London: Longman, 1882–85).

Snedegar, Keith, 'John Ashenden and the *Scientia Astrorum Mertonensis*', unpublished PhD thesis, University of Oxford, 1988.

———, 'The Works and Days of Simon Bredon. A Fourteenth-Century Astronomer and Physician', in *Between Demonstration and Imagination: Essays in the History of Science and Philosophy, presented to John D. North*, ed. by Lodi Nauta and Arjo Vanderjagt (Leiden: Brill, 1999), pp. 285–309.

———, 'Ashenden, John (d. in or before 1368?), Astrologer', in the on-line *Oxford Dictionary of National Biography*, Oxford University Press, 2004 [accessed: 05.10.2020].

Schabel Christopher and Fritz S. Pedersen, 'Miraculous, Natural, or Jewish Conspiracy? Pierre de Ceffons, Question on the Black Death, with Astrological Predictions by Gersonides and Jean de Murs/Firmin de Beauval', *Recherches de Théologie et Philosophie médiévales*, 81 (2014), 137–79.

Stock, Brian, *The Implications of Literacy: Written Language and Models of Interpretation in the Eleventh and Twelfth Centuries* (Princeton: Princeton University Press, 1983).

Stratford, Jenny, Webber, Teresa, 'Bishops and kings: private book collections in medieval England', in *The Cambridge history of libraries in Britain and Ireland*, ed. by Elisabeth Leedham-Green and Teresa Webber (Cambridge: Cambridge University Press, 2006), pp. 178–217.

Sylla, Edith D., *The Oxford Calculators and the Mathematics of Motion, 1320–1350: Physics and Measurements by Latitudes* (New York: Garland, 1991).

———, 'Bradwardine, Thomas', in the on-line *Complete Dictionary of Scientific Biography*, Charles Scribner's Sons, 2008 [accessed: 03.02.2020].

Synan, Edward A., 'Richard of Campsall, an English Theologian of the Fourteenth Century', *Mediaeval Studies*, 14 (1952), 1–8.

Tabule astronomice illustrissimi Alfontii regis Castelle (Venice: Ratdolt, 1483).

Talbot, Charles H, 'Simon Bredon (*c.* 1300–72): Physician, Mathematician and Astronomer', *British Journal for the History of Science*, 1 (1962), 19–30.

Thomson, Rodney M., *The University and College Libraries of Oxford* (London: The British Library, 2015).

———, 'William Reed, Bishop of Chichester (d. 1385) — Bibliophile?', in *The Study of Medieval Manuscripts of England: Festschrift in Honor of Richard W. Pfaff*, ed. George H. Brown and Linda E. Voigts (Tempe-Turnhout: Arizona for Medieval and Renaissance Studies-Brepols, 2010), pp. 281–93.

Thomson, S, 'Fitzjames, Richard (d. 1522), Bishop of London', in the on-line *Oxford Dictionary of National Biography*, Oxford University Press, 2004 .

Thorndike, Lynn, *A History of Magic and Experimental Science*, 8 vols (New York: Columbia University Press, 1923–58).

———, *Latin Treatises on Comets, 1238–1368 A.D.* (Chicago: Chicago University Press, 1950).

———, and Pearl Kibre, *A Catalogue of Incipits of Mediaeval Scientific Writings in Latin* (Cambridge: The Medieval Academy of America, 1963).

Vuillemin-Diem Gudrun, and Carlos Steel, *Ptolemy's Tetrabiblos in the Translation of William of Moerbeke, Claudii Ptolemaei liber iudicialium* (Leuven: Leuven University Press, 2015)

Watson Andrew G., 'A Merton College Manuscript Reconstructed: Harley 625; Digby 178, ff. 1–14, 88–115; Cotton Tiberius B. IX, ff. 1–4, 225–35', *Bodleian Library Record*, 9 (1976), 207–17.

Weisheipl, James A., 'Ockham and the Mertonians', in *The History of the University of Oxford. I: The Early Oxford Schools*, ed. J. I. Catto (Oxford: Oxford University Press, 1984), pp. 607–58.

———, 'Repertorium Mertonense', *Mediaeval Studies*, 31 (1969), 174–224.

Wickersheimer, Ernest, *Dictionnaire biographique des médecins en France au Moyen Âge*, new ed. (Geneva: Droz, 1979).

Zepeda, Henry, '*The Medieval Latin Transmission of the Menelaus Theorem*', unpublished PhD Dissertation, University of Oklahoma, 2013.

———, *The First Latin Treatise on Ptolemy's Astronomy: The Almagesti minor (c. 1200)* (Turnhout: Brepols, 2018).

RICHARD L. KREMER

Exploring a Late-Fifteenth-Century Astrologer's Toolbox: British Library Add MS 34603

Introduction: What were the Alfonsine Tables?

The corpus of Alfonsine manuscripts, as currently defined by the ALFA Project, contains roughly 170 separate copies of the Parisian Alfonsine Tables. Perhaps half of these copies are what I refer to as 'well identified', meaning they include a discrete set of individual tables explicitly delimited by the original scribes. Of the sixty-two copies I have seen, forty-two are defined as 'Alfonsine' in the incipits or explicits; only fifteen copies include both an incipit and an explicit. For example, one of the earliest copies, Erfurt CA 2° 360, folios 1–23, dated 1360 in its explicit, begins and ends with the same phrase: 'tabulae illustris principis Alfonsii regis Castelle'. All of the twenty-seven explicits I have found end the Parisian Alfonsine Tables at the conclusion of the Mercury equations. These, plus the fifteen copies with both incipits and explicits, identically delimit the Parisian Alfonsine Tables to three subsets of tables: calendrical, defining the number of days from various epochs; mean motions and radices for those epochs; and equations for the eighth sphere, apogees, luminaries, and planets. Hence, roughly half of the manuscripts I have examined present a well-identified version of what their scribes called the *tabulae [...] Alfonsii*.

In his 1984 edition, translation, and commentary of *Les Tables alphonsines avec les canons de Jean de Saxe*, Emmanuel Poulle reprinted twenty-seven canons by John of Saxony and twenty related tables from the 1483 *editio princeps*, an imprint entitled by its printer Erhard Radolt as *Alfontii regis castellae illustrissimi caelestium motuum tabulae* (GW 1257).[1] Poulle defined the twenty tables, which include the exact three subsets I described

* This work was supported by the ERC project ALFA: Shaping a European scientific scene, Alfonsine astronomy, CoG 723085, PI Matthieu Husson. For their criticisms and suggestions, I thank Jean-Patrice Boudet, Jonathan Green, Alena Hadravová, Stephan Heilen, Nicholas Jacobson, David Juste, Laure Miolo, C.P.E. Nothaft, and participants of the SAW Workshop at Paris Diderot University, organized by Karine Chemla, Christine Proust, and Agathe Keller, where I first discussed this manuscript in November of 2013.

Richard L. Kremer • Dartmouth College

Alfonsine Astronomy: The Written Record, ed. by Richard L. KREMER, Matthieu HUSSON, José CHABÁS, Turnhout, 2022 (*Alfonsine Astronomy: Studies and Sources*, 1), pp. 107-141
© BREPOLS PUBLISHERS 10.1484/M.ALFA.5.124925

above, as the 'basic group' or the 'solid core' of the 'tables properly Alfonsine'. Tables in the 1483 edition that are not explicitly mentioned by John's canons were relegated by Poulle to the *membra adjuncta*: collateral material including tables for velocities of the luminaries, planetary latitudes, fixed stars, ascensions of the signs (i.e. trigonometric material), eclipses, and geographic locations. Of the seventy-four folios of tables printed in the *editio princeps*, Poulle included exactly half, or thirty-seven folios among the 'tables properly Alfonsine'.[2]

Poulle's decision to define the 'proper' Alfonsine Tables by John of Saxony's canon might offer insight into how Parisian astronomers in the 1320s viewed the boundaries of their tables. However, as my somewhat random survey of the extant manuscript copies illustrates, later scribes (presumably many scribes were also users) drew those boundaries differently or even eschewed labels or boundaries altogether. For these scribes, the quire or the codex was the working unit for practicing Alfonsine astronomy, not the named 'set' of tables. For example, of the thirty-three copies without explicits that I have examined, nine present, without a break, additional tables after Mercury's equations, often an equation of time, velocity, or eclipse tables from John of Lignères' Tables of 1322. These nine scribes chose not to define the end of the table set, even though six of them opened the set with the Alfonsine incipit.

I shall refrain from drawing firm conclusions until more of the 170 manuscripts have been examined. But this preliminary survey suggests that, for many Alfonsine astronomers, the contents of the 'Alfonsine Tables' were not canonically fixed, but pragmatically flexible.[3] Different combinations of individual tables could be copied, in different sequences; a different (or no) title could be given to the same individual table; incipits and explicits referring to King Alfonso were inconsistently inserted. Such variety in scribal copying suggests that many Alfonsine practitioners took a pragmatic, modular approach to their manuscript tables. Rather than collecting rigidly defined sets of tables, they freely arranged different tabular materials, often copied by quires and only later bound into codices. We might thus think of the bound manuscript codex as a particular astronomer's toolbox, containing a plethora of individual tables that could be used separately or in combination, just as a medieval craftsman's toolbox might have contained individual hand tools for shaping wood, cutting stone, or working leather. Thus, if we allow the scribes and manuscript culture to define the 'Alfonsine Tables', we will undoubtedly learn more about Alfonsine astronomy over the *longue durée* than we would if we defined those tables simply by John of Saxony's canon.[4]

---

1 Gesamtkatalog der Wiegendrucke, accessible at www.gesamtkatalogderwiegendrucke.de, lists 127 extant copies of this edition. Ratdolt's mathematical imprints tend to be well preserved. His 1482 Euclid exists in 338 copies, his 1482 Sacrobosco in 132 copies, and his 1488 *Astrolabium planum* in ninety-nine copies.

2 Emmanuel Poulle, ed., *Les Tables alphonsines avec les canons de Jean de Saxe: Édition, traduction et commentaire* (Paris: Éditions du Centre national de la recherche scientifique, 1984), pp. 6–26. Interestingly, the table of 'Radices motuum ad eram incarnationis ad loca diversa' (radices for 0 to 30 degrees east of Toledo, at one-degree intervals of longitude), included in the *editio princeps* and Poulle (Table 10), is not present in any of the sixty-two manuscripts I have examined.

3 Poulle, *Tables alphonsines*, p. 16 also recognized the variety among manuscript witnesses.

4 For another perspective on the difficulty of defining 'sets' of astronomical tables, see José Chabás, *Computational Astronomy in the Middle Ages: Sets of Astronomical Tables* (Madrid: Consejo Superior de Investigaciones Científicas, 2019), pp. 14–15.

One such toolbox that places Alfonsine material in a unique context is a fifteenth-century codex purchased in 1894 by the British Museum and kept there ever since (shelfmark: Add MS 34603). The current British Library description, repeating the late-nineteenth-century catalogue, describes the codex as an 'astrological manual' of 305 leaves, 'compiled' *c.* 1500 by Marcus Schinnagel and consisting of eight parts:[5]

- Rules for constructing tables: seventeen leaves
- Tabula [de] vrina non visa per me magistrum Marcum Schynagel … 1500: two leaves
- Signs of the zodiac with associated diseases: seven leaves
- Symbolical drawings, perhaps representing constellations: twenty-nine leaves
- Medical 'tract in four chapters' on critical days: seven leaves
- Hypocratis libellus de medicorum astrologia … a Petro de Albano [sic] in latinum traductus: six leaves
- A large collection of tables for astrological use: 219 leaves
- Another copy of 'De medicorum astrologia', unfinished: two leaves

In 1953, art historians Fritz Saxl and Hans Meier presented a more codicologically inflected description of Add MS 34603, identifying several watermarks (Bavarian paper), and, given their interest in astrological illumination, tersely describing each of the images in the twenty-nine leaves of 'symbolical drawings'. They reproduced three of these images but cited no comparative material and offered no explanation for the iconography. For example, folio 33^r illustrates 'an enthroned king or emperor under a canopy, holding a chain in each hand, under the left one kneels a knight, under the right a mountain; three arms on the pedestal'. Furthermore, folio 33^v shows 'a naked man standing between two dogs'. The second part of the manuscript, they noted, contains, 'only tables that are listed in the [British] catalogue of additions'. Saxl and Meier did not inquire about the significance of this combination of material in the codex.[6]

More recently, another art historian, Heidrun Franz, connected Add MS 34603 to a large, multi-winged painting now in Stuttgart, dated 1489 and completely covered with astronomical and astrological tables and several short texts (Fig. 1). Franz found that several texts and tables copied in the manuscript are also featured on the polyptych. But she could not deduce an 'exact relationship' between the painted panels and the codex, referring to the latter as an 'astrological manual' that 'apparently served as a personal workbook' for Schinnagel.[7]

In sum, as far as I am aware, no previous scholarship has sought to understand the Alfonsine toolbox represented by Add MS 34603. My goal here is to consider what a

5 http://searcharchives.bl.uk/IAMS_VU2:IAMS032-002087557 (accessed 8 May 2020); *Catalogue of Additions to the Manuscripts in the British Museum in the Years 1894–1899* (London: Clowes and Sons, Ltd., 1901), pp. 15–16.

6 Fritz Saxl and Hans Meier, *Verzeichnis astrologischer und mythologischer illustrierter Handschriften des lateinischen Mittelalters, Bd. 3, Handschriften in englischen Bibliotheken*, ed. Harry Bober, 2 vols (London: Warburg Institute, 1953), II, pp. 74–79, Tafel LXXXVI.

7 Heidrun Franz, *Das Hauptwerk des Astrologen Marcus Schinnagel von 1489: Alltagsmanagement und Zukunftsdeutung an der Schwelle zur Neuzeit* (Hamburg: Kovac, 2014), pp. 109, 290, 295. Cf. Richard L. Kremer, 'Marcus Schinnagel's Winged Polyptych of 1489: Astronomical Computation in a Liturgical Format', *Journal for the History of Astronomy*, 43 (2012), 321–45. Franz does not elaborate what she means by 'workbook' but obviously is suggesting that the codex was not intended to be 'read' as a 'text'.

focused analysis of a single manuscript can tell us about the 'Alfonsine Tables' and the practice of Alfonsine astronomy around 1500 in the southwest German region of Swabia. We begin by examining the scribe and presumably the compiler of the codex, Markus Schinnagel, and the manuscript's relation to another artifact he designed, the polyptych now located at the Württembergisches Landesmuseum in Stuttgart. We then consider the codex as a physical object and review its contents as a toolbox. Finally, we will speculate about the identity of its patron or intended user, a puzzle to which the manuscript itself offers few clues.

1. Context for Add MS 34603 and its compiler

Schinnagel signed the Stuttgart polyptych prominently across the central panel: 'Ego magister marcus schynagel alme universitatis crackoviensis / 1489'. His name also appears several times in Add MS 34603, folios 21^{v}–22^{r}, 303^{v}: Tabula [de] urina non visa, per me magistrum Marcum Schynagel alme Vniversitatis Crakouiensis tunc temporis plebanus in landtsperg anno 1500'; or 'Tabula de urina non visa per me magistrum Marcum Schynagel'. Yet our *Magister* was more widely known than these brief explicits might suggest.

Additional details of Schinnagel's biography recently have been summarized by Franz and historian Klaus Graf.[8] In the early fifteenth century, a Schinnagel family is documented in the town of Waiblingen in Württemberg (15 km north of Stuttgart). But our Marcus was apparently born circa 1448 in Košice, about 100 km northwest of Budapest. He matriculated at the Cracow University in 1466, became Bachelor of Arts in 1470. The university's archives do not record him becoming a master, but his earliest extant imprint, a broadside almanac for 1487, is signed, 'in arte astronomia magistrum'. By 1490, he referred to himself as 'der freyen kunsten vnnd in sonder der astronomy doctor'.

At the end of the fifteenth century, Schinnagel became known through his fifteen incunabula editions of broadside almanacs or multi-leaved annual practica, both extremely popular genres of early printing.[9] As can be seen from Table 1, most of his imprints appeared in Swabia, a territory extending from Augsburg to Basel to Strasbourg to Ulm. These calendrical and medical (bloodletting) tracts, offering predictions and astrological advice for a single year, were pragmatic commodities, designed for use in a given year and then to be replaced by the following year's edition. Most survive in very few copies. Schinnagel's imprints are typical of the genre.

8 Klaus Graf, 'Marcus Schinnagel, ein Astrologe in der Zeit Maximilians I., Schöpfer des astronomisch-astrologischen Kompendiums aus Petershausen', http://frueheneuzeit.hypotheses.org/1615, 2014 (accessed 8 May 2020).

9 For an introduction to this periodical literature of annual calendars with astrological/medical predictions, see Richard L. Kremer, 'Incunable Almanacs and Practica as Practical Knowledge Produced in Trading Zones', in *The Structures of Practical Knowledge*, ed. Matteo Valleriani (Cham: Springer, 2017), pp. 333–69.

Table 1. Schinnagel's extant printed editions, as documented by the *Gesamtkatalog der Wiegendrucke* (GW).

Date	GW	Genre (Latin or German)	Location	Printer	Extant copies
1487	M40840	Almanac (L)	Augsburg	Ratdolt	4
1487	M40839	Practica (L)	Strassburg	Grüninger	1
1487	M40842	Almanac (G)	Augsburg	Ratdolt	4
1488	M40843	Almanac (G)	Augsburg	Ratdolt	3
1489	M4084320	Practica (G)	Ulm	Zainer d.Ä	8
1489	M4084330	Practica (G)	Ulm	Zainer d.Ä	1
1490	GW 1447	Almanac (G)	Basel	Furter	4
1491	M40847	Practica (L)	Basel	Amerbach	1
1491	M40848	Practica (L)	Basel	Furter	2
1491	M40846	Practica (G)	Ulm	Zainer d. Ä	2
1493	M40850	Practica (L)	Vienna	Winterburg	1
1493	M40849	Practica (L)	Leipzig	Landsberg	2
1500	M4084020	Practica (G)	Freiburg	Riederer	1
1500	M4085030	Practica (G)	Ulm	Zainer d.J.	2
1500	M35180	Practica (G)	Tübingen	Otmar	1

His first known broadside combines the genres of practica and almanac. This sheet, one of the most complex of the 500 extant incunable almanac editions and surely the finest ever issued by the well-known Augsburg printer, Erhard Ratdolt, must have launched Schinnagel's astronomical career in Swabia. In his 1491 practica, Schinnagel further polished his local image:

Für war den spruch hat gemacht	Indeed, Master Marcus Schinnagel,
gepracticiert vnd auß grund erdacht	as he is known, made the text,
Maister marx schinagel ist er genant	prognosticated and thoroughly conceived it.
jn schwaben wol erkant	Well known in Swabia,
Ain astronomum thüt er sich nennen	he calls himself an astronomer;
ain astrologiam gar wol erkennen	he is well recognized as an astrologer;
Ain arismetricus auch dabey	he is also a mathematician,
mit seinen kunsten ist er frey.[10]	who is generous with his arts.

Schinnagel's astrological activities bore consequences. After predicting a war with the Swiss and cold weather that would kill the grape harvest, he was driven out of Constance in 1490 by 'farmers and the common people'. Seeking patronage, he dedicated his 1491 practica to the outgoing Archduke of Tirol, Sigismund, and to the incoming King Maximilian I, apparently persuading the king to grant him a prebend in the small town of Sulzberg near Kempten, about sixty kilometres east of Constance. In 1494 while visiting Kempten,

10 Marcus Schinnagel, *Prognostikon auf das Jahr 1491* (Ulm: Johann Zainer d.Ä, 1491).

Figure 1. Schinnagel's astronomical compendium, 330 × 140 cm, Stuttgart, Landesmuseum Württemberg Inv. 1995–323. P. Frankenstein, H. Zwietasch, Creative Commons.

Maximilian named Schinnagel, now described as *astronomicae scientiae peritia celebrem*, one of his court chaplains, a minor post with unspecified responsibilities. In 1495, Schinnagel drafted a horoscope and introduction to astrology for the Bavarian Duke Albrecht IV, preserved in autograph in HAB Cod. Guelf. 22.1 Aug. 4°.[11] Several years later, Albrecht appointed Schinnagel to another church post in Landsberg, twenty kilometres south of Augsburg. In 1496, Schinnagel corresponded with Duke René II of Lorraine, offering astrological predictions about the death of the King of Naples, a territory long claimed by René. Clearly, Schinnagel was an influential astrologer and calendar maker in Swabia. Indeed, a 1543 practica by the Zürich city physician vaguely refers to what had happened to Schinnagel after his predictions came true, apparently a reference to his banishment from Constance some fifty years earlier.[12]

We learn even more about Schinnagel from the Stuttgart polyptych, a complex set of painted panels covered in calendrical, astronomical, and astrological texts, tables, diagrams, images, and even instruments (now lacking the moveable vovelles).[13] The central panel measures 140 × 130 cm; when both wings are opened, the overall width extends more than 330 cm (see Fig. 1). Schinnagel's polyptych is undoubtedly the largest physical object in the Alfonsine corpus. Since the details of the polyptych have recently been studied, here we need only consider some links between the polyptych and BL Add MS 34603.

11 Like Add MS 34603, this codex also contains many unwritten folios.

12 Schinnagel's 1493 practica is dedicated to King Johann I Albert of Poland, a gesture that apparently did not result in patronage. Cf. Schinnagel to Duke René II, 3 March 1496, Paris, BnF Dept. of Manuscripts, Lorraine 7, f. 203^{r} (autograph); Christophor Clauser, *Practica Tütsch vff das MDXLIII Jar* (Zurich: Christoph Froschauer d.Ä, 1543), sig. a2^{r-v}; Franz, *Schinnagel*, pp. 57–83; Graf, 'Marcus Schinnagel'; William Monter, *A Bewitched Duchy: Lorraine and Its Dukes, 1477–1736* (Geneva: Droz, 2007), pp. 21–37. I thank Jonathan Green for helping me transcribe the letter to René.

13 Kremer, 'Schinnagel's Polyptych'; Franz, *Schinnagel*.

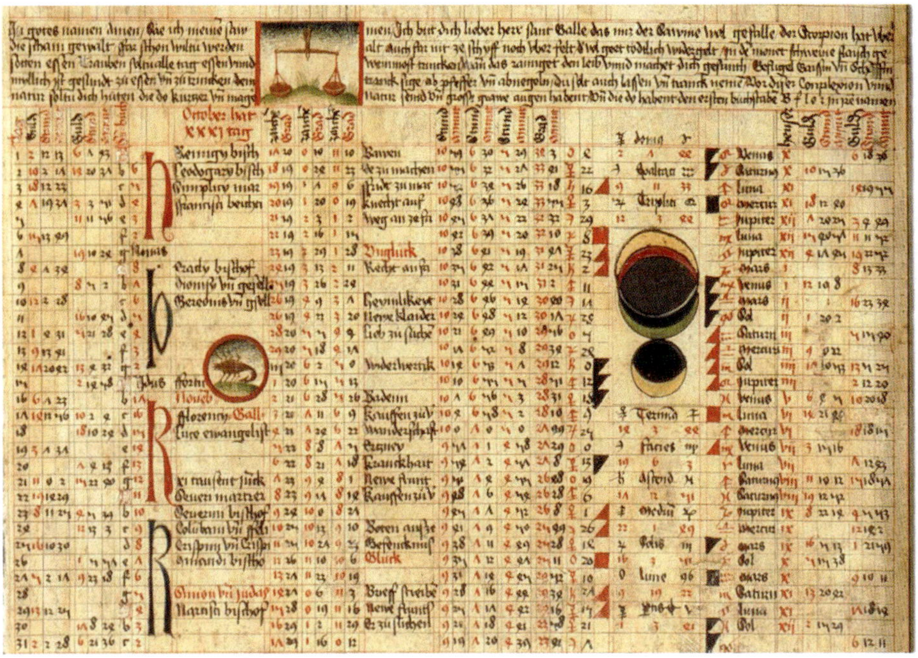

Figure 2. Schinnagel's astronomical compendium, panel 3, detail October. Stuttgart, Landesmuseum Württemberg Inv. 1995–323. Author's photograph.

Both the polyptych and manuscript present many texts and tables copied verbatim from incunable imprints. And many of the individual tables found in both are identical in format and content:[14]

- House tables for latitude of 45° (panel 2a): Table 30
- Astrological triplicities, terms, and decans of the zodiacal signs (panel 2a): Table 30
- Mean motions, by days, for one year (panel 4a): Table 47
- Mean motion of the planetary apogees, collected years (panel 4a): Table 38
- Planetary equations, to degrees (panel 5a):Table 48
- Geographical places (panel 5b): Table 35
- Radices by year, for 1489–1526 (panel 4a): Table 47, for 1441–1561
- Moveable feasts (panel 2a): Table 32 (the manuscript includes more data)

The iconography of coloured spheres and sickle-moons, found in some unfinished, coloured miniature pages of the codex, also appears on the central calendrical panel of the polyptych (cf. Figs 2 and 3). None of the manuscript images are exactly reproduced, however, in the polyptych.

Nonetheless, the London codex and Stuttgart compendium are closely related in content, design, and iconography. The identity of the patron for the Stuttgart polyptych remains unknown. Franz speculated that a wealthy diplomat and advisor at the Stuttgart

14 I denote panels on the polyptych as designated by Franz and tables in the codex as listed in Appendix 2.

court, Baron Hans von Reischach zum Reichenstein (d. 1492), may have commissioned the compendium. But documentation surrounding this official is slight and none indicates any interest in astrology or astronomy.[15] Furthermore, since the London codex is clean (i.e. not obviously used), we might speculate that Schinnagel prepared it for the patron of the Stuttgart piece, who then never used it seriously, or perhaps Schinnagel made the codex for a potential second patron, after having completed the Stuttgart piece. The unfinished miniatures in the manuscript would have allowed the future owner to help design the iconography for his yet-to-be-constructed compendium, or perhaps the patron simply wanted a codex and not a massive polpytych. As we lack sources, we can only speculate about such patrons.

In any case, to prepare all his printed annual astrological predictions, Schinnagel must have computed many horoscopes for the 'revolutions of the year' (the Sun's entry into the four tropical points), the two-dozen true syzygies each year, and the eclipses and auspicious planetary aspects. Such a working professional would have needed a toolbox of astronomical and astrological implements to enable him to perform the many computational and astrological steps involved in horoscope-making. Add MS 34603, we shall argue, was either Schinnagel's toolbox or a toolbox designed by him for a patron still unknown to us.

2. The codex as a physical object

Upon taking the manuscript codex to hand, we immediately notice two features not emphasized in the previous descriptions, namely the early binding and the considerable heft of the book, which is more than ten-centimetres thick. The 1901 catalogue reports that it is 'bound in white pigskin, tooled with a running-pattern of deer'. Subsequently this rolled stamp has been identified as 'Hirsch-Rolle 3', which was used by the Ingolstadt bookbinder, Heinrich Trencker, and is found today on more than fifty bindings in south German libraries, usually on printed books dating from 1485 to 1517.[16] The current physical configuration of the codex was thus fixed either contemporaneously with, or shortly after, the quires were written. A printed ex-libris pasted on the inside front cover indicates that in 1588 the codex was donated to the library of the Augustian Chorherrenstift Heiligkreuz in Augsburg by the provost of that city's cathedral.[17] Our codex apparently originated in Swabia and spent its first century in that region.

Secondly, Add MS 34603 contains 440 leaves of rather thick paper, of which 134, or nearly one third, carry no text, drawings, or tables. Our codex appears to be unfinished.

---

15 Franz, *Schinnagel*, pp. 88–94.

16 Ernst Kyriß, *Verzierte gotische Einbände im alten deutschen Sprachgebiet*, 4 vols (Stuttgart: Max Hettler Verlag, 1951–58), #50.

17 For the ex-libris, see Saxl and Meier, *Verzeichnis*, I, p. 74. The provost, Wolfgang Andreas Rem (1511–88), was a well-known collector of books, mathematical instruments, sundials, and globes. Upon his death, more than 1400 books and instruments went to the monastery library. For a manuscript copy of his will, see the verso of the broadside 'Wahre Contrafactur des … Wolfgang Andreas Rhem' (Augsburg, 1588), Munich, BSB 2 Inc.c.a. 1320#Vorderdeckel. The monastery library was 'secularized' in 1802; some manuscripts went to the Bavarian Court Library in Munich, others to the Augsburg City Library. Our manuscript somehow passed into the hands of the Edinburgh bookseller and founder of that city's Bibliographical Society, George Pyper Johnston, who in 1894 sold it to the British Museum. Stephan Kellner and Annemarie Spethmann, *Historische Kataloge der bayerischen Staatsbibliohek München, Münchner Hofbibliothek und andere Provenienzen* (Wiesbaden: Harrassowitz Verlag, 1996), pp. 137–38.

Perhaps the scribe or the early owner who bound it intended to add additional material. Given the cost of writing support in medieval Europe, it is unusual to find a manuscript of this size containing so many blank leaves. Whoever bound Add MS 34603 apparently wanted to leave space in the toolbox so that more implements could be added after the quires were bound.[18]

Looking more deeply at the physical object, we notice the several foliations and two codicological parts. A later (probably nineteenth-century) foliation in pencil marks 306 text-bearing leaves in the manuscript (my citations refer to this pencil hand). The first codicological part consists of eight quires with a total of 106 leaves, spanning the pencil-foliated leaves 4 to 85; twenty-three leaves are blank and unfoliated. An early (probably sixteenth-century) hand in brown ink has for the second codicological part numbered quires 2, 3, 5, 6 …, 31, spanning pencil-foliated sheets 86–306. Another early foliation in brown ink numbers the leaves in the second codicological part from 1 (fourth leaf of quire 3) to 335 (final leaf of quire 30). The early numbered quires 1 and 4 have disappeared. The second codicological part thus contains 221 pencil-foliated leaves that are filled with text and 105 unfoliated leaves that, generally, are not blank but lined for a tabular text block that the scribe did not enter.[19]

An analysis of the paper confirms this division of the codex into two parts. The first codicological part contains five units of content, each found in its own quire or quires of distinct paper. The first text (Th/K 1313; see Appendix 1), a practical canon for astronomical and astrological computation in the hand of Schinnagel, fills quire 1 and is written on paper with three watermarks dated to Swabia in the late 1490s.[20] The second text, a medical tract *Tabula urina non visa*, also an autograph by Schinnagel and dated Landsberg (in Swabia) 1500, requires most of the second quire of paper dating to Augsburg 1500.[21] The third section, comprising the third quire, features a set of delicately painted miniatures, one per page, of human figures below, and coloured spheres and sickle-moons (vaguely resembling eclipse diagrams) above, surrounded by text blocks lined for thirty-one rows and obviously intended to serve as monthly calendars but bearing no text (see Fig. 3). The codex includes twenty-six such pages, perhaps intended to represent one month per opening (with February doubled for regular and leap years?). The paper of this quire is thicker and stiffer than the other paper in the codex and carries a watermark I have not been able to identify.[22] The fourth section, found in the fourth, fifth, and first five leaves of the sixth quires, also presents miniatures, one per page, much more crudely drawn in ink, each at the centre of the outlines of a square horoscope showing only the house divisions but no

18 The final quire, numbered '31' by the sixteenth-century hand, consists of six completely blank leaves; quires 27–30, of ten or twelve leaves each, have the text block outlined in brown ink, but contain no textual content except for two small tables in Schinnagel's hand, 'Tabula de urina non visa per me magister Marcum Schynagel' (f. 303^{v}), 'De significatione planetarum in signis secundum membra hominis' (f. 304^{r-v}), and a verbatim copy of the opening of the ps.-Hippocratic text (ff. 305^{v}–306^{r}) found in the first part of the codex (ff. 80^{v}–81^{r}).

19 Blank leaves, of course, allow users freedom to add any content; lined leaves might constrain users to insert only tabular content. We might also speculate that the prevalence of blank leaves in the codex implies that Schinnagel was a collector, always open to adding more tools to his toolbox.

20 Wasserzeichen-Informationssystem (https://www.wasserzeichen-online.de/wzis/index.php), AT3800-PO-53287 (Hohenstein 1496), AET800-PO-35291 (Augsburg 1499), DE7710-PO-15181 (Lindau 1497).

21 WZIS AT3800-PO-129445 (Augsburg 1500).

22 Six-pointed star in sickle-moon, on a two-contoured support extending above a 'Dreiburg'.

Figure 3. Unfinished calendar pages with coloured miniatures. London, British Library Add MS 34603, folios 34v–35r. Reproduced with permission of the British Library.

other data (see Fig. 4). Fifty-six horoscope images fill twenty-three folios. These quires, using the same paper found in quire 2 plus two other papers from 1498, must have been written contemporaneously with quire 2.[23] The final unit, containing short excerpts from recently printed texts on medical astrology, fills quire 7 and the first leaf of quire 8, both containing the two 1498 papers.[24] Presumably, each of these five text sections now in the first codicological unit was copied independently and may have circulated independently until the codex was bound sometime between 1500 and 1517.

The second codicological part of the codex is much more homogeneous. It contains paper with the same two watermarks across all twenty-nine quires and tables that freely range across quire boundaries. Clearly, these quires (originally thirty-one) were prepared and written as a single unit; they undoubtedly circulated in this format before being bound into the codex. The same hand, presumably Schinnagel's, wrote most, if not all, of the material in the second part. The paper is the same found in Schinnagel's autograph in the first quire of the first codicological part, as well as in one of the flyleaves before the opening quire.[25] Hence, it seems highly likely that Schinnagel himself had the codex bound, providing flyleaves from his own personal paper stock.

23 WZIS AT3800-PO-72661 (Freiburg i.Br. 1498), AT3800-PO-71980 (Freiburg i.Br. 1498).

24 WZIS DE2910-PO-72023 (no location, 1499) and type DE81200-PO-72238 (Cologne 1485).

25 WZIS AT3800-PO-53287, AT3800-PO-53291.

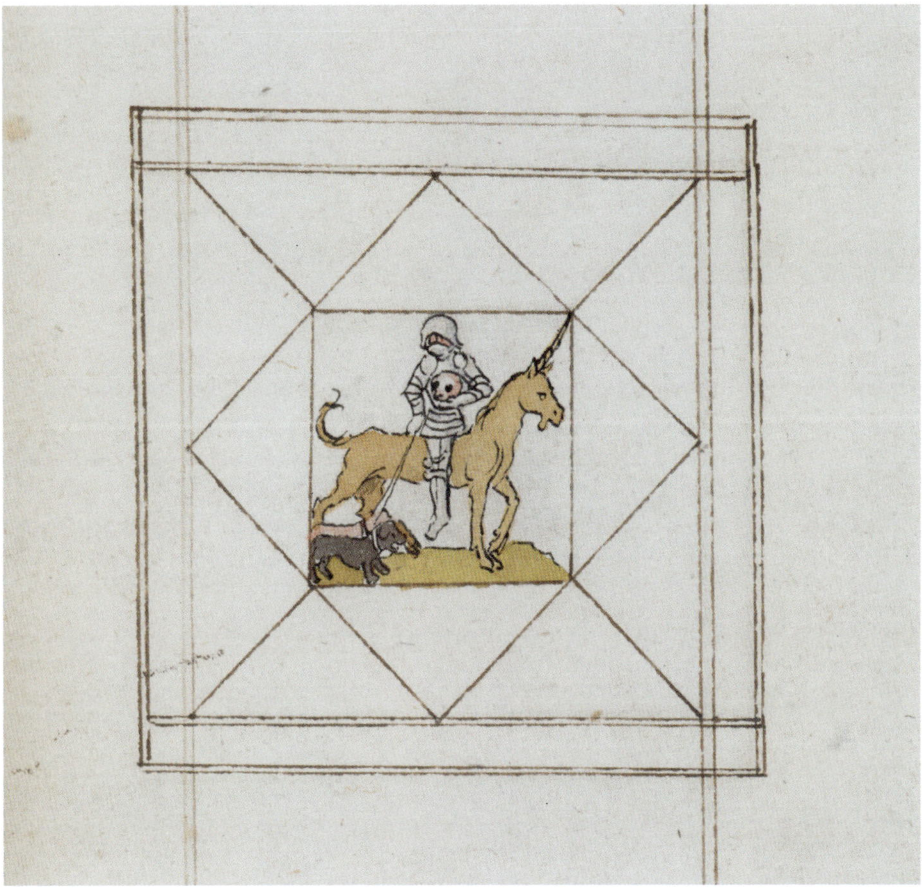

Figure 4. Unfinished horoscopes with ink miniatures. London, British Library Add MS 34603, folio 44r. Reproduced with permission of the British Library.

From this physical analysis, we conclude that Schinnagel was directly involved in writing and compiling most of Add MS 34603; he probably also had the codex bound in its present configuration. As we shall see later, sources for the texts and tables of the codex can be identified. The two sections Schinnagel claimed to have authored are explicitly identified, '... per me magistrum Marcum Schynagel', a flourish we might expect only if the codex were intended for some other user or owner. Schinnagel thus prepared a toolbox, but for whom, we cannot say. A closer look at the tools in the box may shed light on the interests of this unknown person.

3. Contents of the second codicological part

We begin with the larger and more homogenous section of the codex. Filling nearly three-quarters of the foliated (i.e. text-bearing) manuscript, the second codicological unit of twenty-nine quires contains primarily tables. Both the early brown ink quire numbers and folio numbers suggest that the order of the material did not change between copying and binding. Add MS 34603 thus gives us the individual tools and their sequential arrangement as selected by Schinnagel himself. Similar to the first codicological section, the second was not finished. Quire 2 has lined text blocks but no text. Quire 3 has text blocks and tabular lines but also lacks text. The tables begin in quire 5 (quire 4 is missing) and generally fill the remaining quires through to the twenty-sixth. The end of this unit (quires 26–31) also has marked text blocks but no text (nearly fifty folios). At the time of binding, Schinnagel must have envisioned someone adding more tools to the toolbox. In Appendix 2, I transcribe titles of the more than fifty individual tables and identify their sources. This content tells us something about Schinnagel or the practicing astronomer for whom the codex was designed.

We notice first that Schinnagel did not arrange his tools into named 'sets'. He was not interested in Bianchini's tables, the Oxford Tables, the *Tabulae resolutae*… or even the Alfonsine Tables. Indeed, I find the term 'Alfonsine' only once in the entire codex, in the title of a table that Schinnagel copied verbatim from his source, namely *Tabula coniunctionum Saturni et Jovis post incarnationem Christi secundum medium motum per tabulas Alfonsi notate* (f. 121^r^). Yet Schinnagel was surely aware of the existence of the Alfonsine Tables, since he copied nearly half of his tables (twenty-four of the fifty-six) directly (preserving both format and contents) from the 1492 edition of the Parisian Alfonsine Tables printed in Venice by Johann Hamann and edited by Johannes Santritter. Unlike many incunabula, this imprint has a crisply formatted title page: *Tabule Astronomice / Alfonsi * Regis*. However, nowhere in Add MS 34603 did Schinnagel seek to ascribe unity to the individual tables he copied or to identify them as Alfonsine.

This does not mean that Schinnagel's fifty-six tables lack structure. He did order the tables according to common tasks that a late-fifteenth-century astronomer/astrologer would have performed:

- Tables for planetary motions in longitude and latitude
- Entry of the Sun into the first point of Aries and the other zodiacal signs
- Eclipses, parallax, and true syzygies
- Velocities of the Sun and Moon
- Directions and profections for horoscopes
- Spherical trigonometry, including house boundaries and half-day lengths
- Computus and calendar of moveable feasts
- Lists of geographical locations
- Mean motions for collected years, months, and days
- Tables for duration of pregnancy and medical diagnosis
- Equations of time
- Radices for years 1441–1531, but only to degrees
- Astrological tables of the zodiacal signs and lunar mansions
- Star catalogue of John Vimond 'verified' for '1313' (see n. 68 below)

This list represents my conceptual ordering of Schinnagel's tools; he simply copied the tables one after the other, offering no apparent rationale for their sequential order. A comparison of my list with the recent survey of medieval tables by Chabás and

Goldstein suggests, however, that Schinnagel included most of the major topics found in that literature.[26]

Given his lack of interest in 'sets' of tables, it is perhaps not surprising that only two short canons appear amidst the more than fifty individual tables in the second codicological part.[27] Medieval authors often wrote canons to accompany 'sets' of tables. Neither appear in Add MS 34603. Perhaps Schinnagel or his intended user had canons available in other codices in his personal library and saw no need to duplicate them in Add MS 34603. Or perhaps as experienced astronomical practitioners they had no need to refer to written instructions while performing astronomical computations; that is to say, perhaps he envisioned the codex as a working tool, not a pedagogical textbook. On the other hand, the tables section of the manuscript is very 'clean' and shows little, if any, signs of use (unlike some other manuscript copies of the Parisian Alfonsine Tables that show soiled, heavily thumbed leaves, underlines, marginal notes). For whatever reason, Schinnagel's toolbox was not often opened by computers.

Nonetheless, its tools do contain some surprises, even for users (myself included) well practiced in the computational procedures of Alfonsine astronomy. For example, the first copy of the equations (Table 1 in Appendix 2) formats the material as does the vast majority of the manuscript copies of the Parisian Alfonsine Tables and the first two printed editions, giving the entries to minutes of arc for arguments of the mean centre and true argument. The second copy of the equations (Table 48) gives entries to degrees, using a double-argument format, and shifts the argument for the equation of the centre from the mean centre to the mean longitude (no text points to this shift). That is, the longitude of apogee for each planet (which moves over time with the eighth sphere) has been added to the mean centre. I do not recall seeing another copy of the Parisian Alfonsine Tables' equations of centre which formats the arguments for mean longitude. Only an experienced practitioner could use Table 48 in the absence of a canon explaining its idiosyncratic format.

At no point does Schinnagel's toolbox include radices to minutes, which is the standard of precision of the *Almagest*, most Arab zijes, the Toledan Tables, the Parisian Alfonsine and related tables. Although most of Schinnagel's tables are extended to minutes, one could not compute positions to that precision if working only with the tools of Add MS 34603. Radices to degrees, however, also appear on Schinnagel's polyptych, which makes us wonder whether the codex was designed to use in conjunction with that object.

Schinnagel's codex repeats not only the equations. It includes two tables of the equation of time (Tables 44 and 45), one labelled 'old', the other 'modern'; several tables of geographical places (Tables 35, 46, 50); and two tables for finding the time of true syzygy, by the unnamed methods of John of Lignères and Nicholaus de Heybech (Tables 6 and 39, respectively). Redundancy, not just efficiency, can be seen in Schinnagel's toolbox.

Perhaps most striking, however, is how extensively Schinnagel borrowed his tools from printed books. Only thirty years after printers in southern Germany began issuing astronomical tables, Schinnagel had gained close access to the new medium, as both a publishing author and a reading consumer. Yet in Add MS 34603, he transferred content

26 José Chabás and B. R. Goldstein, *A Survey of European Astronomical Tables in the Middle Ages* (Leiden: Brill, 2012).

27 Short canons accompany the Tabula more infantium (f. 263^{r-v}) and the Tabula latitudinis et longitudinis civitatum (f. 265^{r}).

from the world of print 'back' into the world of manuscript. Of course, some of the tables he copied may have been available to him in either manuscript or print. Yet, in many cases, Schinnagel's handwritten titles and formats match exactly those found in printed sources; I am convinced that most of the tables in Add MS 34603 were copied from imprints, from the following sources, listed chronologically:[28]

- Regiomontanus, *Kalendarium* (Nürnberg: Regiomontanus, 1474).
- [Beham, Lazarus], *Notamdum quod in isto libello habentur duo kalendarij* (Cologne: Nikolaus Götz, *c.* 1476).
- Pflaum, Jacob, *Kalendarium* (Ulm: Johann Zainer, 1478).
- Alfonso X, *Tabulae astronomicae* (Venice: Ratdolt, 1483).
- Firmin of Beauval, *Opusculum reportorii pronosticon in mutationes aeris tam via astrologica quam metheorologica* (Venice: Ratdolt, 1485).
- Angelus, Joannes, *Astrolabium planum in tabulis* (Augsburg: Ratdolt, 1488).
- Regiomontanus, *Tabulae directionum profectionumque* (Augsburg: Ratdolt, 1490).
- Alfonso X, *Tabule astronomice Alfonsi regis*, edited by Johannes Lucilius Santritter (Venice: Johannes Hamann, 1492).
- Zacut, Abraham, *Almanach perpetuum* (Leira: Abraham Ben-Samu'el d'Ortas, 1496).

Most of Schinnagel's sources were printed in or close to Swabia. Most were among the first printed editions to feature calendrical and astronomical material in tabular formats. Indeed, the only type of printed tables that Schinnagel did not place in his codex was Regiomontanus's ephemerides (Nuremberg 1474), namely, daily positions of the planets for time spans extending over years. Mathematical astronomers, since the Babylonians, had calculated daily ephemerides; Schinnagel's toolbox lacks them.

As noted above, Schinnagel inserted himself as author into the codex only twice. In the first codicological part, he claimed a table for making medical diagnoses strictly by astrological methods without physically examining the patient, borrowing the title from a well-known thirteenth-century text attributed to William the Englishman, *De urina non visa*.[29] At the very end of the second codicological part of the codex, separate from the earlier tables, Schinnagel added a 'Tabula de urina non visa per me magistrum Schynagel' (f. 303^{v}), written in the same format as the mean motions tables for collected years, but based on a rate of motion that I do not understand (slightly more than 107°/day) and do not find explained in the earlier short text.

Another text, apparently authored by Schinnagel, appears in the second codicological part among the astrological tables for finding 'profections' and 'directions' of horoscopes.

---

28 For other examples of copying from print to manuscript, see Ann Blair, 'Reflections on Technological Continuities: Manuscripts Copied from Printed Books', *Bulletin of the John Rylands Library*, 91 (2015), 7–33; Almuth Märker, 'Inkunabelnabschriften in Handschriften aus dem Leipziger Universitätsbetrieb', *Wolfenbütteler Notizen zur Buchgeschichte*, 42 (2018), 151–71. Book historians have long rejected the teleological claim that print quickly vanquished manuscript production in the fifteenth century.

29 Laurence Moulinier-Brogi, *Guillaume l'Anglais, le frondeur de l'uroscopie médiévale (XIII*^{e} *siècle): Édition commentée et traduction du De urina non visa* (Geneva: Droz, 2011); Hilary M. Carey, 'Medieval Latin Astrology and the Cycles of Life: William English and English Medicine in Cambridge, Trinity College MS 0.5.26', in *Astro-Medicine: Astrology and Medicine, East and West*, ed. by A. Akasoy, Charles Burnett, and R. Yoeli-Tlalim (Florence: Sismel Edizioni del Galluzzo, 2008), pp. 33–74.

Most of this material Schinnagel copied from a recently published set of tables for these purposes compiled by Regiomontanus.[30] But a four-page 'Tabula directionum per me marcus Schinagel' (ff. 139^{v}–141^{r}) is not found in Regiomontanus. Prepared for the sixth and seventh climates, meaning latitudes in Swabia, these tables may well have been invented by Schinnagel (see Fig. 5). Nothing similar to their format is listed in Chabás and Goldstein's survey of European astronomical tables and I do not recall ever seeing anything similar. The tables provide 'directions' for intervals of the ecliptic divided into the astrological 'Egyptian terms'; for each of those longitudes, Schinnagel simply computed the right and oblique ascensions and converted the latter into time spans using Ptolemy's scheme: one degree = one year, five minutes = one month, and one minute = six days.[31]

Interestingly, amidst the astrological tables of the second codicological part, Schinnagel copied a short text that presents birth horoscopes and some of their directions for a 'Suffraganii' (1439, sixty-eight time minutes east of Toledo) and a 'Doctori Nicolai' (1463, eighty-five time minutes east of Toledo). A nearly identical copy of this text appears in Vat. pal. lat. 1444, folios 1^{v}–6^{r} (including one leaf similarly copied in landscape format), labelling the former chart 'episcopi Danielis'. The subjects of these charts were probably Daniel Zehender (d. 1500), suffragan Bishop of Constance from 1473 and patron of a celestial globe and the well-known ephemerides by the Tübingen astronomer, Johannes Stöffler; and Nicolaus Pol (d. 1532), bibliophile, correspondent (usually regarding astrology) of the Augsburg humanist Veit Bild, and personal physician, on call in Innsbruck from 1495, of the King and then Emperor Maximilian I. It is not clear who prepared these horoscopes; birthdates for neither subject are otherwise known. The meridian of Pol's horoscope might support the speculation that he was from Poland.[32] In any case, the appearance of these two nativities, exemplars for computing 'directions', further anchors Schinnagel's toolbox to Swabia.

---

30 Mathematical techniques for predicting future events from a birth horoscope, including life expectancy. For details and the history of this doctrine, see Regiomontanus, *Tabulae directionum profectionumque* (Augsburg: Ratdolt, 1490); Stephan Heilen, *Hadriani genitura: Die astrologischen Fragmente des Antigonos von Nikaia*, 2 vols (Berlin: de Gruyter, 2015), pp. 992–1003.

31 F. E. Robbins, ed., *Ptolemy, Tetrabiblos* (Cambridge: Harvard University Press, 1940), pp. 294–95; Chabás and Goldstein, *Survey of Astronomical Tables*, pp. 205–26. Recomputation suggests that Schinnagel prepared the right ascensions using a table with an obliquity of 23;35 (al-Battani or the Toledan Tables; see C. A. Nallino, ed. *Al-Battani sive Albatenii opus astronomicum [1899–1907]*, reprint ed., 3 vols (Hildesheim: Georg Olms Verlag, 1977), II, p. 58; Fritz S. Pedersen, *The Toledan Tables: A Review of the Manuscripts and the Textual Versions with an Edition* (Copenhagen: C. A. Reitzels Forlag, 2002), p. 970). For the oblique ascensions, however, his obliquity is closer to 23;30 (Regiomontanus, *Tabulae directionum profectionumque*, sig. d2^{r}.), for the sixth climate at a latitude of 45;20. For Egyptian terms and their ruling planets, also marked in Schinnagel's table, see Charles Burnett, Keiji Yamamoto, and Michio Yano, *Al-Qabīṣī (Alcabitius): The Introduction to Astrology, Editions of the Arabic and Latin Texts and an English Translation* (London: Warburg Institute, 2004), p. 29. Unlike Regiomontanus's tables of directions, which yield the values directly, Schinnagel's version requires users to sum sequences of intervals whenever the direction extends beyond one term; i.e., Regiomontanus's tables were more 'user friendly' but Schinnagel's would have been easier to prepare.

32 Johann Stöffler and Jacob Pflaum, *Almanach nova plurimis annis venturis inservientia* (Ulm: Joannis Reger, 1499), sig. a2^{r-v}; Günther Oestmann, *Schicksalsdeutung und Astronomie: Der Himmelsglobus des Johannes Stoeffler von 1493* (Stuttgart: Württembergisches Landesmuseum Stuttgart, 1993); Max H. Fisch and Dorothy M. Schullian, *Nicolaus Pol Doctor 1494, with a Critical Text of His Guaiac Tract* (New York: Reichner, 1947); Alfred Schröder, 'Der Humanist Veit Bild, Mönch bei St Ulrich: Sein Leben und sein Briefwechsel', *Historischer Verein für Schwaben und Neuburg*, 20 (1893), 173–227. I thank Stephan Heilen for identifying Zehender.

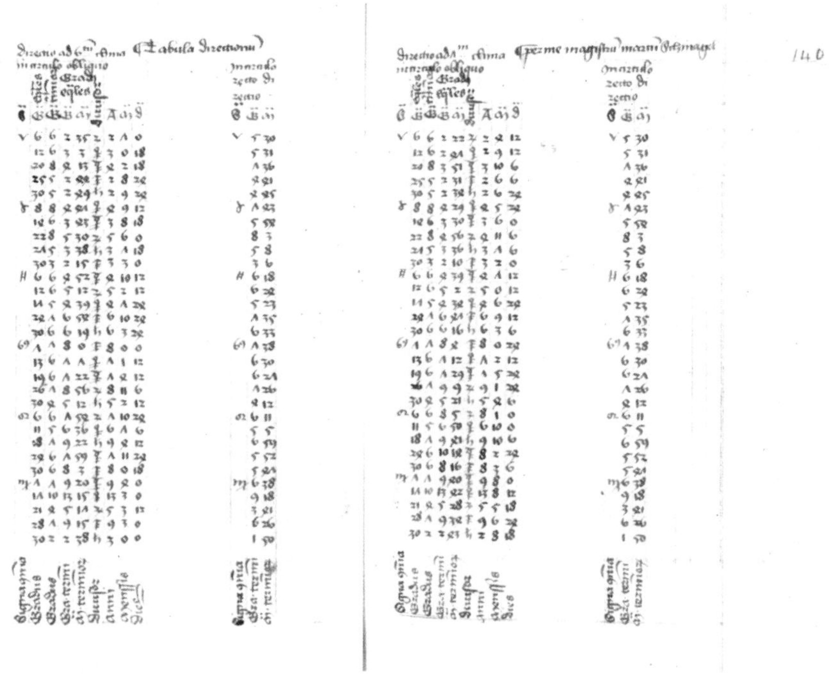

Figure 5. Schinnagel's *Tabula directionum*. London, British Library Add MS 34603, folios 139v–140r. Reproduced with permission of the British Library.

Hence, Schinnagel claimed intellectual authorship for only three of the nearly sixty tables copied into Add MS 34603, a rather modest contribution compared to better-known Alfonsine astronomers like John of Murs in the fourteenth century or Giovanni Bianchini in the fifteenth. On the other hand, the number of medieval Latin astronomers who claimed to author new tables is very small, no more than perhaps several dozen. On the basis of Add MS 34603, we must assign Schinnagel to this group.[33]

4. Contents of the first codicological part

In contrast to the second codicological part, the first contains no quantitative, tabular material. And its individual components are much more heterogenous in content: some are very unusual, perhaps unparalleled in the Alfonsine corpus.

33 Schinnagel's expression of authorship is quite similar to those found in earlier Alfonsine texts, e.g. in the canons of John of Lignères ('Hic incipient canones magistri Johannes de Lyneriis' for the *Cujuslibet arcus propositi sinum* or 'Expliciunt canones … per magistrum Johannem Pychardum de Lynerii et completi Parisius [*sic*] … 1322' for the *Priores astrologi motus corporum cesestium diligentissimis considerationibus observantes*) or John of Gmunden ('Expliciunt canones … compilati et conscripti Wyenne per magistrum Iohannem de Gmunden anno Domine 1437' for the *Quedam notabilia ad practicam tabularum*). See Marie-Madeleine Saby, 'Les Canons de Jean de Lignères sur les tables astronomiques de 1321: Édition critique, traduction et étude', unpublished thesis, Paris, École Nationale des Chartes, 1987, pp. 171, 277; Beatriz Porres de Mateo, 'Les Tables astronomiques de Jean de Gmunden: Édition et étude comparative', unpublished PhD thesis, Paris, École Pratique des Hautes Études, 2003, p. 501.

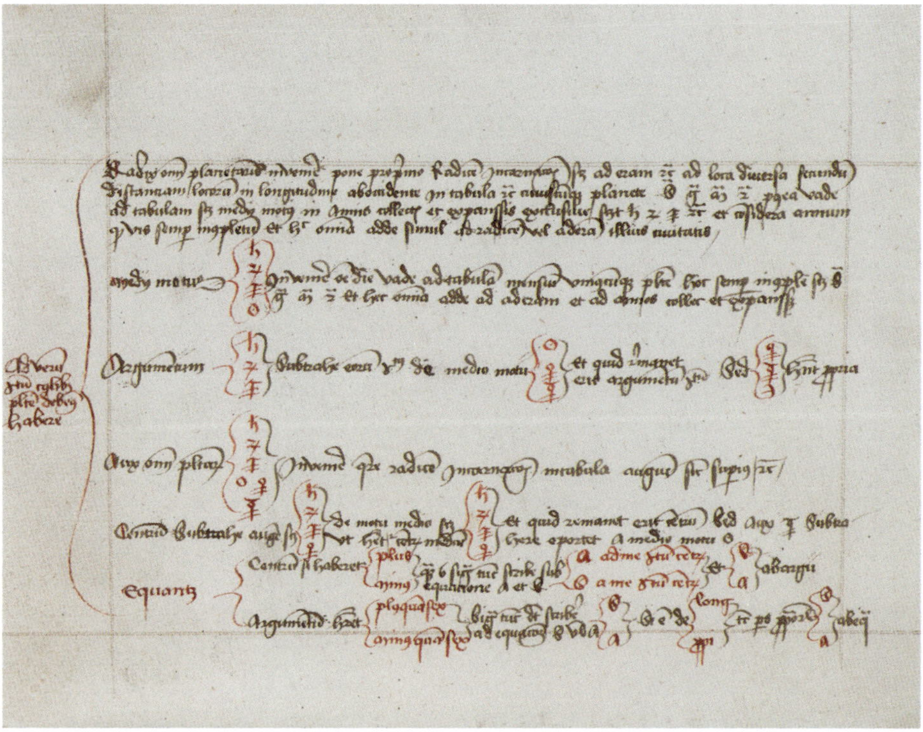

Figure 6. Schinnagel's algorithmic canon, 'Ad verum motum cuiuslibet planete debemus habere', in landscape format. See Appendix 1. London, British Library Add MS 36403, folio 4^v^. Reproduced with permission of the British Library.

The first quire contains what I refer to as an 'algorithmic canon', written in Schinnagel's hand and probably authored by him (see Appendix 1). The first two folios provide a terse, graphical 'flow diagram' of the individual steps required to compute true longitudes for the planets with the Parisian Alfonsine Tables. Interestingly, these sheets are written in what we now call 'landscape format' (only one additional landscape folio appears in the codex), apparently to enhance the visual integrity of the step-by-step instructions.[34] The steps in the procedure flow from left to right and top to bottom, with computational options listed for given steps. Not all of the steps, however, are specified; this canon seems to have been designed for an expert user who needed only brief reminders, especially to carry algebraic signs through the intermediate values. The remaining fifteen leaves of the canon step through several computations, finding true positions of Mars and the Moon for noon, 1 January 1496, for a meridian of sixty-five minutes east of Toledo, and the times of mean conjunctions of the luminaries, at twenty-four-year intervals, from 1440 to 1560 for seventy minutes east.

34 In Appendix 1, my non-diplomatic transcription of this canon into a 'portrait' format requires more line breaks than in the original.

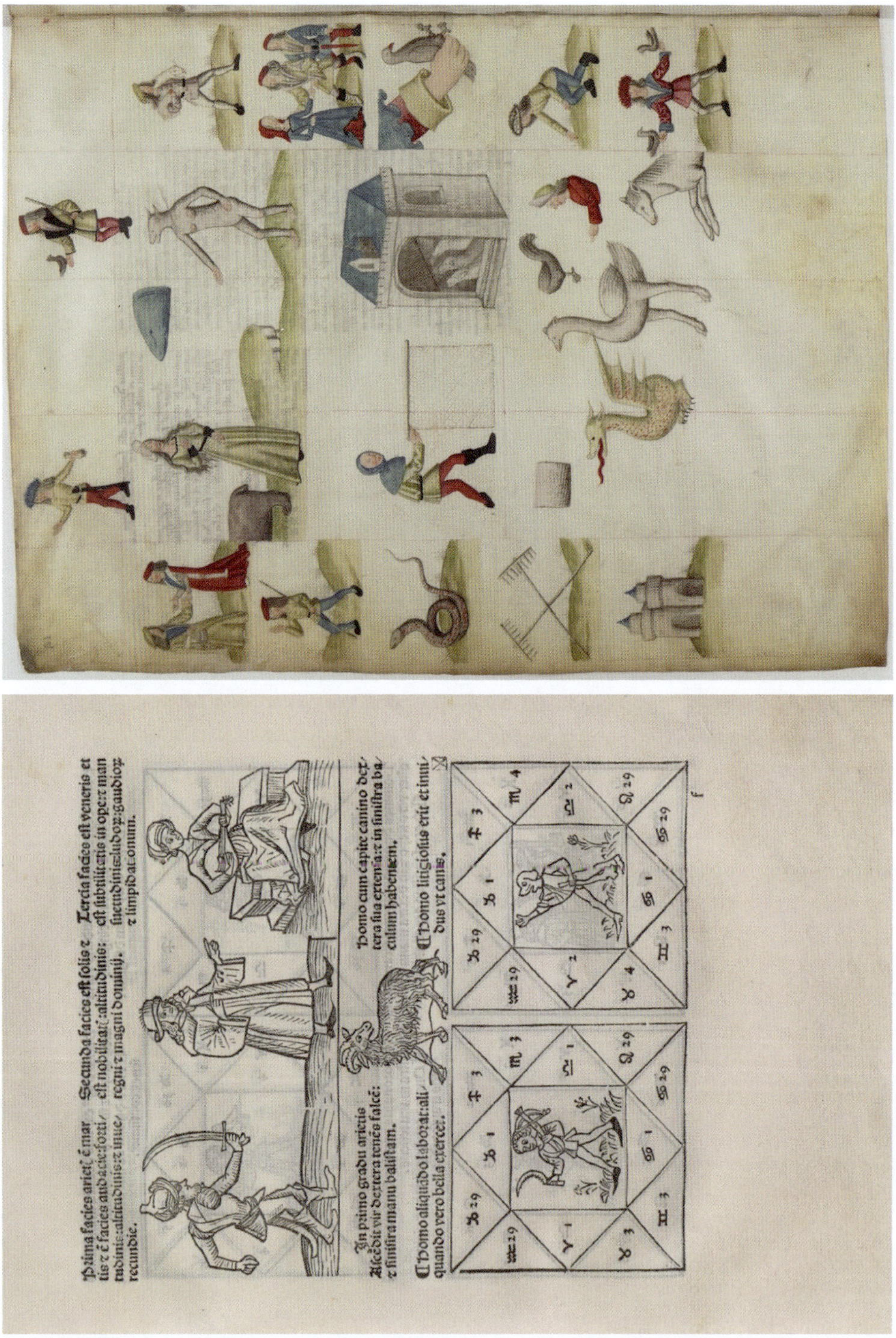

Figure 7. Paranatellonta as depicted in Angelus, *Astrolabium planum* (Augsburg 1488), sig. 1^{r}, and Paris, BnF lat. 7344, folio 19^{v} (Simon de Phares, between 1488 and 1490). Cf. Figure 4. Sources www.digitale-sammlungen.de and gallica.bnf.fr.

The second quire of miniatures embedded in unfinished calendrical pages is obviously linked to the Stuttgart polyptych. The third, fourth, and partial fifth quires, showing cruder miniatures placed in unfinished horoscopes, seem related to one of the printed sources we encountered in the second codicological part, Angelus's *Astrolabium planum* of 1488. In addition to some tabular material (see Tables 30 and 31 in Appendix 2), this imprint includes images of 360 horoscopes, one for each degree of the ascendant (see Fig. 7), specifying house cusps for a geographical latitude of forty-five degrees. Images for the three decans of each zodiacal sign are also indicated. A short phrase ascribes, for each degree of the ascendant, astrological meaning to a person born under that degree. This idea of a distinct interpretation for every degree of the zodiac and every decan went back at least to Manilius's *Astronomica* and Firmicus Maternus's *Mathesis*, Bk VIII, and prompted various traditions of medieval manuscript illumination of the paranatellonta or sub-constellation sets of stars that rise at the same time as a given degree of the zodiac. Angelus's 1488 imprint stimulated the creation of a new group of finely illuminated manuscripts (e.g. BnF lat. 7344) of which Schinnagel's codex surely was a part.[35] However, Add MS 34603 includes only fifty-six miniatures in the horoscopes; the manuscript provides no clues as to how those images might relate to the 360° of the zodiac. Likewise, I cannot explain why Schinnagel did not complete the horoscopes that surround the miniatures.

The fourth section of the first codicological part of Add MS 34603 contains copies of two classical medical texts that had recently been printed. Four excerpts on critical days of diseases were copied verbatim from Angelus's edition of Bonatus, printed in 1491 by Ratdolt in Augsburg, and from Pietrus de Abano's short *Hypocratis libellus de medicorum astrologia*, an appendix to the 1485 edition of Firmin of Beauval's *De mutatione aeris*, also printed by Ratdolt but in Venice.[36] Both texts offer practical advice for medical astrology. As he did in the tabular section of Add MS 34603, Schinnagel here copied printed material into his toolbox. Yet the medical material is slight, filling only thirteen leaves of the large codex. Schinnagel's interests were more astronomical than medical, it would seem.

Conclusion

Add MS 34603 belongs to the Alfonsine corpus. It contains twenty-two tables from the 1492 printed edition of the Parisian Alfonsine Tables. However, the London codex does

35 Angelus's book was reprinted twice in Venice, in 1494 by Johann Emerich de Spira, and in 1502 by Lucas Antonius de Giunta. Only several of Schinnagel's miniatures (e.g. 14 Aries, 20 Aries, 17 Taurus, or 9 Gemini) resemble those found in the Angelus edition; most of Schinnagel's iconography seems independent, even if their designs are close to those in the imprint. For a sample of the massive literature on Angelus and the paranatellonta, see Bernhard D. Haage, 'Astrolabium planum deutsch', *Sudhoffs Archiv*, 65 (1981), 117–43; Eberhard Knobloch, 'Astrologie als astronomische Ingenieurkunst des Hochmittelalters: Zum Leben und Wirken des Iatromathematikers und Astronomen Johannes Engel (vor 1472–1512)', *Sudhoffs Archiv*, 67 (1983), 129–44; Wolfgang Hübner, *Manilius, Astronomica, Buch V*, 2 vols (Berlin: de Gruyter, 2010); Dieter Blume and Wolfgang Metzger, *Sternbilder des Mittelalters und der Renaissance: Der gemalte Himmel zwischen Wissenschaft und Phantasie*, 5 vols (Berlin: de Gruyter, 2012–2016), II/2, pp. 173–82, 889–906.

36 Guido Bonatus, *Decem tractatus astronomiae* (Augsburg: Erhard Ratdolt, 1491), sig. P6^r-Q1^r; Firmin of Beauval, *Opusculum repertorii pronosticon in mutationes aeris tam via astrologica quam metheorologica* (Venice: Ratdolt, 1485), ff. 46^r–49^v.

not, by incipits or explicits, anywhere identify itself as Alfonsine. It does not preserve the sexagesimal treatment of the mean motions as designed by Parisian astronomers in the 1320s. Add MS 34603 is 'Alfonsine' not because it matches a particular 'set' of tables. It is Alfonsine because it exemplifies how at least one Alfonsine practitioner worked at the end of the fifteenth century. Hardly caring about sets of tables, these astronomer-astrologers pragmatically assembled computational tools from whatever sources they had at hand, after 1470 increasingly from printed sources. They personalized their codices, combining disparate tables or texts for their own astronomical, astrological, medical, computistic, or pedagogical purposes. That is, they created toolboxes for themselves.

I am certainly not the first historian of astronomy to offer such a claim. A modern editor of Greek mathematical texts has observed that 'texts were often edited by [ancient] scholars who were themselves practitioners, or teachers, of the fields that the texts transmitted, and who took the scope of their role to include a correction of the words of text based on their own understanding of the ideas that these words conveyed'. Another scholar of Byzantine astronomy noted that 'texts tended to be copied by professionals interested more in gathering useful information than in preserving the *verba ipsa* of any author except the most authoritative'.[37] Similar scribal practices, I argue, can be found in the Alfonsine era.

Hence, the surviving Alfonsine codices tell us something about the 'Alfonsine Tables'... but they tell us more about how Alfonsine astronomers worked and arranged their computational practices and their toolboxes.

37 David Pingree, 'From Alexandria to Baghdād to Byzantium: The Transmission of Astrology', *International Journal of the Classical Tradition*, 8 (2001), 3–37 (p. 3); Nathan Sidoli, 'Research on Ancient Greek Mathematical Sciences, 1998–2012', in *From Alexandria, through Baghdad: Surveys and Studies in the Ancient Greek and Medieval Islamic Mathematical Sciences in Honor of J.L. Berggren*, ed. Glen Van Brummelen (Berlin: Springer, 2014), pp. 25–50 (p. 28).

Appendix 1: Schinnagel's algorithmic canon*

Edition

Ad verum motum cuiuslibet planete debemus habere:[38]

<1> Radix omnium planetarum: invenire pone pro primo radicem Incarnationis, scilicet ad eram etc. ad loca diversa secundum distanciam locorum in longitudine ab occidente in tabula etc. cuiuscumque planete s[igna] g[radus] m[inuta] s[ecunda] postea vade ad tabulam scilicet medii motus in annis collectis et expanssis exclusive, scilicet ♄ ♃ ♂ etc. et considera annum quamvis semper incompletum et hec omnia adde simul ad radicem vel ad eram illius civitatis.

<2> Medii motus {♄ ♃ ♂ ☉}: Invenire omne die vade ad tabulam mensium uniuscumque planete hoc semper incomplete scilicet s[igna] g[radus] m[inuta] s[secunda] et hec omnia adde ad {ad} eram et ad annos collectos et expanssos.[39]

<3> Argumentum {♄ ♃ ♂}: Subtrahe eorum motus de medio motu {☉ ♀ ☿}, et quid remanet erit argumentum motum.[40] Sed {♀ ☿ ☽} habent propria.

<4> Aux omnium planetarum {♄ ♃ ♂ ☉ ♀ ☿}: Invenire quere radicem Incarnationis in tabula augium, sicut superius etc.

<5> Centrum: Subtrahe augem scilicet {♄ ♃ ♂ ♀} de motu medio scilicet ut habetur centrum medium {♄ ♃ ♂ ♀}, et quid remanet erit centrum. Sed aux ☿ subtrahere oportet a medio motu ☉.

* Add MS 34603, ff. 4^{v}–5^{v} (landscape format); Th/K 1313 (this manuscript only); Saxl and Meier, *Verzeichnis*, p. 74. Editorial note: < > angle brackets mark editorial additions, { } curly brackets athetize redundant characters, [ ] square brackets indicate lacuna or illegible characters. See Figure 6 for ink colours.

38 A single curly bracket encompasses sentences <1-6>.

39 recte: incompleti

40 recte: motus

<6> Equantem

- centrum: si haberet { plus / minus } quam 6 signa, tunc scribe sub equacione
  a<dde> et s<ubtrahe> { a<dde> ad me<dium> motum centrum / s<ubtrahe> a me<dio> motum centrum }
  et { s / a } ab argumento.[4]
- argumentum: <si> haberet { plus quam sex / minus quam sex } signa, tunc debet scribere
  ad equacionem s<ubtrahe> vel a<dde> { s / a }, si est de { longiori / propiori }
  tunc pars proportionalis { s / a } ab equatione. <4v/5r>

<7> Argumentum ☉: quere augem ☉, quem subtrahe a medio motu ☉ erit argumentum ☉ et centrum ♀, minuta proporciona{bi}lia debemus querere cum centro vero.

<8> Diversitas diametri scilicet de { longitudine <longiori> / <longitudine> propiori etc. } : debemus querere cum argumento vero.

<9> Si autem intrabis cum duplici introitu. Si centrum vel argumentum habent minuta hec [?]ela valet scilicet.

<10> Si prima equacio est

maior quam secunda, tunc sub<trahe> partem proporcionalem a maiore equacione, tunc erit vera equacio, tunc scribe desuper A si centrum habet plus sex signa, vel S si centrum habet minus sex signa. Et econverso in <vero> argumento si plus sex signa tunc scribe S, si minus sex tunc A debet scribi.[41]

minor quam secunda, tunc adde partem proporcionalem ad minorem equacionem, tunc erit vera equatio, tunc scribe desuper S si centrum habet minus sex signa, vel A si habet plus. Et econverso in <vero> argumento, si minus sex signa tunc debet scribi A, si maius sex tunc scribi debet S.

<11> Si minuta proporcionalia sint { de longitudine longiori / de longitudine propiori }, tunc { sub<trahe> / adde } pars

proportionalis a prima equatione argumenti tunc erit vera <equatio argumenti>.[42]

41 Recte: motu centri
42 recti: partem proportionalem

<12> Nota: si iste equaciones, scilicet

- centrum A / argumentum S et econverso { s / a }, tunc sub<trahe> minore de maiore, et quod remanet, A<dde> vel S<ubtrahe> ad medium motum.
- centrum A / argumentum A tunc adde ad simul et adde postea ad medium motum.
- centrum S / argumentum S tunc adde adinvicem et hoc, quod remanet, subtrahe a medio motu.

Translation

For the true motion of any planet, we must have:

<1> The radix of all the planets: to find, first take the radix of Incarnation, that is, to the era, etc., to various places according to the distance of the place in longitude from the west in the table, etc., for any planet in signs, degrees, minutes, seconds.[43] Then go to the table of mean motion in collected and expanded years only, for Saturn, Jupiter, Mars, etc., and consider the year, which is always incomplete, and add all these also to the radix or to the era of that city.[44]

<2> The mean motions of Saturn, Jupiter, Mars, Sun: to find every day, go to the table of months, always incomplete, for whatever planet, in signs, degrees, minutes, and seconds, and all these add to [the mean motions for] the era and for the collected and expanded years.

<3> The <mean> argument of Saturn, Jupiter, Mars: subtract their <mean> motion from the mean motion of the Sun, Venus, Mercury, and what remains will be the <mean> argument of motion. But Venus, Mercury, Moon have their own <mean arguments of motion>.

<4> The aux of all planets (Saturn, Jupiter, Mars, Sun, Venus, Mercury): to find, seek the radix at Incarnation in the table of apogees, as above.[45]

<5> The <mean> centre: subtract the aux of Saturn, Jupiter, Mars, Venus from the mean motion of these planets to have the mean centre of Saturn, Jupiter, Mars, Venus, and what remains will be the <mean> centre. But the aux of Mercury must be subtracted from the mean motion of the Sun.

<6> Equated centre: if it <mean centre> has (more/fewer) than six signs, then write under the equation <of centre> 'a' and 's' ('a' add to the mean motion of the centre/'s'

43 '… to the era … from the west' is quoted verbatim from the header to Table 10 of the Parisian Alfonsine Tables; cf. Poulle, *Tables alphonsines*, p. 127.

44 The Parisian Alfonsine Tables, as formulated in the 1320s, presented mean motions in tables entered with sexagesimal numbers of days; tables of mean motions in collected and expanded years soon begin appearing in manuscripts containing copies of the Parisian Alfonsine Tables. Both formats are found in Add MS 34603 (Appendix 2, Tables 1, 38).

45 The canon does not discuss the need to precess the apogees from the epoch to the given date of the computation.

subtract from the mean motion of the centre) <to find the true centre>. And (subtract/add) <the equation of centre> from <or to> the mean argument <to find the true argument>.

Equated argument: if it <true argument> has (more than six/fewer than six) signs, then write above the equation <of argument> 's' or 'a' ('s' / 'a'); if it is (farther/nearer), then (subtract/add) the proportional part from <or to> the <first> equation <of argument>.[46]

<7> The <mean> argument of the Sun: subtract the solar aux from the mean motion of the Sun to find the <mean> solar argument and the [mean] centre of Venus; to find the proportional minutes <for Venus> we must <enter> with the true centre.

<8> The variation of diameter of (farther/nearer, etc.): <to find> we must enter with the true argument.

<9> If, however, you enter with double entry <and> if the centre or argument has minutes, away with them.[47]

<10> If the first equation <of argument> is larger than the second, then subtract the proportional part from the larger equation, which will be the true equation <of argument>; then write 'A' over it if the <mean> centre has more than six signs, or 'S' if the <mean> centre has fewer than six signs. Conversely, if the <true> argument has more than six signs, then write 'S', if fewer than six then must be written 'A' <above the equation of argument>.

<If the first equation of argument is> smaller than the second, then add the proportional part to the smaller equation, which will be the true equation <of argument>, and write above it 'S' if the centre has fewer than six signs, or 'A' if it has more. And conversely, if the <true> argument has fewer than six signs, then must be written 'A', if more than six then must be written 'S' <above the equation of argument>.

<11> If the proportional minutes are (of farther distance/of nearer distance), then (subtract/add) the proportional part <from/to> the first equation of the argument, which will be the true <equation of argument>.

<12> Note: if those equations

(Centre A/Argument S) or conversely (<Centre> S/ <Argument> A), then subtract the smaller from the larger and a<dd> or s<subtract> what remains to the mean motion.

(Centre A/Argument A), then add together and add to the mean motion what remains.

(Centre S/Argument S), then add in turn and subtract what remains from the mean motion.

Commentary

This canon provides a highly abridged description of most, but not all, of the steps required to compute planetary longitudes with the Parisian Alfonsine Tables, or tables closely related to them. Unlike most canons for medieval astronomical tables, this algorithmic text does not define terms, explain arithmetical procedures, refer to Ptolemaic geometrical

46 The final step of sentence <6> is repeated in sentence <11>.

47 Perhaps words are missing in this sentence; its meaning is not clear to me. Add MS 34603 includes no double-entry tables for planetary equations. See Chabás, *Computational Astronomy*, Chapters 9 and 12.

models, or mention any external texts. In does not consider more complicated tasks such as calendrical conversions or calculating motions of the eighth sphere, lunar motions, syzygies, eclipses, planetary stations, retrogradations, or latitudes.

As can be seen from a facsimile of the manuscript (Fig. 6), the manuscript witness combines several written and grammatical formats. The tall curly brackets in the canon, a graphical device I have not previously noticed in canons to medieval astronomical tables, serve two functions.[48] In sentences <1–5>, they list the relevant bodies to be considered for those steps. In the subsequent sentences, they describe parallel branches in the algorithms and introduce if-then conditionals: for example, if an equation is greater than six signs (or 180°), then make its algebraic sign negative, if fewer than six signs, make the sign positive. The canon instructs readers with verbs sometimes in infinitive forms, sometimes in imperative forms. But throughout, the instructions are exceedingly terse. In the translation, I have not infrequently added technical terms to clarify the computational steps being described.

Many of the sentences manage the algebraic signs of intermediate quantities found while computing planetary equations. Following a format initiated by Ptolemy, the Parisian Alfonsine Tables (like most medieval astronomical tables) exploit symmetries in the geometrical models and present entries for arguments ranging only from zero to 180 degrees, moving from top to bottom of a column. For arguments from 180 to 360 degrees, one moves from the bottom to the top of a column and changes the algebraic sign of the entries. The printed edition of Parisian Alfonsine Tables 1492 conveniently marks the columns of equations with 'Adde' and 'Minue' to help users remember the signs; many manuscript versions (as well as Parisian Alfonsine Tables 1483) do not mark signs. This algorithmic canon, although composed presumably just when printed editions of the Parisian Alfonsine Tables were becoming available, seems designed for expert table users needing quick, graphical reminders as they determined planetary longitudes.[49]

As can be seen below in my attempt to reduce the instructions to modern algebraic representation, the algorithmic canon does not specify every step required to compute a true planetary longitude. For example, the canon assumes one knows how to compute the Alfonsine motion of the eighth sphere for a given date. It describes how to compute the mean centre and mean argument, but not how to enter the table of equations to look up the equations of centre and argument or to find their algebraic signs. It does not explain rationales behind the conventions for the signs. Moreover, I cannot understand the exceedingly terse sentence 10, which treats the trickiest step in the computation, adjusting

48 For a similar use of curly brackets in this codex, see 'Canon de mora infantis', f. 263^r, in the same hand and with similar use of red and black ink. A left curly bracket is commonly found in medieval diagrams of 'distinctio' or subdivisions, grouping classifications that become more differentiated in moving across or down a written page. Using double curly brackets to denote algorithmic branching options is less well attested. Cf. John E. Murdoch, *Album of Science: Antiquity and the Middle Ages* (New York: Charles Schribner's Sons, 1984), pp. 42–46; Mary Carruthers, *The Book of Memory: A Study of Memory of Medieval Culture*, 2nd ed. (Cambridge: Cambridge University Press, 2008), pp. 106–13.

49 Another short Alfonsine canon, but at 7000 words, far more expansive than this one, has been aptly described as being written for 'a more experienced audience, who were already familiar with some of the more basic moves in astronomical computation and the terminology involved, and who merely looked for some quick reminders concerning the details of particular operations'. Cf. C. P. E. Nothaft, 'Jean des Murs's Canones tabularum alfonsii of 1339', *Erudition and the Republic of Letters*, 4 (2019), 98–122 (p. 105).

for varying distances of the epicycle from the Earth by applying proportional minutes and the proportional part to the 'first', or provisional, equation of the argument.[50] Indeed, the canon reads like an idiosyncratic checklist of reminders crafted for a particular calculator, not like a general set of instructions for university students.

Interestingly, in sentence 9, the canon briefly notes that, rather than using the single-entry equations of the Parisian Alfonsine Tables, one might instead use double-entry equations. Such tables greatly ease the bookkeeping of the algebraic signs. The canon offers no instructions for employing such tables apart from asserting that minutes can be ignored in the double-entry arguments. Such a move would simplify interpolation in the double-entry context, but it would also degrade the accuracy of the computation. Nowhere else does this canon advise such approximation.

The following translation of the sentences from prose to algebra seeks to enhance the visibility of the algorithmic steps in the canon. With only one exception (sentence 5), these steps concisely and correctly depict the computational requirements of the Parisian Alfonsine Tables, as we understand them from other, more lengthy canons such as John of Saxony's *Tempus est mensura motus* (1327) or from modern recomputation of medieval computations using the Parisian Alfonsine Tables.[51]

<1, 2, 4> $\text{radix} = \text{radix}_o + \Delta t_{\textit{from epoch}} {}^{*}\omega_{radix} + \Delta t_{\textit{from meridian}} {}^{*}\omega_{radix}$

<3> $a_m = \bar{\lambda}_{Sun} - \bar{\lambda}$, for superior planets

<5> $c_m = \bar{\lambda}_{planet} - \lambda_{aux\ planet}$; for Mercury $c_m = \bar{\lambda}_{Sun} - \lambda_{aux\ Sun}$[52]

<6> if $c_m > 180$, then $c_v = c_m + q$ and $a_v = a_m - q$
if $c_m < 180$, then $c_v = c_m - q$ and $a_v = a_m + q$
$p = p_o - \frac{\Delta\Pi_l}{60}$ or $p = p_o + \frac{\Delta\Pi_p}{60}$
if $a_v > 180$, then $p < 0$
if $a_v < 180$, then $p > 0$

<7> $a_m = \bar{\lambda}_{Sun} - \lambda_{aux\ Sun}$, for Sun; for Venus $c_m = \bar{\lambda}_{Sun} - \lambda_{aux\ Sun}$
Lookup $\Pi(c_v)$

<8> Lookup $\Delta\ (a_v)$

<10> If p_o > 'second equation', then $p = p_o - \frac{\Delta\Pi}{60}$[53]

---

50 Cf. Poulle, *Tables alphonsines*, pp. 207–8.

51 My notation follows Olaf Pedersen, *A Survey of the Almagest* (Odense: University Press, 1974) and J.D. North, ed. *Richard of Wallingford: An Edition of His Writings, with Introductions, English Translation and Commentary*, 3 vols (Oxford: Clarendon Press, 1976), III, pp. 175–76: Δt = time since epoch, ω = mean motion, λ = true longitude, $\bar{\lambda}$ = mean longitude, a_m = mean argument, a_v = true argument, c_m = mean centre, c_v = true centre, p_o = provisional equation of argument, tabulated for the value of c_m where the distance from the centre of the epicycle to the equant point is sixty parts, p = equation of argument, q = equation of centre, Δ = *diversitas diametri*, Π = proportional minutes.

52 Venus (not Mercury) uses the solar apogee.

53 I do not understand what is meant here by the 'second' equation. The algebraic sign of the proportional part [$\frac{\Delta\Pi}{60}$] is determined by the sign of the proportional minutes [Π], which in turn is determined by the true centre.

$$\text{If } p_{\circ} < \text{'second equation' then } p = p_{\circ} + \frac{\Delta\Pi}{60}$$

In both cases:
if $c_v > 180$, then $p > 0$
if $c_v < 180$, then $p < 0$
if $a_v < 180$, then $p > 0$
if $a_v > 180$, then $p < 0$

<11> Repeats <6> final step

<12> If $q > 0$ and $p < 0$, then $\lambda = \bar{\lambda} + (q - |p|)$
If $q < 0$ and $p > 0$, then $\lambda = \bar{\lambda} - (|q| - p)$
If $q > 0$ and $p > 0$, then $\lambda = \bar{\lambda} + (q + p)$
If $q < 0$ and $p < 0$, then $\lambda = \bar{\lambda} - (q + p)$

Appendix 2: Add MS 34603's second codicological unit

I transcribe here titles of the successive tables and describe their extent in the modern foliation of the codex. Blank unfoliated leaves, often scattered among the tables, are indicated only when they comprise more than three or four folios. Sources of the individual tables are listed in parentheses; full citations of the printed sources are listed above (Section 3); 'PAT' refers to the Parisian Alfonsine Tables.

Twenty-five blank folios

1 (ff. 86^{r}–107^{r}): Tabula equationis Solis, Lune ... Saturni (PAT 1492, sig. d3^{r}-g4^{v})[54]

2 (ff. 108^{r}–110^{r}): Tabula visionum et occultationum (visibilities, stations, daily velocities, latitudes for Venus, Mercury, Mars, Jupiter and Saturn, PAT 1492, sig. e5^{r}, f1^{r}, f5^{r}, g1^{r}, g5^{r})

3 (ff. 110^{v}–116^{r}): Tabula tabularum ad omnes calculationes (sexagesimal multiplication table, PAT 1492, sig. g5^{v}-h3^{r})

4 (ff. 117^{r}–118^{r}): Tabula motus solis (lune) in uno minuto diei (PAT 1492, sig. h4^{r}-h5^{r})

5 (ff. 118^{v}–119^{r}): Tabula veri motus solis et lune in una hora (PAT 1492, sig. h5^{v}-h6^{r})[55]

6 (ff. 119^{v}–120^{v}): Tabula inventionis temporis inter coniunctionem et oppositionem veram et mediam (two double-argument tables for finding times of true syzygy, PAT 1492, sig. h6^{v}-h7^{v})[56]

54 PAT 1492 interleaves the mean motions and equations for each successive planet; Add MS 34603 includes only the equations, but in the same order as in the printed edition.

55 These values differ slightly from those in the PAT *editio princips* 1483, with its otherwise unattested lunar velocities from 0;30,21–0;36,25°/hr. See Bernard R. Goldstein, 'Lunar Velocity in the Middle Ages: A Comparative Study', in *From Baghdad to Barcelona: Studies in the Islamic Exact Sciences in Honour of Prof. Juan Vernet*, ed. Josep Casulleras and Julio Samsó (Barcelona: Instituto Millás Vallicrosa de Historia de la Ciencia Árabe, 1996), pp. 181–94 (p. 190).

56 These simple tables, performing one division, were formulated by Ibn al-Kammād in the twelfth century and can be found in Alfonsine manuscripts bearing tables compiled by John of Lignères. See José Chabás and B.R. Goldstein, 'Computational Astronomy: Five Centuries of Finding True Syzygy', *Journal for the History of Astronomy*, 28 (1997),

7 (ff. 121^{r}–123^{v}): Tabula coniunctionum Saturni et Jovis, Saturni et Martis, Jovis et Martis, post Incarnationem Christi secundum medium motum per tabulas Alfonsi notate (PAT 1492, sig. h8^{r}-i2^{v}, title quoted verbatim from the printed edition)

8 (f. 124^{r}): Tabula ad inveniendum tempus distantie Sol a principio Arieti (PAT 1492, sig. i3^{r})

9 (ff. 125^{r}–128^{r}): Tabula diversitatis aspectus Lune (lunar parallax for climates 1–7, PAT 1492, sig. i4^{r}-k1^{r})

10 (f. 128^{v}): Tabula semidiametrorum solis et lune et umbre (PAT 1492, sig. k1^{v})

11 (f. 129^{r}): Tabula equationis diversitati aspectus sive tabula attacium (PAT 1492, sig. k2^{r})

12 (f. 129^{v}): Tabula eclipsis solis (PAT 1492, sig. k2^{v})

13 (f. 130^{r}): Tabula eclipsis lune ad longitudinem longiorem (PAT 1492, sig. k3^{r})

14 (f. 130^{v}): Tabula proportionis augmentata per duos gradus (PAT 1492, sig. k3^{v})

15 (f. 131^{r}): Tabula eclipsis solis ad longitudinem longiorem (PAT 1492, sig. k4^{r})

16 (f. 131^{v}–132^{r}): Tabula eclipsis lune ad longitudine longiorem/propiorem (PAT 1492, sig. k4^{v}-k5^{r})

17 (ff. 132^{v}–133^{r}): Tabula radicum temporum in introitum Solis in principiis signorum, ad meridianum Venetijs cum equatione dierum ad annos Christi 1371 (PAT 1492, sig. k5^{v}-k6^{r})[57]

18 (f. 133^{v}): Introitus et exitus lune in 12 signis (incomplete, source unknown)

19 (f. 134^{v}–135^{r}): Nota de directionibus; Inc: Accipe per quolibet gradu annum unum … (source unknown)[58]

20 (f. 136^{v}): Tabella mensium profectionalium ac usualium (Regiomontanus, *Tabulae directionum* 1490, sig. s5^{v})

21 (f. 137^{r}): Tabule profectionum (source unknown)

22 (f. 137^{v}): Tabula profectionis mensurne (Regiomontanus, *Tabulae directionum* 1490, sig. s4^{v})

23 (f. 138^{r}): Tabula profectionis diurne (Regiomontanus, *Tabulae directionum* 1490, sig. s5^{r})

24 (ff. 139^{v}–141^{r}): Tabula directionum per me marcus Schinagel (for climates 6 and 7)[59]

25 (ff. 142^{v}–147^{r}): Directiones hilegiorum secundum doctrinam Alkabilii [*sic*] (with a nativity and directions for 'Nativitatas Suffragancij' [Daniel Zehender], dated 30 March 1439, at 15;32 equal hours, and another for 'Doctori Nicolai' [Nicolaus Pol], dated 10 October 1463, at 17;50 equal hours, followed by four blank folios)[60]

26 (ff. 148^{r}–149^{r}): Tabula elevationum signorum in circulo directo (with equation of time, Pedersen, *Toledan Tables*, pp. 972–75)

27 (ff. 149^{v}–159^{v}): Tabula elevationum signorum in primo climate (through climate 7, Pedersen, *Toledan Tables*, pp. 1037–1070)

93–105 (pp. 94–95); Chabás and Goldstein, 'The Medieval Moon in a Matrix: Double Argument Tables for Lunar Motion', *Archive for History of Exact Sciences*, 73 (2019), 335–59 (pp. 347–49).

57 These radices are computed according to the PAT (including the 1483 equation of time) for a meridian of about eighty-three time minutes east of Toledo, although the Table of places in the *editio princeps* (this table is not included in the 1492 edition) lists Venice at ninety time minutes east.

58 For a copy of the first third of this text, see Vat. pal. lat. 1444, f. 1^{v}.

59 See n. 31 above.

60 Much of this text is found in Vat. pal. lat. 1444, ff. 2^{r}–6^{r}. See n. 32 above.

28 (ff. 160^{v}–161^{v}): Tabula conversionis horarum et suarum fractionum in gradus et suas fractiones (PAT 1492, sig. c5^{r-v})
29 (ff. 162^{v}–186^{r}): Tabula ascensionum obliquarum (for latitudes 30–53° in 1° intervals; Regiomontanus, *Tabulae directionum* 1490 offers same table for latitudes 1–60° in 1° intervals)
30 (ff. 187^{v}–210^{r}): Tabula signi et gradis ascendentis qualibet hora atque minuto, Tabula equationis domorum (for climates 5–7; the latter also gives astronomical information for each degree of the ascendant, i.e. triplicitas, termini, facies, etc., Angelus, *Astrolabium planum* 1488, sig. c4^{v}-e3^{r})
31 (ff. 211^{v}–223^{r}): Tabula horarum inequalium diei artificialis (climates 5–7, Angelus, *Astrolabium planum* 1488, sig. C1^{v}-D5^{r})
32 (ff. 224^{v}–225^{v}): Tabula festorum mobilium (incomplete, only January-June (source unknown)
33 (f. 226^{r}): Tabula declinationis solis secundum Albategni (obliquity 23;33;30, probably Pedersen, *Toledan Tables*, p. 964)[61]
34 (f. 226^{v}–227^{r}): Radices motuum ad eram Incarnationis ad loca diversa secundum distantiam locorum in longitigudine ab occidente cum radice Toletana (for places zero to thirty degrees east of Toledo, PAT 1483, sig. d1^{v}-d2^{v}, not in PAT 1492)[62]
35 (f. 227^{v}–228^{r}): untitled composite list of sixty places, listing longitudes in minutes of time from Nuremberg and Ulm and latitudes in degrees (Regiomontanus, *Kalendarium* 1474, which includes two additional places) and longitudes in degrees and minutes from Toledo at 11;00 east (PAT 1483, sig. m5^{r}-m6^{r}, which lists 146 places)[63]
36 (ff. 228^{v}–229^{r}): Tabula quantitatis dierum (for latitudes 44–55°, PAT 1492, sig. c3^{r-v})
37 (f. 229^{v}): Tabula equationis accessus et recessus (PAT 1492, sig. b3^{r})
38 (f. 230^{r}–239^{v}): mean motions in collected and expanded years (to 1000), months, days, hours, including:[64]

Tabula continens motum augium[65]
Tabula accessus et recessus octavi circuli[66]
Tabula continens medium motum augium et stellarum fixarum
Tabula continens medium motum Solis/Mecurii/Veneris
Tabula continens medium motum Lune

61 Albattani's zij uses an obliquity of 23;35,00; the value 23;33,30 is found in the Toledan Tables. See Nallino, *Albattanii opus*, II, p. 58.

62 Compare n. 2 above.

63 The table also includes, for each place, the difference in time of sunrise and length of day at summer solstice from Paris/Passau/Salzburg (latitude 48°); the difference in latitude from Nuremberg (latitude 49°); the difference in ascendant from a reference at latitude 52°; and the difference in the maximal or minimal solar altitude from a reference at latitude 52°. A planet or luminary and a star is also associated with each place. Cf. Chabás and Goldstein, *Survey of Astronomical Tables*, pp. 201–3.

64 These tables include no radices, except for 'Radices coniunctionum Solis et lune' (236^{r}), which are computed for 12 January, Julian year 1 (first mean conjunction after Incarnation), for Vienna (noted on 236^{v}) at eighty time minutes east of Toledo, to a precision of four sexagesimal digits of hours. Some of the material for mean syzygies is repetitive, arranged in formats for collected years and months that I have not previously seen.

65 This table specifies a motion of 1;00,02/100 years which is Ptolemy's rate of precession. The remaining mean motion tables are all Alfonsine but arranged in formats not found in PAT 1494.

66 I note that no table of equations for the access and recess of the eighth sphere is provided. These tables could not be used to compute the Alfonsine motion of trepidation.

Tabula continens medium argumentum Lune
Tabula continens medium motum capitis draconis
Tabula continens medium motum Saturni
Tabula continens medium motum Jovis
Tabula continens medium motum Martis
Tabula continens medium argumentum Veneris
Tabula continens medium argumentum Mercury
Radices coniunctionum Solis et lune
Tabula proventionum coniunctionum luminarium (mean synodic months)
Tabula medii motus solis et lune finibus tempore conjunctionis et oppositionis (mean synodic months)
Tabula medii argumenti lune ad tempore conjunctionis et oppositionis solis et lune
Tabula argumenti medii latitudinis lune ad tempore conjunctionis et oppositionis solis et lune
Tabula continens medium motum elongationis solis aut lune
Tabula continens verum motum capitis draconis
39 (ff. 240^{v}–241^{r}): Tabula equationis solis et lune tempore coniunctionis (Nicholas de Heybech's tables for true syzygy, never printed)[67]
40 (ff. 242^{v}–243^{v}): Tabula more infantis in vtero matris (Zacut, *Almanach perpetuum* 1496, ff. 161^{v}–162^{v})
41 (ff. 244^{r}–245^{v}): Untitled calendar of saints' days (Zacut, *Almanach perpetuum* 1496, ff. 165^{r}–166^{v})
42 (f. 246^{r}): Tabula latitudinis lune (Pedersen, *Toledan Tables*, pp. 1253–1257)
43 (f. 246^{v}): Tabula latitudinis quinque planetas (Pedersen, *Toledan Tables*, pp. 1325–1326)
44 (f. 247^{r}): Tabula equationis dierum cum noctibus suis vetus (Peter of St Omer, Pedersen, *Toledan Tables*, p. 985; PAT 1492, sig. e[first]4^{r})[68]
45 (f. 247^{v}): Tabula equationis dierum cum noctibus moderna verificata ad annos Christi (PAT 1492, sig. e[first]4^{v})[69]
46 (f. 248^{v}–249^{r}): Tabula regionum provinciarum europe et que in vicinijs singularum iacent ad scriptas tanque primarias reducentur pauxillo quopiam intervallo nullam differentiam notatu dignam importante (times from meridian of Toledo, latitudes and columns for time shifts and degrees from meridian of Nuremberg, PAT 1492, e[first]1^{v})

67 Edited by José Chabás and B. R. Goldstein, 'Nicholaus de Heybech and His Table for Finding True Syzygy', *Historia mathematica*, 19 (1992), 265–89.

68 PAT 1492, as described by GW 1258, includes a quire inserted between the end ('Finis') of the opening canon (sig. A1^{r}-D8^{v}) and the opening of the Parisian Alfonsine Tables (sig. a1^{r}-k6^{r}) that is marked e1–6 and gives six small tables with canons (ending 'Finis') that seem separate from the content of the Parisian Alfonsine Tables that follows. By 'first', I distinguish this inserted quire from the e-quire in the PAT 1492. I note that some copies of GW 1258 (Barcelona, Wolfenbuttel, Florence 2) do not include the 'inserted' e6 quire. Apparently, some early binders doubted whether that quire belonged to this title.

69 Schinnagel did not copy the date from the title given in the PAT 1492: '... moderna verificata ad annos Christi 1456'. This table is taken from Giovanni Bianchini's *Tabulae primi mobilis*. See Chabás, *Computational Astronomy*, pp. 272, 358.

47 (ff. 249^{v}–257^{r}): untitled radices, by year for 1441–1561 and incomplete for 1532–61, by day for the months, to signs and degrees (Schinnagel's polyptich gives only the years 1489–1526[70])

48 (ff. 258^{r}–261^{v}): untitled planetary equations and stations to degrees, with 2°-intervals for equations of centre and 1°-intervals for equations of argument; Mars, Venus and Mercury include tables of proportional minutes and 'longitudo longior/propior' to degrees (Beham, *Kalendarium* 1476, folio 13^{v}; Schinnagel polyptych, panel 5a[71])

49 (ff. 262^{r}–263^{v}): Tabula more infantium, with canons and example for 12 November 1452 (related to ps.-Ptolemy, *Centiloquium*, verb. 51, on rectifying a horoscope when the time of birth is not known)[72]

50 (ff. 264^{r}–265^{v}): Tabula latitudinis et longitudinis civitatum ab occidente habitato, with canon, inc: Nam hec tabula continet de civitatibus in ea nominates longitudiones … (128 places, only longitude and latitudes in degrees and minutes, PAT 1483, sig. m5^{r}-m6^{r})

21 blank folios

51 (ff. 266^{v}–269^{v}): untitled, times of true syzygy, 1498–1539 (Pflaum, *Kalendarium* 1478, ff. [11^{r}–13^{r}])

19 blank folios

52 (ff. 270^{v}–271^{r}): Tabula qualitates 12 signorum (source unknown)

53 (ff. 272^{r}–273^{r}): Tabula mansionum lune (source unknown)

54 (ff. 274^{r}–277^{r}): Tabula stellarum fixarum que sunt prope viam solis verificate ad annun domini 1313 (John Vimond's star catalog, Firmin of Beauval, *De mutatione aeris*, 1485, ff. 13^{r}–15^{v}, '… ad domini 1312')[73]

55 (ff. 278^{v}–302^{r}): untitled, giving for each hour of a year the ascendant in signs and degrees, for an unspecified year and geographical latitude, starting with 9 Gemini for the first hour of 1 January (source unknown)

18 blank folios

56 (f. 303^{v}): Tabula de urina non visa per me magistrum Schynagel (mean motions for completed years, months and days, rate of 105;15,44,12°/day,[74] with radix for Incarnation

70 Franz, *Schinnagel*, pp. 239–40.

71 Franz, *Schinnagel*, p. 248. Unlike the polyptych, the manuscript presents the equations of centre and proportional minutes, for Mars, Venus, and Mercury, with arguments of mean longitude rather than mean centre.

72 On f. 262^{v}, under the title Tabula more, verbum 51 is quoted in the translation by Plato of Tivoli: 'Locus lune in nativitate est tempore <sic> gradus ascendens in circulo hora casus spermatis, et locus lune hora casus spermatis est gradus ascendens nativitatis'. Cf. Jean-Patrice Boudet, 'Naissance et conception: Autour de la proposition 51 du Centiloquium attribué à Ptolémée', in *De l'homme, de la nature et du monde: Mélanges d'histoire des sciences médiévales offerts à Danielle Jacquart* (Geneva: Droz, 2019), pp. 165–78 (p. 168). Chapter 50 of the canons accompanying the astronomical tables of Giovanni Bianchini deal with 'Moram nati in utero materno'; but Schinnagel does not appear to have copied from Bianchini. See Giovanni Bianchini, *Tabulae celestium motuum earumque canones* (Venice: Simon Bevilaqua, 1495), sig. C7^{v}-C9^{r}.

73 According to Lynn Thorndike, *A History of Magic and Experimental Science*, 7 vols (New York: Macmillan, 1923–58), III, p. 274, the correct date '1321' was probably mistakenly replaced at the print shop by the incorrect '1312'. Schinnagel miscopied the incorrect date; see José Chabás and B.R. Goldstein, 'Early Alfonsine Astronomy in Paris: The Tables of John Vimond (1320)', *Suhayl*, 4 (2004), 207–94 (pp. 267–90).

74 Entries tabulated in units of 360° counted up to 7, 30° signs, degrees, minutes, and seconds. I cannot explain the significance of a unit of 7 × 360° and do not recall ever having seen such a unit in medieval astronomical or astrological manuscripts. At a rate of 105+°/day, it takes 23;50 days to realize 7 × 360°.

of Christ, a column for additional radices for the years 1489–1506 is not completed, source unknown)

57 (f. 304^{r}-v) De significatione planetarum in signis secundum membra hominis (tabulated by planet and zodiacal sign, Giovanni Calderia, *Liber canonum astrologiae*, II, 8, unpublished)[75]

58 (ff. 305^{v}–306^{r}): Inc: Cum legere libros Hippocratis medicorum optimi diei … (Firmin of Beauval, *De mutatione aeris*, 1485, ff. 46^{r-v}; copy, in another hand, of f. 80^{v} abov)[76]

25 blank folios

Manuscript sources

Erfurt, Universitätsbibliothek, CA 2° 360
London, British Library, Add Ms 34603
Paris, Bibliothèque nationale de France, Lorraine 7
Paris, Bibliothèque nationale de France, lat. 7344
Wolfenbüttel, Herzog August Bibliothek, Cod. Guelf. 22.1 Aug. 4°

Bibliography

Alfonso X, *Tabulae astronomicae* (Venice: Ratdolt, 1483).

———, *Tabule astronomice Alfonsi regis*, ed. by Johannes Lucilius Santritter (Venice: Johannes Hamann, 1492).

Angelus, Joannes, *Astrolabium planum in tabulis* (Augsburg: Ratdolt, 1488).

[Beham, Lazarus], *Notamdum quod in isto libello habentur duo kalendarij* (Cologne: Nikolaus Götz, *c.* 1476).

Bianchini, Giovanni, *Tabulae celestium motuum earumque canones* (Venice: Simon Bevilaqua, 1495).

Blair, Ann, 'Reflections on Technological Continuities: Manuscripts Copied from Printed Books', *Bulletin of the John Rylands Library*, 91 (2015), 7–33.

Blume, Dieter, and Wolfgang Metzger, *Sternbilder des Mittelalters und der Renaissance: Der gemalte Himmel zwischen Wissenschaft und Phantasie*, 5 vols (Berlin: de Gruyter, 2012–16).

Bonatus, Guido, *Decem tractatus astronomiae* (Augsburg: Erhard Ratdolt, 1491).

Boudet, Jean-Patrice, 'Naissance et conception: Autour de la proposition 51 du Centiloquium attribué à Ptolémée', in *De l'homme, de la nature et du monde: Mélanges d'histoire des sciences médiévales offerts à Danielle Jacquart* (Geneva: Droz, 2019), pp. 165-78.

Burnett, Charles, Keiji Yamamoto, and Michio Yano, *Al-Qabīṣī (Alcabitius): The Introduction to Astrology, Editions of the Arabic and Latin Texts and an English Translation* (London: Warburg Institute, 2004).

Carey, Hilary M, 'Medieval Latin Astrology and the Cycles of Life: William English and English Medicine in Cambridge, Trinity College MS 0.5.26', in *Astro-Medicine: Astrology and*

75 Thorndike, *Magic and Experimental Science*, IV, pp. 163–66, 668.

76 See n. 36 above.

Medicine, East and West, ed. by A. Akasoy, Charles Burnett, and R. Yoeli-Tlalim (Florence: Sismel Edizioni del Galluzzo, 2008), pp. 33–74.

Carruthers, Mary, *The Book of Memory: A Study of Memory of Medieval Culture*, 2nd ed. (Cambridge: Cambridge University Press, 2008).

Catalogue of Additions to the Manuscripts in the British Museum in the Years 1894–1899 (London: Clowes and Sons, Ltd., 1901).

Chabás, José, *Computational Astronomy in the Middle Ages: Sets of Astronomical Tables* (Madrid: Consejo Superior de Investigaciones Científicas, 2019).

———, and Bernard R. Goldstein, 'Nicholaus de Heybech and His Table for Finding True Syzygy', *Historia mathematica*, 19 (1992), 265–89.

———, and ———, 'Computational Astronomy: Five Centuries of Finding True Syzygy', *Journal for the History of Astronomy*, 28 (1997), 93–105.

———, and ———, 'Early Alfonsine Astronomy in Paris: The Tables of John Vimond (1320)', *Suhayl*, 4 (2004), 207–94.

———, and ———, *A Survey of European Astronomical Tables in the Middle Ages* (Leiden: Brill, 2012).

———, and ———, 'The Medieval Moon in a Matrix: Double Argument Tables for Lunar Motion', *Archive for History of Exact Science*, 73 (2019), 335–59.

Clauser, Christophor, *Practica Tütsch vff das MDXLIII Jar* (Zurich: Christoph Froschauer d.Ä, 1543).

Firmin of Beauval, *Opusculum repertorii pronosticon in mutationes aeris tam via astrologica quam metheorologica* (Venice: Ratdolt, 1485).

Fisch, Max H., and Dorothy M. Schullian, *Nicolaus Pol Doctor 1494, with a Critical Text of His Guaiac Tract* (New York: Reichner, 1947).

Franz, Heidrun, *Das Hauptwerk des Astrologen Marcus Schinnagel von 1489: Alltagsmanagement und Zukunftsdeutung an der Schwelle zur Neuzeit* (Hamburg: Kovac, 2014).

Goldstein, Bernard R, 'Lunar Velocity in the Middle Ages: A Comparative Study', in *From Baghdad to Barcelona: Studies in the Islamic Exact Sciences in Honour of Prof. Juan Vernet*, ed. by Josep Casulleras and Julio Samsó (Barcelona: Instituto Millás Vallicrosa de Historia de la Ciencia Árabe, 1996), pp. 181–94.

Graf, Klaus, 'Marcus Schinnagel, Ein Astrologe in der Zeit Maximilians I., Schöpfer des astronomisch-astrologischen Kompendiums aus Petershausen' (http://frueheneuzeit.hyotheses.org/1615, 2014).

Haage, Bernhard D, 'Astrolabium planum deutsch', *Sudhoffs Archiv*, 65 (1981), 117–43.

Heilen, Stephan, *Hadriani genitura: Die astrologischen Fragmente des Antigonos von Nikaia*, 2 vols (Berlin: de Gruyter, 2015).

Hübner, Wolfgang, *Manilius, Astronomica, Buch V*, 2 vols (Berlin: de Gruyter, 2010).

Kellner, Stephan, and Annemarie Spethmann, *Historische Kataloge der bayerischen Staatsbibliohek München, Münchner Hofbibliothek und andere Provenienzen* (Wiesbaden: Harrassowitz Verlag, 1996).

Knobloch, Eberhard, 'Astrologie als astronomische Ingenieurkunst des Hochmittelalters: Zum Leben und Wirken des Iatromathematikers und Astronomen Johannes Engel (vor 1472–1512)', *Sudhoffs Archiv*, 67 (1983), 129–44.

Kremer, Richard L, 'Marcus Schinnagel's Winged Polyptych of 1489: Astronomical Computation in a Liturgical Format', *Journal for the History of Astronomy*, 43 (2012), 321–45.

———, 'Incunable Almanacs and Practica as Practical Knowledge Produced in Trading Zones', in *The Structures of Practical Knowledge*, ed. by Matteo Valleriani (Cham, Springer, 2017), pp. 333–69.

Kyriß, Ernst, *Verzierte gotische Einbände im alten deutschen Sprachgebiet*, 4 vols (Stuttgart: Max Hettler Verlag, 1951–58).

Märker, Almuth, 'Inkunabelnabschriften in Handschriften aus dem Leipziger Universitätsbetrieb', *Wolfenbütteler Notizen zur Buchgeschichte*, 42 (2018), 151–71.

Monter, William, *A Bewitched Duchy: Lorraine and Its Dukes, 1477–1736* (Geneva: Droz, 2007).

Moulinier-Brogi, Laurence, *Guillaume l'Anglais, Le frondeur de l'uroscopie médiévale (XIIIe siècle): Édition commentée et traduction du De urina non visa* (Geneva: Droz, 2011).

Murdoch, John E, *Album of Science: Antiquity and the Middle Ages* (New York: Charles Schribner's Sons, 1984).

Nallino, C. A., ed. *Al-Battani sive Albatenii opus astronomicum [1899–1907]*, reprint ed. 3 vols (Hildesheim: Georg Olms Verlag, 1977).

North, J. D., ed. *Richard of Wallingford: An Edition of His Writings, with Introductions, English Translation and Commentary*, 3 vols (Oxford: Clarendon Press, 1976).

Nothaft, C. P. E, 'Jean des Murs's Canones tabularum alfonsii of 1339', *Erudition and the Republic of Letters*, 4 (2019), 98–122.

Oestmann, Günther, *Schicksalsdeutung und Astronomie: Der Himmelsglobus des Johannes Stoeffler von 1493* (Stuttgart: Württembergisches Landesmuseum Stuttgart, 1993).

Pedersen, Fritz S, *The Toledan Tables: A Review of the Manuscripts and the Textual Versions with an Edition* (Copenhagen: C. A. Reitzels Forlag, 2002).

Pedersen, Olaf, *A Survey of the Almagest* (Odense: University Press, 1974).

Pflaum, Jacob. *Kalendarium* (Ulm: Johann Zainer, 1478).

Pingree, David, 'From Alexandria to Baghdād to Byzantium: The Transmission of Astrology', *International Journal of the Classical Tradition*, 8 (2001), 3–37.

Porres de Mateo, Beatriz, 'Les Tables astronomiques de Jean de Gmunden: Édition et étude comparative', unpublished PhD thesis, Paris, École Pratique des Hautes Études, 2003.

Poulle, Emmanuel, ed. *Les Tables alphonsines avec les canons de Jean de Saxe: Édition, traduction et commentaire* (Paris: Éditions du Centre national de la recherche scientifique, 1984).

Regiomontanus, *Kalendarium* (Nürnberg: Regiomontanus, 1474).

———, *Tabulae directionum profectionumque* (Augsburg: Ratdolt, 1490).

Robbins, F. E., ed. *Ptolemy, Tetrabiblos* (Cambridge: Harvard University Press, 1940).

Saby, Marie-Madeleine, 'Les Canons de Jean de Lignères sur les tables astronomiques de 1321: Édition critique, traduction et étude, unpublished thesis, Paris, École Nationale des Chartes, 1987.

Saxl, Fritz, and Hans Meier, *Verzeichnis astrologischer und mythologischer illustrierter Handschriften des lateinischen Mittelalters, Bd. 3, Handschriften in englischen Bibliotheken*, ed. by Harry Bober, 2 vols (London: Warburg Institute, 1953).

Schinnagel, Marcus, *Prognostikon auf das Jahr 1491* (Ulm: Johann Zainer d.Ä, 1491).

Schröder, Alfred, 'Der humanist Veit Bild, Mönch bei St Ulrich: Sein Leben und sein Briefwechsel', *Historischer Verein für Schwaben und Neuburg*, 20 (1893), 173–227.

Sidoli, Nathan, 'Research on Ancient Greek Mathematical Sciences, 1998–2012', in *From Alexandria, through Baghdad: Surveys and Studies in the Ancient Greek and Medieval Islamic*

Mathematical Sciences in Honor of J. L. Berggren, ed. by Glen van Brummelen (Berlin: Springer, 2014), pp. 25–50.

Stöffler, Johann, and Jacob Pflaum, *Almanach nova plurimis annis venturis inservientia* (Ulm: Joannis Reger, 1499).

Thorndike, Lynn, *A History of Magic and Experimental Science*, 7 vols (New York: Macmillan, 1923–58).

Zacut, Abraham, *Almanach perpetuum* (Leira: Abraham Ben-Samu'el d'Ortas 1496).

ALEXANDRE TUR

From Computus Material to Preacher's Toolbox: Manufacturing a Bat-Book Almanac in the Fifteenth Century

Introduction

In the entire collection of 40 000 medieval manuscripts at the Bibliothèque nationale de France, only three qualify as 'bat books' or, as we used to say before Johann Grumbert's 2016 study on this particular format, books *en forme de cliquette de ladre* (in the form of a leper's clapper), or *en forme de crécelle de pestiféré* (in the form of a pestiferous rattle).[1] All three of them are rather recent acquisitions; the last two, BnF NAL 375 and NAL 482 were bought in the nineteenth century by librarians out of curiosity for their odd format.[2] Both are fourteenth-century almanacs adapted from Peter of Dacia's *Kalendarium*, a widespread cyclic calendar used to obtain reliable syzygy times for liturgical purposes, in a folded form that proved to be, if rare (or at least rarely preserved), not exceptional at the time. Because of their specific format and delicacy, both have been preserved in the Department of Manuscripts' *Réserve* since their acquisition, although they are not the most precious items in the library.

The third BnF bat book entered the library at an earlier time, and, probably because it was part of a large donated collection, it does not seem to have attracted much attention upon its arrival. Manuscript Latin 7478 had been acquired by the bibliophile Roger of

* This work was supported by the ERC project ALFA: Shaping a European scientific scene, Alfonsine astronomy, CoG 723085, PI Matthieu Husson. The author thanks the project's members for their valuable feedback and comments over the course of this research, and particularly José Chabás, Richard L. Kremer, and Eric Ramirez-Weaver for their detailed review of this final paper.

1 Johan Peter Gumbert, *Bat Books: A Catalogue of Folded Manuscripts Containing Almanacs or Other Texts* (Turnhout: Brepols, 2016).

2 Léopold Delisle, *Manuscrits latins et français ajoutés aux fonds des nouvelles acquisitions pendant les années 1875–1891* (Paris: H. Champion, 1891), I, pp. 91–92. The acquisition of these bat books took place in a context of scientific interest for portable books (even those not considered as bat books today, such as the pocket breviary, BnF Latin 10479. See also, Hercule Géraud, 'Calendrier perpétuel portatif dressé l'an 1381', *Bibliothèque de l'École des chartes*, 2 (1841), 272–80; 'Calendrier portatif du XIVe siècle', *Bibliothèque de l'École des chartes*, 44 (1883), 569; 'Calendrier portatif du XVe siècle', *Bibliothèque de l'École des chartes*, 45 (1884), 136–37.

Alexandre Tur • Bibliothèque nationale de France

Alfonsine Astronomy: The Written Record, ed. by Richard L. Kremer, Matthieu Husson, José Chabás, Turnhout, 2022 (*Alfonsine Astronomy: Studies and Sources*, 1), pp. 143-198
© BREPOLS PUBLISHERS 10.1484/M.ALFA.5.124926

Gaignières (1644–1715), before he gifted his manuscript collection to King Louis XIV in 1711. After entering the royal collection, the manuscript was rebound in what was at the time the customary 'Royal Morocco binding' for the King's manuscripts.[3] While much of the codicological information on how it was composed as a bat book was lost in the process, it is striking that it was done quite competently and not in a way that would have entirely obliterated its bat book past: the folding remains, as well as the order of the leaves (for the most part). The procedure also strengthened the manuscript, leaving it less fragile than, for instance, NAL 375, which is in its original binding. Included in the scientific section of the royal library, MS Latin 7478 remained mostly unstudied, except for a few exhibitions or presentations of 'unusual manuscripts'.[4]

Looking more closely at its organization and content, however, these appear very uncommon for a bat-book almanac. This paper aims to reclassify it more precisely within the contexts of late medieval bat books and of fifteenth-century 'enhanced calendars'. With this rare case of a non-English Alfonsine bat book almanac, we endeavour to submit a few hypotheses on the role of such an object in the Alfonsine framework and late-medieval culture.

1. A bat book or folded almanac?

1.1. From content to form: a codicological notion

The concept of a bat book is relatively recent, as it was forged by late Leiden University codicology professor, Johan Peter Gumbert, in his 2016 published catalogue of folded manuscripts.[5]

In Gumbert's sense, a bat book is a codicological type of manuscript, alternative to *codex* or *rotulus*. That means that a bat book is built in a manner entirely alien to that which we are accustomed with codex manuscripts. A bat book is made of individual folded leaves sewn together on a marginal tab, which means that there are no quires as there are in codices,

3 It is described as unbound in the 1711 inventory of the Gaignières collection ('Ancien almanach avec des cartes sur velin, in-12° non relié', BnF Clairambault 1032, #73, f. 347r). On the 1717 register of the entry of the collection in the Royal Library, an early-nineteenth-century librarian added a #73/2 item as 'Anonymi tabulæ variæ ad cyclum solarem et lunarem accomodatæ, accedunt canones, sæc. XV°', with the actual shelfmark (attributed 1744), '[Latin] 7478' (BnF NAF 5738, p. 10). As both descriptions match the object of this paper, and there is no other manuscript from the Gaignières collection matching #73[/1], it is safe to assume, as hypothesized by Gumbert, that both lines refer to the same Latin 7478. On this subject, see also the COLLECTA database on the Gaignières collection and its record for item #73: https://www.collecta.fr/permalien/COL-OUV-00078 [Accessed 28 July 2020].

4 Although it has been identified as a bat book for some time, Latin 7478 remains less highlighted than the other two, and less altered. It was, for instance, mentioned in 'Le Livre' exhibition as a relation to the displayed NAL 375 (*Le Livre* [May 17–Oct. 31, 1972], ed. by Roger Pierrot and Marcel Thomas [Paris: Bibliothèque nationale, 1972], p. 74, #222). See also Hermann Degering, 'Der Bucheinband: Ein calendarium pugillare mit Computus aus dem Jahre 1294', in *Buch und Bucheinband, Aufsätze und graphische Blätter zum 60. Geburtstage von Hans Loubier*, ed. by Max Joseph Husung (Leipzig: Hiersemann, 1923), pp. 79–88, mostly pp. 81–83 and pl. 7. Latin 7478 is mentioned in *Mappemondes AD 1200–1500: Catalogue préparé par la Commission des cartes anciennes de l'Union géographique internationale*, ed. by Marcel Destombes (Amsterdam: N. Israel, 1964), p. 61 nb. 23 for the map of the world displayed in f. 17D, maybe a later adjunct (see below).

5 Gumbert, *Bat Books*.

Figure 1. Bat book; *en forme de crécelle de pestiféré or en forme de cliquette de ladre* (shaped like a leper's clapper or rattle).

(a) A pestiferous rattle. Paris, BnF, Arsenal, MS-5080, f. 373^{r}.
(b) A leper's clapper. Paris, BnF, Français 9140, f. 151^{v}.
(c) Bat book. Paris, BnF, NAL 482, tail edge. All reproduced with permission of the BnF.

and these leaves are to be unfolded to be read, from the outer edge to the stub, sometimes with several compartments for each step of the unfolding process.

While Gumbert was not the first to notice such books (there have been regular mentions since the nineteenth century, and in the case of English bat book almanacs, we have Hilary Carey's 2003 and 2004 illuminating papers), he is the first to give such a strong physical definition that goes back to the core of the book and is disconnected from the content, which enables a better typological view.[6]

Since the release of Gumbert's catalogue, even the name, bat book, has been widely adopted, in part because of the appealing metaphor, 'For myself I have gradually grown accustomed to calling them "bat books", because when in rest they hang upside-down and all folded up, but when action is required they lift up their heads and spread their wings wide', but also because previous English denominations were not really satisfactory.[7] Even 'folded book' or 'folded manuscript' can be misleading, as a traditional codex can obviously contain folded leaves without being a bat book. *A fortiori*, 'folded almanac' confuses form and content, weakening the coherence of the notion.

The French denominations mentioned above (books shaped like a leper's clapper or rattle) are somewhat more interesting, because they relate to the appearance given by the

6 Hilary M. Carey, 'What is the Folded Almanac? The Form and Function of a Key Manuscript Source for Astro-medical Practice in Later Medieval England', *Social History of Medicine*, 16 (2003), 481–509; Hilary M. Carey, 'Astrological Medicine and the Medieval English Folded Almanac', *Social History of Medicine*, 17 (2004), 345–63. See also the nineteenth-century references mentioned above, and, more recently, André Vernet, 'Le Calendrier portatif de Mamert Fichet (1440)', *Bulletin de la Société Nationale des Antiquaires de France*, 1959 (1961), 243–44; Monique-Cécile Garand, 'Livres de poche médiévaux à Dijon et à Rome', *Scriptorium*, 25 (1971), 18–24; Chelsea Silva, 'Opening the Medieval Folding Almanac', *Exemplaria*, 30 (2018), 49–65.

7 Gumbert, *Bat Books*, p. 19.

main codicological characteristic, the tab sewing, that may evoke a clapper or a rattle, and mostly because, as Gumbert points out, the expression is found in the 1472 catalogue of the Clairvaux library, which seems to prove that it was already used with that meaning in the fifteenth century.[8] For twenty-first-century readers, however, we must admit (and appreciate) that clappers and rattles are not as evocative.

1.2. Bat books in the late Middle Ages

In his catalogue, Gumbert identifies sixty-three bat book manuscripts, either currently preserved in public libraries, or that were attested precisely enough in the last century. He gathers them in seven groups, depending on their date and content, which gives an interesting picture of the history of this type of book.[9]

Bat books seem to have been invented in the middle of the thirteenth century as a solution to, as Gumbert puts it, 'make something that is small outside but large inside'.[10] The first bat books were not specifically designed to be astronomical tools: they carried texts as diverse as historical chronicles, legal texts, liturgical or philosophical treaties, as well as medical recipes, and, early as it was, calendars and computistical tables. They were easy to transport as they were smaller than codices with the same text, but also thicker and a good choice for scholars or practitioners wanting to travel with lots of texts.

On that market, however, they would soon face competition from what we now call 'pocket bibles', which are small-dimension codices featuring very thin parchment and small script, thus containing a lot of text in a small and light book. They were easier to make and easier to use.[11]

The earliest bat book known to Gumbert is a partially preserved compendium of historical texts on Glastonbury Abbey, today in the Bodleian Library, dated around 1265. He identifies six more of the oldest, non-astronomical bat books from the thirteenth century.[12]

Around 1300, however, such textual bat books had disappeared in the face of the competition. They only remained the preferred alternative for tabular content, in fact, mostly calendars or astronomico-computistical tables. Calendars need a large surface for content that cannot be divided over several leaves nor written too small as offered by the pocket bible technology, but in return, they do not need many leaves, usually about fifteen if we count one leaf per month plus a few extras. In the fourteenth and fifteenth centuries, Gumbert identifies fifty-six bat books, only three of which are not calendars or almanacs in tabular form: a Bosnian collection of prognoses, following a calendar sequence but in textual form, dating from the 1450s, now in Parma, and two 'late breviaries' dating to the

8 André Vernet, *La Bibliothèque de l'abbaye de Clairvaux du XIIe au XVIIIe siècle* (Paris: Editions du Centre national de la recherche scientifique, 1979), I, p. 334, #2221: 'Item une autre Journel fait en manière d'une cliquette de ladre, et y sont les leçons et respons d'apres Pasques et Pentecoste et le commun des sainctz, commençant on second feuillet d'apres le kalendier || nostre configuratum et finissant on penultime pedibus marinos || Ainsi signé &55'. See Gumbert, *Bat Books*, p. 19, note 8.

9 Most of this description is borrowed from Gumbert, *Bat Books*, passim.

10 Gumbert, *Bat Books*, p. 23.

11 Gumbert, *Bat Books*, pp. 24–25.

12 Gumbert, *Bat Books*, p. 25.

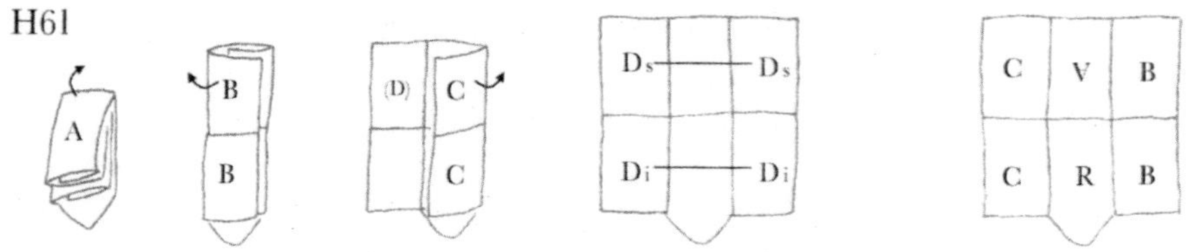

Figure 2. Gumbert's figure of the H6l pattern.

© Johan Peter Gumbert, *Bat books: A catalogue of folded manuscripts containing almanacs or other texts* (Turnhout: Brepols, 2016), p. 20. Reproduced with permission.
Latin 7478 has a H6r pattern: its leaves open on the top as well, but the first compartment to be unfolded (B) is then on the right, and only the last one (C) is on the left.

late-fifteenth or even early-sixteenth century and seeming to be witnesses of a surprising Cistercian trend that might be related to the invention of printing.[13]

All of the other fifty-three bat books from the fourteenth and fifteenth centuries can be described as 'calendars' or 'almanacs', two categories for which a dividing line is not always very easy to determine. Gumbert distinguishes between 'Continental Calendars' (nine + Parma prognoses), 'Continental Almanacs' (14), and three types of 'English Almanacs'; 'Early' (10), 'Mid-[Fifteenth] Century' (5) and 'Late' (15).[14] This census gives a new scope to a phenomenon long believed to be specifically English. The distinction Gumbert draws between calendars and almanacs is described as astrological; while calendars might encompass some basic precalculated astrological data such as lucky days, they are mostly filled with liturgical content. Some are perpetual calendars, sometimes with basic computistical tables, mostly to find the golden number and the date of Easter. On the other hand, an almanac gives enough astronomical information to allow an astrologer to make his own predictions in a variety of subjects.[15]

1.3. Latin 7478: A codicological description

Gumbert classes manuscript BnF Latin 7478 in the 'Continental Almanacs' category. He dates it, convincingly enough, to 1456 Italy, as we shall see later.

Latin 7478 has seventeen leaves, measuring 10 × 7 cm folded and 19 × 22 cm unfolded, making it rather large for a continental bat book. All leaves are folded to the same pattern, categorized by Gumbert as a 'H6r' horizontal pattern (cf. Fig. 2). This pattern, one of the main seven known to Gumbert, is somewhat uncommon; most bat books used rather a vertical pattern — the last fold being vertical, which means that each leaf first opens on

13 Gumbert, *Bat Bbooks*, pp. 24 and 207–8.
14 Gumbert, *Bat Books*, pp. 40, 73, 121, 153, 160 (respectively).
15 We qualify this distinction in the following paragraphs. In fact, astronomical information related to almanacs tends to be limited to a more effective indication of the dates and hours of syzygies than just the golden number, which does not allow much direct astrological use without further calculations. It is important to note that Gumbert's educated distinction does not necessarily reflect any self-designation as 'kalendarium' or 'almanach' in the bat books' titles or canons.

the right or left, whereas the leaves of Latin 7478 open on the top. The H6 pattern is rare but not distinctive: Gumbert found instances in every time period and place, whereas a V8 pattern, for instance (with eight compartments instead of six), would have been characteristic of the late English almanacs.[16]

Except for the first leaf, which is probably sewn backwards, maybe due to a rebinding mistake, the folded part is first visible on the recto while looking at the manuscript like a codex; however, it has to be turned to be read, so that the tabs (or now the binding) always remain at the bottom. Most of the content, and especially the tabular content, is written on the unfolded face of the leaves. In most cases, the folded recto and verso are mainly the support for a repeated title, and all intermediary compartments are empty. However, this leaves very little space for canons and commentary, and on some leaves, part of the text goes over the edges of the folded compartments. On the first leaf, the copyist seems to have made use of this possibility without careful consideration; the text initiated on the unfolded face extends onto an intermediary compartment. It is thus only readable with the leaf half-folded. When completely folded, it appears as a partial (and unusable) text beside the title. By contrast, one of the last leaves, the *tabola ad inveniendum ciclum solarem et lunarem* (f. 15) exploits all compartments, allowing different texts, schemas, and tables to be read in any position. This could be a clue that the quality of the codicological features of the copyist may have improved through experience while manufacturing this copy.[17]

While sharing the H6r pattern, the last leaf, copied on a different, less refined sort of parchment — but seemingly by the same hand — might be a later adjunct. Describing the earthly and heavenly spheres with scarce verses giving the rotation period of each planet, it is also less in tune with the rest of the almanac, which we shall now consider.

2. A rare copy of a widespread *Kalendarium*

2.1. A late anonymous almanac

Like most bat books, Latin 7478 is mostly a seventy-six-year lunations calendar with some astronomical and astrological additions. The first leaf is called 'Tabula quattuor cyclorum decemnovenalium' (*tabula 4^or^ ciclorum 19^m^*) and contains a diagram that is used to obtain the golden number of each of the seventy-six plus two years included, 1456–1533. Additionally, there are brief canons explaining how to use it, together with the following calendar, to find the precise date and time (to the minute) of each of the luminaries' mean conjunctions and oppositions in those seventy-eight years.[18] The next twelve leaves, folios 2–13, are dedicated to the calendar, one per month. Each displays a table with a line for each day of the month, and four sets of columns including:

16 Gumbert, *Bat Books*, p. 20.

17 This would mean that Latin 7478 is the first bat book manufactured by its copyist. We revisit this hypothesis in the last part of this paper.

18 Cycles of seventy-six years were widely used throughout the medieval era as they combined the four-year solar cycle and the nineteen-year lunar cycle, thus making them particularly accurate in describing and predicting luni-solar phenomena, both astronomical and liturgical (see below).

- dates and times of mean conjunctions and oppositions for each cycle, depending on the year's golden number;
- core calendar data (golden number, Sunday letter, Roman date, and number of the day);
- a textual qualification (mostly liturgical feasts and rites, but occasionally brief astronomical or astrological information);
- an additional sub-table gathering the Sun longitude in degrees, the lunar letter (*littera signorum*: allowing the calculation of the Moon longitude in signs with the help of the table on f. 14), length of the day, time of sunrise, and time of sunset.[19]

The next leaf, folio 14, is a *tabula signorum lune per totum annum*, using the golden number and lunar letter (found in the calendar) to equate the zodiacal position of the moon at any time of the nineteen-year cycle. It is completed with short canons and basic warnings on what to do and not to do while the moon is in each zodiacal sign, which would allow someone to calculate auspicious periods.

These first fourteen leaves work as a whole, as if they could be distributed separately without any lack of information. The next three are additional and allow further liturgical calculations.

Folio 15 is the most complicated one as it features several folded compartments. It is entitled 'Tabula ad inveniendum ciclum solarem et lunarem', and it gives computistical data, with two diagrams on the unfolded face: on the left, the first one equates a golden number (*numerus aureus*), an epact (*pacta*), and liturgical keys (*claves terminorum*) for the years 1456–74; on the right, the second diagram gives the dominical letters for the years 1456–83. Short canons are given underneath and continue on the unfolded 'B' face, beginning with 'si vis invenire lunam cotidie'. On the unfolded 'C' face, a more astrologically oriented table helps us to find the ruling planet for each hour of the week. Finally, on the folded verso, a drawing of two hands explains how to manually count the liturgical keys to find the date of the moveable feasts. On this leaf, we also find the copyist's signature and date: *1456, 25a die 7tembris, frater Paulus de Kignin*, above some additional mnemonics for the liturgical keys.[20]

Folio 16 is a classical *tabula ad inveniendum festa mobilia et ebdomadas*, making it possible to calculate the date of the liturgical feasts with only the golden number (or epact) and the Sunday letter (even when we know nothing of the liturgical keys).

As we have already seen, the final leaf, folio 17, is somewhat different to the preceding ones. In all likelihood written by the same hand but on a different, cruder sort of parchment, it might be a later adjunct, or a sign that the bat book maker was a little short on raw material. The content is also very peculiar: without a title, it contains a sort of combination of a didactic cosmographical system diagram and a small geographical *mappa mundi* (Fig. 3a). All earthly and heavenly spheres are depicted as concentric circles, one for each of the four elements, water coloured in green, air in blue, and fire in red, and one for each of the planets, including the duration of its revolution. While this sort of diagram is fairly

19 These astronomical data and their source are investigated in the next section. See also their transcription in the Appendix.

20 We further investigate the identity and role of Paulus de Kignin (or 'Szignin') in the final section of this paper.

Figure 3. A cosmographical system or *mappa mundi*? (a) Paris, BnF, Latin 7478, folio 17D.

(b) "Planisphere found in a geographical poem manuscript of the fifteenth century" (facsimile from Santarém, *Atlas*, XCIX, pl. 26, #1, BnF Ge BB 248). Both reproduced with permission of the BnF.

common in didactic literature in the last centuries of the Middle Ages, its 'Earth sphere' is usually very schematic as well, often just a 'T-O' figure. This *mappa mundi* drew historians' attention because of its detailed geographical depiction.[21] Although too small to be as precise as the large and meticulous contemporary maps, this figure is not as schematic as most *mappae mundi*, usually presenting continents as compartments (T-O and Y-O maps, for instance) or zones (such as the ptolemaic *climata*). On the contrary, it depicts a realistic geographical description centred on the Mediterranean Sea, with north on top and labels such as 'Francia', 'Spania' (?), 'Italia', 'Turchia', 'Caucasia' or 'Egyptus', and a clear focus on central Europe ('Polonia', 'Ungaria', 'Scla[via]', 'Alba').[22]

Except for the copyist's name, the date, and some conclusions we can draw from the choices made for the cycles and for the sanctorale, manuscript Latin 7478 does not contain much information about its origin and the work from which it was copied. Gumbert observes that the almanac is neither Peter of Dacia's *Kalendarium* — the most popular in his 'Continental Almanacs' category of bat books — nor one of John Somer's or Nicholas of Lynn's (mostly copied in the fifteenth century) English Almanac bat-books.[23] This is probably due to the fact that, having been copied in 1456, Latin 7478 is a very late example of a continental almanac, and at that time, Peter of Dacia's 1292 *Kalendarium*, even updated, was mainly obsolete. Even John Somer's and Nicholas of Lynn's 1397–1462 almanacs would not have been useful for very long in 1456. In fact, a study of the cycles chosen by Paul of Kignin (or his patron) shows that he updated a 1439–1514 76-year calendar. In order to begin in 1456 without 'wasting' almost an entire cycle, he changed the initial diagram (f. 1D, see Fig. 4) to switch the first cycle from 1439–57 to 1514–33. The last two years, 1532 and 1533, are given a marginal equivalence to 1456 and 1457. The offhandedness with which this 'update' is presented above the original layout, not to mention the neglected consequences on the syzygy times, leads us to think that this decision did not originate from a very skilled astronomer (or even calendar-maker).

---

21 *Mappemondes AD 1200–1500*, p. 61, #23.

22 *Mappemondes AD 1200–1500* categorizes this sort of mappemond as a 'type D', or 'ecumenical mappemond with nomenclature and configurations'. In his 1852 *Atlas composé de mappemondes et de cartes hydrographiques et historiques depuis le XIe jusqu'au XVIIe siècle… devant servir de preuves à l'ouvrage sur la priorité de la découverte de la cote occidentale d'Afrique au dela du Cap Bojador par les Portugais et à l'histoire de la géographie du Moyen Âge*, the viscount of Santarém, Manuel Francisco de Barros e Sousa (1791–1856), reproduces two such mappemonds (sadly with a very vague quotation of the manuscripts in which he found them): this one, *monumento LV* (plate 7, #13), as a 'geographical system found in a manuscript of the beginning of the 14th century' [*sic*], and another one, very similar, labelled: 'planisphere found in a geographical poem manuscript of the fifteenth century' (mon. XCIX, pl. 26, #1, see Fig. 3b). I have not been able to identify this other manuscript, but it seems highly probable that it is either the direct source of Latin 7478 mappemond or the copy of a close common source. I have consulted the Santarém Atlas through its 1989 facsimile edition: Manuel Francisco de Barros e Sousa de Mesquita de Macedo Leitão e Carvalhosa Santarém, *Atlas du vicomte de Santarem*, facsimile edition of the final maps (Lisbon: Administração do Porto de Lisboa sous les auspices de la commission nationale pour les commémorations des découvertes portugaises, 1989). The textual description in Manuel Francisco de Barros e Sousa de Mesquita de Macedo Leitão e Carvalhosa Santarém, *Essai sur l'histoire de la cosmographie et de la cartographie pendant le Moyen Âge et sur les progrès de la géographie après les grandes découvertes du xve siècle*, 3 vols (Paris: impr. de Maulde et Renou, 1848–1852), III, pp. 227–29 (LXXXVII), quoted by *Mappemondes AD 1200–1500*, is similar to, but not consistent with, either of these two mappemonds, both with north on top, and it could be a third, more distant witness.

23 Gumbert, *Bat Books*, pp. 74–75.

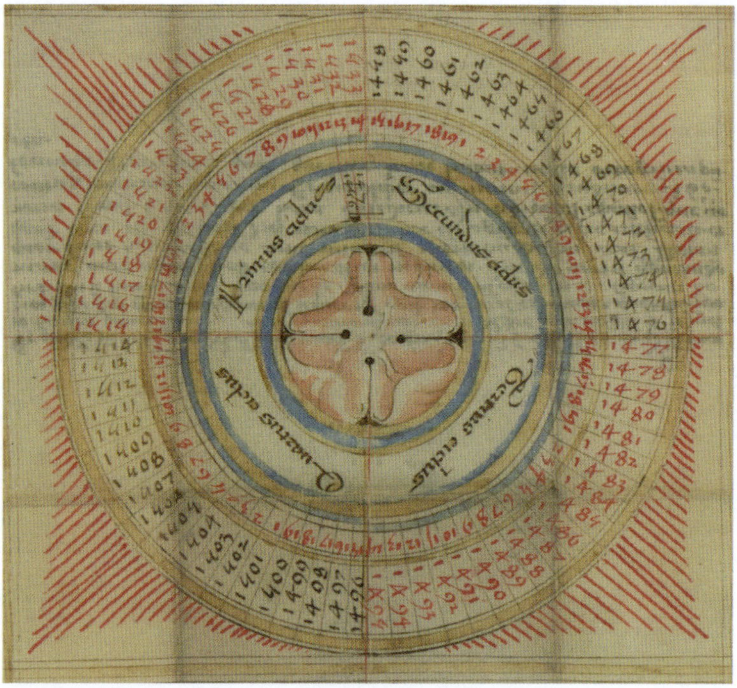

Figure 4. Diagram of the years constituting the four cycles, Paris, BnF, Latin 7478, folio 1Ds. Source gallica.bnf.fr.

On the other hand, the unusual composition of each cycle, the non-updated astronomical data of the calendar, and even some of the canons help point to another fifteenth-century, influential seventy-six-year calendar: John of Gmunden's calendar for 1439–1514.[24] Indeed, it turns out that most (but not all) of the content of Latin 7478 originates from Gmunden's calendar, although it is, to our knowledge, the only bat book for which this is the case.

2.2. *Works (mostly) copied from John of Gmunden*

John of Gmunden (before 1385–1442) was a skilled astronomer and prominent figure at the University of Vienna. Besides his lectures on mathematics, physics, and astronomy, he introduced the Alphonsine Tables in Vienna and computed syzygy calendars, first for 1420–38, and then 1439–1514. These belong to the last generation of what Philipp Nothaft

24 Unlike most bat-book almanacs, Latin 7478's nineteen-year cycles do not match the Metonic cycles: the first year of each cycle has a golden number of fifteen (and the Metonic cycle starts back on the sixth year of these cycles). Although this is astronomically correct, it does not seem the most obvious choice, and the calendars of Peter of Dacia, John Somer, and Nicholas of Lynn all used a Metonic cycle, even when they did not date from the first year of the cycle. This is another hint that the original author of this almanac was mastering the astronomical science behind the calendar cycles and could have easily changed them to update the almanac, rather than naively switching years.

calls 'enhanced calendars'.[25] According to Nothaft, the production of such calendars from the late-thirteenth to the fifteenth century was a reaction to the inactivity of the Church when it came to addressing the growing gap between the 'official syzygies' (as they could be deduced from the golden number according to the Nicaea instructions) and the observed ones, and the liturgical consequences in computing the moveable feasts. In this sense, Gmunden's calendars, improving on those of predecessors such as the pseudo-Grosseteste, pseudo-Bacon, Peter of Dacia, William of Saint-Cloud, John Somer, or Nicholas of Lynn, proved at least as influential as his astronomical tables, with more than 150 remaining copies in Latin and German, and even printed adaptations.[26] His successors in Vienna, Georg Peuerbach and Regiomontanus, continued his almanac work, not without bringing their own influence.

In this sense, it is unsurprising that a mid-fifteenth-century Italian bat book maker like Paul of Kignin (or his patron) would turn to Gmunden's calendar; why he would not proclaim it, placing himself under the moral authority of the renowned astronomer, is more of a surprise.

It must be underlined that, unless it was copied onto an already heavily altered copy, the manuscript Latin 7478 is not an exact reproduction of Gmunden's calendar, as it is found in most codices.[27] While it usually takes no more than twenty leaves (including twelve for the calendar), Paul of Kignin seems to have tried to save as much space as possible, mostly by cutting or abbreviating canons and auxiliary tables. On the other hand, he added some related content not originating from Gmunden (at least, not from the copies we have seen). Table 1 summarizes the main differences.

Table 1. Comparison between Latin 7478 and Gmunden's calendar (Cod. 2440).

Gmunden (Cod. 2440)	**Latin 7478**
F. 1^{v}–13^{r}. Calendar.	F. 2D–13D.[28] Calendar. [*Mostly identical*]
F. 13^{v}. Rota: golden number and liturgical key for a 19-years cycle, adaptable for the years 1439–1818 ('Aureus numerus 15 et clavis festorum mobilium 22 ad annos hic in parva tabula descriptos'). Canon: 'Si vis scire aureum numerum alicujus anni …'	F. 15D (left). Rota: golden number, epact and liturgical key for a 19-years cycle, 1456–1474. Canon: 'Queras in hac rota aureus numerus, claves terminorum et pactam …' [*Adapted*]

25 C. Philipp E. Nothaft, *Scandalous Error: Calendar Reform and Calendrical Astronomy in Medieval Europe* (Oxford: Oxford University Press, 2018), see, in particular, chapter 5.1.

26 See Rudolf Klug, *Der Astronom Johannes von Gmunden und sein Kalender* (Linz: Pirngruber, 1912), and, more recently, Kathrin Chlench, Beatriz Porres de Mateo, and Rudolf Simek, 'Johannes von Gmunden: Personalbibliographie und Handschriftenverzeichnis', in *Johannes von Gmunden (ca. 1384–1442): Astronom und Mathematiker (2006)*, ed. by Simek and Chlench (Vienna: Fassbaender, 2006), pp. 183–223.

27 Unfortunately, Gmunden's *Kalendarium* lacks a critical edition. In comparing it to BnF Latin 7478, we have used the following manuscripts: Vienna, ÖNB, Cod. 2440, ff. 1^{v}–13^{r}, which is usually considered an early copy or even an autograph of John of Gmunden; Munich, BSB, Clm 14504, ff. 1^{v}–20^{r} (digitized online: http://mdz-nbn-resolving.de/urn:nbn:de:bvb:12-bsb00046291-3); Dresden, SLUB, Mscr. F.95.

28 When needed, we used Gumbert's way of referring to each compartment of a bat book leaf, as shown in Fig. 2. We caution that Latin 7478 is in fact a H6r bat book, which means that compartment B unfolds first on the right, then C on the left (symmetrically to what is represented in Fig. 2 for a H6r bat book).

Gmunden (Cod. 2440)	**Latin 7478**
F. 14^{r}. Rota: dominical letter for a 28-years cycle, adaptable for the years 1439–1831 ('Cyclus solaris 20 littera dominicalis 7 ad annos hic in parva tabula descriptos'). Canon: 'Si litteram dominicalem et numerum cycli solaris in aliquo anno scire volueris…'	F. 15D (right). Rota: dominical letter for a 28-years cycle, 1456–1483. Canon: 'Queras in hac rota litteram dominicalem ab annis Domini 1456…' [*Adapted*]
F. 14^{v}. Tabula intervalli. Canon: 'Si vis scire intervallum, id est numerum septimanarum que sunt a festo Nativitatis Christi…'	[*Missing*]
F. 14^{v}–15^{r}. Canon: 'Si vis scire quandoque festa mobilia, scilicet 70me, 40^{e}, Pasce, Rogationum et Penthecostes, vide primo in kalendario loca clavium istorum festorum…'	[*Replaced*]
	F. 15B. Canon: 'Si vis invenire lunam cotidie, id est quot dies habet et quando facit revolutionem…'
	F. 15B. De clavibus. 'Prima igitur Jan[uari]i jacet 70 [Setpuagesime] clavis…'
	F. 15R. Diagrams and short canons to compute manually epacts and liturgical keys: 'Idem quando est bisextus cuilibet clavi…'
F. 15^{r}–16^{v}. Canon: 'Si per has duas tabellas in aliquo anno volueris scire intervallum, et quandoque festa mobilia…'	F. 16D. Canons: 'Omnia festa mobilia et ebdomadas cum diebus superfluis invenies super litteram dominicalem…' [*Adapted and shortened*]
F. 15^{v}. Table of Easter letters depending on golden number and dominical letter.	[*Missing*]
F. 16^{r}. Table of moveable feasts and intervals depending on Easter letter, golden number and dominical letter.	F. 16D. Table of moveable feasts and intervals depending on golden number and dominical letter.
F. 16^{v}–17^{r}. Canon: 'Ad sciendum per hoc kalendarium tempus conjunctionum et oppositionum solis et lune…'	F. 1Ds. Table of the years constituting the four 19-years cycles.[29] F.~1Di–1BRC. Canon: 'Ad sciendum per kalendarium sequens tempus conjunctionum et oppositionum solis cum lune…' [*Abbreviated*]
F. 17^{v}. Tabula continuationis conjunctionum et oppositionum. Canon: 'Transactis igitur 76 annis, quod erit anno Domini 1515 currente, si post hoc vis scire tempus conjunctionis vel oppositionis…'	[*Missing*]
F. 18^{r}. Canon: 'Si vis scire in quo signo zodiaci et quotto gradu ipisus sit sol quocumque die anni…'	[*Missing*]

29 See Fig. 4. While this table is clearly derived (and adapted) from Gmunden's cycle choices, there is no equivalent table in any of the Gmunden manuscripts we have been able to check.

Gmunden (Cod. 2440)	**Latin 7478**
F. 18v. Tabula signorum lune etc. F. 19r. Canon: 'Si scire desideras quocumque die anni in quo signo sit luna, intra kalendarium precedens cum die mensis prout dictum est de sole…'	F. 14D (left). Tabula signorum lune per totum annum. [*Identical*] F.~14Di. Canon: 'Si scire desideras quocumque die anni in quo signo sit luna, intra kalendarium precedens cum die mensis…' [*Abbreviated*]
F. 19v. Canon: 'Ad habendum autem bonum tempus seu malum pro flebotomia seu minutione…'	[*Replaced*]
	F. 14D (right). Verses giving the medical qualities of each sign: 'Aries: nil capiti facias, Aries cum luna refulget. Brachia tunc minuas…'[30]
	F. 15C. Table of the hourly ruling planets. Canon: 'Si vis scire dominationem planetarum in omni hora…'
F. 20r. Canon: 'Si vis scire quantitatem, id est numerum horarum et minutorum diei artificialis et etiam noctis per kalendarium precedens…'	[*Missing*]
F. 20v. Canon: 'Si vis scire quotta hora et quotto minuto oriatur sol vel occidat aliquo die…'	[*Missing*]
	F. 17D. Diagram: earthly and heavenly spheres, with a mappemond and the duration of each planet's revolution.

While Latin 7478 is globally adapted from Gmunden's calendar, significant differences remain both on large and smaller scales. Most canons are abbreviated or adapted, even when they have the same incipit and/or address the same topic. Diagrams and tables are also affected. There is in fact much ingenuity in the way Gmunden's two tables for moveable feasts are very practically condensed into one with a very limited loss of information. In a similar way, Gmunden's rotas for finding the golden number and dominical letter of any given year (up to the nineteenth century) are adapted to match only one cycle, sparing one table each while losing mostly useless information. It is noteworthy that the canons are also adapted, and their new version, coherent with the new diagrams, remains very intelligible, avoiding a rather common flaw in the late Middle-Ages process of adapting diagrams.

These are clues that Paul of Kignin was probably using an intermediary source between Gmunden (as handed over in ÖNB Cod. 2440) and his own copy. In fact, manuscript Latin 7478 abounds in hints that its copyist was not a skilled astronomer himself. The curious Latin spelling of technical words, for instance, already noticed by Gumbert, or the

30 These are popular verses from the *Salernitan Regimen sanitatis*, (mostly) as edited in: *Collectio Salernitana ossia documenti inediti, e trattati di medicina appartenenti alla scuola medica Salernitana*, ed. by Salvatore de Renzi (Naples: Filiatre-Sebezio, 1852), I, pp. 486–87. See also the transcription in the Appendix.

Table 2. Tables of mean syzigies for the first cycle in June (Cod. 2440 and Latin 7478).

	Gmunden (Cod. 2440)						Latin 7478					
	Conjunctio			Oppositio			Conjunctio			Oppositio		
		h	m		h	m		h	m		h	m
1	5	10	54				5	10	54			
2	13	23	35	9	5	15	13	23	35	9	5	15
3												24
4	2	19	44	17	1	24	2	19	44	17	1	5
5				6	14	5				6	14	46
6	10	8	25				10	8	25			
7				14	2	46				14	2	55
8	18	4	34	3	22	55	18	4	34	3	22	36
9	7	17	15				7	17	15			
10				11	11	36				11	11	45
11	15	13	25				15	13	25			
12				19	7	45				19	7	26
13	4	2	6	8	20	26	4	2	6	8	20	35
14	12	14	47	16	16	35	12	14	47	16	16	16
15	1	10	56				1	10				
16	9	23	37	5	5	16	9	23	56	5	5	57
17				13	17	57				13	17	6
18	17	19	48				17	19	37			
19				2	14	6				2	14	47
20	6	8	27				6	8	46			
21	14	21	8	10	2	47	14	21	27	10	2	56
22				18	22	56				18	22	37
23	3	17	17				3	17	8			
24				7	11	37				7	11	47
25	11	5	58				11	5	58			
26				15	7	47				15	7	28
27	19	2	7	4	20	28	19	2	7	4	20	9
28	8	14	48				8	14	48			
29	16	10	57	12	9	9	16	10	57	12	9	9
30	5	23	38	1	5	18	5	23	38	1	5	18

dubious grammatical 'corrections' that appear when comparing the text of the canons is one such hint. His treatment of the astronomical data in the calendar is even more conclusive.

The twelve monthly tables of Latin 7478 are mostly identical with the examined Gmunden manuscripts. The structure is preserved; only the Easter letters, which appear in March and April on ms. ÖNB 2440 are missing in Latin 7478, which was to be expected as the auxiliary table they feature in has also been removed. The data itself is mostly identical, except for what seem to be copy mistakes. Out of the 9108 numbers displayed in the twelve month tables, we count 271 'deviant' numbers, different from ÖNB 2440 (one in thirty-four). While they mostly concern minutes, they could, albeit rarely, be computation approximations.[31] Some are obvious misreadings (some even corrected afterwards), and in certain cases an entire column is displaced, involving a chain of errors (cf. Table 2). We incidentally learn that tabular data was often copied by column and not by row.[32]

Even more noteworthy is the initiative to shift the first nineteen-year cycle. As described above, Paul of Kignin decided to abandon Gmunden's first nineteen-year cycle (1439–57), which was already almost over when he copied the manuscript in 1456, to replace it with a new cycle for 1515–33 at the end of the original timespan. In doing so, he seems to have forgotten that seventy-six-year calendars are not, in fact, perpetual calendars, and while a simple addition is enough to go from one cycle's syzygy times to the fifth, it is still a mandatory operation. In fact, Gmunden gives specific instructions to do so, even providing a shift table to allow the computing of new cycles up to the nineteenth century.[33] The astronomer who helped Kignin prepare his material (or maybe wrote the copy he drew from) would have known he had to subtract five hours, fifty-two minutes, and thirteen seconds from the 1439–57 cycle's syzygy times. Kignin only changed the initial diagram associating years with each cycle, thus placing inaccurate times in the calendar.

3. How was a bat-book almanac useful in the mid-fifteenth century?

3.1. A medical-astrological companion or a liturgical tool?

In spite of Hilary Carey's and Johann Gumbert's serial studies on folded almanacs and bat books as a genre, the context of how they were used remains largely hypothetical. Analysing twenty-nine English folded almanacs, Carey postulates a mostly astrological use for medical practitioners, and a real influence in diffusing astrological medicine in fifteenth-century England.[34]

Based on bat books adapted from John Somer's and Nicholas of Lynn's almanacs, these conclusions do not apply very well to Latin 7478. Similarly to the *Kalendarium* it derives from, Latin 7478 only offers very limited astrological or medical data.[35] Except for syzygy times, the calendar only gives an approximation of the solar longitude to the degree, and the way to calculate the lunar longitude to the decan using the dedicated table. Interpretation is limited to well-known brief verses (more concise than the original Gmunden canon) supplying advice for each sign. For instance, the only astrological/medical instruction while the Moon is in Taurus is the following:

31 In rare cases, however, there is room for doubt. See the Appendix below, where the differences between Latin 7478 and ÖNB 2440 are highlighted.

32 This is discussed in Richard L. Kremer's forthcoming edition of John of Murs's *Tabulae permanentes*.

33 ÖNB Cod. 2440, f. 17ᵛ. This table appears in the (few) copies we have checked, but may have not been present in the exemplar Paulus was copying.

34 Carey, 'Astrological Medicine and the Medieval English Folded Almanac'.

35 While it is uncertain whether (and in what extent) Gmunden was involved with astrology in Vienna, very little appears in his *Kalendarium*. See *Johannes von Gmunden - zwischen Astronomie und Astrologie*, ed. by Rudolf Simek and Manuela Klein (Vienna: Fassbaender, 2012).

> Trees are to be planted while the Moon have Taurus. You may build, and spread Earth seeds. The physician is not to touch neck with iron.[36]

Latin 7478 does not contain any medical or astrological addition to the calendar, as is often the case in English almanacs: no Zodiac Man or Vein Man, no information on planetary astronomy, not even bloodletting tips or Egyptian days in the calendar, and no medical canons like those of Nicholas of Lynn (or astronomical ones, like those of John Somer).

Of course, syzygy dates and times have in themselves astrological and medical value, and any portable calendar can prove a useful tool for a medical practitioner, if only for the timing of the remedies, which was crucial in medieval medicine.[37] However, this possible use does not seem to be highlighted in Latin 7478.

On the contrary, liturgy is clearly an object of interest in this almanac. This comes as no surprise, as the first objective of 'enhanced calendars' like Gmunden's *Kalendarium* was to correct flaws in the official liturgical calendar.[38] As a matter of fact, the *Kalendarium* itself provides more tools to compute the temporal liturgy than astrological or medical ones. These tools, such as the table of moveable feasts, are improved in Latin 7478, and some related tools are added, such as the canons and diagrams on liturgical keys.

This new content is not in itself a major innovation: hand diagrams, for instance, have a widespread tradition going back (at least) to Bede, revitalized by the *computus manualis* literature in the thirteenth and fourteenth centuries.[39] Likewise, *claves terminorum* are a dated way of computing liturgical feasts for the fifteenth century, and partly redundant with the table of moveable feasts. By compiling them, however, one seems to provide a portable toolbox for liturgical *computus*.

3.2. *A Franciscan* vade mecum

The calendar tables themselves bear a very useful piece of liturgical information: the sanctorale. Unlike the numerical data of these tables, Latin 7478's sanctorale is very distinct from Gmunden's model.

Of the 377 feasts mentioned in at least one of the two sanctorales (Latin 7478 and Cod. 2440), only 154 are common to both. Latin 7478 is more precise, providing 280 feasts in 198 days (of the 365), against 232 feasts in 214 days in Cod. 2440. That means that Paul of Kignin — at least if he copied from an exemplar using a sanctorale similar to Cod. 2440's — did not only add new festivals, more suited to his context of production, but he also erased some he found useless, even though he was usually willing to write several festivals for one day.

36 'Taurus — Arbor plantetur cum luna Thaurus habebit. Edificare potes, et sperges [*sic* for *spergas*] semina terre. Et medicus caveat cum fero tangere collum.' (Latin 7478, f. 14; see the entire transcription in the Appendix and ed. *Collection Salernitana*, p. 486).

37 Carey, 'Astrological Medicine and the Medieval English Folded Almanac', pp. 353–54.

38 See above and Nothaft, *Scandalous Error*.

39 See Nothaft, *Scandalous Error* for an overview of the subject. Many thanks to Laura Fernández Fernández for sharing with me her soon to be published paper on computus hand diagrams in RBME Ms. O-II-10, including an extensive bibliography: Laura Fernández Fernández, 'La mano como herramienta visual mnemotécnica: los diagramas de cómputo del Ms. O-II-10, RBME', forthcoming.

Latin 7478 also gives liturgical precisions in the style of breviaries' calendars that did not appear in usual copies of Gmunden, such as 'duplex majus', 'duplex minus', 'semi-duplex', 'non tenetur nisi in dominica' (to be celebrated on Sundays only), 'dicitur *Credo* in missa', '*Te deum laudamus* dicitur' etc. (all heavily abbreviated, as to be read by accustomed clerics).

This new sanctorale appears clearly as Franciscan. Of the 128 celebrations added in Latin 7478, fewer than ten do not belong to the classical fifteenth-century Franciscan sanctorale.[40] On the contrary, some are highly characteristic: St Francis of Assisi (5 October, with octave, translation on 25 May, stigma on 17 September); St Antonius of Padua (13 June, with octave and translation on 15 February); St Clare of Assisi (12 August, with solemn octave and translation on 2 October), St Bernardine of Sienna, recently canonized (20 May).[41] Some others, while not as crucial for the Franciscan liturgy, are known to appear almost exclusively in Franciscan calendars at that time: St Gilbert de Sempringham (4 February), St Patrick of Armagh (17 March), Pope John I (27 May), St Rufina and Secunda (10 July), Popes Anacletus (13 July), Victor (28 July) and Evaristus (26 October), Dedication of the Basilicas of the Apostles Peter and Paul (18 November), etc. Recent canonizations, such as St Nicholas of Tolentino (1446), St Bernardine of Sienna (1450), and St Vincent Ferrer (1455) are also up to date.

This very convincingly points towards a Franciscan context of production, and leads us to believe that Frater Paulus de Kignin, otherwise unknown, was either an active Franciscan Friar in Northern Italy, or working for one.[42]

3.3. Authorship of Latin 7478: The Case of Paulus de Kignin

To our current knowledge, bat book Latin 7478 seems to be a special case. In order to better understand how it was produced, and to what end, we need to investigate who intervened in this process. We have seen so far that while unacknowledged, most of the content was derived from Gmunden's *Kalendarium*. Some of it, however, was either adapted or replaced, sometimes by other circulating material such as the *Salernitan Regimen sanitatis*. There seems to be at least two players involved in this adaptation: one relatively skilled almanac-maker, able to abbreviate diagrams and tables with a minimal loss of information, and a more naive adapter, involved enough to decide to shift the nineteen-year cycles to fit the current date, but not skilled enough to do so without distorting the syzygy times. It seems logical to identify the latter player with the signing and dating of folio 15B: 'frater Paulus de Kignin' (Fig. 5). It would also seem safe to assume that this apparent friar would be the link with the Franciscan community the sanctorale appears to designate as the final user.

40 Some of those exceptions can in fact be found in the *Franciscano-Tridientinum sanctorale*, or are just displaced (usually celebrated the day before in the classical sanctorale). For this analysis, we used the Grotefend corpus through the IRHT *Calendoscope* database (http://calendoscope.irht.cnrs.fr/).

41 St Francis of Assisi on 5 Oct is probably a mistake for 4 October, especially since the octave is to be celebrated on 11 October. The celebration of St Francis on 5 October is only witnessed in a few breviaries from the north of France (Leroquais). This one-day interval occurs for other festivals. See, for instance, the celebration of Frederick III's coronation, 18 June instead of 17 June (1442). St Bernadine of Sienna's feast is annotated directly in the sanctorale: 'Obiit ex [h]ac vita 1444 anno; canonicatus fuit gloriosus confessor 1450 anno' (Latin 7478, f. 6, 20 May).

42 Probably Umbria, maybe Assisi. Particular thanks go to Laura Albiero for confirming this localization, consistent with the presence of Roman and papal feasts in conjunction with others only found in Francisco-Tridentine sanctorales.

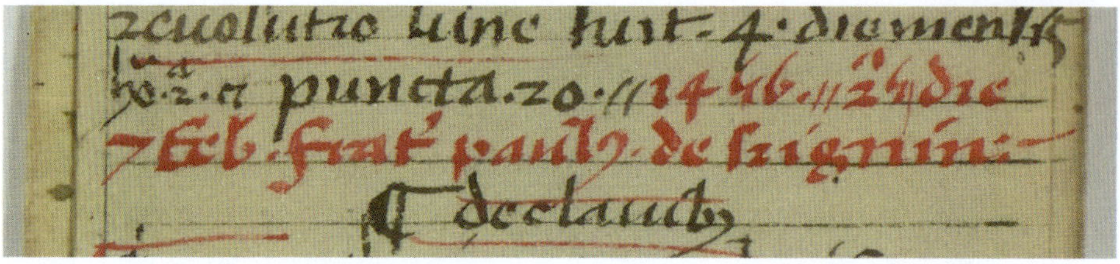

Figure 5. Paulus de Kignin's signature, Paris, BnF, Latin 7478, folio 15B. Source gallica.bnf.fr.

(a) Ad sciendū per kłm [kalendarium] (f. 1Di)

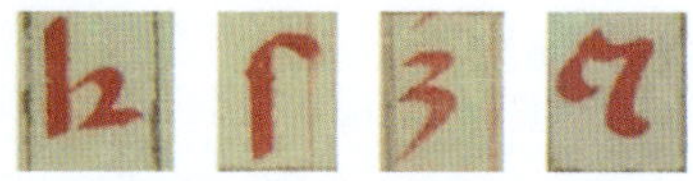

(b) Letters 'k', 's', 'z', and 'et' in the table of the signs of the moon (f. 14D)

The name Paulus de Kignin itself is not without suspicion. We have adopted thus far the reading suggested by Johann Gumbert[43]; while he is probably right to discard Marcel Destombes' 'Paulus de Frignin'[44], a reasonable alternative would be Paulus de Szignin. Unfortunately, palaeographical evidence is scarce, as this would be the only occurrence of an in-word 'z' in the text, and the letter is indeed formed differently in the calendar tables (cf. Fig. 5).

In any case, we have not been able to find any biographical evidence of the existence of either Paulus de Kignin or Paulus de Szignin in the fifteenth century. Both names would 'sound' Central-European rather than Italian, and 'Szignin' in fact became an attested Hungarian family name a few centuries later. It would obviously be unwise to draw a conclusion from this vague clue; however, there are other hints that Latin 7478 could have originated from Central Europe.

The most obvious is the copy of Gmunden's *Kalendarium* itself. While Gmunden had a European reputation, this particular copy is very close to the 'original' ÖNB Cod. 2440. As we have seen previously, while the canons have definitely been altered, the numerical data remain very close to the Vienna manuscript, and most of the changes seem to be copy mistakes.[45] There seems to be no attempt, for instance, to shift the syzygies' astronomical hours (counted from noon) into Italian hours (counted from sunset); likewise, daylight duration and sunrise and sunset times, also unchanged, are indeed valid for Vienna and localities with the same geographical latitude, but not for any place in Italy.

In the same way, it is probably not accidental that it reminds us of Holy Roman Emperor Frederick III's coronations (both as King of the Romans and Emperor), but no other sovereign's, not even Pope Callixtus III.[46] All this would seem in favour of an attempt to export a product of the astronomical network that (even before Matthias Corvinus) linked Vienna, Hungary, and the Imperial Court. The involvement of Georg Peuerbach, successor

43 Gumbert, *Bat Books*, pp. 111–12.
44 *Mappemondes AD 1200–1500*, p. 61.
45 See above p. 156 and Table 2.
46 These records are somewhat approximate: King of the Romans on 18 June (instead of 17 June 1442) and Emperor on 12 May (instead of 19 March 1452). See below, p. 177 and footnote 50.

of Gmunden at Vienna University and frequent traveller through Central Europe and Italy (particularly from 1448–51), astrologer of King Ladislas V of Bohemia and Hungary (before joining the service of Frederick III in 1457) should not be discarded.

Conclusion: Is Latin 7478 a prototype of a personalized toolbox for Franciscans?

With a canonical Franciscan sanctorale and ways to easily compute dates of moveable feasts, bat book Latin 7478 proves to be a very useful compendium of official Franciscan liturgy. Portable, with a small but adequate map of the known world, it is suited to the life of a travelling preacher without access to astronomical books. In fact, however it might have been used, this design would at least have been consistent with the Franciscan vocation. This also raises questions of a larger scale production of such bat books by the mendicant orders.

Is Latin 7478 a surviving copy of a serial production? There is, of course, very little evidence to support it. Bat books proved to be fragile materials, and if they were produced in great quantities, not enough survived to attest to that. Even so, Latin 7478 appears to be a special case among the sixty-three bat books known to Gumbert.

Friar Paul of Kignin (or Szignin), who signed the copy, was clearly not an astrologer. This is not to say that he did not play a significant role in the confection of this bat book; he may very well have been behind the adaptation of the sanctorale to the Franciscan liturgy, and perhaps at the origin of some of the structural changes in the composition as well. He seemingly took it upon himself to update the nineteen-year cycles, even if this decision proved somewhat unfortunate.

That raises the question of the exemplar of Gmunden calendar he was working with. Why did he not follow the express canon on how to update the cycles? John of Gmunden saw to it to be easy enough with a little arithmetic, and if Paul of Kignin knew it, he could have at least tried. It seems more probable that he did not cut himself that canon but worked with an exemplar already deprived of it. Similarly, some astute changes, for instance, in condensing the table for moveable feasts, would tend to suggest that he used an already adapted exemplar (but still with the original Gmunden cycles). That may also be a reason why he did not credit Gmunden.

Could this exemplar, probably in circulation in Hungary, have been a bat book, now lost to us? Kignin worked in 1456, fifteen years after the composition of Gmunden's *Kalendarium*. That would have been time enough to adapt it, especially given the (possible) involvement of Georg Peuerbach. Studying Somer's almanacs, Hilary Carey suggests that bat books may have been produced almost at the same time as the codices of the same work.[47] In itself, the adaptation of the same almanac is not unusual; in the case of John Somer's and Nicholas of Lynn's *Kalendarii*, it happened regularly, both in codex and bat book format. Obviously, the sanctorale was particularly easy to change according to local customs, mostly without updating the astronomical data.[48] Imagining a mendicant initiative to produce bat books

47 Carey, 'Astrological Medicine and the Medieval English Folded Almanac', p. 354.

48 *The Kalendarium of Nicholas of Lynn*, ed. by Sigmund Eisner (Athens: University of Georgia Press, 1980). For bat books, cf. Carey, 'Astrological Medicine and the Medieval English Folded Almanac', p. 355. Carey knows of an Oxford-calculated almanac (as attested by the height of the Sun at noon) featuring saints of Northern France, which was probably used there.

based on Nothaft's 'enhanced calendars' as a liturgical tool is appealing; however, it must be stated that it does not explain the massive loss of such bat books. To our knowledge, Latin 7478 would be, in fact, the sole surviving bat book to attest it. Even if this initiative proved less successful and short-lived compared to the astrological and medical one in England, a few other sources to corroborate it would definitely be welcome.

A more humble hypothesis would be that Franciscan Friar Paul of Kignin, a native of Central Europe, came into contact with some material brought by astronomers of the Peuerbach network, and realizing it was unknown in Italy, took the liberty (or maybe was given the task) of preparing a prototype to present to the Assisi convent. In doing so, he might have been idealizing the life and needs of a friar on a faraway preaching mission in *Tartaria* or *Egyptis*. For that reason, or maybe because of the cost involved in producing a bat book for a fragile result, his prototype would have been declined by the Assisi Franciscans, or he might have produced it but never delivered. That could explain why we preserve the sole copy in the Bibliothèque nationale de France.

Manuscript sources

Dijon, Bibliothèque municipale, 115 (*Breviarium*, bat book)

Dresden, Sächsische Landesbibliothek — Staats- und Universitätsbibliothek, F. 95 (John of Gmunden, *Kalendarium*)

Munich, Bayerische Staatsbibliothek, Clm 14504 (John of Gmunden, *Kalendarium*)

Oxford, Bodleian Library, Laud. Misc. 750 (compendium for the Abbey of Glastonbury, bat book)

Paris, Bibliothèque nationale de France, Clairambault 1032 (inventory of the Gaignières collection)

Paris, Bibliothèque nationale de France, lat. 7478 (*Kalendarium*, bat book)

Paris, Bibliothèque nationale de France, lat. 10479 (*Breviarium*)

Paris, Bibliothèque nationale de France, NAF 5738 (inventory of the Gaignières collection)

Paris, Bibliothèque nationale de France, NAL 375 (*Kalendarium*, bat book)

Paris, Bibliothèque nationale de France, NAL 482 (*Kalendarium*, bat book)

Parma, Biblioteca Palatina, 1993 (prognoses in Bosnian, bat book)

Solothurn, Zentralbibliothek, S 839 (Cistercian *Breviarium*, bat book)

Vienna, Österreichische Nationalbibliothek, Cod. 2440 (John of Gmunden, *Kalendarium*)

Printed sources

'Calendrier portatif du XIVe siècle', *Bibliothèque de l'École des chartes*, 44 (1883), 569.

'Calendrier portatif du XVe siècle', *Bibliothèque de l'École des chartes*, 45 (1884), 136–37.

Carey, Hilary M., 'What is the Folded Almanac? The Form and Function of a Key Manuscript Source for Astro-medical Practice in Later Medieval England', *Social History of Medicine*, 16 (2003), 481–509.

———, 'Astrological Medicine and the Medieval English Folded Almanac', *Social History of Medicine*, 17 (2004), 345–63.

Chlench, Kathrin, Beatriz Porres de Mateo, and Rudolf Simek, 'Johannes von Gmunden: Personalbibliographie und Handschriftenverzeichnis', in Simek and Chlench (eds), *Johannes von Gmunden (ca. 1384–1442): Astronom und Mathematiker* (2006) (Vienna: Fassbaender, 2006), 183–223.

Degering, Hermann, 'Der Bucheinband: Ein calendarium pugillare mit Computus aus dem Jahre 1294', in Max Joseph Husung (ed.), *Buch und Bucheinband, Aufsätze und graphische Blätter zum 60. Geburtstage von Hans Loubier* (Leipzig: Hiersemann, 1923), pp. 79–88.

Delisle, Léopold, *Manuscrits latins et français ajoutés aux fonds des nouvelles acquisitions pendant les années 1875–1891* (Paris: H. Champion, 1891).

Destombes, Marcel (ed.), *Mappemondes AD 1200–1500: Catalogue préparé par la Commission des cartes anciennes de l'Union géographique internationale* (Amsterdam: N. Israel, 1964).

Eisner, Sigmund (ed.), *The Kalendarium of Nicholas of Lynn* (Athens: University of Georgia Press, 1980).

Fernández Fernández, Laura, 'La mano como herramienta visual mnemotécnica: los diagramas de cómputo del Ms. O-II-10, RBME', forthcoming.

Garand, Monique-Cécile, 'Livres de poche médiévaux à Dijon et à Rome', *Scriptorium*, 25 (1971), 18–24.

Géraud, Hercule, 'Calendrier perpétuel portatif dressé l'an 1381', *Bibliothèque de l'École des chartes*, 2 (1841), 272–80.

Gumbert, Johan Peter, *Bat Books: A Catalogue of Folded Manuscripts Containing Almanacs or Other Texts* (Turnhout: Brepols, 2016).

Klug, Rudolf, *Der Astronom Johannes von Gmunden und sein Kalender* (Linz: Pirngruber, 1912).

Kremer, Richard L. 'Cracking the *Tabulae permanentes* of John of Murs and Firmin of Beauval with Exploratory Data Analysis', forthcoming.

Nothaft, C. Philipp E., *Scandalous Error: Calendar Reform and Calendrical Astronomy in Medieval Europe* (Oxford: Oxford University Press, 2018).

Pierrot, Roger and Marcel Thomas (eds), *Le livre ([Exposition] 17 mai - octobre 1972)* (Paris: Bibliothèque nationale, 1972).

Renzi, Salvatore de (ed.), *Collectio Salernitana ossia documenti inediti, e trattati di medicina appartenenti alla scuola medica Salernitana*, 5 vols (Naples: Filiatre-Sebezio, 1852).

Santarém, Manuel Francisco de Barros e Sousa de Mesquita de Macedo Leitão e Carvalhosa, *Essai sur l'histoire de la cosmographie et de la cartographie pendant le Moyen Âge et sur les progrès de la géographie après les grandes découvertes du XVe siècle*, 3 vols (Paris: impr. De Maulde et Renou, 1848–52).

———, *Atlas du vicomte de Santarem*, facsimile edition of the final maps (Lisbon: Administração do Porto de Lisboa sous les auspices de la commission nationale pour les commémorations des découvertes portugaises, 1989).

Silva, Chelsea, 'Opening the Medieval Folding Almanac', *Exemplaria*, 30 (2018), 49–65.

Simek, Rudolf and Manuela Klein (eds), *Johannes von Gmunden - zwischen Astronomie und Astrologie* (Vienna: Fassbaender, 2012).

Vernet, André, 'Le calendrier portatif de Mamert Fichet (1440)', *Bulletin de la Société Nationale des Antiquaires de France*, 1959 (1961), 243–44.

———, *La Bibliothèque de l'abbaye de Clairvaux du XIIe au XVIIIe siècle* (Paris: Editions du Centre national de la recherche scientifique, 1979).

Appendix: A transcription of BnF Latin 7478

Manuscript BnF Latin 7478 has a reference bibliographical record on the *BnF Archives et Manuscrits* online catalogue under the digital id ark:/12148/cc124800v. It is also available as an online facsimile on the *Gallica* digital library (ark:/12148/btv1b100215083).[49]

In the following pages, we intend to offer an extensive transcription of the tables and canons found in Latin 7478. Although some readings remain unsure and some sources unidentified, we hope this material could prove useful to researchers to identify other bat books derived from John of Gmunden's *Kalendarium*, and maybe at some point help produce a critical edition of that work. We caution readers, however, that at this point it is not something that can be expected of this appendix; as a rule, it does not provide a critical or historical apparatus.

For practical reasons, we transcribe the calendar tables (ff. 2–13) and the canons (ff. 1 and 14–17) separately. The calendar itself distinguishes, somewhat artificially, the numerical tables per se (syzygy times, calendar data, Sun and Moon longitude, duration of the day sunrise and sunset times), the sanctorale, and other astronomical information copied textually inside the sanctorale column (for instance, astromedical advice).

In the calendar tables, numbers different from the supposed original manuscript for Gmunden's *Kalendarium*, Österreichsiche Nationalbibliothek Cod. 2440, are highlighted. Only in the sanctorale section do we provide, in comparison, a transcription of the feasts mentioned in Cod. 2440. Diagrams embedded in the canons are reproduced here as straight tables. While this helps us to understand the operation of said diagrams, we strongly suggest studying their shape as well (as they can be found in the online facsimile) and replace them not only according to a textual tradition but also a 'diagrammatic tradition'.

Likewise, while semantic entities have been formatted as well as possible, we do not render the original red and black ink alternation.

Figure 6. In this transcription, the calendar's lunar letters (*littere signorum*) are displayed as follows: a b c d e f g h i k l m n o p q r ſ s t u v x y ʒ 3 ȥ. Compare, especially, the last three to the way they are formed by the copyist in the reproduction. Paris, BnF, Latin 7484, folio 14D (left). Source gallica.bnf.fr.

Calendar

Calendar tables

49 Bibliothèque nationale de France, 'Latin 7478', *BnF Archives et Manuscrits* (2020), <https://archivesetmanuscrits.bnf.fr/ark:/12148/cc124800v> [Accessed 28 July 2020]. The 2018 digital facsimile is available at https://gallica.bnf.fr/ark:/12148/btv1b100215083.

Annus habet dies 365 et 6 horas, habet 12 menses et 52 ebdoma[da]s.
Habet dies 31m Januarius, luna vero ejus 30.

Primus ciclus						Secundus ciclus						Tertius ciclus						Quartus ciclus						KL					[JANUARIUS]		Littere signorum	Quantitas dierum		Ortus solis		Occasus solis	
Conjunctio	h	m	[Oppositio]	h	m	Conjunctio	h	m	[Oppositio]	h	m	Conjunctio	h	m	[Oppositio]	h	m	Conjunctio	h	m	[Oppositio]	h	m									h	m	h	m	h	m
																													Capricornus								
			4	16	3				4	8	35	8	19	28	4	1	7				4	17	39	3	a		K	1		20	a	8	24	7	49	4	11
8	10	24				8	2	56	12	21	16				12	13	48	8	12	0					b	4	N	2		21	b	8	27	7	48	4	12
			12	4	44	16	23	5				16	15	37				16	8	9	12	6	20	11	c	3	N	3		22	c	8	29	7	47	4	13
16	6	33							1	17	25				1	9	57	5	20	50	1	2	29		d	2	N	4		23	d	8	31	7	46	4	14
5	19	14	1	0	53	5	11	46				5	4	18	9	22	38				9	15	10	19	e		N	5		24	e	8	33	7	45	4	15
			9	13	34				9	6	6	13	16	59				13	9	31				8	f	8	I	6		25	f	8	35	7	44	4	16
13	7	55	17	9	44	13	0	27							17	18	47				17	11	19		g	7	I	7		26	g	8	37	7	43	4	17
2	4	4	6	22	25	2	20	36	17	2	15	2	13	8				2	40	40				16	a	6	I	8		27	h	8	39	7	42	4	18
10	16	54							6	14	56				6	7	28	10	18	21	6	0	0	4	b	5	I	9		28	i	8	41	7	41	4	19
			14	11	6	10	9	17	14	3	38	10	1	49	14	20	9				14	12	41		c	4	I	10		29	k	8	44	7	40	4	20
18	12	54				18	5	26	3	23	47	18	21	58				18	14	30				13	d	3	I	11	Aquarius		l	8	47	7	38	4	22
			3	7	15	7	8	7							3	16	19				3	8	50	2	e	2	I	12		1	m	8	49	7	37	4	23
7	1	35	11	19	56				11	13	28	7	10	39				7	3	11	11	21	32		f		I	13		2	n	8	51	7	36	4	24
15	21	44				15	16	16				15	6	48	11	5	0	15	23	20				10	g	19	K	14		3	o	8	45	7	35	4	25
			19	16	5				19	8	36	4	19	29	19	1	9				19	17	40		a	18	K	15		4	p	8	56	7	33	4	27
4	10	25				4	2	54	8	21	18				8	13	50	4	12	1				18	b	17	K	16		5	q	8	54	7	34	4	28
12	23	6	8	4	46	12	15	38				12	8	10				12	0	42	8	6	22	7	c	16	K	17		6	r	8	2	7	31	4	29
									16	17	27				16	9	59	1	20	51	16	2	31		d	15	K	18		7	f	9	5	7	29	4	31
1	19	15	16	0	55	1	11	47				1	4	19	5	22	40				5	15	2	15	e	14	K	19		8	s	9	8	7	28	4	32
			5	13	36				5	6	8	9	17	0				9	9	31				4	f	13	K	20		9	t	9	10	7	26	4	34
9	7	56	13	2	17	9	0	28	13	18	49				13	11	20				13	3	53		g	12	K	21		10	u	9	13	7	25	4	35
17	4	6	2	22	26	17	20	37				17	13	9				17	5	42				12	a	11	K	22		11	v	9	16	7	25	4	36
6	16	47							2	14	58				2	7	31	6	18	22	2	0	2	1	b	10	K	23		12	x	9	19	7	22	4	38
			10	11	7	6	9	18	10	3	29	6	1	50	10	12	11				10	12	43		c	9	K	24		13	y	9	22	7	21	4	39
14	5	28				14	22	0	18	23	58	14	14	31				14	7	3				9	d	8	K	25		14	ʒ	9	28	7	19	4	41
			18	7	16	3	18	9							18	16	20				18	8	52		e	7	K	26		15	ꝫ	9	32	7	16	4	44
3	1	37	7	19	57				7	12	29	3	10	41				3	3	13	7	21	33	17	f	6	K	27		17	z	9	35	7	14	4	46
11	14	18				11	6	50				11	23	22	7	5	1	11	15	54				6	g	5	K	28		18	a	9	38	7	13	4	47
			15	16	6				15	8	38	19	19	31	15	1	10				15	17	42		a	4	K	29		19	b	9	41	7	11	4	49
19	10	27				19	2	59	4	21	19				4	13	51	19	12	3				14	b	3	K	30		20	c	9	49	7	10	4	50
8	23	8	4	4	47	8	15	40				8	8	12							4	6	23	3	c	2	K	31		21	d	9	45	7	8	4	52

Februarius habet dies 28, quando est bixestus habet dies 29.
Luna vero illo tunc habet dies XXX.

Dies	Primus ciclus Conjunctio	h	m	Primus ciclus [Oppositio]	h	m	Secundus ciclus Conjunctio	h	m	Secundus ciclus [Oppositio]	h	m
1				12	17	28				12	10	0
2	16	19	17				16	11	49			
3				1	13	37				1	6	9
4	5	7	59	9	2	18	5	0	30	9	18	50
5	13	20	39	17	22	28	13	13	11			
6	2	17	48							17	15	0
7				6	11	9	2	9	20			
8	10	5	29	14	23	50	10	22	1	6	3	41
9							18	18	10	14	16	22
10	18	1	38	3	19	40				3	12	31
11	7	14	19				7	6	51			
12				11	8	49				11	1	12
13	15	10	28				15	3	0	19	21	21
14	4	23	9	19	4	30	4	15	41			
15				8	17	39				8	10	2
16	12	11	50				12	4	22			
17				16	13	30				16	6	11
18	1	7	59				1	0	31	5	18	52
19	9	20	40	5	3	20	9	13	12			
20	17	16	50	13	15	1				13	7	33
21				2	11	10	17	9	22			
22	6	5	31	10	23	50	6	22	33	2	3	42
23	14	18	12							10	16	23
24				18	20	0	14	10	44	18	12	32
25	3	14	21				3	6	53			
26				7	8	41	11	19	34	7	1	13
27	11	3	2							15	21	22
28	19	23	11	15	10	50	19	15	43			

Dies	Tertius ciclus Conjunctio	h	m	Tertius ciclus [Oppositio]	h	m	Quartus ciclus Conjunctio	h	m	Quartus ciclus [Oppositio]	h	m
1				12	2	32	8	0	44	12	19	4
2	16	4	21	1	22	41	16	20	53	1	5	13
3	5	17	2				5	9	34			
4				9	11	22	16	22	15	9	3	54
5	13	5	43									
6				17	7	31	2	18	24	17	0	3
7	2	1	52	6	20	12				6	12	44
8	10	14	33				10	7	5			
9				14	8	54				14	1	25
10	18	10	42				18	3	14	3	21	35
11	7	23	23	3	5	3	7	15	55			
12	15	19	32	11	17	44				11	10	16
13				19	13	53	15	12	4			
14	4	8	13							19	6	25
15	12	20	54	8	2	34	4	0	45	8	19	6
16				16	22	43	12	13	26	16	15	15
17	1	17	3				1	9	35			
18				5	11	24	9	22	16	5	3	56
19	9	5	44							13	16	27
20				13	0	5	17	18	25			
21	17	1	53	2	20	14				2	12	46
22	6	14	35				6	7	6			
23				10	8	55	14	19	47	10	1	27
24	14	3	16	18	5	4				18	21	37
25	3	23	25	7	17	47	3	15	57			
26	11	12	6							7	10	15
27				15	13	14	11	4	38	15	6	26
28	19	8	15				19	0	47	4	19	7

KL					[FEBRUARIUS]	Littere signorum	Quantitas dierum h	m	Ortus solis h	m	Occasus solis h	m
					Aquarius							
	d		K	1	22	e	9	48	7	6	4	54
11	e	4	N	2	23	f	9	41	7	5	4	55
19	f	3	N	3	24	g	9	44	7	3	4	57
8	g	2	N	4	25	h	9	57	7	2	4	58
	a		N	5	26	i	10	0	7	0	5	0
16	b	8	I	6	27	k	10	4	6	58	5	2
4	c	7	I	7	28	l	10	7	6	57	5	3
	d	6	I	8	29	m	10	10	6	55	5	5
13	e	5	I	9	Pisces	n	10	13	6	53	5	7
2	f	4	I	10	1	o	10	17	6	52	5	8
	g	3	I	11	2	p	10	21	6	48	5	10
10	a	2	I	12	3	q	10	24	6	46	5	12
	b		I	13	4	r	10	28	6	45	5	14
18	c	16	K	14	5	ſ	10	31	6	43	5	15
7	d	15	K	15	6	s	10	35	6	41	5	17
	e	14	K	16	7	t	10	39	6	40	5	19
16	f	13	K	17	8	u	10	41	6	38	5	20
4	g	12	K	18	9	v	10	45	6	36	5	22
	a	11	K	19	10	x	10	48	6	34	5	24
12	b	10	K	20	11	y	10	53	6	33	5	26
1	c	9	K	21	12	ʒ	10	55	6	31	5	27
	d	8	K	22	13	ꝫ	10	59	6	29	5	29
9	e	7	K	23	14	z	11	3	6	27	5	31
	f	6	K	24	15	a	11	6	6	25	5	33
17	g	5	K	25	16	b	11	10	6	24	5	35
6	a	4	K	26	17	c	11	13	6	28	5	36
	b	3	K	27	18	d	11	17	6	22	5	38
14	c	2	K	28	19	e	11	20	6	20	5	40

Martius habet dies XXXI.
Luna vero ejus habet dies XXX.

Primus ciclus						Secundus ciclus						Tertius ciclus						Quartus ciclus						KL					[MARTIUS]	Littere [signorum]	Quantitas dierum		Ortus solis		Occasus solis	
Conjunctio			[Oppositio]			Conjunctio			[Oppositio]			Conjunctio			[Oppositio]			Conjunctio			[Oppositio]															
	h	m		h	m		h	m		h	m		h	m		h	m		h	m		h	m						Pisces		h	m	h	m	h	m
			4	17	31				4	10	3	8	10	56	4	2	35	8	13	28				3	d		K	1	20	f	11	24	6	18	5	42
8	11	52				8	4	24	12	22	44				12	15	16				12	7	48		e	6	N	2	21	g	11	28	6	16	5	44
16	8	1	12	6	12							16	17	5				16	9	37				11	f	5	N	3	22	h	11	31	6	15	5	45
5	10	42	1	2	21	16	0	33	1	18	53				1	11	25	5	22	18	1	3	57		g	4	N	4	23	i	11	35	6	13	5	47
			9	15	3	5	13	14				5	5	46							9	16	38	19	a	3	N	5	24	k	11	39	6	11	5	49
13	9	23							9	7	34	13	18	27	9	0	6	13	10	59				8	b	2	N	6	25	k	11	43	6	9	5	51
			17	11	12	2	22	4	17	3	44				17	20	16				17	12	47		c		N	7	26	l	11	46	6	7	5	53
2	5	32	6	23	53				6	16	25	2	14	36				2	7	8				16	d	8	I	8	27	m	11	49	6	6	5	54
10	18	13				10	10	45				10	3	17	6	8	57	10	19	49	6	1	28	4	e	7	I	9	28	n	11	53	6	4	5	56
			14	12	39				14	5	6	18	23	26	14	21	38				14	14	10		f	6	I	10	29	o	11	56	6	2	5	58
10	14	22				18	6	54							3	17	47	18	15	58				13	g	5	I	11	Aries	p	12	0	6	0	6	0
			3	8	43	7	19	35	3	1	15	7	12	7							3	10	19	2	a	4	I	12	1	q	12	0	5	59	6	1
7	3	3	11	21	24				11	13	56				11	6	28	7	4	39	11	23	0		b	3	I	13	1	r	12	3	5	58	6	2
15	23	12				15	15	44				15	8	16				15	0	48	19	19	9	10	c	2	I	14	2	ſ	12	7	5	57	6	3
			19	17	33				19	10	5	4	10	57	19	2	37	4	13	29					d		I	15	3	s	12	10	5	55	6	5
4	11	53				4	4	25	8	22	44				8	15	18				8	7	40	18	e	17	K	16	4	t	12	14	5	53	6	7
			8	6	14	12	17	6				12	9	38				12	2	10				7	f	16	K	17	5	u	12	17	5	52	6	8
12	0	34	16	2	53				16	18	55				16	11	27	1	22	19	16	3	59		g	15	K	18	6	v	12	21	5	50	6	10
1	20	43	5	15	4	1	13	15				1	5	47							5	16	40	15	a	14	K	19	7	x	12	25	5	48	6	12
9	9	25				9	1	56	5	7	36	9	18	28	5	0	8	9	11	0				4	b	13	K	20	8	y	12	29	5	46	6	14
			13	3	45	17	22	5	13	20	17				13	12	49				13	5	21		c	12	K	21	9	ȝ	12	32	5	44	6	16
17	5	34	2	23	54				2	16	26	17	14	38				17	7	9				12	d	11	K	22	10	ꝫ	12	36	5	42	6	18
6	18	15				6	10	47							2	8	58	6	19	50	2	1	30	1	e	10	K	23	11	ƶ	12	39	5	41	6	19
			10	12	35	14	23	28	10	5	7	6	3	19	10	21	39				10	14	11		f	9	K	24	12	a	12	42	5	39	6	21
14	6	56										14	16	0	18	17	48	14	8	32				9	g	8	K	25	13	b	12	46	5	37	6	23
			18	8	44	3	19	37	18	1	16	3	12	9							10	10	20		a	7	K	26	14	c	12	50	5	35	6	25
3	3	5	7	21	25				7	13	57				7	6	29	3	4	41	7	23	1	17	b	6	K	27	15	d	12	54	5	33	6	27
11	15	46				11	8	18				11	0	50				11	17	22	15	19	10	6	c	5	K	28	16	e	12	57	5	32	6	28
			15	17	34				15	10	6	19	20	59	15	2	38	19	13	31					d	4	K	29	17	f	13	1	5	30	6	30
19	2	55				19	4	27	4	22	47				4	15	19				4	7	51	14	e	3	K	30	18	g	13	4	5	28	6	33
			4	6	15	8	17	7				8	9	40				8	2	12	12	20	32	3	f	2	K	31	19	h	13	8	5	26	6	34

Aprilis habet dies XXX.
Luna vero habet XXIX.

Primus ciclus						Secundus ciclus						Tertius ciclus						Quartus ciclus						KL					[APRILIS]	Littere [signorum]	Quantitas dierum		Ortus solis		Occasus solis	
Conjunctio			[Oppositio]			Conjunctio			[Oppositio]			Conjunctio			[Oppositio]			Conjunctio			[Oppositio]															
	h	m		h	m		h	m		h	m		h	m		h	m		h	m		h	m						Aries		h	m	h	m	h	m
8	0	36	12	18	56				12	11	28				12	4	0	16	22	21					g		K	1	20	i	13	12	5	24	6	36
16	20	45	1	15	6	16	13	17				16	5	49							1	16	41	11	a	4	N	2	21	k	13	15	5	23	6	37
5	9	26							1	7	37	5	18	30	1	0	9	5	13	2					b	3	N	3	22	l	13	18	5	21	6	39
13	22	7	9	3	47	5	1	58	9	20	19				9	12	50	13	23	43	9	5	22	19	c	2	N	4	23	m	13	22	5	19	6	41
			17	23	58	13	14	39	17	16	28	13	7	11										8	d		N	5	24	n	13	25	5	17	6	45
2	18	57				2	10	48							17	9	0	2	19	52	17	1	31	16	e	8	I	6	25	o	13	29	5	15	6	43
			6	12	37	10	23	29	6	5	9	2	3	20	6	21	41				6	14	13	5	f	7	I	7	26	p	13	32	5	14	6	46
10	6	6							14	17	50	10	16	1				10	8	33					g	6	I	8	27	q	13	26	5	12	6	48
			14	1	18	18	19	38				18	12	10	14	10	22				14	2	54	13	a	5	I	9	28	r	13	29	5	10	6	50
18	3	47	3	21	27				3	13	59				3	6	31	18	4	42	3	22	30	2	b	4	I	10	29	ſ	13	43	5	8	6	52
7	1	16				7	8	19				7	0	51	11	19	12	7	17	23					c	3	I	11	Taurus	s	13	46	5	7	6	53
			11	10	8				11	2	40	15	21	0				15	13	32	11	11	44	10	d	2	I	12	1	t	13	49	5	5	6	54
15	11	56				15	4	28	19	22	49				19	5	21				19	7	53		e		I	13	2	u	13	53	5	3	6	57
			19	6	17	4	17	9				4	9	41				4	2	13	8	20	34	18	f	18	K	14	3	v	13	56	5	2	6	58
4	0	37	8	18	58				8	11	30	12	22	22	8	4	2	12	14	54				7	g	17	K	15	4	x	14	0	5	0	7	0
12	13	18	16	15	7	12	5	50													16	16	43		a	16	K	16	5	y	14	3	4	58	7	2
1	9	28							16	7	39	1	18	31	16	0	11	1	11	3				15	b	15	K	17	6	ʒ	14	6	4	57	7	3
9	22	9	5	3	48	1	2	59	5	20	20				5	10	52	9	23	44	5	5	28	4	c	14	K	18	7	ɜ	14	9	4	55	7	5
			13	16	29	9	14	41				9	7	12							13	18	5		d	13	K	19	8	z	14	12	4	54	7	6
17	18	18				17	10	50	13	9	1				13	31	33	17	19	54				12	e	12	K	20	8	a	14	13	4	53	7	7
			2	12	38	6	23	31	2	5	10	17	3	22	2	21	42				2	14	14	1	f	11	K	21	9	b	14	15	4	52	7	8
6	6	59							10	17	51	6	16	3				6	8	35					g	10	K	22	10	c	14	18	4	51	7	9
14	1	40	10	1	19	14	12	12							10	10	23	14	21	16	10	2	55	9	a	9	K	23	11	d	14	22	4	49	7	11
			18	21	28				18	14	0	14	4	44	18	6	32				18	23	34		b	8	K	24	12	e	14	25	4	47	7	13
3	15	49				3	8	21				3	0	53	7	19	13	3	17	55				17	c	7	K	25	13	f	14	29	4	45	7	15
			7	10	9	11	21	2	7	2	41	11	13	34							7	11	45	6	d	6	K	26	14	g	14	31	4	44	7	16
11	4	30							15	22	50				15	15	22	11	6	6	15	7	54		e	5	K	27	15	h	14	35	4	42	7	18
			15	6	18	19	17	11				19	9	43				19	2	15	4	20	35	14	f	4	K	28	16	i	14	38	4	41	7	19
19	0	9	4	18	59				4	11	31	8	22	24	4	4	3	8	14	56				3	g	3	K	29	17	k	14	41	4	39	7	21
8	13	20				8	5	52							12	16	44				12	9	16		a	2	K	30	18	l	14	44	4	38	7	22

Madius habet dies XXXI.
Luna vero habet dies XXX[a].

Primus ciclus						Secundus ciclus						Tertius ciclus						Quartus ciclus						KL					[MADIUS]	Littere [signorum]	Quantitas dierum		Ortus solis		Occasus solis	
Conjunctio			Oppositio			Conjunctio			[Oppositio]			Conjunctio			Oppositio			Conjunctio			[Oppositio]															
	h	m		h	m		h	m		h	m		h	m		h	m		h	m		h	m								h	m	h	m	h	m
																													Taurus							
16	9	29	12	7	41				12	0	12	16	18	32				16	11	4				11	b		K	1	19	m	14	46	4	37	7	23
5	22	10	1	3	50	16	2	1	1	20	22				1	12	53	5	23	46	1	5	25		c	6	N	2	20	n	14	49	4	35	7	25
			9	16	31	5	14	42				5	7	14							9	18	6	19	d	5	N	3	21	o	14	52	4	34	7	26
13	10	51				13	3	23	9	9	3	13	19	55	9	1	35	13	12	27				8	e	4	N	4	22	p	14	55	4	32	7	28
			17	12	40	2	23	32	17	5	12				17	21	44				17	14	16		f	3	N	5	23	q	14	58	4	31	7	29
2	7	2							6	17	53	2	16	4				2	8	36				16	g	2	N	6	24	r	15	1	4	29	7	31
10	19	41	6	1	21	10	12	13							6	10	25	10	21	17	6	2	57	5	a		N	7	25	ſ	15	3	4	28	7	32
			14	14	2				14	6	34	10	4	45	14	23	6				14	15	38		b	8	I	8	26	s	15	6	4	27	7	33
18	15	50				18	8	22				18	0	54	3	19	15	18	17	26				13	c	7	I	9	27	t	15	8	4	26	7	34
			3	10	11	7	21	3	3	2	43	7	13	35							3	11	47	2	d	6	I	10	28	u	15	11	4	24	7	36
7	4	31	11	22	52				11	15	24				11	7	56	7	6	7	11	0	20		e	5	I	11	29	v	15	13	4	23	7	37
						15	17	12				15	9	44				15	2	16	19	20	37	10	f	4	I	12	Geminy	x	15	16	4	22	7	38
15	0	40	19	19	1				19	11	33	4	22	25	19	4	5	4	14	57					g	3	I	13	1	y	15	18	4	21	7	39
4	13	21				4	5	53							8	16	46				8	9	18	18	a	2	I	14	2	ʒ	15	20	4	20	7	40
			8	7	42	12	18	34	8	0	14	12	11	6				12	3	38				7	b		I	15	3	ꝫ	15	21	4	19	7	41
12	2	3	16	3	51				16	20	23				16	12	55	1	23	47	16	5	27		c	17	K	16	3	ƶ	15	23	4	18	7	42
1	22	12	5	16	32	1	14	44				1	7	16							5	18	8	15	d	16	K	17	4	a	15	25	4	17	7	43
9	10	53				9	3	25	5	9	4	9	19	57	5	1	36	9	12	28				4	e	15	K	18	5	b	15	26	4	16	7	44
			13	5	13	17	23	34	13	21	45				13	14	17				13	6	49		f	14	K	19	6	c	15	29	4	15	7	45
17	7	2							2	17	54	17	16	6				17	8	38				12	g	13	K	20	7	d	15	31	4	14	7	46
6	19	43	2	1	22	6	12	15							2	10	26	6	21	19	2	2	58	1	a	12	K	21	8	e	15	33	4	13	7	47
			10	14	3				10	6	35	6	4	47	10	23	7				10	15	39		b	11	K	22	9	f	15	35	4	12	7	48
14	8	24				14	0	56				14	17	28	18	19	16	14	10	0				9	c	10	K	23	10	g	15	37	4	11	7	49
			18	10	12	3	21	5	18	2	44	3	13	37							18	11	48		d	9	K	24	11	h	15	39	4	10	7	50
3	4	23	7	22	53				7	15	25				7	7	57	3	6	9	7	0	29	17	e	8	K	25	12	i	15	40	4	10	7	50
11	17	14				11	9	46				11	2	18				11	18	50	15	20	38	6	f	7	K	26	13	k	15	42	4	9	7	51
			15	19	2				15	11	34	19	22	27	15	4	6	19	14	59					g	6	K	27	14	k	15	43	4	8	7	52
19	13	23				19	5	55							4	16	47				4	9	19	14	a	5	K	28	15	l	15	45	4	7	7	53
8	2	4	4	7	44	8	18	36	4	0	15	8	11	8				8	3	49	12	22	0	3	b	4	K	29	16	m	15	46	4	7	7	53
16	22	13	12	10	25				12	12	57				12	5	28	16	23	49					c	3	K	30	17	n	15	47	4	6	7	54
			1	16	34	16	14	45				16	7	17							1	18	9	11	d	2	K	31	18	o	15	48	4	6	7	54

Junius autem habet dies XXX.
Luna vero ejus habet dies XXIX.

Primus ciclus						Secundus ciclus						Tertius ciclus						Quartus ciclus						KL					[JUNIUS]	Littere signorum	Quantitas dierum		Ortus solis		Occasus solis	
Conjunctio			[Oppositio]			Conjunctio			[Oppositio]			Conjunctio			[Oppositio]			Conjunctio			[Oppositio]															
	h	m		h	m		h	m		h	m		h	m		h	m		h	m		h	m								h	m	h	m	h	m
																													Gemini							
5	10	54							1	9	6	5	19	58	1	1	38	5	12	30					e		K	1	19	p	5	49	4	5	7	55
13	23	35	9	5	15	5	3	26	9	21	47				9	14	19				9	6	51	19	f	4	N	2	20	q	5	50	4	5	7	55
					24	13	16	7	17	17	56	13	8	39				13	1	11				8	g	3	N	3	21	r	5	51	4	4	7	56
2	19	44	17	1	5	2	12	16							17	10	28	2	21	20	17	3	0	16	a	2	N	4	21	ſ	5	51	4	4	7	56
			6	14	46				6	6	37	2	4	48	6	23	9				6	15	41	5	b		N	5	22	s	5	52	4	4	7	56
10	8	25				10	0	57	14	19	14	10	17	29				10	10	1					c	8	I	6	23	t	5	52	4	4	7	56
			14	2	55	18	21	6				18	13	38	14	11	50				14	4	22	13	d	7	I	7	24	u	5	53	4	3	7	57
18	4	34	3	22	36				3	15	27				3	7	59	18	6	10				2	e	6	I	8	25	v	5	54	4	3	7	57
7	17	15				7	9	47				7	2	19	11	20	40	7	18	51	3	0	31		f	5	I	9	26	x	5	54	4	3	7	57
			11	11	45				11	4	8	15	22	28				15	1	0	11	13	12	10	g	4	I	10	27	y	5	55	4	2	7	58
15	13	25				15	5	56							19	16	49				19	9	21		a	3	I	11	28	ʒ	5	55	4	2	7	58
			19	7	26	4	18	37	19	0	17	4	11	9				4	3	41	8	22	2	18	b	2	I	12	29	ꝫ	5	56	4	2	7	58
4	2	6	8	20	35				8	12	58	12	23	50	8	5	30	12	16	22				7	c		I	13	Cancer	z	5	56	4	2	7	58
12	14	47	16	16	16	12	7	19													16	18	11		d	18	K	14	1	a	5	55	4	2	7	58
1	10								16	9	7	1	20	0	16	1	39	1	12	31				15	e	17	K	15	2	b	5	55	4	2	7	58
9	23	56	5	5	57	1	3	28	5	21	48				5	14	20				5	6	52	4	f	16	K	16	3	c	5	55	4	2	7	58
			13	17	6	9	16	9				9	8	41				9	1	23	13	19	33		g	15	K	17	4	d	5	54	4	3	7	57
17	19	37				17	12	18	13	10	29				13	3	1	17	21	22				12	a	14	K	18	5	e	5	54	4	3	7	57
			2	14	47				2	6	38	14	4	50	2	23	10				2	15	42	1	b	13	K	19	6	f	5	53	4	3	7	57
6	8	46				6	0	59	10	19	19	6	17	31				6	10	3					c	12	K	20	7	g	5	54	4	3	7	56
14	21	27	10	2	56	14	13	40							10	11	15	14	22	45	10	4	29	9	d	11	K	21	7	h	5	52	4	4	7	56
			18	22	37				18	15	28	14	6	12	18	8	0								e	10	K	22	8	i	5	52	4	4	7	55
3	17	8				3	9	49				3	2	21	7	20	41	3	18	53	18	0	32	17	f	9	K	23	9	k	5	51	4	4	7	55
			7	11	47	11	22	30	7	4	9	11	15	2							7	13	13	6	g	8	K	24	10	l	5	50	4	5	7	54
11	5	58													15	16	50	11	7	34	15	9	22		a	7	K	25	11	m	5	49	4	6	7	54
			15	7	28	19	18	39	15	0	18	19	11	11				19	3	43	4	22	3	14	b	6	K	26	12	n	5	48	4	6	7	53
19	2	7	4	20	9				4	13	0	8	23	52	4	5	31	8	16	24				3	c	5	K	27	13	o	5	47	4	7	7	52
8	14	48				8	7	20							12	18	12				12	10	44		d	4	K	28	14	p	5	46	4	7	7	52
16	10	57	12	9	9				12	1	41	16	20	1				16	12	33				2	e	3	K	29	15	q	5	44	4	8	7	52
5	23	38	1	5	18	16	3	29	1	21	50				1	14	22				1	6	54		f	2	K	30	16	r	5	43	4	8	7	52

Julius habet dies XXXI.
Luna vero habet 29.

Primus ciclus						Secundus ciclus						Tertius ciclus						Quartus ciclus						KL					[JULIUS]		Littere [signorum]	Quantitas dierum		Ortus solis		Occasus solis	
Conjunctio			[Oppositio]			Conjunctio			[Oppositio]			Conjunctio			[Oppositio]			Conjunctio			[Oppositio]																
	h	m		h	m		h	m		h	m		h	m		h	m		h	m		h	m							Cancer		h	m	h	m	h	m
			9	17	59	5	16	10				5	8	42				5	1	14	9	19	35	19	g		K	1		17	f	15	42	4	9	7	51
13	12	19							9	10	31	13	21	23	9	3	3	13	13	55				8	a	6	N	2		18	s	15	40	4	10	7	50
			17	14	8	13	4	51	17	6	40				17	23	12				17	15	44		b	5	N	3		19	t	15	39	4	11	7	49
2	8	28				2	1	0	6	19	21	2	17	32				2	10	40				16	c	4	N	4		20	u	15	37	4	12	7	48
10	21	9	6	2	49	10	13	41							6	11	53	10	22	45	6	4	25	5	d	3	N	5		21	v	15	35	4	13	7	47
			14	15	30				14	8	2	10	6	13	14	0	34				14	17	6		e	2	N	6		22	x	15	33	4	14	7	46
18	17	18				18	9	50				18	2	22	3	20	43	18	18	54				13	f		N	7		23	y	15	31	4	15	7	45
			3	11	39	7	22	31	3	4	11	7	15	3							3	13	15	2	g	8	I	8		24	ʒ	15	28	4	16	7	44
7	5	59							11	16	52				11	9	24	7	7	35	11	1	56		a	7	I	9		25	ɜ	15	27	4	17	7	43
			11	0	20	15	18	40				15	11	12				15	3	44	19	22	5	10	b	6	I	10		26	ƶ	15	25	4	18	7	42
15	2	9	19	20	29				19	13	1	4	23	53	19	5	33	4	16	25					c	5	I	11		27	a	15	23	4	19	7	41
4	14	59				4	7	22							8	18	14				8	10	46	18	d	4	I	12		28	b	15	20	4	20	7	40
12	3	31	8	9	10	12	20	3	8	1	42	12	12	35				12	5	6				7	e	3	I	13		29	c	15	18	4	21	7	39
1	23	40	16	5	19				16	21	51				16	14	23				16	6	55		f	2	I	14		Leo	d	15	16	4	22	7	38
			5	18	0	1	16	52				1	8	44				1	1	16	5	19	36	15	g		I	15		1	e	15	14	4	23	7	37
9	12	21							5	10	32	9	21	25	5	3	4	9	13	57				4	a	17	K	16		1	f	15	13	4	24	7	36
			13	6	41	9	4	53	13	33	13				13	14	45				13	8	17		b	16	K	17		2	g	15	11	4	25	7	35
17	8	30				17	1	2	2	19	22	17	17	34				17	10	6				12	c	15	K	18		3	h	15	8	4	26	7	34
6	21	11	2	2	50	6	13	43							2	11	54	6	22	47	2	4	26	1	d	14	K	19		4	i	15	6	4	27	7	33
			10	15	31				10	8	3	6	6	15	10	0	34				10	17	7		e	13	K	20		5	k	15	3	4	28	7	30
14	9	52				14	2	24				14	18	56	18	20	44	14	11	28				9	f	12	K	21		6	l	15	0	4	30	7	29
			18	11	40	3	22	33	18	4	12	3	15	5							18	3	16		g	11	K	22		7	m	14	58	4	31	7	29
3	6	1							7	16	53				7	9	25	3	7	37	7	1	57	17	a	10	K	23		8	n	14	55	4	32	7	28
11	18	42	7	0	21	11	11	14				11	3	46				11	10	18	15	22	6	6	b	9	K	24		9	o	14	52	4	34	7	26
			15	20	31				15	13	3	19	23	55	15	5	34	19	16	27					c	8	K	25		10	p	14	49	4	35	7	25
19	14	51				19	7	23							4	18	16				4	10	47	14	d	7	K	26		11	q	14	46	4	37	7	23
8	3	32	4	9	12	8	20	4	4	1	44	8	12	36				8	5	8	12	23	28	3	e	6	K	27		12	r	14	43	4	39	7	21
			12	21	53				12	14	25				12	6	57								f	5	K	28		13	ſ	14	41	4	40	7	20
16	23	41	1	18	2	16	16	13				16	8	45				16	1	17	1	19	38	11	g	4	K	29		14	s	14	38	4	41	7	19
5	12	22							1	10	34	5	21	26	1	3	6	5	13	58	9	8	19	19	a	3	K	30		15	t	14	35	4	43	7	17
			9	6	43	5	4	54	9	23	55				9	15	47								b	2	K	31		16	u	14	31	4	45	7	15

Augustus habet dies XXXI.
Luna vero ejus habet XXX.

Primus ciclus						Secundus ciclus						Tertius ciclus						Quartus ciclus						KL					[AUGUSTUS]	Littere signorum	Quantitas dierum		Ortus solis		Occasus solis	
Conjunctio			[Oppositio]			Conjunctio			[Oppositio]			Conjunctio			[Oppositio]			Conjunctio			[Oppositio]															
	h	m		h	m		h	m		h	m		h	m		h	m		h	m		h	m								h	m	h	m	h	m
																													Leo							
13	1	3				13	17	35	17	19	24	13	10	7				13	2	39				8	c		K	1	17	v	14	28	4	46	7	14
2	21	12	17	2	52	2	13	44							17	11	56	2	23	48	17	4	28	16	d	4	N	2	18	x	14	25	4	48	7	12
			6	15	33				6	8	5	2	6	16							6	17	9	5	e	3	N	3	19	y	14	22	4	49	7	11
10	9	53				10	2	25	14	20	46	10	18	57	6	0	37	10	11	29					f	2	N	4	20	ʒ	14	18	4	51	7	9
			14	4	14	18	22	34				18	15	6	14	13	18				14	5	50	13	g		N	5	21	ȝ	14	15	4	53	7	7
18	6	2							3	16	55				3	9	27	18	7	38				2	a	8	I	6	22	z	14	12	4	54	7	6
7	18	44	3	0	23	7	11	15				7	3	47	11	22	8	7	20	19	3	1	59		b	7	I	7	23	a	14	10	4	55	7	5
			11	13	4				11	5	36	15	23	56				15	16	28	11	14	40	10	c	6	I	8	23	b	14	9	4	56	7	4
15	14	53				15	7	25							19	18	17				19	10	49		d	5	I	9	24	c	14	6	4	57	7	3
			19	9	13	4	20	6	19	1	45	4	12	38				4	5	9	8	23	30	18	e	4	I	10	25	d	14	3	4	59	7	1
4	3	34	8	21	54				8	14	26				8	6	58	12	17	51				7	f	3	I	11	26	e	14	0	5	0	7	0
12	16	15	16	18	3	12	8	47				12	1	19							16	19	39		g	2	I	12	27	f	13	56	5	2	6	58
1	12	24							16	10	35	1	21	28	15	3	7	1	14	0				15	a		I	13	28	g	13	53	5	4	6	56
			5	6	44	1	4	56	5	23	16				5	15	48				5	8	20	4	b	19	K	14	29	h	13	49	5	6	6	54
9	1	5	13	19	55	9	17	37				9	10	9				9	2	41	13	21	1		c	18	K	15	Virgo	i	13	46	5	7	6	53
17	21	14				17	13	46	13	11	57				13	4	29	17	22	50				12	d	17	K	16	1	k	13	43	5	9	6	51
			2	15	34				2	8	6	17	6	18							2	17	10	1	e	16	K	17	2	l	13	39	5	11	6	49
6	9	55				6	2	27	10	20	47	6	18	59	2	0	38	6	11	31					f	15	K	18	3	m	13	37	5	13	6	47
14	22	36	10	4	15	14	15	8							10	13	19				10	5	51	9	g	14	K	19	4	n	13	32	5	15	6	45
									18	16	56	14	7	40	18	9	28	14	0	12					a	13	K	20	5	o	13	29	5	16	6	44
3	18	45	18	0	25	3	11	17				3	3	49	7	22	9	3	20	21	18	2	0	17	b	12	K	21	6	p	13	25	5	18	6	42
			7	13	6	11	23	58	7	5	37	11	16	30							7	14	41	6	c	11	K	22	7	q	13	22	5	19	6	41
11	7	26													15	18	19	11	9	2	15	10	50		d	10	K	23	8	r	13	18	5	21	6	39
			15	9	15	19	20	7	15	1	47	19	12	30				19	5	11	4	23	32	14	e	9	K	24	9	r	13	15	5	23	6	37
19	3	35	4	21	25				4	14	28				4	7	0	8	15	42				3	f	8	K	25	10	ſ	13	11	5	25	6	35
8	16	16				8	8	48				8	1	20	12	13	41				12	12	13		g	7	K	26	11	s	13	8	5	26	6	34
16	12	25	12	10	37				12	3	9	16	21	29				16	14	1				11	a	6	K	27	12	t	13	4	5	28	6	32
			1	6	46	16	4	57	1	23	18				1	15	50				1	8	22	19	b	5	K	28	13	u	13	1	5	30	6	30
5	1	6	9	19	27	5	17	38				5	10	10				5	2	42	9	21	3		c	4	K	29	14	v	12	57	5	32	6	28
13	13	47							9	11	59	13	12	51	9	4	31	13	15	23				8	d	3	K	30	15	x	12	54	5	33	6	27
			17	15	36	13	6	19	17	8	8										17	17	12		e	2	K	31	15	y	12	50	5	35	6	25

September habet diex XXX, luna 29.

[Day]	Primus ciclus: Conjunctio	h	m	[Oppositio]	h	m	Secundus ciclus: Conjunctio	h	m	[Oppositio]	h	m
1	2	9	56				2	2	28	6	20	49
2	10	22	37	6	4	17	10	15	9			
3				14	16	58				14	9	30
4	18	18	47				18	11	19			
5				3	13	7				3	5	59
6	7	7	28				7	0	0	11	18	20
7				11	1	48	15	20	9			
8	15	3	37	19	21	57				19	14	29
9	4	16	18				4	8	50			
10				8	10	38	12	21	51	8	3	10
11	12	4	59	16	6	47				16	23	19
12	1	1	8	5	19	28	1	17	40			
13	9	13	49							5	12	0
14				13	8	9	9	6	21	13	0	41
15	17	9	58				17	2	30	2	20	50
16	6	22	30	2	4	18	6	15	11			
17				10	16	59				10	9	31
18	14	11	20				14	3	52			
19				18	13	9				18	5	40
20	3	7	29				3	0	1	7	18	22
21	11	20	10	7	1	50	11	12	42			
22				15	21	59				15	14	31
23	19	16	19				19	8	52			
24				4	10	40	8	21	32	4	3	12
25	8	5	0	12	23	21				12	15	53
26	16	1	9	1	19	30	16	17	41			
27	5	13	50							1	12	2
28				9	8	11	5	6	22	9	0	43
29	13	2	31				13	19	3	17	20	53
30	2	22	40	17	4	20	2	15	12			

[Day]	Tertius ciclus: Conjunctio	h	m	[Oppositio]	h	m	Quartus ciclus: Conjunctio	h	m	[Oppositio]	h	m
1	2	19	0	17	0	40	2	11	32			
2				6	13	21				6	5	53
3	10	7	41	14	2	2	10	0	13	14	18	34
4	18	3	50	3	22	11	18	20	22			
5	7	16	31							3	14	43
6				11	10	52	7	9	3	11	3	24
7	15	12	41				15	5	12	19	23	33
8				19	7	1	4	13	54			
9	4	1	22	8	19	42				8	12	14
10	12	14	3				12	6	35			
11				16	15	51				15	8	23
12	1	10	12				1	2	44	5	21	4
13	9	22	53	5	4	33	9	15	25			
14				13	17	13				13	9	45
15	17	19	2				17	11	34			
16				2	13	22				3	5	54
17	6	7	43	10	2	3	6	0	15	10	18	34
18	14	20	24	18	22	12	14	12	56			
19	3	16	33							18	14	44
20				7	10	53	3	9	5	7	3	25
21	11	5	14				11	21	46	15	23	35
22				15	7	3	19	17	55			
23	19	1	23	4	19	44				4	12	16
24	8	14	4				8	6	36			
25				12	8	25				12	0	57
26	16	10	13				15	2	45	1	21	6
27	5	22	54	1	4	34	5	15	26			
28				9	17	15				9	9	47
29	13	11	35				13	4	7			
30				17	13	24				17	5	56

KL					[SEPTEMBER]	Littere signorum	Quantitas dierum h	m	Ortus solis h	m	Occasus solis h	m
16	f		K	1	Virgo 17	ʒ	12	47	5	37	6	23
5	g	4	N	2	18	ꝫ	12	43	6	29	6	21
	a	3	N	3	19	z	12	39	6	41	6	19
13	b	2	N	4	20	a	12	36	6	42	6	18
2	c		N	5	21	b	12	32	6	40	6	16
	d	8	I	6	22	c	12	28	6	46	6	14
10	e	7	I	7	23	d	12	24	6	46	6	12
	f	6	I	8	24	e	12	21	6	50	6	10
18	g	5	I	9	25	f	12	17	6	52	6	8
7	a	4	I	10	26	g	12	14	6	53	6	7
	b	3	I	11	27	h	12	10	6	55	6	5
15	c	2	I	12	28	i	12	7	6	57	6	3
4	d		I	13	29	k	12	3	6	59	6	1
	e	18	K	14	Libra	l	12	0	6	0	6	0
12	f	17	K	15	1	m	11	58	6	1	5	59
1	g	16	K	16	1	n	11	58	6	2	5	58
	a	15	K	17	2	o	11	53	6	4	5	56
9	b	14	K	18	3	p	11	49	6	6	5	54
	c	13	K	19	4	q	11	46	6	7	5	53
17	d	12	K	20	5	r	11	42	6	9	5	51
6	e	11	K	21	6	ſ	11	39	6	11	5	49
	f	10	K	22	7	s	11	35	6	13	5	47
14	g	9	K	23	8	t	11	31	6	15	5	45
3	a	8	K	24	9	u	11	28	6	16	5	44
	b	7	K	25	10	v	11	24	6	18	5	42
11	c	6	K	26	11	x	11	21	6	20	5	40
19	d	5	K	27	12	y	11	17	6	21	5	39
	e	4	K	28	13	ʒ	11	13	6	23	5	37
8	f	3	K	29	14	ꝫ	11	10	6	25	5	35
	g	2	K	30	15	z	11	6	6	27	5	33

October habet dies XXXI.
Luna vero ejus 30.

Primus ciclus Conjunctio	h	m	Primus ciclus Oppositio	h	m	Secundus ciclus Conjunctio	h	m	Secundus ciclus [Oppositio]	h	m	Tertius ciclus Conjunctio	h	m	Tertius ciclus Oppositio	h	m	Quartus ciclus Conjunctio	h	m	Quartus ciclus [Oppositio]	h	m	KL				[OCTOBER]	Littere [signorum]	Quantitas dierum h	m	Ortus solis h	m	Occasus solis h	m
																												Libra							
			6	17	1				6	9	33	2	7	44				2	0	16	6	18	37	16	a		K 1	16	a	11	3	6	29	5	31
10	11	21				10	3	53	14	22	14	10	20	25	6	2	5	10	12	57				5	b	6	N 2	17	b	10	59	6	31	5	29
			14	5	42							18	16	34	12	14	46				14	7	18	13	c	5	N 3	18	c	10	55	6	33	5	27
18	7	31				18	0	3	3	18	23				3	10	55	18	9	6				2	d	4	N 4	19	d	10	52	6	34	5	26
7	20	12	3	1	51	7	12	44				7	5	16	11	23	36	7	21	47	3	3	27		e	3	N 5	20	e	10	48	6	36	5	24
			11	14	32				11	7	4							15	17	57	11	16	8	10	f	2	N 6	21	f	10	45	6	38	5	22
15	16	21				15	8	53				15	1	25	19	19	45				19	12	17		g		N 7	22	g	10	41	6	40	5	20
			19	10	41	4	21	34	19	3	13	4	14	6				4	6	38				18	a	8	I 8	23	h	10	38	6	41	5	19
4	5	2	8	2	21				8	15	54				8	8	26	12	19	19	8	0	58	7	b	7	I 9	24	i	10	35	6	43	5	17
12	17	43	16	19	31	12	10	17				12	2	47							16	21	7		c	6	I 10	25	k	10	31	6	45	5	15
1	13	52							16	12	3	1	22	56	16	4	35	1	15	28				15	d	5	I 11	26	l	10	28	6	46	5	14
			5	8	12	1	6	24							5	17	16				5	9	58	4	e	4	I 12	27	m	10	24	6	48	5	12
9	2	33	13	20	53	9	19	5	5	0	44	9	11	37				9	4	9	13	22	29		f	3	I 13	28	n	10	21	6	50	5	10
17	22	42				17	15	14	13	13	22				13	5	57							12	g	2	I 14	29	o	10	18	6	52	5	8
			2	17	2				2	9	34	17	7	46				17	0	18	2	18	38	1	a		I 15	Scorpio	p	10	14	6	53	5	7
6	13	23				6	3	55	10	22	15	6	20	27	2	2	6	6	12	59					b	17	K 16	1	q	10	10	6	55	5	5
			10	5	44	14	16	36							10	14	47				10	7	19	9	c	16	K 17	2	r	10	7	6	57	5	3
14	0	4							18	18	25	14	9	8	18	10	56	14	1	40					d	15	K 18	3	f	10	4	6	58	5	2
3	20	13	18	1	53	3	12	45				3	5	17	7	23	38	3	21	49	18	3	28	17	e	14	K 19	4	s	10	0	7	0	5	0
			7	14	34				7	7	6	11	17	58							7	16	9	6	f	13	K 20	5	t	9	57	7	2	4	58
11	8	54				11	1	26							15	19	47	11	10	39	5	12	19		g	12	K 21	6	u	9	54	7	3	4	57
			15	10	43	19	21	35	15	3	15	19	14	7				19	6	39				14	a	11	K 22	7	v	9	51	7	5	4	55
19	5	3	4	23	24				4	15	56				4	8	28	8	19	20	4	1	0	3	b	10	K 23	8	x	9	48	7	6	4	54
8	17	44				8	10	16				8	2	48	12	21	9				12	13	41		c	9	K 24	9	y	9	45	7	8	4	52
16	13	53	12	12	5				12	4	37	16	22	57				16	15	29				11	d	8	K 25	10	ʒ	9	41	7	10	4	50
			1	8	14	16	6	25							1	17	18				1	9	50	19	e	7	K 26	11	ꝫ	9	38	7	11	4	49
5	2	34	9	20	55	5	19	6	1	0	46	5	11	38				5	4	10	9	22	31		f	6	K 27	12	z	9	35	7	13	4	47
13	15	15							9	13	27				9	5	59	13	16	51				8	g	5	K 28	13	a	9	31	7	15	4	45
			17	17	4	13	7	47	17	9	36	13	0	19							17	18	40		a	4	K 29	14	b	9	28	7	16	4	44
2	11	24				2	3	56	6	22	17	2	20	28	17	2	2	2	13	0				16	b	3	K 30	15	c	9	25	7	18	4	42
			6	5	45	10	16	27							6	14	49				6	7	21	5	c	2	K 31	16	d	9	22	7	19	4	41

November habet dies XXX, luna vero 29.

Primus ciclus						Secundus ciclus						Tertius ciclus						Quartus ciclus						KL					[NOVEMBER]		Littere signorum	Quantitas dierum		Ortus solis		Occasus solis	
Conjunctio			[Oppositio]			Conjunctio			[Oppositio]			Conjunctio			[Oppositio]			Conjunctio			[Oppositio]																
	h	m		h	m		h	m		h	m		h	m		h	m		h	m		h	m									h	m	h	m	h	m
10	0	6	14	18	26				14	10	58	10	9	9	14	3	30	10	1	40	14	20	2		d		K	1	Scorpius	17	e	9	19	7	21	4	39
18	20	15				18	12	47				18	5	18	3	23	39	18	21	50				13	e	4	N	2		18	f	9	15	7	23	4	37
			3	14	35				3	7	7	7	18	0							3	16	11	2	f	3	N	3		20	g	9	10	7	25	4	35
7	8	56				7	1	28	11	19	48				11	12	20	7	10	34					g	2	N	4		21	h	9	8	7	26	4	34
			11	3	16	15	21	37				15	14	9				15	6	41	11	4	52	10	a		N	5		22	i	9	5	7	28	4	32
15	5	5	19	23	25				19	15	57				19	8	29	4	19	24	19	1	1		b	8	I	6		23	k	9	2	7	29	4	31
4	17	46				4	10	18				4	2	50	8	21	10				8	13	42	18	c	7	I	7		24	l	8	59	7	31	4	20
			8	12	6	12	22	59	8	4	38	12	15	31				12	8	3				7	d	6	I	8		25	m	8	56	7	32	4	28
12	6	27	16	8	15										16	17	19				16	9	51		e	5	I	9		26	n	8	54	7	33	4	27
1	2	36	5	20	56	1	19	8	16	0	47	1	11	40				1	4	12	5	22	32	15	f	4	I	10		27	o	8	51	7	35	4	25
9	15	17							5	13	28				5	6	0	9	16	53				4	g	3	I	11		28	p	8	49	7	36	4	24
			13	9	37	9	7	49	13	2	9	9	0	21	13	18	41				13	11	13		a	2	I	12		29	q	8	46	7	37	4	23
17	11	26				17	3	58	2	22	18	17	20	30				17	13	2				12	b		I	13	Sagittarius		r	8	44	7	38	4	22
			2	5	47	6	16	39							2	14	50				2	7	22	1	c	18	K	14		1	r	8	42	7	39	4	21
6	0	7	10	18	28				10	11	0	6	9	11	10	3	31	6	1	43	10	20	3		d	17	K	15		2	ſ	8	39	7	41	4	19
14	12	48				14	5	20				14	21	52	18	23	41	14	14	24				9	e	16	K	16		3	s	8	37	7	42	4	18
			18	14	37				18	7	9	3	18	1							18	16	12		f	15	K	17		4	t	8	35	7	43	4	17
3	8	17				3	1	29	7	19	50				7	12	22	3	10	33				17	g	14	K	18		5	u	8	33	7	44	4	16
11	21	38	7	3	18	11	14	10				11	6	42				11	23	14	7	4	54	6	a	13	K	19		6	v	8	31	7	45	4	15
			15	23	27				15	15	59				15	8	31	19	19	23	15	1	3		b	12	K	20		7	x	8	29	7	46	4	14
19	17	47				19	10	19				19	20	51	4	21	12				4	13	44	14	c	11	K	21		8	y	8	27	7	47	4	13
			4	12	8	8	23	0	4	4	40	8	15	32				8	8	4				3	d	10	K	22		9	ʒ	8	25	7	48	4	12
8	6	28	12	0	49				12	17	21				12	9	53				12	2	25		e	9	K	23		10	ꝫ	8	23	7	49	4	11
16	2	37	1	20	58	16	19	9				15	11	41				16	4	13	1	22	34	11	f	8	K	24		11	z	8	21	7	50	4	10
5	15	18							1	13	30				1	6	2	5	16	54				19	g	7	K	25		12	a	8	18	7	50	4	10
			9	9	39	5	7	50	9	2	11	5	0	22	9	18	43				9	11	15		a	6	K	26		13	b	8	17	7	51	4	9
13	3	59				13	20	31	17	22	20	13	13	3				13	5	35				8	b	5	K	27		14	c	8	15	7	52	4	8
			17	5	48	2	16	41							17	14	52				17	7	24		c	4	K	28		15	d	8	16	7	53	4	7
2	0	9	6	18	29				6	11	1	2	9	12				2	1	44	6	30	5	16	d	3	K	29		16	e	8	14	7	53	4	7
10	12	50				10	5	22	14	23	42	10	21	53	6	3	33	10	14	25				5	e	2	K	30		17	f	8	13	7	54	4	6

December habet dies XXXI,
luna vero XXX.

Primus ciclus						Secundus ciclus						Tertius ciclus						Quartus ciclus						KL					[DECEMBER]		Littere [signorum]	Quantitas dierum		Ortus solis		Occasus solis	
Conjunctio			Oppositio			Conjunctio			Oppositio			Conjunctio			Oppositio			Conjunctio			Oppositio																
	h	m		h	m		h	m		h	m		h	m		h	m		h	m		h	m									h	m	h	m	h	m
			14	7	10							18	18	3	14	16	14				14	8	46		f		K	1	Sagittarius	18	g	8	12	7	54	4	6
18	8	59				18	1	31	3	19	51				3	12	23	18	10	35				13	g	4	N	2		19	h	8	11	7	55	4	5
7	21	40	3	3	19	7	14	12				7	6	44				7	23	16	3	4	55	2	a	3	N	3		20	i	8	10	7	55	4	5
			11	16	0				11	8	32				11	1	4	15	19	25	11	17	36		b	2	N	4		21	k	8	9	7	56	4	4
15	17	49				15	10	21				15	2	53	19	21	13				19	13	45	10	c		N	5		22	l	8	8	7	56	4	4
			19	12	9	4	23	2	19	4	41	4	15	34				4	8	6					d	8	I	6		23	m	8	7	7	57	4	3
4	6	30	8	0	50				8	17	22				8	9	54	12	20	47	8	2	26	18	e	7	I	7		24	n	8	7	7	57	4	3
12	19	11	16	20	59	12	11	43				12	4	15							16	22	35	7	f	6	I	8		25	o	8	6	7	57	4	3
1	15	20							16	13	31				16	6	3	1	16	56					g	5	I	9		26	p	8	6	7	57	4	3
			5	9	40	1	7	52				1	0	34	5	18	44				5	11	16	15	a	4	I	10		27	q	8	5	7	58	4	2
9	4	1	13	22	21	9	10	33	5	2	12	9	13	5				9	5	37	3	23	57	4	b	3	I	11		28	r	8	5	7	58	4	2
						17	16	42	13	14	53				13	7	25								c	2	I	12		29	ſ	8	4	7	58	4	2
17	0	10	2	18	31				2	11	3	17	9	14				17	1	48	2	20	6	12	d		I	13	Capricornus		s	8	4	7	58	4	2
6	12	51				6	5	23	10	23	44	6	21	55	2	3	34	6	14	27				1	e	19	K	14		1	t	8	4	7	58	4	2
			10	7	12	14	18	4							10	16	16				10	8	47		f	18	K	15		2	u	8	5	7	58	4	2
14	1	32							18	19	53	14	10	36	18	12	25	14	3	8				9	g	17	K	16		3	v	8	5	7	58	4	2
3	21	41	18	3	21	3	14	13				3	6	45				3	22	17	18	4	57		a	16	K	17		4	x	8	6	7	57	4	3
			7	16	4				7	8	34	11	19	26	7	1	6				7	17	38	17	b	15	K	18		5	y	8	6	7	57	4	3
11	10	22				11	3	54							15	21	15	11	11	58	15	13	47	6	c	14	K	19		7	ʒ	8	7	7	57	4	3
			15	12	11	19	23	3	15	4	43	19	15	35				19	8	7					d	13	K	20		8	ꝫ	8	8	7	56	4	4
19	6	31							4	17	24				4	9	56	8	20	48	4	2	28	14	e	12	K	21		9	z	8	9	7	56	4	4
8	19	12	4	0	52	8	11	44				8	4	16	12	21	37				12	15	9	3	f	11	K	22		10	a	8	10	7	55	4	5
16	15	21	12	13	33				12	6	5							16	16	57					g	10	K	23		11	b	8	11	7	55	4	5
			1	9	43	16	7	53				16	0	25	1	18	46				1	11	18	11	a	9	K	24		12	c	8	12	7	54	4	6
5	4	2	9	22	23	5	20	34	1	2	14	5	13	6				5	5	38	9	23	59	19	b	8	K	25		13	d	8	13	7	54	4	6
13	16	44							9	14	55				9	7	27	13	18	19					c	7	K	26		14	e	8	14	7	53	4	7
			17	18	32	13	9	15	17	11	4	13	1	47							17	20	8	8	d	6	K	27		15	f	8	15	7	53	4	7
2	12	53				2	5	25	6	23	45	2	21	57	17	3	36	2	14	28					e	5	K	28		16	g	8	17	7	52	4	8
			6	7	13	18	18	6							6	16	17				6	8	49	16	f	4	K	29		17	h	8	18	7	51	4	9
10	1	34							14	12	26	10	10	38				10	3	9	14	21	30	5	g	3	K	30		18	i	8	19	7	51	4	9
18	21	43	14	19	54	18	14	15				18	6	47	14	4	58	18	23	19					a	2	K	31		19	k	8	21	7	50	4	10

Textual astronomical data

Jan 6	A festo Stelle numerando, perfice lunam post quadraginta dies et 70^{a} fiet et si bisestus fuerit similiter additur unus; si cadit in lucem Dei, subpone sequentem.
Jan 7	Claves 70^{e}.
Jan 18	Ante diem Prisce nunquam deponitur *Alleluia*.
Jan 26	Nota quod feria secunda post 70as dicitur officium pro defunctis, fratris et benefactoribus.
Jan 28	Claves 40^{e}.
Feb 6	Ver incipit secundum Grecos.
Feb 13	Hic minue contra febres.
Feb 18	Hic Adam cum peno peccavit.
Feb 21	Incipit ver secundum Ysidorum.
Feb 22	Cedit hiems retro Cathedrato Simone Petro.
Feb 23	Ver secundum Ambrosium.
Feb 24	Bixestum sexste martis tenuere calende, posterioraque die celebrantur festa Mathie. Transfertur si venerit in die Cinerum.
Feb 27	Post 2^{am} primam computa duos dies et prima sequenti dominica est Quadragessima. Item post revolutionem lune februarii, prima dies Mercurii est dies Cinerum.
Mar 1	Nota quod numerus pacte semper mutatur prima die martii et durat usque ad alium martium.
Mar 11	Claves Pasce.
Mar 21	Post festum sancti Benedicti, ubicumque 14 luna invenitur [que tenet undenas Aprilis Luna, pacte numerum monstrant per quodlibet annum], dominica proxima celebratur Pasca.
Mar 27	Post 3^{am} primam computa 14 dies et prima sequenti dominica est Pasca Domini nostri Yhesu Christi.
Apr 15	Claves Rogationum.
Apr 21	[Claves Penthecostes.]
May 12	Sacramento [?] Kaesaris hic est.[50]
May 15	Nota quod 8^{a} Ascensionis est semi-duplex et 4^{a} feria post Pentecostes Quatuor Tempora celebrantur, et dominica prima fit officium Trinitatis cum 9^{e} dominice, et 5^{a} feria post Pentecostes fit Officium Corporis.
May 26	Ver fugat Urbanum, estas Symphorianum; illico quod tibi restat autumpni tempora prestart.
Jun 18	Coronatio Kaesaris hic est.[51]
Jul 14	Dies caniculares.
Nov 27	Ultimo die ante dominica de Adventu fit officium pro defunctis: primam et matutinam oratio *Deus qui*.
Dec 13	Vult Crux, Lucia, Cynis et Carismata dia, ut det vota pia quarta sequens feria.
Dec 30	Hic leguntur sermones hore vel castissius quando nativitas Domini venerit 2^{a} feria.

50 Relates probably to the coronation of Holy Roman Emperor Fredrick III in Rome on March 19th, 1452. Reading is not certain, but seems to relate to 'coronatio Kaesaris' on June 18th [*sic* for his coronation King of the Romans on June 17th, 1442].

51 See above.

Sanctorale[52]

Day	Gmunden	Latin 7478
		JANUARIUS
1	Circumcisio Domini	Circuncisio Domini (duplex minus, praeceptum); Basilii episcopi et confessoris; Martine virginis martyris.
2	Octava sancti Stephani	Octava sancti Stephani (semi-duplex minus). Dicitur *Credo* in missa.
3	Octava sancti Johannis	Octava sancti Johannis (semi-duplex minus).
4	Octava sanctorum Innocentum	Octava Innocentum (semi-duplex minus). *Te Deum laudamus* dicitur.
5	Severini episcopi	Vigilia quo ad officium (semi-duplex minus). Hic non dicitur officium dominice.
6	Epyphanie Domini	Epiphania Domini (duplex majus, praeceptum).
7	Valentini episcopi	
8	Erhardi episcopi	
9	Juliani et sociorum	
10	Pauli primi hermite	Sancti Pauli primi heremite.
11		Sancti Ygini pape et martyris.
12		
13	Octava Epyphanie	Octava Epiphanie (semi-duplex minus).
14	Felicis in Pincis	Sancti Felicis in Pincis presbyteri et martyris.
15		Sancti Mauri abbatis.
16	Marcelli pape	Sancti Marcelli pape et martyris. Pro 8° responsorio dicitur *Domin[e] prevenisti eum.*
17	Anthonii abbatis	Sancti Anthonii abbatis (semi-duplex minus). Non tenetur nisi in dominica.
18	Priscce virginis	Sancte Prisce virginis et martyris.
19		Sanctorum Mauri, Marthe, Audifax et Abbacuc.
20	Fabiani et Sebastiani	Sanctorum martyrum Fabiani et Sebastia[ni]. Non tenetur nisi in dominica.
21	Angnetis virginis	Sancte Agnetis virginis et martyris (semi-duplex majus). Non tenetur nisi in dominica 70^{e}.
22	Vincenti martitis	Sanctorum martyrum Vincentii et Anastasii. Non tenetur nisi in dominica.
23	Emerenciane virginis	sancte Emerentiane virginis et martyris.
24	Thymothei apostoli	
25	Conversio sancti Pauli	Conversio sancti Pauli Apostoli (duplex minus). Non tenetur nisi in dominica 70^{e} vel 60^{e}.
26	Poicarpi episcopi	

52 Many thanks to Dr Laura Albiero for her help in several readings and liturgical abbreviation deciphering. (Obviously, any remaining mistakes are my own.) This sanctorale is to be read in conjunction with the calendar on p. 165 and the additional textual data that I decided, maybe mistakenly, was not part of the sanctorale per se (see p. 177).

Day	Gmunden	Latin 7478
27	Johannis Crisotomi	Sancti Johannis Crisostimi et episcopi et confessoris. Oratio: *Deus venie largitor.*
28	Octava Angnetis	Sancte Agnetis secundo. Non tenetur nisi in dominica.
29		
30		
31	Vigili et sociarum ejus	Sanctorum martyrum Ciri et Johannis.
	FEBRUARIUS	
1	Brigide virginis	Sancti Ygnatii episcopi et martyris; sancte Brigide virginis (non tenetur nisi in dominica). Pro 8° responsorio dicitur *Domin[e] prevenisti eum.*
2	Purificatio Marie Virginis	Purificatio Virginis Marie (duplex majus, praeceptum). Non tenetur nisi in dominica 70^e^ vel 60^e^ vel 50^e^.
3	Wlasii episcopi	Sancti Blasii episcopi et martyris.
4		Sancti Gilberti confessoris.
5	Agathe virginis	Sancte Agathe virginis et martyris. Non tenetur nisi in dominica 70^e^, 60^e^ et 50^e^.
6	Dorothee virginis	Sancte Dorothee virginis.
7		
8		
9	Appollonie virginis	Sancte Apolonie virginis et martyris. Non tenetur nisi in dominica.
10	Scolastice virginis	Sancte Scolastice virginis.
11		
12		
13		
14	Vincenti martiris	Sancti Valentini presbyteri et martyris. Obii anno Domini 280.[53]
15		Translatio sancti Antonii de Padua (duplex minus).
16	Juliane virginis	Sancte Juliane virginis.
17		
18		
19		
20		
21		
22	Kathedra sancti Petri	Cathedra sancti Petri (duplex minus).
23	Vigilia	Vigilia (praeceptum).
24	Mathie apostoli	Sancti Mathie apostoli (duplex minus, praeceptum). Transfertur si venerit in die cinerum.
25	Waltpurge virginis	
26		

53 Saint Valentin of Terni is supposed to have been beheaded in 269 AD.

Day	Gmunden	Latin 7478
27		
28		
	MARTIUS	
1		
2		
3	Chunigundis regine	
4	Adriani martiris	
5		
6		
7	Perpetue et Felicitatis	Sancti Tome de Aquino confessoris (non tenetur nisi in dominica); sanctarum virginum Perpetue et Felicitatis.
8		
9		Sanctorum Quadraginta Martirum.
10		
11		
12	Gregorii pape	Sancti Gregorii pape et [con]fessoris (duplex minus). Dicitur in missa *Credo*.
13		
14		
15		Sancti Longini martyris.
16	Ciriaci martiris	
17	Gertrudis virginis	Sancti Patricii martyris et confessoris. Non tenetur nisi in dominica.
18		
19		Sancti Joseph confessoris. Non tenetur nisi in dominica.
20		
21	Benedicti abbatis	Sancti Benedicti abbatis et confessoris (duplex minus).
22		
23		
24		
25	Anunciatio Marie	Annuntiatio Virginis Marie (duplex majus, praeceptum).
26		
27	Ruperti episcopi	
28		
29		
30		
31		
	APRILIS	
1		
2		

Day	Gmunden	Latin 7478
3		
4	Ambrosii episcopi	
5		Sancti Vincentii confessoris. Non tenetur nisi in dominica.
6		
7		
8		
9	Marie Egypciace	
10		
11	Leonis pape	
12		
13	Eufemie virginis	
14	Tiburtii et Valeriani	Sanctorum martyrum Tiburtii, Valeriani et Maximi.
15		
16		
17		Sancti Aniceti pape et martyris.
18		
19		
20		
21		
22	Victoris pape	Sanctorum martyrum Sotheris et Georii pape et martyris.
23		Sancti Georgii martyris.
24	Geor[g]ii martitis	
25	Marci evangeliste	Sancti Marci evangeliste (duplex minus).
26	Cleti pape	Sancti Cleti et Marcelli pape et martyris.
27		
28	Vitalis martiris	Sancti Vitalis martyris.
29		Sancti Petri martyris de Ordine Predicatorum. Non tenetur nisi in dominica.
30		
		MAIUS
1	Philippi et Jacobi	Sanctorum apostolorum Philippi et Jacobi (duplex minus, praeceptum).
2	Sigismundi regis	
3	Inventio sancti Crucis	Inventio sancte Crucis (duplex minus, praeceptum); Alexandri et sociorum ejus.
4	Floriani martiris	
5	Gothardi episcopi	
6	Johannis ante portam latinam	Sancti Johannis ante portam latinam (duplex minus).
7		
8	Victoris martiris	Apparitio sancti Michaelis archangeli (semi-duplex majus). Non tenetur nisi in dominica.

Day	Gmunden	Latin 7478
9		
10	Gordiani episcopi	Sanctorum martyrum Giordani et Epimachi.
11		
12	Nerei, Achilei, Pangrati	Sanctorum martyrum Nerei et Achilei et Pancratii.
13		
14		Sancti Bonifatii martyris.
15	Sophye virginis	
16		
17		
18		
19	Potenciane virginis	Sancti Petri de Morone pape (non tenetur nisi in dominica); sancte Pote[n]tiane virginis.
20		Sancti Bernardi confessoris (duplex majus). Obiit ex ac vita 1444 an[n]o; canonicatus fuit gloriosus confessor 1450 anno.
21		
22	Helene regine	
23		
24		
25	Urbani pape	Translatio sancti Francisci (duplex majus); sancti Urbani pape martyris.
26		Sancti Eleuterii pape martyris.
27		Sancti Johannis pape et martyris.
28		
29		
30		Sancti Felicis pape et martyris.
31	Petronelle virginis	Sancte Petronille virginis.
		JUNIUS
1	Nicomedis martiris	
2	Marcelli[ni] et Petri	Sanctorum martyrum Marcellini, Petri atque Herasmi.
3	Erasmi episcopi	
4		
5	Bonifacii et sociorum ejus	
6		
7		
8		
9	Primi et Feliciani	Sanctorum martyrum Primi et Feliciani.
10		
11	Barnabe apostoli	Sancti Bernabe apostoli (minus duplex).
12	Basilidis [et] Cirini	Sanctorum martyrum Basilidis, Cirini, Narboris et Nazarii.
13		Sancti Antonii confessoris (duplex majus).
14		

Day	Gmunden	Latin 7478
15	Viti et Modesti	Sanctorum martyrum Viti et Modesti atque Crescentie.
16		
17		
18	Marci et Marcelli	Sanctorum martyrum Marci et Marcelliani.
19	Gervasii [et] Prothasii	Sanctorum martyrum Gervasii et Protasii. Pro 8° responsorio dicitur *Hec est vera.*
20		Octava sancti Antonii (semi-duplex minus); sancti Si[l]verii pape et martyris. Pro 8° responsorio dicitur *Domine prevenisti.*
21	Albani martiris	
22	Achatii et sociorum ejus	Sancti Paulini episcopi et confessoris.
23	Vigilia	Vigilia (praeceptum).
24	Johanis Baptiste	Nativitas sancti Johannis Baptiste (duplex majus, praeceptum).
25		
26	Johannis et Pauli	Sanctorum martyrum Johannis et Pauli (non tenetur nisi in dominica). 8^m^ responsorium dicitur *Hec est vera fraternitas.*
27	Septeni dormientium	
28	Leonis pape	Sancti Leonis pape et confessoris (non tenetur nisi in dominica); Vigilia (praeceptum).
29	Petri et Pauli apostolorum	Apostolorum Petri et Pauli (majus duplex, praeceptum).
30	Commemoratio Pauli	Commemoratio sancti Pauli apostoli (duplex minus).
	JULIUS	
1	Octava Johannis Baptiste	Octava sancti Johannis Baptiste (semi-duplex majus).
2	Visitacio Marie	Visitatio Virginis Mari (majus duplex); sanctorum martyrum Processi et Martiniani.
3		
4	Udalrici episcopi	
5		
6	Octava Petri et Pauli	Octava apostolorum Petri et Pauli (duplex minus).
7	Wilibaldi episcopi	
8	Kiliani et sociorum ejus	
9	Translatio Nicolai	Octava Visitationis Virginis Marie (duplex minus).
10	Septem fratrum	Sanctorum martyrum 7^em^ fratrum et sanctarum Rufine et Secunde virginum. Pro 8° responsorio *Hec est vera.*
11	Translatio Benedicti	Sancti Pii pape et martyris.
12	Margarethe virginis	Sanctorum martyrum Naboris et Felicis.
13	Hamrici regis	Sancti Anacleti pape et martyris; sancte Margarite virginis et martyris (non tenetur...).
14		

Day	Gmunden	Latin 7478
15	Divisio apostolorum	Sanctorum martyrum [C]irici et Julite.
16		
17	Allexii confessoris	Sancti Alexii confessoris.
18	Arnolphi martiris	Sanctorum martyrum Symphorose cum 7^em^ filiis suis. Pro 8° responsorio dicitur *Hec est.*
19		
20		
21	[P]raxedis virginis	Sancte Praxedis virginis; Officium pro defunctis fratribus et benefactoribus.
22	Marie Magdalene	Sancte Marie Magdalene (semi-duplex majus). Non tenetur nisi in dominica.
23	Appollinaris martiris	Sancti Apolinaris episcopi et martyris.
24	Christine virginis	Sancte Cristine virginis et martyris; Vigilia (praeceptum).
25	Jacobi apostoli, Kristofori	Sancti Jacobi apostoli (duplex minus, praeceptum); sancti Cristofori martyri.
26	Anne matris Marie	Sancte Anne matris Virginis Marie (non tenetur nisi in dominica); sancti Pastoris presbyteri.
27		Sancti Pantaleonis martyris.
28	Panthaleonis martiris	Sanctorum martyrum Nazarii, Celsi, Victoris et Innocentii pape et martyris.
29	Felicis et sociorum ejus	Sancte Marthe virginis (non tenetur nisi in dominica); sancti Symplicii et sociorum ejus.
30	Abdon et Senne[n]	Sanctorum martirum Abdon et Senen.
31		
	AUGUSTUS	
1	Ad Vincula sancti Petri	Sancti Petri ad Vincula (duplex minus); sanctorum Machabeorum martyrum. Responsorium *Hec est.*
2	Stephani pape	Dedicatio Portiuncule (duplex minus); sancti Stephani pape et martyris.
3	Inventio sancti Stephani	Inventio sancti Stephani prothomartiris. Non tenetur nisi in dominica.
4	Translatio Valentini	Sancti Justini presbyteri et martyris.
5	Oswaldi regis	Festum nivis (majus duplex); sancti Dominici confessoris (non tenetur nisi in dominica).
6	Sixti pape	Sancti Systi pape et martyris; F[el]icissimi et Agapiti martyrum.
7	Afre martiris	Sancti Donati episcopi et martyris.
8	Ciriaci et sociorum ejus	Sanctorum martyrum Ciriaci, Largi et Sma[ra]gdi.
9	Romani martiris	Sancti Romani martyris; Vigilia (praeceptum).
10	Laurentii martiris	Sancti Laurencii martyris (praeceptum, majus duplex).
11	Tyburcii martiris	Sanctorum martyrum Tiburcii et Susanne. Non tenetur nisi in dominica.

Day	Gmunden	Latin 7478
12	Clare virginis	Sancte Clare virginis de Assisio (duplex majus).
13	Ypoliti et soc[i]orum ejus	Sanctorum martyrum Ipoliti et sociorum ejus.
14	Eusebii confessoris	Sancti Eusebii presbyteri et confessoris; Vigilia (praeceptum).
15	Assumptio Marie Virginis	Assumptio sancte Marie virginis (duplex majus, praeceptum).
16		
17	Octava Laurentii	Octava sancti Laurentii (semi-duplex minus).
18	Agapiti martiris	Sancti Agapiti martyris.
19		Sancti Ludovici episcopi et confessoris (duplex majus); et de octava sancte Clare fit confessio.
20	Bernhardi abbatis	Sancti Bernardi abbatis et martyris. Fit statim post 8^{am} Assumptionis.
21		
22	Thymothei [et] Simphoriei	Octava Assumptionis (duplex minus); sanctorum martyrum Timotei et sociorum ejus.
23	Vigilia	
24	Bartholomei apostoli	Vigilia (preceptum).
25		Sancti Bartholomei apostoli (duplex minus, fit secundum morem patrie, praeceptum).
26		Octava sancti Ludovici episcopi (semi-duplex minus); sancti Ceserini pape martyris.
27	Rufi martiris	Sancti Ludovici confessoris Regis Francie. Fit statim post 8^{am} Ludocivi episcopi.
28	Augustini episcopi	Sancti Augustini episcopi et confessoris; dicitur *Credo* in Missa; duplex minus; Hermetis martyris.
29	Decolatio Johanis	Decolatio sancti Johannis Baptiste (semi-duplex majus, non tenetur nisi in dominica); Sabine martyris.
30	Felicis et Adaucti	Sanctorum martyrum Felicis et Audacti.
31		
	SEPTEMBER	
1	Egidii abbatis	Sanctorum martyrum 12 fratrum; pro 8° responsorio dicitur *Hec vera fraternitas*; sancti Egidii abbatis et confessoris.
2		Sancti Anthonini martyris.
3		
4		
5		
6	Magni confessoris	
7		
8	Nativitas Marie	Nativitas Virginis Marie (duplex majus, praeceptum); sancti Andriani martyris.
9	Gorgonii martiris	Sancti Gorgonii martyris.
10		Sancti Nicolai Tolentini confessoris.

Day	Gmunden	Latin 7478
11	Prothy et Jacincti	Sanctorum martyrum Prothi et Iacinti.
12		
13	Septem dormientium	
14	Exaltatio sancti Crucis	Exaltatio sancte Crucis (minus duplex, praeceptum); sanctorum Cornelii et Cypriani.
15	Nicomedis martiris	Octava Virginis Marie (semi-duplex majus); sancti Nicomedi fide martyris.
16	Eufemie virginis	Sanctorum martyrum Lucie, Germiniani et Eufemie virginis.
17	Lamperti episcopi	Stimata beati Francisci (majus duplex). Non tenetur nisi in dominica.
18		
19	Januarii et sociorum ejus	[*Scraped*].[54]
20	Vigilia	Sanctorum martyrum Eustachii et sociorum ejus (non tenetur nisi in dominica); Vigilia (praeceptum).
21	Mathiei apostoli	Sancti Mathei apostoli et evangeliste (minus duplex, praeceptum). Non tenetur nisi in dominica.
22	Mauricii et sociorum ejus	Sanctorum martyrum Mauricii et sociorum ejus.
23	Tecle virginis	Sancti Lini pape et martyris.
24	Translatio Ruperti	
25		[*Unreadable*].
26	Translatio sancti Virgilii	[*Unreadable*].
27	Cosme et Damiani	[*Unreadable*] (semi-duplex majus, non tenetur nisi in dominica); [*Unreadable*].
28	Wenzeslai regis	Officium pro benefactoribus (oratio *Deus venie*); [*Unreadable*]. 8^{m} responsorium dicitur *Hec est*.
29	Michaelis archangeli	Dedicatio Michaelis (duplex majus, non tenetur nisi in dominica, praeceptum).
30	Jeronimi confessoris	Jeronimi presbyteris (duplex minus). Dicitur *Credo* in missa; non tenetur nisi in dominica.
	OCTOBER	
1	Remigii episcopi	Sancti Remigii episcopi et confessoris.
2	Leodegarii episcopi	Translatio sancte Clare virginis (duplex minus). Non tenetur nisi in dominica.
3		
4	Francisci confessoris	
5		Sancti Francisci confessoris (majus duplex). Non tenetur nisi in dominica.

54 The scribe seems to have been correcting a one-line-shift; hence the difficulties to decipher either the scraped or the replacing text (especially in black ink).

Day	Gmunden	Latin 7478
6		
7	Sergii et Bachii martirum	Sanctorum martyrum Sergi, Bachii, Marcelli et Apulei; sancti Marci pape et confessoris.
8		
9	Dyonisii et sociorum ejus	Sanctorum martyrum Dionisii, Rusticii et Eleutherii.
10	Gereonis et sociorum ejus	Sancti Cerboni episcopi et confessoris.
11	Translatio sancti Augustini	Octava sancti Francisci (duplex minus).
12	Maximiliani episcopi	
13	Cholomanni martiris	
14	Calixti pape	Sancti Kalisti pape et martyris.
15	Hedwigis vidue	
16	Galli abbatis	
17	Marthe hospite Christi	
18	Luce ewangeliste	Sancti Luce evangeliste (duplex minus). Non tenetur nisi in dominica.
19	Januarii et sociorum ejus	
20		
21	Undecim milium virginum	Sancte Ursule et sociarum ejus (non tenetur nisi in dominica); sancti Ilarionis abbatis et confessoris.
22	Severi episcopi	
23	Severini episcopi	
24		
25	Crispini et Crispiani	Sanctorum martyrum [Ch]risanti et Darie.
26	Amandi episcopi	Sancti Evaristi pape et martyris.
27	Vigilia	Sancti Yvonis confessoris de tertio ordine (semi-duplex majus, non tenetur nisi in dominica); Vigilia (praeceptum).
28	Symonis et Jude	Aspostolorum Symonis et Jude (duplex majus, non tenetur nisi in dominica, praeceptum).
29	Narcissi episcopi	
30		
31	Wolfgangi episcopi	Vigilia omnium sanctorum (praeceptum).
	NOVEMBER	
1	Omnium sanctorum	Festum omnium sancctorum (duplex majus, praeceptum); Cesarii martyris.
2	Commemoratio animarum	Commemoratio omnium fidelium defunctorum.
3		
4		Sancti Vitalis et Agricole martyrum.
5		
6	Bernhardi confessoris	Sancti Leonardi confessoris.
7		
8	Quatuor cor[o]natorum	Translatio sancti Ludovici episcopi (duplex majus); Sanctorum martyrum 4 coronatorum.

Day	Gmunden	Latin 7478
9	Theodorii martiris	Dedicatio Basilice (duplex majus); sancti Teodori martyris.
10	Martini pape	Sanctorum martyrum Triphonis, Respicii et Nymphe virginis.
11	Martini episcopi	Sancti Martini episcopi (semi-duplex majus, praeceptum); sancti Menne martyris.
12		Sancti Martini pape et martyris. Pro 8^{m} responsorium dicitur *Domine prevenisti eum.*
13	Briccii episcopi	Sancti Britii episcopi et confessoris.
14		
15		
16	Othmari abbatis	
17		
18	Octava sancti Martini	Dedicatio Basilice Petri et Pauli apostolorum (duplex majus).
19	Elysabeth vidue	Sancte Helisabet nec virginis nec martyris (semi-duplex majus, non tenetur nisi in dominica); sancti Pontiani pape et martyris. Pro 8^{m} responsorium dicitur *Domine prevenisti eum.*
20		
21		
22	Cecilie virginis	Sancte Cecilie virginis et martyris (semi-duplex minus). Non tenetur nisi in dominica.
23	Clementis pape	Sancti Clementis pape et martyris (semi-duplex minus). Non tenetur nisi in dominica.
24	Crisogoni martiris	Sancti Grisogoni martyris; sancte Felicitatis martyris.
25	Katherine virginis	Sancte Katherine virginis et martyris (semi-duplex minus). Non tenetur nisi in dominica.
26	Lini pape	Sancti Petri Alexandrini episcopi et martyris.
27	Virgilii episcopi	
28		Vigilia (praeceptum).
29	Vigilia	Sancti Saturnini martyris.
30	Andree apostoli	Sancti Andree apostoli (duplex minus, praeceptum).
	DECEMBER	
1		
2		Sancte Bibiane virginis et martyris.
3		
4	Barbare virginis	Sancte Barbare virginis et martyris.
5		Sancti Sabbe abbatis et confessoris.
6	Nicolai episcopi	Sancti Nicolai episcopi et confessoris (semi-duplex majus). Non tenetur nisi in dominica.

Day	Gmunden	Latin 7478
7	Octava sancti Andree	Sancti Ambrosii episcopi et confessoris (duplex minus). Non tenetur nisi in dominica.
8	Conceptio Marie	Conceptio Virginis Marie (majus duplex). Non tenetur nisi in dominica.
9		
10		
11	Damasii pape	
12		
13	Lucie et Otilie virgin[um]	Sancte Lucie virginis et martyris (semi-duplex majus).
14	Nicasii episcopi	
15		
16		
17	Ignacii episcopi	
18		
19		
20	Vigilia	Vigilia (praeceptum).
21	Thome apostoli	Sancti Thome apostoli (minus duplex, praeceptum).
22		
23		
24	Vigilia	Vigilia (praeceptum).
25	Nativitas Christi	Nativitas Domini Nostri Yhesu Christi (duplex majus, praeceptum); Anastasie martyris; Confessio in missa 2[a].
26	Stephani prothomartiris	Sancti Stephani prothom[arty]ris (praeceptum, duplex majus).
27	Johanis apostoli	Sancti Johannis evangeliste (praeceptum, duplex majus).
28	Sanctorum Innocentum	Sanctorum Innocentum martyrum (semi-duplex majus). *Te Deum laudamus*, praeceptum. Non dicitur nisi venerit in dominica.
29	Thome episcopi	Sancti Thome episcopi et martyris (non tenetur).
30		
31	Silvestri pape	Sancti Silvestri pape et confessoris (non tenetur nisi in dominica, praeceptum).

Canons and ancillary tables

F. 1. Tabula 4^{or} ciclorum 19^{m}

[F. 1Ds]

Primus ciclus			Secundus ciclus		Tertius ciclus		Quartus ciclus	
15	1515		15	1458	15	1477	15	1496
16	1516		16	1459	16	1478	16	1497
17	1517		17	1460	17	1479	17	1498
18	1518		18	1461	18	1480	18	1499
19	1519		19	1462	19	1481	19	1500
1	1520		1	1463	1	1482	1	1501
2	1521		2	1464	2	1483	2	1502
3	1522		3	1465	3	1484	3	1503
4	1523		4	1466	4	1485	4	1504
5	1524		5	1467	5	1486	5	1505
6	1525		6	1468	6	1487	6	1506
7	1526		7	1469	7	1488	7	1507
8	1527		8	1470	8	1489	8	1508
9	1528		9	1471	9	1490	9	1509
10	1529		10	1472	10	1491	10	1510
11	1530		11	1473	11	1492	11	1511
12	1531		12	1474	12	1493	12	1512
13	1532	1456	13	1475	13	1494	13	1513
14	1533	1457	14	1476	14	1495	14	1514

[F. 1Di]

Ad sciendum per kalendarium sequens tempus conjunctionum et oppositionum solis cum lune: nota primo [quod] in eo posui 4^{or} ciclos conjunctionum et oppositionum, quorum quilibet ciclus continet spatium 19 annorum, et ex hiis integratur ciclus perfectus qui est spatium 76 annorum, ut patet hic super in rota in qua posui tibi predictos ciclos cum milessimo. In quarum prima ponitur aureus numerus in directo illius diei in quo fit conjunctio vel oppositio. In secunda linea ponitur numerus [horarum]. Et in tertia linea ponitur numerus minutorum hore, quibus completis, fit tunc conjun[c]tio vel oppositio.

Primo ergo scias aureum numerum illius anni in quo queris, demum quere eundem aureum numerum sub ciclo conjunctionis vel oppositionis et in directo ejusdem aurei numeri, in linea dierum illius mensis in quo queris predicta, habere diem conjunctionis etc.

[F. 1A][55]

Et considera bene sub quo ciclo debes predicta querere 2^{m} doctrinam notabilis premium. Et nota quod si hora[e] invente fuerint pauciores quam 12, tunc conjunctio vel oppositio erit eodem die post meridiem, scilicet inter meridiem ejusdem diei et meridiem noctis sequentis, post tot horas et minuta quot sunt inposita; si vero fuerint plures hore quam 12, tunc conjunctio vel oppositio erit die sequenti de mane, ante meridiem, post tot horas et minuta a medi[o] noctis computando quot sunt hore residue cum minutis ultra duodecim horas. Nota etiam quod hora est 24^{a} pars diei naturalis, hoc est agregati ex die artificiali et nocte simul. Et quelibet hora dividitur in 60^{a} partes equales que vocantur minuta horarium. Et quodlibet minutum [dividitur] in 60^{a} secunda. Item nota quod quelibet hora habet mille octoginta puncta.

F. 14. Tabula signorum lune per totum annum

[F. 14D (left)]: See opposite page.

[F. 14D (right)][56]

Aries — Nil capiti facias, Aries cum luna refulget. Brachia tunc minuas et balneo tutius intras [*sic* for *intres*]. Non tangas aures nec barbam radere cures.

Taurus — Arbor plantetur cum luna Thaurus habebit. Edificare potes, et sperges [*sic* for *spergas*] semina terre. Et medicus caveat cum fero tangere collum.

Gemini — Brachia non minuas cum lucet luna Gemellis. Ungwibus et manibus cum fero cura negetur.

Cancer — Pectus, pulmone, jecur non minuantur. Potio sumatur, securus perge viator. Somnia falsa vites [*sic* for *vides*]; est utilis emptio rerum.

Leo — Cor gravat et stomachum, cum cernit luna Leonem. Non facias vestes sed ad convivia vadas. Et nichil ore vomas, nec sumas tunc medicinas.

Virgo — Luna Virgo tenens, uxorem ducere noli. Unguento caveas. Costas curare cyrogere [*sic*]. Detur agro semen, caveas intrare carinam.

Libra — Libra tenens lunam, nemo genitalia tangat. Aut renes, nates; iter nec capere cures, extremam partem Libre cum luna tenebit.

Scorpio — Augmentat Scorpio morbos in parte pudenda. Vulnera non curas, cave ascendere navem. Et carpes iter, timeas de morte ruinam.

Sagittarius — Luna nocet femori per partes mota [*sic*] Sagitte. Unguues et crines potes prescendere [*prescindere*] tute.

Capricornus — Capra nocet genibus, ipsam cum luna tenebit. Intres aquas maris melius curabitur eger. Fundamenta ruunt, nichil est quod duret in ipsa.

Aquarius — Tangere crura cave cum luna videbit Aquosam. Insere tunc plantes [*plantas*], excelas erige turus. Ac si capis iter, ad finem tardius ibis.

Pisces — Piscis habens lunam, noli curare podagram. Carpe viam tucius, hujusmodi potio sumpta salubris.

---

55 This leaf's tab has been replaced and it is mounted incorrectly; compartment A can be read on the verso of the entirely unfold leaf, tab on top.

56 These verses are derived from the *Salernitan Regimen sanitatis*; compare to the critical edition in *Collection Salernitana*, pp. 486–87.

1456			Tabula 12 signorum lune per totum annum																		
Aureus numerus			1	2	3	4	5	6	7	8	9	10	11	12	13	14	15	16	17	18	19
		/	y	n	c	u	l	ƶ	ſ	h	ʒ	p	e	v	m	a	s	i	3	q	f
Aries	Bonum	—	ʒ	o	d	v	m	a	s	i	3	q	f	x	n	b	t	k	ƶ	r	g
		\	3	p	e	x	n	b	t	k	ƶ	r	g	y	o	c	u	l	a	ſ	h
		/	ƶ	q	f	y	o	c	u	l	a	ſ	h	ʒ	p	d	v	m	b	s	i
Taurus	Malum	—	a	r	g	ʒ	p	d	v	m	b	s	i	3	q	e	x	n	c	t	k
		>	b	ſ	h	3	q	e	x	n	c	t	k	ƶ	r	f	y	o	d	u	l
Gemini	Malum	—	c	s	i	ƶ	r	f	y	o	d	u	l	a	ſ	g	ʒ	p	e	v	m
		>	d	t	k	a	ſ	g	ʒ	p	e	v	m	b	s	h	3	q	f	x	n
Cancer	Indifferentia	—	e	u	l	b	s	h	3	q	f	x	n	c	t	i	ƶ	r	g	y	o
		>	f	v	m	c	t	i	ƶ	r	g	y	o	d	u	k	a	ſ	h	ʒ	p
Leo	Malum	—	g	x	n	d	u	k	a	ſ	h	ʒ	p	e	v	l	b	s	i	3	q
		\	h	y	o	e	v	l	b	s	i	3	q	f	x	m	c	t	k	ƶ	r
		/	i	ʒ	p	f	x	m	c	t	k	ƶ	r	g	y	n	d	u	l	a	ſ
Virgo	Indifferentia	—	k	3	q	g	y	n	d	u	l	a	ſ	h	ʒ	o	e	v	m	b	s
		>	l	ƶ	r	h	ʒ	o	e	v	m	b	s	i	3	p	f	x	n	c	t
Libra	Bonum	—	m	a	ſ	i	3	p	f	x	n	c	t	k	ƶ	q	g	y	o	d	u
		>	n	b	s	k	ƶ	q	g	y	o	d	u	l	a	r	h	ʒ	p	e	v
Scorpio	Indifferentia	—	o	c	t	l	a	r	h	ʒ	p	e	v	m	b	ſ	i	3	q	f	x
		>	p	d	u	m	b	ſ	i	3	q	f	x	n	c	s	k	ƶ	r	g	y
Sagittarius	Bonum	—	q	e	v	n	c	s	k	ƶ	r	g	y	o	d	t	l	a	ſ	h	ʒ
		\	r	f	x	o	d	t	l	a	ſ	h	ʒ	p	e	u	m	b	s	i	3
		/	ſ	g	y	p	e	u	m	b	s	i	3	q	f	v	n	c	t	k	ƶ
Capricornus	Malum	—	s	h	ʒ	q	f	v	n	c	t	k	ƶ	r	g	x	o	d	u	l	a
		>	t	i	3	r	g	x	o	d	u	l	a	ſ	h	y	p	e	v	m	b
Aquarius	Indifferentia	—	u	k	ƶ	ſ	h	y	p	e	v	m	b	s	i	ʒ	q	f	x	n	c
		>	v	l	a	s	i	ʒ	q	f	x	n	c	t	k	3	r	g	y	o	d
Pisces	Indifferentia	—	x	m	b	t	k	3	r	g	y	o	d	u	l	ƶ	ſ	h	ʒ	p	e
		\	y	n	c	u	l	ƶ	ſ	h	ʒ	p	e	v	m	a	s	i	3	q	f
Aureus numerus			1	2	3	4	5	6	7	8	9	10	11	12	13	14	15	16	17	18	19

[F. 14Di]

Si scire desideras quocumque die anni in quo signo sit luna, intra kalendarium precedens cum die mensis cum directo [*in directo*] illius diei versus dextram in linea sub qua est scriptum 'littere signorum', invenies unam litteram de litteris alphabeti quam bene considera, et ejusdem littere figuram. Demum eandem litteram et ejus [figuram] quere in tabula supra scripta signorum lune directe sub aureo numero illius anni. Descend[end]o usque illam litteram — quem aureum numerum invenies in superiori vel inferiori parte istius tabule, scilicet in prima linea superius vel inferius transversaliter posita — habita autem litera predicta directe sub aureo numero descendeda [*discendente*] inventa, tunc in directo ejusdem littere invenies a sinistris nomen signi sub quo est luna die illa aut eadem die ad minus intrabit vel exsibit [*exibit*] idem signum.

F. 15. Tabola ad inveniendum ciclum solarem et lunarem

[F. 15B]

Si vis invenire lunam cotidie, id est quot dies habet et quando facit revolutionem. Oportet tibi habere tres numeros, scilicet pactam, et menses, incipiendo a martio usque ad mensem in quo es, et dies mensis illius in quo es. Et omnes istos tres numeros pone in simul. Et vide si superat 30^{a}. Si vero superat 30^{a}, abjectis 30^{a}, illud quod remanet est luna quam queris. Si vero habueris 30^{a} vel 60^{a}, abjectis 30^{a} de 60^{a}, remanet 30^{a}: illo die est revolutio luna (30^{a} vel etiam 60^{a} non removendo 30^{a} a 60^{a}).

Si vero desideras qua hora et quo puncto luna facit revolutionem, teneas istam regulam: primo oportet te scire quo die mensis fuit revolutio lune, 4 vel 6 et cetera. Si vero fuerit 4^{a} divide illos 4^{or} dies per medium: prima pars, scilicet 2, sunt horae, secunda pars, scilicet que remanserunt, sunt puncta multiplicando pro quolibet uno die decem. Verbi gratia luna fecit revolutionem 4^{a} die mensis: divide 4^{or} per medium; prima pars, scilicet 2°: sunt due hore; secunda pars multiplicando quolibet unum per decem, sunt puncta horarum, ergo revoutio lune fuit 4^{a} die mensis hora 2^{a} et puncta 20.

1456 25^{a} die 7tembris frater Paulus de Kignin.

De clavibus

Prima igitur Jan[uari]i jacet 70 [*Setpuagesime*] clavis.
Ultima igitur Jan[uari]i 40 [*Quadragesime*] inspice clavem.
Secundaque igitur Martis Pasce dat tibi clavem.
Tertia igitur Aprelis Rogationum dat tibi clavem.
Ultima igitur ejusdem Pentecostes dat tibi clavem.

Katholicus pro Sept[u]agessima	**Bonus** pro Quadragessima	**Offert** pro Pasca
Vinum pro Rogationes		**Delicatum** pro Pemtecoste

[**F. 15C**]

	Sol	Venus	Mercurius	Luna	Saturnus	Jupiter	Mars
Feria 1^a	**1**	**2**	**3**	**4**	**5**	**6**	**7**
	8	**9**	**10**	**11**	**12**	**13**	**14**
	15	**16**	**17**	**18**	**19**	**20**	**21**
Feria 2^a	**22**	**23**	**24**	1	2	3	4
	5	6	7	8	9	10	11
	12	13	14	15	16	17	18
Feria 3^a	19	20	21	22	23	24	**1**
	2	**3**	**4**	**5**	**6**	**7**	**8**
	9	**10**	**11**	**12**	**13**	**14**	**15**
	16	**17**	**18**	**19**	**20**	**21**	**22**
Feria 4^a	**23**	**24**	1	2	3	4	5
	6	7	8	9	10	11	12
	13	14	15	16	17	18	19
Feria 5^a	20	21	22	23	24	**1**	**2**
	3	**4**	**5**	**6**	**7**	**8**	**9**
	10	**11**	**12**	**13**	**14**	**15**	**16**
	17	**18**	**19**	**20**	**21**	**22**	**23**
Feria 6^a	**24**	1	2	3	4	5	6
	2	3	4	5	6	7	8
	14	15	16	17	18	19	20
Feria 7^a	21	22	23	24	**1**	**2**	**3**
	4	**5**	**6**	**7**	**8**	**9**	**10**
	11	**12**	**13**	**14**	**15**	**16**	**17**
	18	**19**	**20**	**21**	**22**	**23**	**24**

Si vis scire dominationem planetarum in omni hora: Accipe quadrantem et mitte intrare per ambo, foramina radium solis posita primo margarita super diem mensis, 1 si est prima die vel 2^a, 3 vel 4^a aut 20^a et quota die aut feria. Dabit tibi certitudinem quam tu petis. Ex hoc apparebit quis planetarum regnat in qualibet hora. Nam in prima hora diei dominice regnat sol, in 2^a Venus, in 3^a Mercurius, et cetera. Et sic fac de aliis diebus sicut de die dicto fecisti.

Vale in Domino.

[F. 15D]

	Aureus Numerus	Claves	Pacta
1456	13	14	23
1457	14	33	4
1458	15	22	15
1459	16	11	26
1460	17	30	7
1461	18	19	18
1462	19	38	29
1463	1	26	11
1464	2	15	22
1465	3	34	3
1466	4	23	14
1467	5	12	26
1468	6	31	6
1469	7	20	17
1470	8	39	28
1471	9	28	9
1472	10	17	20
1473	11	36	1
1474	12	25	12

– Pacta
– Menses incipiendo a Martio semper
– Dies mensis

1456	d	c
1457	b	
1458	a	
1459	g	
1460	f	e
1461	d	
1462	c	
1463	b	
1464	a	g
1465	f	
1466	e	
1467	d	
1468	c	b
1469	a	
1470	g	
1471	f	
1472	e	d
1473	c	
1474	b	
1475	a	
1476	g	f
1477	e	
1478	d	
1479	c	
1480	b	a
1481	g	
1482	f	
1483	e	

Queras in hac rota aureum numerum, claves terminorum et pactam ab annis Domini 1456 ut aparet hic supra in paragrafis positis.
Et semper ab ip[s]is numeris inclusive pro quolibet anno unum numerum computa pro tempore futuro, pro preterito vero retrogradando. Et ubi terminatur ultimus annus, ibi sunt supra dicti numeri. Ante vel retro numerabis.
Et sic faciendo nu[n]quam herabis.
Item habes in qualibet casella milessimum ut posis melius atque citius quod cupis invenire. Amen.

Queras in hac rota litteram dominicalem ab annis Domini 1456, in quibus annis curebant littere dominicales d et c, ut patet supra in paragrafo. Et ab ipsis litteris pro quolibet anno unam litteram computa pro tempore futuro — pro preterito vero retrogradando — et habebis litteram dominicalem.
Quod si annus bixestus erit, duas litteras dominicales habebit. Quarum altera scilicet in majori circulo erit dominicalis a Januario usque ad festum sancti Matthie apostoli [24/02]. Reliqua vero per residuum anni.
Item posui tibi milessium in qualibet cassella ut citius et levius posis invenire.

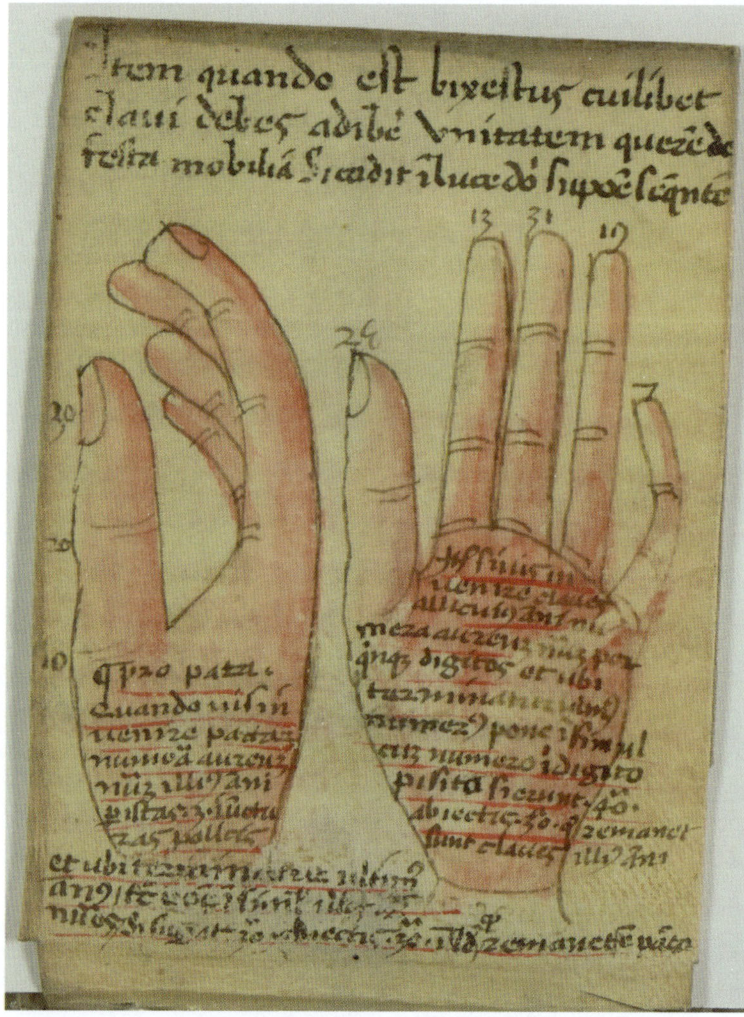

Figure 7. Paris, BnF, Latin 7478, f. 15R (folded). Source gallica.bnf.fr.

[F. 15R][57]

Idem quando est bixestus cuilibet clavi debes adibere unitatem querendo festa mobilia: si cadit in luce Domini supone sequentem.

Pro Pacta: quando vis invenire pactam, numera aureum numerum illius anni per istas 3 juncturas policis, et ubi terminatur ultimus annus [*sic* for *numerus*?] [...] pone insimul illos extremos numeros [?]. Si superat 30ª, abjectis 30ª illud quod remanet est pacta.

57 On this compartment, Paulus of Kignin has drawn two hands with the numerical value associated with each joint and digit. Paragraphs beginning with *Pro Pacta*... and *Item si vis invenire*... are respectively written over these drawings, that I am omitting here (compare to Fig. 7).

Item si vis invenire claves allicujus anni, numera aureum numerum per quinque digitos et ubi terminatur aureus numerus pone insimul cum numero in digito pisito [*sic*]: si erunt 40ª, abjectis 30ª, qui remanet sunt claves illius anni.

F. 16. Tabula ad inveniendum festa mobilia et ebdomadas

[F. 16D]

See table on opposite page.

Nota quod pacta, claves terminorum et littera martilogi vel dactarii, isti tres numeri semper sunt in linea ubi est aureus numerus, ut pate[t] hic supra scriptum.

Omnia festa mobilia et ebdomadas cum diebus superfluis invenies super litteram dominicalemn primam post aureum numerum. Et si dictus aureus numerus cadit super litteram dominicalem, accipe dominica[le]m sequentem et in linea illius littere omnia festa mobilia et ebdomadas reperies pro illo anno in quo queris, etc.

F. 17. [Earthly and heavenly spheres]

Compare this transcription with the reproduction in Fig. 3a.

[Heavenly spheres]
Saturnus facit cursum suum in triginta annis.
Jupiter complet cursum suum in duodecim annis.
Mars complet cursum suum in 2 annis.
Sol facit cursum suum in tres centis sessaginta quinque diebus et sex horis, et ex istis horis fit bixestus.
Venus complet cursum in uno [anno] minus 17 diebus.
Mercurius complet cursum suum in uno anno minus quadraginta diebus.
[Luna] complet cursum suum in viginti octo diebus et congiungit se soli.

Ignis
Aer
Aqua
Terra

[Mappemond]

Teoria (?)	Polonia	Tartaria
Francia	Ungaria	Caucasia
Italia	Scla[via]	Turchia
Spania (?)	Alba	
Tunisi		Arabia
Barbaris	Egyptus	India
	Etiopia	

Numerus pacte	Claves terminorum	Littera martilogii	Aureus numerus	[Littera dominicalis]	Septuagessima	Carnis privium	Prima dominica in 40e	Pascha Domini Yhesu Christi	Ascensio Domini	Pentecostes	Trinitas	Festum corporis Christi	Dominice inter 8am Pentecostis et Adventum	Inter Nativitatem et 70am ebdomade		Inter Nativitatem et Pascam		Inter Nativitatem et Pentecostem		Quando est Adventus et de quo mense, nota hic inferius		Ab Adventu usque ad Nativitatem	
26	11	Q	16		Jan.	Feb.	Feb.	Mar.	Apr.	Mai.	Mai.	Mai.	*Ebd.*	*Ebd.*	*Dies*	*Ebd.*	*Dies*	*Ebd.*	*Dies*	*Adventus*		*Ebd.*	*Dies*
25	12	E	5	d	18	3	8	22	30	10	17	21	28	3	3	12	3	19	3	29	november	3	5
				e	19	4	9	23	Mai. 1	11	18	22	28	3	4	12	4	19	4	30	november	3	4
23	14	N	13	f	20	5	10	24	2	12	19	23	28	3	5	12	5	19	5	1	december	3	3
22	15	B	2	g	21	6	11	25	3	13	20	24	28	3	6	12	6	19	6	2	december	3	2
				a	22	7	12	26	4	14	21	25	27	4	0	13	0	20	0	3	december	3	1
20	17	K	10	b	23	8	13	27	5	15	22	26	27	4	1	13	1	20	1	27	november	4	0
				c	24	9	14	28	6	16	23	27	27	4	2	13	2	20	2	28	november	3	6
18	19	S	18	d	25	10	15	29	7	17	24	28	27	4	3	13	3	20	3	29	november	3	5
17	20	G	7	e	26	11	16	30	8	18	25	29	27	4	4	13	4	20	4	30	november	3	4
				f	27	12	17	31	9	19	26	30	27	4	5	13	5	20	5	1	december	3	3
15	22	P	15	g	28	13	18	Apr. 1	10	20	27	31	27	4	6	13	6	20	6	2	december	3	2
14	23	D	4	a	29	14	19	2	11	21	28	Jun. 1	27	5	0	14	0	21	0	3	december	3	1
				b	30	15	20	3	12	22	29	2	26	5	1	14	1	21	1	27	november	4	0
12	25	M	12	c	31	16	21	4	13	23	30	3	26	5	2	14	2	21	2	28	november	3	6
11	26	A	1	d	Feb. 1	17	22	5	14	24	31	4	26	5	3	14	3	21	3	29	november	3	5
				e	2	18	23	6	15	25	Jun. 1	5	26	5	4	14	4	21	4	30	november	3	4
9	28	I	9	f	3	19	24	7	16	26	2	6	26	5	5	14	5	21	5	1	december	3	3
				g	4	20	25	8	17	27	3	7	26	5	6	14	6	21	6	2	december	3	2
7	30	R	17	a	5	21	26	9	18	28	4	8	26	6	0	15	0	22	0	3	december	3	1
6	31	F	6	b	6	22	27	10	19	29	5	9	25	6	1	15	1	22	1	27	november	4	0
				c	7	23	28	11	20	30	6	10	25	6	2	15	2	22	2	28	november	3	6
4	33	O	14	d	8	24	Mar. 1	12	21	31	7	11	25	6	3	15	3	22	3	29	november	3	5
3	34	C	3	e	9	25	2	13	22	Jun. 1	8	12	25	6	4	15	4	22	4	30	november	3	4
				f	10	26	3	14	23	2	9	13	25	6	5	15	5	22	5	1	december	3	3
1	36	L	11	g	11	27	4	15	24	3	10	14	25	6	6	15	6	22	6	2	december	3	2
				a	12	28	5	16	25	4	11	15	25	7	0	16	0	23	0	3	december	3	1
29	38	T	19	b	13	Mar. 1	6	17	26	5	12	16	24	7	1	16	1	23	1	27	november	4	0
28	39	H	8	c	14	2	7	18	27	6	13	17	24	7	2	16	2	23	2	28	november	3	6
				d	15	3	8	19	28	7	14	18	24	7	3	16	3	23	3	29	november	3	5
				e	16	4	9	20	29	8	15	19	24	7	4	16	4	23	4	30	november	3	4
				f	17	5	10	21	30	9	16	20	24	7	5	16	5	23	5	1	december	3	3
				g	18	6	11	22	31	10	17	21	24	7	6	16	6	23	6	2	december	3	2
				a	19	7	12	23	Jun. 1	11	18	22	24	8	0	17	0	24	0	3	december	3	1
				b	20	8	13	24	2	12	19	23	23	8	1	17	1	24	1	27	november	4	0
				c	21	9	14	25	3	13	20	24	23	8	2	17	2	24	2	28	november	3	6

ERIC RAMÍREZ-WEAVER

Bohemian King Wenceslas IV's Copy of the Alfonsine Tables and Their Place Within his Astronomical and Astrological Corpus

Introduction

Working among the creative teams that compiled, composed, and illuminated the seven canonical manuscripts associated with Wenceslas IV, a band of Bohemian courtiers supplied their king with three refined astronomical and astrological codices.[1] These manuscripts include the *Astronomical Anthology for Wenceslas IV* in Munich (Bayerische Staatsbibliothek, Clm 826), prepared in Prague after 1400 by two late medieval painters and a historical astrologer, who is usually designated in historical scholarship by a courtly sobriquet, Terzysko, the moniker he used to sign his diagrammatic summary of astrological tradition on folio 8^{r} (Fig. 1).[2]

* Research presented in this chapter was supported by the ERC project ALFA: Shaping a European scientific scene, Alfonsine astronomy, CoG 723085, PI Matthieu Husson. I wish to thank the University of Virginia for institutional support including a summer research stipend in 2020, and the invaluable insights received from the anonymous referee, Matthieu Husson and Richard L. Kremer.

1 Barbara Boehm and Jiří Fajt, eds, *Prague, The Crown of Bohemia, 1347–1437* (New York: Metropolitan Museum of Art, 2005), 220; Ulrike Jenni and Maria Theisen, *Mitteleuropäische Schulen, Die Illuminierten Handschriften und Inkunabeln der Österreichischen Nationalbibliothek*, ser. 1, vol. 13 (Vienna: Verlag der Österreichischen Akademie der Wissenschaften, 2014), IV, 13 identifies the canonical set: *Willehalm* Romance (Vienna, Österreichische Nationalbibliothek Ser. n. 2643, 1387–1400); Wenceslas *Bible* in German vernacular translation (Vienna, Österreichische Nationalbibliothek Cod. 2759–64, *c.* 1389–95, late 1390s?); Astrological and Astronomical Anthology with Alfonsine Planetary Tables (Vienna, Österreichische Nationalbibliothek Cod. 2352, *c.* 1392–93); *Psalter* of Wenceslas IV with the German edition of Nicholas of Lyra's *Commentary* (Salzburg, Universitätsbibliothek M.III.20, *c.* 1395); *de luxe* edition of the Golden *Bull* (Vienna, Österreichische Nationalbibliothek Cod. 338, 1400); 'Alī ibn Riḍwān's *Commentary* on Ptolemy's *Quadripartitum* in Latin translation (Vienna, Österreichische Nationalbibliothek Cod. 2271, *c.* 1400); Astronomical Anthology for Wenceslas IV (Munich, Bayerische Staatsbibliothek Clm 826, after 1400). For a general historical background, see Miloslav Polívka, 'The Expansion of the Czech State during the Era of the Luxemburgs (1306–1419)', in *A History of the Czech Lands*, 2nd ed. by Jaroslav Pánek and Oldřich Tůma, trans. by Justin Quinn, Petra Key, and Lea Bennis (Prague: Karolinum P, 2018), 123–56.

2 Eric Ramírez-Weaver, 'Reading the Heavens: Revelation and Reification in the Astronomical Anthology for Wenceslas IV', *Gesta*, 53 (2014), 73–94; a dated but helpful overview is supplied by Karel Stejskal and Josef Krása, 'Astralvorstellungen in der mittelalterlichen Kunst Böhmens', *Sborník Prací Filosofické Fakulty Brněnske University*,

Eric Ramírez-Weaver • University of Virginia

Alfonsine Astronomy: The Written Record, ed. by Richard L. KREMER, Matthieu HUSSON, José CHABÁS, Turnhout, 2022 (*Alfonsine Astronomy: Studies and Sources*, 1), pp. 199-240
© BREPOLS PUBLISHERS 10.1484/M.ALFA.5.124927

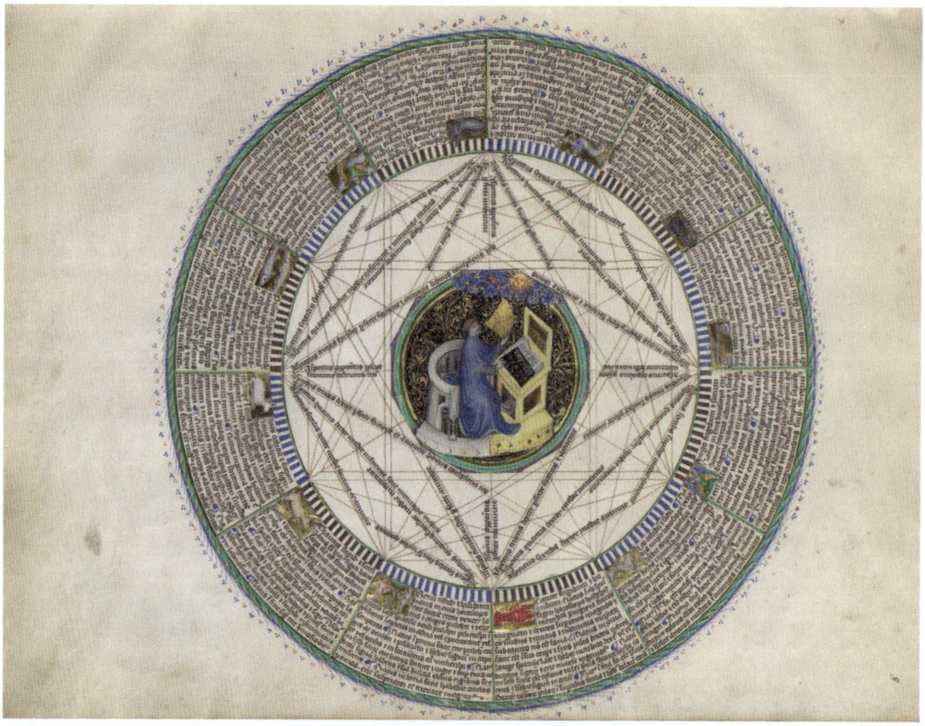

Figure 1. Terzysko, *Astronomical Anthology for Wenceslas IV*, Munich, Bayerische Staatsbibliothek, Clm 826, folio 8r. From Prague, after 1400, reproduced with permission.

As a testimonial to the longstanding significance of the Wenceslas astronomical codices in Central Europe, this manual, which remained forever a work in progress, was updated in 1501, a century after its original manufacture in Prague, by scribes and artists at work for Wilhelm IV Haller (d. 1534) in Nürnberg.[3] One way to assess the sorts of resources vital to the study of astronomy and astrology in Prague *c.* 1400 is to carefully review the contents of the three major manuscripts that document Wenceslas IV's celestial concerns. Assessing the significance of the élite Bohemian copy of the Parisian Alfonsine Tables (Österreichische Nationalbibliothek, Cod. 2352) will assume a central role, permitting a reappraisal of the various functions—iconographic and pedagogical— that these astronomical and astrological manuscripts performed at the court of Wenceslas IV. Although most details concerning the life and activity of the historic Terzysko remain a total mystery,

8 (1964), 61–85; Josef Krása, *Die Handschriften König Wenzels IV*, trans. Herta Soswinski (Vienna: Forum Verlag, 1971), 56–57, 211–14, esp. 277n356; updating Josef Krása, 'Astrologické Rukopisy Václava IV', in *České Iluminované Rukopisy 13./16. Století*, ed. by Jiří Dvorský and Jiří Kropáček (Prague: Odeon, 1990), 180–203; Boehm and Fajt, *Prague*, 223; Anton Legner, ed., *Die Parler und der Schöne Stil 1350–1400*, 5 vols (Cologne: Schnütgen Museum, 1978–80), III (1978), 104.

3 Dieter Blume, Mechthild Haffner, and Wolfgang Metzger, *Sternbilder des Mittelalters und der Renaissance: Der gemalte Himmel zwischen Wissenschaft und Phantasie*, 5 vols (Berlin: de Gruyter, 2012–16), II/1 (2016), 373 with earlier bibliography; Legner, ed., III, 104–5; also see Regina Cermann, 'Beschreibung einer problematischen Corvine: Cod. Guelf. 69.9 Aug. 2°', in *Corvina Augusta: Die Handschriften des Königs Matthias Corvinus in der Herzog August Bibliothek Wolfenbüttel*, ed. by Edina Zsupán (Budapest: Bibliotheca Nationalis Hungariae, 2014), 123–51.

it is at least anecdotally significant that at the hub of courtly activity in Prague *c.* 1400 an astronomically minded astrologer is depicted. Hence, an analysis of the significance of the three *de luxe* astronomical and astrological manuscripts, prepared during the life of Wenceslas IV, reveals the prevalence and importance of the science of the stars in Bohemia and one late medieval court's and university's efforts at summoning the aid of the heavens within the lands of the Bohemian crown.

On the one hand, within a courtly milieu the Alfonsine corpus of manuscripts possessed symbolic cachet by virtue of association with Alfonso X el Sabio (King of León and Castile, r. 1252–84) and the compilation of the planetary tables he had originally sponsored.[4] Wenceslas IV (King of the Romans, r. 1376–1400, King of Bohemia, r. 1363/78–1419) attempted to appropriate aspects of the earlier sage sovereign's mantle, studying astrological and astronomical pursuits apposite for a late-fourteenth-century royal. On the other hand, copying the Alfonsine planetary tables, as recast in Paris during the first and second quarters of the fourteenth century, and the concomitant canons supplied by professorial exegetes such as John of Saxony (often referred to as Johannes Dank), performed an alternative variety of symbolic and scientific work at the Prague court and within the Faculty of Liberal Arts at Charles University (founded 1348).[5]

Astronomical courtly prestige *c.* 1350–1400 was not only established by a link to the legacy of liberal arts pursuits. Advances in astronomical study had become a *cause célèbre* in which the university centres such as Bologna, Cracow, Paris, Prague, or Vienna vied for pre-eminence and sought to cultivate the most up-to-date texts and traditions, often with a focus upon iatromathematical applications, including the ubiquitous medicinal intervention, phlebotomy.[6] Philosophical, astronomical, and astrological study in the fourteenth century at the courts of Wenceslas IV and Charles V of France (r. 1364–80) generated new knowledge and translations, rather than merely relaying classical texts from cherished *auctores*. In Paris, Nicole Oresme undertook novel translations of Aristotelian classics, such as the *Politiques* and *Éthiques*, accompanied by original visual and textual commentaries for Charles V, but he also probably translated (into French from Latin) Ptolemy's *Quadripartitum* following the Aegidius de Tebaldis redaction of the text, complete with the influential commentary of 'Alī ibn Riḍwān. A Latin copy was likewise prepared for Wenceslas IV (Österreichische Nationalbibliothek, Cod. 2271).[7] It has become standard in art historical scholarship since the postmodern revisionism of the 1990s to seek examples

4 José Chabás, and Bernard R. Goldstein, *The Alfonsine Tables of Toledo* (Dordrecht: Kluwer, 2003), 1–8.

5 Charles Burnett, 'Teaching the Science of the Stars in Prague University in the Early Fifteenth Century: Master Johannes Borotin', *Aithér (AIΘHP)*, 2 (2014), 9–16; František Šmahel, 'The Faculty of Liberal Arts 1348–1419', in *Die Präger Universität im Mittelalter/The Charles University in the Middle Ages* (Leiden: Brill, 2007), 213–315.

6 Richard Lemay, 'The Teaching of Astronomy in Medieval Universities, Principally at Paris in the Fourteenth Century', *Manuscripta*, 20 (1976), 197–217; Harry Bober, 'The Zodiacal Miniature of the *Très riches heures* of the Duke of Berry—Its Sources and Meaning', *Journal of the Warburg and Courtauld Institutes*, 11 (1948), 1–34; also see, Nancy Siraisi, *Medieval and Early Renaissance Medicine: An Introduction to Knowledge and Practice* (Chicago: University of Chicago Press, 1990), 97–109.

7 Claire Richter Sherman, *Imaging Aristotle: Verbal and Visual Representation in Fourteenth-Century France* (Berkeley: University of California Press, 1995), 6–9, 15–21, 30–38. Max Lejbowicz, 'Guillaume Oresme, traducteur de la *Tetrabible* de Claude Ptolemée', *Pallas*, 30 (1983), 107–33 (pp. 107–8, 117) argues persuasively for Oresme's use of the same Aegidius de Tebaldis text which appears in the Viennese Cod. 2271 discussed here. Lejbowicz's attribution of the French translation to Guillaume, Nicole's brother, remains however contested and unlikely.

of intervisuality, modelled after the studies of cultural historians such as Madeline Caviness working on Hildegard of Bingen or the text-image analyses associated with scholars such as Michael Camille.[8] This approach seeks moments of complex and even inventive creative intervention documenting verifiable moments in which medieval manuscripts contain purposeful, internal self-reference or meaningfully refer to visual precursors and comparanda. Hence, Carolingian manuscripts could be viewed as having established the scientific precedent for later Central European astronomical traditions through complex strategies of exegetical emendation or scientific soteriology and sacralization (manipulating astronomical images for religious and scientific reasons).[9]

Scientific prestige could be demonstrated *inter alios* by acquisition of the most sophisticated treatises and observational equipment.[10] Nowhere is this connection and proposed mode of patronage made more explicit for the Bohemian court of Wenceslas IV than in the patronage of the Old Town Hall tower Astronomical Clock (Fig. 2), made in *c.* 1410 and linked to the scholarly activity of Iohannes Šindel (d. 1455–58, also called Jan Ondřejův). Work on the clock itself required the manufacturing skills of Mikuláš of Kadaň. In fact, Šindel succeeded Utraquist reformer Jan Hus to become rector upon appointment in 1410 and as one of Hus's loyal supporters composed texts about the astrolabe. The complementary role of academics affiliated with Charles University and courtly affairs is underscored by the influential treatises on the use and manufacture of the astrolabe composed by the rector of Charles University, Křišťan of Prachatice, in 1407, which continued to be copied for the next 150 years.[11]

In France, by comparison, Charles V's courtier and translator, Pèlerin de Prusse, also was busy rendering the astrolabe accessible to a royal patron in his *Practique de astrolabe* of 1362. In that vernacular treatise derived from John of Seville's Latin translation, the *Compositio et operatio astrolabii*, and from an influential treatise anonymously authored by pseudo-Māshā'allāh, Pèlerin de Prusse literally modified his text to relate directly to the astrolabes recorded in the comprehensive inventory of Charles V's possessions.[12] An interesting comparison emerges between intellectual kings such as Charles and Wenceslas. Both attempted to consolidate their troubled lands, whether it be through the reconquest

---

8 Madeline Caviness, 'Hildegard as the Designer of the Illustrations to her Works', in *Hildegard of Bingen: The Context of her Thought and Art*, ed. by Charles Burnett and Peter Dronke (London: Warburg Institute, 1998), 29–62; Michael Camille, *Image on the Edge* (London: Reaktion, 1992), 9–55.

9 Eric Ramírez-Weaver, *A Saving Science: Capturing the Heavens in Carolingian Manuscripts* (University Park: Pennsylvania State University Press, 2017), 6–14, 193–98, 209–11, with succinct summaries of earlier and relevant art historical literature pertaining to the Warburgian tradition and the iconographic method associated with Erwin Panofsky and Fritz Saxl.

10 Lys Ann Shore, 'A Case Study in Medieval Nonliterary Translation: Scientific Texts from Latin to French', in *Medieval Translators and Their Craft*, ed. Jeanette Beer (Kalamazoo: Western Michigan University, Medieval Institute Press, 1989), 307–10.

11 Křišťan z Prachatic, *Stavba a Užití Astrolábu*, ed. and trans. by Alena Hadravová and Petr Hadrava (Prague: Filosofia, 2001), 477–80; Zdislav Šíma, *Astronomy and Clementinum*, 2nd ed., trans. by Hana Vajnerová (Prague: Národní knihovna České republiky, 2006), 91; Anna Novotná, *Die Prager Rathausuhr*, trans. by Artlingua (Prague: Práh, 2015); Boehm and Fajt, *Prague*, 99; Alena Hadravová and Petr Hadrava, 'Astronomy in Prague: From the Past to the Present', *Highlights of Astronomy*, 14 (2006), 10–12; Jaroslav Folta, 'Clockmaking in Medieval Prague', *Antiquarian Horological Society*, 23 (1997), 405–17.

12 Edgar Laird and Robert Fischer, *Pèlerin de Prusse on the Astrolabe: Text and Translation of his* Practique de astralabe (Binghamton: SUNY Binghamton Press, 1995), 1–17; Křišťan z Prachatic, *Stavba a Užití Astrolábu*, 477.

Figure 2. Prague, the Astronomical Clock, Old Town Hall, *c.* 1410 and subsequent additions. Photo: Eric Ramírez-Weaver. First published as Figure 3 in 'Reading the Heavens: Revelation and Reification in the Astronomical Anthology for Wenceslas IV', *Gesta* 53 (2014); reproduced here with permission of *Gesta* and the University of Chicago Press. © 2014 by *Gesta*.

of the territories of France during the Hundred Years's War or through quelling the internecine strife that vexed Wenceslas within Bohemia. Simultaneously, they advanced their erudition, consulting, using, and sponsoring the creation of astronomical or astrological tools and manuscripts.[13] Alfonso was more than the historic supplier of suitably packaged planetary tables apposite for study and diachronic revision; he also offered a courtly *exemplum bonum*, as both an active patron and willing participant.[14]

1. The trouble with Terzysko and astrological courtiers

In order to conceptualize how Wenceslas IV and his loyal supporters sought to enable him to adopt and adapt the Alfonsine model to a Bohemian context, it is useful to assess briefly that which can and cannot be known about the enigmatic figure, Terzysko. Those familiar with medieval studies grapple with the perpetual problem of authorial anonymity, so a self-identified working author of any sort who receives an author portrait in a diagram, such as that on folio 8^r, attracts attention. That said, minimal textual documentation clarifies the living conditions of a court astrologer in Prague *c.* 1400. There are, however, two specific references to Terzsyko worthy of mention.[15] As is well known, following the foundation of the Nové Město, or Prague New Town, and Charles University in 1348 by Wenceslas IV's father, the Holy Roman Emperor Charles IV[16], an influx of craftspeople—including specialists in parchment preparation—flocked to the area near the New Town Hall. The area currently occupied by Charles Square formerly supplied the site for the cattle market, local workshops, and annual relic displays that melded religious fervour with the daily performance of bourgeois activity in medieval Prague.[17] Directly up the 'Old Beltway' from the cattle market, past Saint Henry Street and on towards the parish church of Saints Henry and Kunigunde, is situated the hay market, where Terzysko's family property was located (today Senovážné Naměstí). In this context, a document confirming Terzysko's immediate family's acquisition of property by 1405 within the burgeoning New Town is evidence of success and access. The family had a home situated far enough away from the hub of artisanal activity at Prague's cattle market to arguably avoid some of the distracting late medieval cacophonous din. Perhaps this permitted Terzysko to engage in his scholarly astronomical pursuits while maintaining a domicile in comfortable proximity to the resources requisite for the production of a

---

13 Lemay, 'The Teaching of Astronomy', 200–4; Sherman, *Imaging Aristotle*, 3–22. For more on the lives of astronomically minded learned kings, see the magisterial text by Dieter Blume, *Regenten des Himmels: Astrologische Bilder in Mittelalter und Renaissance* (Berlin: Akademie Verlag, 2000), 52–63. See also the succinct overview: Dieter Blume, 'Picturing the Stars: Astrological Imagery in the Latin West, 1100–1550', in *A Companion to Astrology in the Renaissance*, ed. by Brendan Dooley (Leiden: Brill, 2014), 333–98.

14 Chabás and Goldstein, *The Alfonsine Tables of Toledo*, 2.

15 Alena Hadravová and Petr Hadrava, 'Astronomy in Prague', esp. 5.

16 Maria Theisen, *History Buech Reimenweisz: Geschichte, Bildprogramm und Illuminatoren des Willehalm-Codex König Wenzels IV. von Böhmen, Wien, Österreichische Nationalbibliothek, ser. nov. 2643* (Vienna: Verlag der Österreichischen Akademie der Wissenschaften, 2010), 89; Boehm and Fajt, *Prague*, 59–63.

17 Theisen, *History Buech Reimenweisz*, 89; Zoë Opačić, 'Architecture and Religious Experience in 14th-Century Prague', in *Kunst als Herrschaftsinstrument: Böhmen und das Heilige Römische Reich unter den Luxemburgern im europäischen Kontext*, ed. Jiří Fajt and Andrea Langer (Berlin: Deutscher Kunstverlag, 2009), 136–49.

manuscript such as the *Astronomical Anthology* in Munich. In addition, this site was near the eastern border of the Old Town, conferring him direct access to the Town Hall via the Customs Officer Gate.[18] We might assume that this anecdote about real estate is indicative of Terzysko's familial successes to date, although not much else can be gleaned from the reference.

A second reference from the Moravian Dietrichstein Library is also informative, although the Latin is ambiguous: [In 1407] *Dominus Rex de Karlstein recepit… In Tocznyk… Item duos Attlas integros Therzysskoni et bysconi* (In 1407 'at Točník, the Lord King [Wenceslas IV] recovered from Karlstein … two complete [star] atlases by Terzysko [known alternatively as Těříško or Teříšek] and Bušek').[19] This would indicate that Terzysko worked alongside Bušek in addition to plying his trade independently for King Wenceslas IV, and that the king evidently sought out their reading materials. This is all that is certain.

It is, however, reasonable to link the King's preference for the work of Terzysko to the creation of the *Astronomical Anthology*. Although the manuscript in Munich is incomplete, the diagram on folio 8ʳ (Fig. 1) offers introductory insights concerning the science of nativities and configurations of planetary influence. Moving from cursory details in the diagram to the texts that follow, such as the Latin translation of the *Introductorium* by Abraham ibn Ezra, robust explanations or discussions of genethlialogy could ensue.[20] The diagram was a heuristic tool designed to introduce its royal audience, and perhaps here exclusively Wenceslas IV, to the basic properties and traits discernibly linked to decans and zodiac signs, conceits derived from the *Great Introduction to the Science of Astrology* (the *Kitāb al-madkhal al-Kabīr 'alā 'ilm aḥkām alnujūm*) of Abū Ma'shar (d. 886).[21]

Josef Krása identified the visual portions of the introductory diagrams as the work of the first master painter in the *Astronomical Anthology*. The erudite nature of the material required the sort of close collaboration for which Terzysko evidently had a penchant. A second late medieval painter added the eighteen al-Ṣūfī star pictures discussed below, before the aforementioned sixteenth-century additions.[22] Krása and Gerhard Schmidt have proposed that the primary painter in the *Astronomical Anthology* (who painted the diagrams, images of the decans or paranatellonta, and historiated initials such as Fig. 3) also participated in the manufacture of the lavish Beautiful Style historiated initial introducing

---

18 Josef Krása, *Die Handschriften*, 56, 258n100, reports Tomek's original discovery; Ramírez-Weaver, 'Reading the Heavens', 79; Theisen, *History Buech Reimenweisz*, 89; the best map of medieval Prague during the reign of Wencelas IV is in Jiří Fajt, ed., *Karl IV.: Kaiser von Gottes Gnaden, Kunst und Repräsentation des Hauses Luxemburg 1310–1437* (Munich: Deutscher Kunstverlag, 2006), 199.

19 The reference was catalogued in Vladislav Dokoupil, *Soupis Rukopisů Mikulovské Dietrichsteinské Knihovny*, Catalogi codicum manu scriptorum in Bibliotheca Universitatis Brunensis asservatorum, 2 (Brno: Státní pedagogické nakladatelství, 1958), 45, cat. Mk 20, folio 1 f. Legner, ed., III, 104 reads as referring to one co-authored volume, rejecting Krása's earlier interpretation emphasizing one book apiece by Bušek and Terzysko in Krása, *Die Handschriften*, 52, 56. The Latin suggests that the partnership resulted instead in two finished works. It is altogether unclear what their roles or the division of labour might have been. I wish to thank Gregory Hays and Matthieu van der Meer for their input, confirming in my mind the translation of *attlas* as a 'star atlas'; any error on this issue rests solely with the author.

20 Ramírez-Weaver, 'Reading the Heavens', 92.

21 Ramírez-Weaver, 'Reading the Heavens', 84.

22 Josef Krása, *Die Handschriften*, 214, 277n356.

Johann von Saaz's *Officium St Hieronymus* in Prague, Knihovna Národní Muzeum, MS XII A 18, 1404.[23]

Methodologically speaking, in the diagram of the system of aspects, Terzysko and his painter created a visualization of possible planetary configurations before the signs. They made explicit the flexibility and adaptability of the system to future situations and its utility for present applications. This open-ended rhetorical pliability is one hallmark of medieval diagrams, as argued by Mary Carruthers. The fact that the introductory cycle of diagrams and schematics from the *Astronomical Anthology* can be used differently at every reading is in fact one of its advantages and, arguably, a sign of purposeful creative activity in a diagram such as folio 8^{r}.[24] Such a genethlialogical diagram is never intended to be mastered. On the contrary, meaning is perpetually deferred to the next consultation, when information about births yet again finds novel utility in a courtly reading, as a new princess or prince is born. In my estimation, this suggests that serious use of such diagrams requires a mentor, or a tutor, to guide the reader, royal or otherwise, through the adduced properties of children born under the various decans or signs. Granted, the brief notice in the Moravian Dietrichstein Library might indicate nothing more than Wenceslas IV's desire to fill his bookshelves. It is more reasonable, in my opinion, to suspect that Wenceslas IV actually sought out his star atlases for some purposeful reason and that, following Occam's razor, their astronomical study supplies the simplest explanation. Dieter Blume has argued that a similar emphasis upon astrology in the service of imperial identity formation and readiness to rule, at the court of Frederick II, participated in a veritable and concerted *Staatsideologie*.[25]

Likewise, Hilary Carey has argued that similar interests were cultivated at the courts of Charles V of France and, albeit to a far lesser degree, of Richard II Plantagenet of England (who for a time was also the brother-in-law of Wenceslas IV through Anne of Bohemia).[26] According to Carey, a king such as Richard II, however, did not involve astrology or astrologers in the life of the court. Given the extant manuscript evidence, even accounting for extraordinary losses, it appears as if Wenceslas IV and his loyal courtiers took more supportive approaches to astronomical and astrological patronage.[27] Charles V, on the other hand, sponsored iatromathematical, or astrological medical, study at the University of Paris by 1377, fulfilling thereby—albeit in a rather limited manner—Gervais Chrestien's

23 Josef Krása, *Die Handschriften*, 212–20; Gerhard Schmidt, 'Malerei bis 1450: Tafelmalerei—Wandmalerei—Buchmalerei', in *Gotik in Böhmen: Geschichte, Gesellschaftsgeschichte, Architektur, Plastik, und Malerei*, ed. by Karl Swoboda (Munich: Prestel Verlag, 1969), 230, 239, 434n297, 437n357–58.

24 Mary Carruthers, *The Book of Memory: A Study of Memory in Medieval Culture*, 2nd ed. (Cambridge: Cambridge University Press, 2008), 336.

25 Blume, *Regenten des Himmels*, 50–51.

26 Hilary Carey, *Courting Disaster: Astrology at the English Court and University in the Later Middle Ages* (London: Macmillan, 1992), 22, 52, 92–116; at p. 22 Carey emphasizes that at least in the English context, '… for these great lords, astrology was not a tool for statecraft, but an exotic game, a pretty ornament, something to pass the time, like listening to romances, or playing chess.' The situation in Bohemia was altogether different as attested by the three luxury astrological codices in Munich and Vienna, as well as, the books personally consulted by Wenceslas IV.

27 Hilary M. Carey, 'Astrology at the English Court in the Later Middle Ages', in *Astrology, Science, and Society: Historical Essays*, ed. by Patrick Curry (Woodbridge: Boydell, 1987), 41–56 (p. 48), writes that 'geomancy and astrology were probably recognized as having a useful, but nevertheless peripheral place in the fashionable and innovative court life he tried to foster.' Carey essentially reduces astrology in Richard II's England to vogue dandyism.

original hopes for an astrological research centre under the administration of the University of Paris, as Jean-Patrice Boudet has elucidated.[28] These courtly comparisons supply a useful framework in which to now assess the significance of the astronomical and astrological codices prepared for Wenceslas IV.

2. The corpus of Wenceslas IV's celestial books and their contents

Previous scholars investigating the late medieval style of artworks manufactured throughout the lands of the Bohemian Crown, from Otto Pächt to Josef Krása, Gerhard Schmidt, Jiří Fajt, and Maria Theisen, have explored, from an art historical perspective, the key stylistic traits of the Bohemian form of the so-called International Style of 1400, known as the Beautiful Style (krásný sloh).[29] In one apposite succinct formulation, Schmidt identified three salient features of this style, all of which are, to varying degrees, present in the books made in Prague in the late fourteenth and early fifteenth centuries. More importantly, this is the manner of painting associated with the manuscripts linked to Wenceslas IV discussed here, whether outright religious or scientific in theme and content. Schmidt proposed that the three features can be found, to varying degrees, in all period works of art, regardless of their medium; they are particularly relevant for the miniatures, historiated initials, and diagrams from manuscripts discussed here. For Schmidt, illustrations in the Beautiful Style display, first of all, a characteristic tendency to offer vacant expressions in any depicted figures, such as Terzysko in his author portrait. The second feature is that representations are curiously enlivened by animated draperies that swirl and undulate about their figures in dramatic folds of freely falling fabric. These artistic and painterly motifs emphasize the linear quality of such compositional strategies, rendering, as a third feature, the palpable presence of the surface, rich yet flattened by the exaggerated emphasis upon ornamentation at the expense of three-dimensionality or illusionism.[30]

In terms of iconography (the art historical study of symbols as synecdoche capable of informing about larger cultural trends as well as isolated creative novelties), there has been a tendency to foreground, or to attempt to decipher, the dynastic and heraldic insignia or individual iconographic motifs, such as bath maidens, Wild Men, and knotted veils associated with Wenceslas IV's emblems and found throughout the canonical set of his

---

28 Carey, 'Astrology at the English Court', 48; Carey, *Courting Disaster*, 52; Jean-Patrice Boudet, 'A "College of Astrology and Medicine"? Charles V, Gervais Chrétien, and the Scientific Manuscripts of Maître Gervais's College', *Studies in History and Philosophy of Biological and Biomedical Sciences*, 41 (2010), 99–102, 107, derived from Jean-Patrice Boudet, 'Charles V, Gervais Chrétien et les manuscrits scientifiques du collège de Maître Gervais', *Médiévales*, 52 (2007), 15–38.

29 Otto Pächt, 'Die Gotik der Zeit um 1400 als gesamteuropäische Kunstsprache', in *Europäische Kunst um 1400*, ed. by Vinzenz Oberhammer (Vienna: Kunsthistorisches Museum, 1962), 53–54; Krása, *Die Handschriften*, 23–63, 115–222; Gerhard Schmidt, 'Kunsthistorischer Kommentar', in *Die Wenzelsbibel, vollständige Faksimile-Ausgabe der Codices Vindobonenses 2759–2764 der Österreichischen Nationalbibliothek Wien*, ed. by Hedwig Heger and others (Graz: Akademische Druck- und Verlaganstalt, 1998), 125–250; Jiří Fajt, 'Magister Theodoricus-Court Artist to Emperor Charles IV', in *Magister Theodoricus, Court Painter to Emperor Charles IV: The Pictorial Decoration of the Shrines at Karlštejn Castle*, ed. by Jiří Fajt (Prague: National Gallery, 1998), 217–77; Theisen, *History Buech Reimenweisz*, 81–97; Jenni and Theisen, *Mitteleuropäische Schulen*, IV, 14–22; Boehm and Fajt, *Prague*, 104–11.

30 Boehm and Fajt, *Prague*, 105–11 (pp. 109–10).

seven surviving codices.[31] The rich contributions of Milena Bartlová, Gerhard Schmidt, Ewa Śnieżyńska-Stolot, Milada Studničková, and Maria Theisen to this scholarship are beyond the scope of the current essay.[32] Instead, building upon their astute insights, it remains useful to explore how the scribes and painters who crafted Wenceslas IV's astronomical or astrological manuscripts drew upon these normative emblems and symbolic iconographies to present an idealized vision of Bohemian astronomical sophistication into which Wenceslas IV was invited, even compelled, to insert himself. In other words, a manuscript like the *Astronomical Anthology* summoned as well as served the king, intellectually inviting him to join Terzysko, at the very least visually and intellectually, at the centre of the genethlialogical diagram on folio 8ʳ, looking outwards at a panoramic vista activating the system of aspects and applying ideas about planetary influence to individual decans and their corresponding nativities.[33]

In general, the Bohemian king's symbols have served to link the distressed Wenceslas IV, in 1400 deposed as King of the Romans, to spiritually motivated beliefs in the necessary role of kings as sacrificial ambassadors of mercy and deliverance, rendering Wenceslas a second Moses or persecuted Christ figure on the Vltava, who bled for the kingdom he wed, Bohemia.[34] This focus upon ideas of restoration, purification, and renewal supplies an increasingly canonical approach to the exegesis of the limited number of surviving manuscripts associated with the library of Wenceslas IV.[35] If, instead, we consider the manuscripts linked to the king for their astronomical and astrological sophistication, a different image of the troubled court in Prague emerges.

In other words, the present paper recommends a methodological reversal, interrogating the contents, illustration, and use of the astronomical or astrological texts contained within the Wenceslas library on their own terms as emic evidence for the various roles such books could play in Prague courtly life during the fourteenth and fifteenth centuries. Rather than operating within a vacuum of sustainable patronage, courtiers such as Terzysko maintained personal scholarly interests which they cultivated alongside those of their king. The texts contained within Wenceslas IV's library offer indelible records of such astrological or astronomical pursuits. Contents of the books illuminate the diverse themes and topics deemed essential for a late medieval student of the skies—someone

---

31 Boehm and Fajt, *Prague*, 220–23; Schmidt, 'Kunsthistorischer Kommentar', 126–30; Jenni and Theisen, *Mitteleuropäische Schulen*, IV, 5–12.

32 Milena Bartlová, *Skutečná přítomnost: Středověký obraz mezi ikonou a virtuální realitou* (Prague: Argo, 2012), 381–91 for an English summary of this important volume's chief arguments; Schmidt, 'Kunsthistorischer Kommentar', 125–74; Ewa Śnieżyńska-Stolot, *Ikonografia znaków zodiaku i gwiazdozbiorów w rękopisie monachijskim Abrahama ibn Ezry* (Cracow: Wydawnictwo Uniwersytetu Jagiellońskiego, 1998); Ewa Śnieżyńska-Stolot, 'Christian Interpretation of the Zodiac in Mediaeval Psalters', *Umění*, 37 (1989), 97–109; Milada Studničková, '*Gens Fera*: The Wild Men in the System of Border Decoration of the Bible of Wenceslas IV', *Umění*, 62 (2014), 214–39; Milada Studničková, 'Drehknoten und Drachen: Die Orden Wenzels IV. und Sigismunds von Luxemburg und die Polysemantik ihrer Zeichen', in *Kunst als Herrschaftsinstrument: Böhmen und das Heilige Römische Reich unter den Luxemburgern im europäischen Kontext*, ed. by Jiří Fajt and Andrea Langer (Berlin: Deutscher Kunstverlag, 2009), 377–87; Jenni and Theisen, *Mitteleuropäische Schulen*, IV, 1–14.

33 Ramírez-Weaver, 'Reading the Heavens', 73–94.

34 Studničková, '*Gens Fera*', 220–22; Jenni and Theisen, *Mitteleuropäische Schulen*, IV, 5, 10–11; Milada Studničková, 'Drehknoten und Drachen', 381–85.

35 Milada Studničková, 'Drehknoten und Drachen', 382; Jenni and Theisen, *Mitteleuropäische Schulen*, IV, 11; Krása, *Die Handschriften*, 97–99.

who could be meaningfully labelled an 'otherworldly courtier'—in partial fulfilment of the functional need to discern and interpret heavenly phenomena within a courtly setting such as Prague.

As coeval pieces of evidence, the contents and illustrations in Wenceslas IV's three astronomical and astrological manuscripts will be interrogated for their joint ability to assert a set of educational priorities considered worthy of the King's attention and to reveal complementary information about the interests of the stargazers who surrounded him in medieval Bohemia. As I argue in greater detail below, Wenceslas IV's father, the Holy Roman Emperor Charles IV, had been raised in fourteenth-century Paris. The reverberations of Charles' formation in Paris after 1323 can be felt in the luxury copy of the Alfonsine corpus contained within the *Astronomical Anthology with Alfonsine Planetary Tables* (now Cod. 2352 in Vienna), compiled in 1392–93 for the Bohemian King.[36] The connection to Alfonso el Sabio and the inclusion of the recast Parisian tables associated with the canons of John of Saxony attest to historic diplomatic and cultural exchange between Toledo, Paris, and Prague, as well as to the personal predilections of Charles IV, who cultivated notoriously Francophile tastes in the lands of the Bohemian crown.[37]

In the *Astronomical Anthology for Wenceslas IV*, the author portrait of Terzysko located at the centre of the diagrammatic *rota* supplies a privileged representation of what we could label an 'otherworldly courtier', pivotally positioned where his science meshes with the entanglements of unstable courtly activity in Prague *c.* 1400. Such a diagram includes, in the radial sections, information pertaining to birth horoscopes, that is nativities. Each zodiac sign designated by one of the twelve pie-shaped wedges along the circumference contains an important summary of texts about the psychological and physical attributes pertaining to each of that sign's three decans (or ten-degree radial zones). The texts inscribed within the wedges inform about the genethlialogical (or horoscopic) significance of zodiac signs for those born under that respective sign, with the caveat that those 'born at the end of' a sign like Taurus evidently are all hermaphrodites.[38]

These concise notes summarize a text included later in the *Astronomical Anthology*, that with creative and calculated precision extracts the essence of the descriptions of the signs from the Latin translation of Abraham ibn Ezra's *Introductorium quod dicitur principium sapientiae*, described in greater detail below.[39] For now, it is important to underscore that this constitutes a marvellous example of both intertextuality and intervisuality in late medieval astronomical codicology. The purposeful layout of the book informs us of the specific and special nature of this particular commission, whatever else can be said about the *Astronomical Anthology*. Its purpose was unique, as was the inventive freedom

36 Jenni and Theisen, *Mitteleuropäische Schulen*, IV, 89–122; Boehm and Fajt, *Prague, The Crown of Bohemia*, 4; Emmanuel Poulle, 'Les Astronomes parisiens au XIVe siècle et l'astronomie alphonsine', *Histoire littéraire de la France*, 43 (2005), 1–2; Emmanuel Poulle, *Les Tables alphonsines avec les canons de Jean de Saxe* (Paris: Éditions du Centre National de la Recherche Scientifique, 1984), 3; Chabás, and Goldstein, *The Alfonsine Tables of Toledo*, 243–45.

37 Boehm and Fajt, *Prague*, 3–33, 59–73.

38 Ramírez-Weaver, 'Reading the Heavens', 73–83. For more theoretical information about diagrams, now see Faith Wallis, 'What a Medieval Diagram Shows: A Case Study of Computus', *Studies in Iconography*, 36 (2015), 1–40, alongside the more classic article, Michael Evans, 'The Geometry of the Mind', *Architectural Association Quarterly*, 12 (1980), 32–55 (pp. 42–44).

39 Ramírez-Weaver, 'Reading the Heavens', 84–90.

exemplified by its makers. The diagram was created from the text but in its clever presentation it supersedes, even truly supplements the text and establishes a new schematic framework through which the original text by ibn Ezra can in fact be assessed, as certain details are highlighted at the expense of others, within the economy of space and form in the extracted diagrammatic version.

On the one hand, the neologism 'otherworldly courtier' has more general applicability to any court astrologer. It designates the intertwining but versatile skill set of the late medieval astronomer or astrologer, who could coordinate his observation of the night sky using an astrolabe or quadrant as in the illustration, anticipate planetary longitudes, reckon horoscopes, and undertake physical observation in the service of heavenly interpretation or prognostication, often with iatromathematical (medical) application.[40] On the other hand, there is an ironic use of the term 'courtier' within the Bohemian context. For in the decade under discussion here, 1393–1403, Wenceslas IV was nearly poisoned and suffered incarceration twice from his burghers, aspirational nobility, and his usurping half-brother, King Sigismund of Hungary. In 1400, Wenceslas IV was divested of his title with more imperial pretensions, losing the right to be King of Romans, although death alone concluded his reign as King of Bohemia in 1419.[41] During this chaotic period, it is perhaps more interesting that Wenceslas IV's astrologically minded courtiers perpetuated late medieval courtly models of prestige patronage, cultivating their own hybrid brands of Prague-based ateliers, even during the lapses of centralized power experienced during the 1390s. The manuscripts discussed here are the best evidence for this atypical late-medieval situation. As the political realities elucidated the fantasy of control, otherworldly courtiers such as Terzysko asserted the promise of control in their own visualizations of their astrological system in the *Astronomical Anthology*.

Analysing the range of sources likely consulted in the creation of such a diagram permits a telling foray into the intellectual process of compilation employed by Terzysko in Prague *c.* 1400. It is often overlooked that source material is itself highly informative about an author's formation and position within society. In other words, much can be learned about a mysterious figure such as Terzykso through his library or, at least, the range of sources consulted in the creation of his compilation. Rather than dismiss this figure to obscurity, it is actually a fruitful exercise to recuperate as much of this individual's identity as possible, especially when such a dearth of source material remains for the Bohemian context. In any case, Terzysko relied upon a range of likely or verifiable sources, including, but hardly limited to, John of Seville's *c.* 1134 translation of al-Qabīṣī's mid-tenth-century Arabic *Book of the Introduction to the Craft of Astrology*.[42] This book was a vital resource for medieval medical faculty and one copy is found in the so-called *Celestial Atlas of the Bohemian Kings* (Cusanus 208) of roughly 1311.[43] Interestingly, in

40 Carey, *Courting Disaster*, 22, 25–27; Krása, *Die Handschriften*, 56.

41 Boehm and Fajt, *Prague*, 91–92; Jenni and Theisen, *Mitteleuropäische Schulen*, IV, 2–5.

42 Charles Burnett, Keiji Yamamoto, and Michio Yano, *Al-Qabīṣī (Alcabitius): The Introduction to Astrology* (London: Warburg Institute, 2004), 1–2, 200–02.

43 Burnett, Yamamoto, and Yano, *Al-Qabīṣī (Alcabitius): The Introduction to Astrology*, 1–2, 159; the best description of the manuscript to date is the catalog entry by Lenka Panušková in Klára Benešovská, ed., *A Royal Marriage: Elisabeth Premyslid and John of Luxembourg – 1310*, 2 vols (Prague: Muzeum hlavního města Prahy, 2011), I, 350–57. For an

1331 John of Saxony drafted one of the most widely influential and copied medieval commentaries on the *Book of the Introduction to the Craft of Astrology* during a fecund period of Parisian scholarly activity following the composition of his widely known canons explaining the Alfonsine planetary tables. As Charles Burnett has argued, John of Saxony supplied thereby two of the most important scholarly tractates routinely used for instruction and as definitive astronomical resources at the University of Paris. In his words, they belonged to 'the same syllabus'.[44]

John of Saxony's commentary on al-Qabīṣī also attests to a portion of the curriculum apposite for astronomical instruction at Charles University during the career of Johannes de Borotin spanning four decades, *c.* 1412–54, as documented by his extant lecture notes. This demonstrates one verifiable link between the Parisian master and the use of his texts to foster astronomical instruction in Prague. For example, in 1454 Borotin lectured on al-Qabīṣī at the Charles University; it is believed these handwritten notes record his pedagogical preparation and final commentary on the subject. It is worth noting that the earliest portions of this manuscript (Prague Castle Archive, Metropolitan Chapter Library, MS O. 1) document Borotin's tutelage under the same Iohannes Šindel who designed aspects of the aforementioned Prague Astronomical Clock. Borotin routinely drew upon John of Saxony in his exegetical clarifications of al-Qabīṣī.[45]

In any case, John of Seville's version of al-Qabīṣī's text is sufficient to clarify the doctrine that triangular (called *trine*) or hexagonal (labelled *sextile*) planetary configurations betoken beneficent results whereas planets arranged in a square (called *quartile*) or diametrical *opposition* presage disaster.[46] In the trapezoidal sections linked to the zodiac signs, Terzysko summarized the information about birth horoscopes contained within the 1293 Pietro d'Abano Latin translation of Abraham ibn Ezra's mid-twelfth-century treatise, *The Beginning of Wisdom*, in its revised format created 1148, and referred to by its later Latin title as the *Introductorium quod dicitur principium sapientiae*. Terzysko drafted these summaries even though he included the entire treatise later in his compilation on folios 11^{v}–27^{r} (Fig. 3). He embedded this data within a sophisticated presentation of the system of aspects in the diagram on folio 8^{r}, revealing myriad countervailing planetary configurations dispersed through the zodiac signs and assembling a comprehensive creative presentation of astronomical and astrological conceits.[47] However as a possible reflection of the unrest

earlier discussion of this important manuscript and its sale to Nicolaus Cusanus, see Alois Krchňák, 'Die Herkunft der astronomischen Handschriften und Instrumente des Nikolaus von Kues', *Mitteilungen und Forschungsbeiträge der Cusanus-Gesellschaft*, 3 (1963), 109–80.

44 Charles Burnett, 'Al-Qabīṣī's *Introduction to Astrology*: From Courtly Entertainment to University Textbook', in *Studies in the history of culture and science: A tribute to Gad Freudenthal*, ed. by Resianne Fontaine, Ruth Glasner, Reimund Leicht, and Giuseppe Veltri (Leiden: Brill, 2011), 53.

45 Burnett, 'Teaching the Science of the Stars', 9–19.

46 Burnett, Yamamoto, and Yano, *Al-Qabīṣī (Alcabitius): The Introduction to Astrology*, I.18, 234–35. For helpful introductions, see also Roger Beck, *A Brief History of Ancient Astrology* (Oxford: Blackwell, 2007), 20–23, 38–42; Tamsyn Barton, *Ancient Astrology* (New York: Routledge, 2006), 99–102; Ramírez-Weaver, 'Reading the Heavens', 82–83.

47 Raphael Levy, *The Astrological Works of Abraham ibn Ezra: A literary and linguistic study with special reference to the Old French translation of Hagin* (Baltimore: Johns Hopkins P, 1927), 7–46; Raphael Levy and Francisco Cantera, eds, *The Beginning of Wisdom: An Astrological Treatise by Abraham ibn Ezra* (Baltimore: Johns Hopkins P, 1939), 11–15. Also see, Franz Boll, *Sphaera: Neue griechische Texte und Untersuchungen zur Geschichte der Sternbilder* (Leipzig: B. G. Teubner,

Figure 3. Abraham ibn Ezra, *Introductorium quod dicitur principium sapientiae*, *Astronomical Anthology for Wenceslas IV*, Munich, Bayerische Staatsbibliothek, Clm 826, folio 11^v. From Prague, after 1400, reproduced with permission.

in Prague during the turbulent reign of Wenceslas IV, the *Astronomical Anthology* remains, regrettably, unfinished.[48]

Even in its incomplete state, the complementary contents of the manuscript indicate a range of desiderata and compilation practices, revealing what was apposite for inclusion within an astrological florilegium. After the ibn Ezra *Introductorium*, Abu Sa'id Shadan's fictive dialogue on astrological conceits, the *Excerpta de secretis Albumasar*, was included on folios 27ᵛ–33ʳ (Fig. 4).[49] Both are introduced by formulaic author portraits, betraying aspects of the Beautiful Style associated with courtly activity during the reign of Wenceslas IV. In the author portrait of Abraham ibn Ezra on folio 11ᵛ (Fig. 3), there is a historiated initial of the letter 'C' introducing the text, *cum initium sapientiae dei timor existat* ('the fear of God

1903; repr. 1967), 419–25. Also see Ramírez-Weaver, 'Reading the Heavens', 79–92; Eric Ramírez-Weaver, 'Creative Cosmologies in Late Gothic Bohemia: Illuminated Diagrams and Memory Tools for the Court of Wenceslas IV', *Manuscripta*, 54 (2010), 21–48.

48 Legner, ed. III, 104–5. Boll, *Sphaera*, 421–22n3 [sic] indicates that on folios 2ʳ and 3ᵛ additions were introduced during the sixteenth century.

49 David Juste, ed., *Les Manuscrits astrologiques latins conservés à la Bayerische Staatsbibliothek de Munich* (Paris: CNRS, 2011), 86–87 ; Graziella Federici Vescovini, 'La Versio latina degli *Excerpta de Secretis Albumasar di Sadan*', *Archives d'histoire doctrinale et littéraire du Moyen Âge*, 65 (1998), 273–330.

Figure 4. *Excerpta de secretis Albumasar, Astronomical Anthology for Wenceslas IV*, Munich, Bayerische Staatsbibliothek, Clm 826, folio 27ᵛ. From Prague, after 1400, reproduced with permission.

is the beginning of wisdom').[50] The gilded background and resplendent vegetal acanthus in jewel tones such as vibrant emerald, lapis, and ruby red, all complement the grisaille patterned painting of vegetation within the structural curves of the historiated initial. These painterly approaches exemplify Schmidt's second and third features presented above—accentuating decoration and flattening space—typical of Beautiful Style illustration. The ceremonial helmet (top centre) and the bath maiden on the right are indicators of the courtly style promulgated throughout Bohemian ateliers *c.* 1400. The emotionless visages of Abraham ibn Ezra and the bath maiden comically stare past one another like thespians on a proscenium, trained not to truly engage with one another but instead to gaze past one's interlocutor at roughly a 45-degree angle.

The bright hues with similarly cascading watery folds of drapery in the historiated initial for the *Excerpta* on folio 27ᵛ likewise exemplify the Beautiful Style (Fig. 4). The painter celebrates delineated forms at the expense of structure (recalling Schmidt's third feature). He relies upon a decorative sense of line (called ductus) and stays close to the (surface texture or) facture of the parchment (exemplifying Schmidt's features two and three). This manuscript illuminator compartmentalizes the figures and initial into modelled zones rather than integrating them into a coherent scene of scholars engaged

50 Ramírez-Weaver, 'Reading the Heavens', 85.

in debate within a discernible architectural interior (related to condition three). There is no single vantage point that yields a comprehensible illusionistic space (Schmidt's third feature) and the apathetic sages lack facial expression, appearing fatigued by the dialogue (satisfying condition one).[51]

Although Lynn Thorndike described this text as an intellectual fantasy motivated by a desire to preserve the teachings of Abu Ma'shar, Graziella Federici Vescovini has underscored the more contemporary significance of this tractate. The *Excerpta* is probably a late-thirteenth- or early-fourteenth-century collection of anecdotal legends and real references packaged by Henri Bate for a wider readership or preserved and presented in dialogue format by his Italian counterpart, Pietro d'Abano.[52] The codicological evidence therefore argues for a more than circumstantial connection between the *Introductorium* and the *Excerpta*, which could have been selected for their potential link to Pietro d'Abano. Since the authorship of the *Excerpta* remains an open question, this cannot be elevated above the level of a conjecture.

The next section of the book (folios 34–41) contains a cycle of eighteen illustrations depicting constellations derived from Abd al-Raḥmān al-Ṣūfī's tenth-century star catalogue, *Kitāb ṣuwar al-kawākib al-thābita* ('*The Book of the Images of the Fixed Stars*'), accompanying the form of the text known as the Ṣūfī-Latinus (Fig. 5).[53]

These astronomical illustrations are modelled after the northern Italian al-Ṣūfī manuscript in the Strahov Monastery (MS DA II 13), studied *in extenso* by Alena Hadravová and Petr Hadrava and probably made at some point during the last two decades of the fourteenth century.[54] This collection closes, on folios 46^{r}–53^{v}, with an ensemble of astrological aphorisms treating divination, known as the *Liber novem iudicum*.[55] This incidentally supplies yet another circumstantial example of texts that found their way into the Wenceslas collection after they played an important role at the court of Charles V. One of Charles' early commissions from his onetime translator, Robert Godefroy, was a French translation of this text, called *Le livre des neuf anciens juges d'astrologie* (1361).[56]

The *Astrological and Astronomical Anthology with Alfonsine Planetary Tables* from Vienna (Österreichische Nationalbibliothek Cod. 2352) presented Wenceslas IV with complementary sets of materials, enabling astronomical observation and astrological prediction in the

51 Boehm and Fajt, *Prague*, 105–11; Schmidt, 'Kunsthistorischer Kommentar', 126–30; Krása, *Die Handschriften*, 176–215.

52 David Juste, *Manuscrits astrologiques latins conservés à la Bayerische Staatsbibliothek de Munich*, 87; Graziella Federici Vescovini, 'La Versio latina degli *Excerpta de Secretis Albumasar di Sadan*', 273–74, 286–95; Lynn Thorndike, 'Albumasar in Sadan', *Isis*, 45 (1954), 22–23.

53 David Juste, *Manuscrits astrologiques latins conservés à la Bayerische Staatsbibliothek de Munich*, 86–87; Stefano Carboni, *Following the Stars: Images of the Zodiac in Islamic Art* (New York: Metropolitan Museum of Art, 1997), 38–39; Krása, *Die Handschriften*, 57, 211; Blume, Haffner, and Metzger, *Sternbilder des Mittelalters und der Renaissance*, II/1: 59–60, 361–73. Also see Paul Kunitzsch, 'Star Catalogues and Star Tables in Mediaeval Oriental and European Astronomy', in *The Arabs and the Stars: Texts and Traditions on the Fixed Stars, and their Influence in Medieval Europe* (Northampton: Variorum, 1989), 117–18; Paul Kunitzsch, 'Ṣūfī Latinus', *Zeitschrift der deutschen morgenländischen Gesellschaft*, 115 (1965), 65–74.

54 Alena Hadravová, and Petr Hadrava, *Sphaera octava IV, Mýty a věda o hvězdách, Katalogy hvězd a přemyslovský nebeský glóbus* (Prague: Artefactum–Academia, 2013), 17–19, 29–235; Blume, Haffner, and Metzger, *Sternbilder des Mittelalters und der Renaissance*, II/1, 347–54, 361–73; Krása, *Die Handschriften*, 211–12.

55 David Juste, *Manuscrits astrologiques latins conservés à la Bayerische Staatsbibliothek de Munich*, 87; Krása, *Die Handschriften*, 57.

56 Sherman, *Imaging Aristotle*, 17.

Figure 5. Perseus (*Ṣūfī Latinus*), *Astronomical Anthology for Wenceslas IV*, Munich, Bayerische Staatsbibliothek, Clm 826, folio 39^{r}. From Prague, after 1400, reproduced with permission.

late fourteenth century.[57] This anthology contains one of the most important sets of star pictures and planetary depictions originating in the High Middle Ages.[58] It is important to identify the portions of this compilation deemed worthy of sophisticated Beautiful Style illumination and to situate the kinds of iconographic work these images were intended to supply within Wenceslas IV's set of astronomical and astrological books.

The contents of the *Anthology with Alfonsine Planetary Tables* are adduced first. Then individual sections that received greater focus and emphasis through the introduction of cycles of illustration are further developed in the ensuing remarks. The book is divided into three major sections. In priority of place, a medieval astrological classic, the *Liber de signis* of Michael Scot, occupies the opening gatherings (folios 1^{r}–31^{v}). John of Saxony's canons and the Parisian Alfonsine Tables follow (spanning folios 34^{r}–83^{r}), with additions from John of Lignères on folios 75^{v}–80^{v}. Finally, there are assorted tools for prognostication, which were updated in the sixteenth century by a careful reader who was keen to introduce the tables and diagrams as a cohesive set and explain their use (folios 83^{v}–102^{r}).[59]

---

57 Jenni and Theisen, *Mitteleuropäische Schulen*, IV, 89–122.

58 Fritz Saxl, *Verzeichnis astrologischer und mythologischer illustrierter Handschriften des lateinsichen Mittelalters* (Heidelberg: Carl Winters Universitätsbuchhandlung, 1927), II, 15–19, 38–39, 86–90, 99–103.

59 Jenni and Theisen, *Mitteleuropäische Schulen*, IV, 89–91; Saxl, *Verzeichnis*, II, 86–90, at 88.

The section of the *Astrological and Astronomical Anthology with Alfonsine Planetary Tables* which would have appealed to a central and western European popular readership was the three-part *Liber Introductorius* of Michael Scot, placed at the outset of the manuscript. This is probably an intentional arrangement, establishing a conceptual and codicological emphasis upon popular or readily accessible forms of astrological texts. As Dieter Blume has argued, this treatise was crafted with an appreciation for equivalent and coeval work that the images and texts were expected to perform as astronomical pendants in a comprehensive, coordinated, and original display of classically inspired or even new constellations.[60] Michael Scot believed his book to be on the cusp of Christian European knowledge but advanced a classically inspired Aratean outlook—literally consulting routinely a twelfth-century (*c.* 1160?) copy of the *Aratea* in the translation by Germanicus, now located in Madrid, Biblioteca Nacional, MS. 19—gazing holistically outwards and upwards at forty-eight identified constellations either visible from Frederick II's court in Palermo or, if need be, fabricated out of whole cloth. Since the traditional cycle of astronomical illustration inherited from Aratus advanced a canonical set of forty-two collected constellations, there was a deficit of six models for anyone creating a new star catalogue based on Ptolemy's *Almagest*. The upshot is that Michael resorted to invention whenever necessary, because he wanted his catalogue to include the forty-eight constellations and 1022 fixed stars reported by Ptolemy in the *Almagest*, which he knew through the Latin translation by the Toledan scholar, Gerard of Cremona.[61] That the Alfonsine Tables come second in the codex could be taken as a sign of courtly emphasis; arguably, it suggests that the tables themselves in this particular book are perhaps to be considered a scientific relic with courtly cachet by their association with fourteenth-century Paris rather than merely a functional tool intended for routine consultation. The cycle of illustration commences with a historiated initial (Fig. 6) in the standard Beautiful Style of the late fourteenth century, opening the text of the *De noticia ordinum stellarum celi seu ymaginum 48…* which begins the text of the *Liber de signis*: *philosophi quidam multis experimentis noverunt celum esse stellarum ordinabiliter* ('the philosophers [qua natural historians presented as astronomers with a quadrant and wax tablet bearing the date of this section, 1393], who have many experiences, recall that the heavens contain constellations in an orderly manner').[62]

The portion of Michael Scot's *Liber Introductorius*, illustrated and featured within Wenceslas IV's *Anthology with Alfonsine Planetary Tables*, is the *Liber de signis*. Silke Ackermann has reconstructed the contents of Michael Scotus's expansive yet remedial natural historical curriculum that was perhaps reworked over the cleric's lifetime. In the service of Frederick II (d. 1250), Michael Scot added his last sprawling ruminations to the text in advance of his death, prior to 1236. The *Liber Introductorius* is comprised of three

60 Blume, Haffner, and Metzger, *Sternbilder des Mittelalters und der Renaissance*, II/1, 25–40, 215–20.

61 Blume, Haffner, and Metzger, *Sternbilder des Mittelalters und der Renaissance*, II/1, 30–38; Silke Ackermann, 'Habent sua fata libelli—Michael Scot and the Transmission of Knowledge Between the Courts of Europe', in *Kulturtransfer und Hofgesellschaft im Mittelalter: Wissenskultur am sizilianischen und kastilischen Hof im 13. Jahrhundert*, eds Gundula Grebner and Johannes Fried (Berlin: Akademie Verlag, 2008), 273–82.

62 Jenni and Theisen, *Mitteleuropäische Schulen*, IV, 91, 96–97, Abb. 19; for the critical edition with German translation, see Silke Ackermann, *Sternstunden am Kaiserhof: Michael Scotus und sein Buch von den Bildern und Zeichen des Himmels* (Frankfurt: Peter Lang, 2009),106–7, at 435 for notes concerning codicology and the various redactions of the *Liber de signis*. The translations into English are my own, unless otherwise specified.

Figure 6. Astronomers engaged in observation and debate, *Astrological and Astronomical Anthology with Alfonsine Planetary Tables*, from Prague, 1392/93. © Österreichische Nationalbibliothek Vienna, Cod. 2352, folio 1r, reproduced with permission.

Figure 7. Alfonso X el Sabio studies the stars, *Astrological and Astronomical Anthology with Alfonsine Planetary Tables*, from Prague, 1392/93. © Vienna, Österreichische Nationalbibliothek Cod. 2352, folio 34[r], reproduced with permission.

discrete volumes; the first divides, in turn, into four parts and is labelled the *Liber quatuor distinctionum*. Book two of this component is properly referred to as the *Liber de signis*.[63] The second volume of the overarching tripartite series is the intriguing *Liber particularis* in which Michael Scot appears to have elucidated, *inter alia*, an array of cosmological concerns vexing Frederick II.[64] Just as Wenceslas IV arguably required an otherworldly courtier in Prague, so too had Frederick needed Michael Scot to situate his astrological knowledge within the framework of his Christian microcosm in Sicily. They discussed prescient but problematic dogmatic issues, inquiring 'Just where are hell, purgatory, and the heavenly paradise?'[65] It is fascinating to note that such pivotal court astrologers sheltered or preserved themselves, in part, through a cultivated ethos of obscurantism. The oeuvre, ultimately, is the best evidence of the otherworldly courtier's research results and lifelong interests.[66]

After the *Liber de signis* comes the most important portion of the book for our discussion, namely, the Bohemian copy of the Parisian Alfonsine Tables. The standard version of John of Saxony's canons is found on folios 34^{r}–51^{r}. A discussion of planetary winds (*De ventis*) follows on folios 51^{v}–52^{v}, before the tables proper begin on folio 53^{r} and continue until 80^{v}.[67] The canons are introduced by yet another historiated initial presenting Alfonso el Sabio in the astronomical act of star-gazing (Fig. 7). The second date of 1392 (1000+300+92), recorded in cipher on Alfonso's wax tablet depicted on folio 34^{r}, indicates when this Bohemian copy of the tables was made.[68]

One piece of circumstantial evidence that this royal Alfonsine book was in fact consulted is contained within the lower margins of the tables, which are filled with scientifically rich annotations. John of Saxony's canon, chapter 13, notes how helpful it can be to personally introduce notes, or significant marginalia, into one's copies of tables.[69] As explained immediately prior in chapter 12, the *radices*, or roots, of the mean motions in the Alfonsine Tables had been calibrated for the meridian of Toledo. Users located elsewhere had to compensate not only diachronically for precession but also synchronically for the difference

63 Ackermann, *Sternstunden am Kaiserhof*, 20–48, 66–75, 435; as well as the shorter presentation of her views in Ackermann, 'Habent sua fata libelli', 273–84, emphasizing the role of Bohemia in the diffusion of Michael Scot's ideas; Charles Burnett, 'Michael Scot and the Transmission of Scientific Culture from Toledo to Bologna via the Court of Frederick II Hohenstaufen', *Micrologus*, 2 (1994), 101–26; Blume, Haffner, and Metzger, *Sternbilder des Mittelalters und der Renaissance*, II/1, 25–26.

64 Blume, *Regenten des Himmels*, 52–53; Ulrike Bauer, *Der Liber Introductorius des Michael Scotus in der Abschrift Clm 10268 der Bayerischen Staatsbibliothek München: Ein illustrierter astronomish-astrologischer Codex aus Padua, 14. Jahrhundert* (Munich: Tuduv, 1983), 2, 110n22; Ackermann, *Sternstunden am Kaiserhof*, 67–69.

65 Ulrike Bauer, *Der Liber Introductorius des Michael Scotus*, 2, 110n22; Charles Homer Haskins, *Studies in the History of Mediaeval Science*, 2nd ed. (New York: Frederick Ungar, 1960), 266–67, for a full English translation with the citation at p. 266, accompanied by a complete Latin transcription on 292–94. See also Lynn Thorndike, *Michael Scot* (London: Thomas Nelson, 1965), 32–59, 92–109; Blume, *Regenten des Himmels*, 47–53. For important background information about courtly translation projects, see Charles Burnett, 'Translation and Transmission of Greek and Islamic Science to Latin Christendom', in *The Cambridge History of Science*, Vol. 2: Medieval Science, ed. by David Lindberg and Michael Shank (New York: Cambridge University Press, 2013), II, 341–64 (pp. 341–45).

66 Ackermann, *Sternstunden am Kaiserhof*, 13–61.

67 Blume, Haffner, and Metzger, *Sternbilder des Mittelalters und der Renaissance*, II/1, 215–16; Jenni and Theisen, *Mitteleuropäische Schulen*, IV, 89–91.

68 Jenni and Theisen, *Mitteleuropäische Schulen*, IV, 108; Krása, *Die Handschriften*, 208–11.

69 Poulle, *Les Tables alphonsines*, 58–59, for a critical edition and French translation of the relevant passage.

in longitude between Toledo and the user's particular place.[70] In the lower margins of the Alfonsine tables on folios 59^{r}–65^{r}, additional radices have been added for Paris and places in central Europe: Prague, Erfurt, Vienna, Magdeburg, Worms, and Bratislava.[71]

From folio 83^{v} until 102^{r} the book turns to practical mantic interventions that astrologically minded, otherworldly courtiers could have deployed in the service of King Wenceslas IV. Traces of a *Geomantie* as well as a cycle of tables, texts and diagrams organized by the pseudo-Socrates Basileus under the category of *Prognostica* were copied onto folios 84^{r}–95^{r}. This section of the manuscript contains a cycle of sixteen divining sovereigns, such as the kings of Cappadocia, Germany, Lybia, and Wenceslas IV himself on folio 95^{r}, identified as the King of the Romans (Fig. 8). In this representation, prognostication through lots is more than simply an opportunity for a late medieval Christian king; it is normalized and presented as an acceptable activity worthy of a sage ruler. The image and text certainly elevated the role of the court astrologer in Prague, who is obliquely presented as a missing but essential participant within a fully functioning courtly structure. This is interesting evidence that the information contained within the *Astrological and Astronomical Anthology with Alfonsine Planetary Tables* (Cod. 2352) likely reflected the sort of divinatory tools and recommendations for fortune telling contained within a more utilitarian but lost hypothetical primer that could have been utilized by Wenceslas IV for his own prophecies.[72] Of astronomical interest is the description of 87 fixed stars found on folio 100$^{r\text{-}v}$, updating stellar coordinates associated with Alfonso X for the year 1357 through the addition of 18;15 to the classical Ptolemaic longitudinal values (compensating for the effects of precession). As Paul Kunitzsch previously noted, the compilers appended on folio 101 another star catalog of 51 items, adjusting in this instance the Ptolemaic values by 18;18.[73] The manuscript ends with a petition for divine assistance as the reader plumbs these spiritual and arguably occult mysteries: 'Lord God, Almighty Father, of one essence, eternal, and unspeakable, before the age in great wisdom you brought forth the Son… give to me, him, and you a responsive heart for the understanding of wise and elevated matters, and of your holy secrets. Whereby, you will live and reign with God the Father and the Holy Spirit'.[74]

---

70 Poulle, *Les Tables alphonsines*, 54–55, with helpful commentary at 193–95. For an insightful introduction, also see Richard L. Kremer and Jerzy Dobrzycki, 'Alfonsine Meridians: Tradition Versus Experience in Astronomical Practice c. 1500', *Journal for the History of Astronomy*, 29 (1998), 187–99, esp. p. 189.

71 The various cities adduced here are listed by folio: 59^{r}-Prague, Erfurt, Paris, Vienna; 60^{r}-Paris, Magdeburg, Worms, Bratislava, Erfurt, Prague, Vienna; 60^{v}- Prague, Erfurt, Bratislava, Vienna, Paris, Magdeburg, Worms; 61^{r}-Paris, Worms, Magdeburg, Vienna, Prague, Erfurt; 61^{v}-Paris, Magdeburg, Vienna, Prague, Erfurt; 62^{r}-Erfurt, Prague; 62^{v}-Paris, Magdeburg, Erfurt, Prague; 63^{r}-Paris, Magdeburg, Vienna, Prague, Erfurt; 63^{v}-Paris, Magdeburg, Prague, Erfurt; 64^{r}-Paris, Magdeburg, Erfurt, Prague, Vienna; 64^{v}-Paris, Magdeburg, Erfurt, Prague; 65^{r}- Paris, Magdeburg, Prague, Erfurt.

72 Blume, Haffner, and Metzger, *Sternbilder des Mittelalters und der Renaissance*, II/1, 216; Jenni and Theisen, *Mitteleuropäische Schulen*, IV, 91, 109–10, 118–19.

73 Blume, Haffner, and Metzger, *Sternbilder des Mittelalters und der Renaissance*, II/1, 216; Jenni and Theisen, *Mitteleuropäische Schulen*, IV, 91; Paul Kunitzsch, 'The Star Catalogue Commonly Appended to the Alfonsine Tables', *Journal of the History of Astronomy*, 17 (1986), 89–98 (p. 96n10).

74 Derived from the transcription reported in Saxl, *Verzeichnis*, II, 89, and Jenni and Theisen, *Mitteleuropäische Schulen*, IV, 91, but confirmed and expanded with the online version of Cod. 2352 available through the website of the Österreichische Nationalbibliothek: 'Domine deus pater omnipotens qui consubstancialem et coeternum et

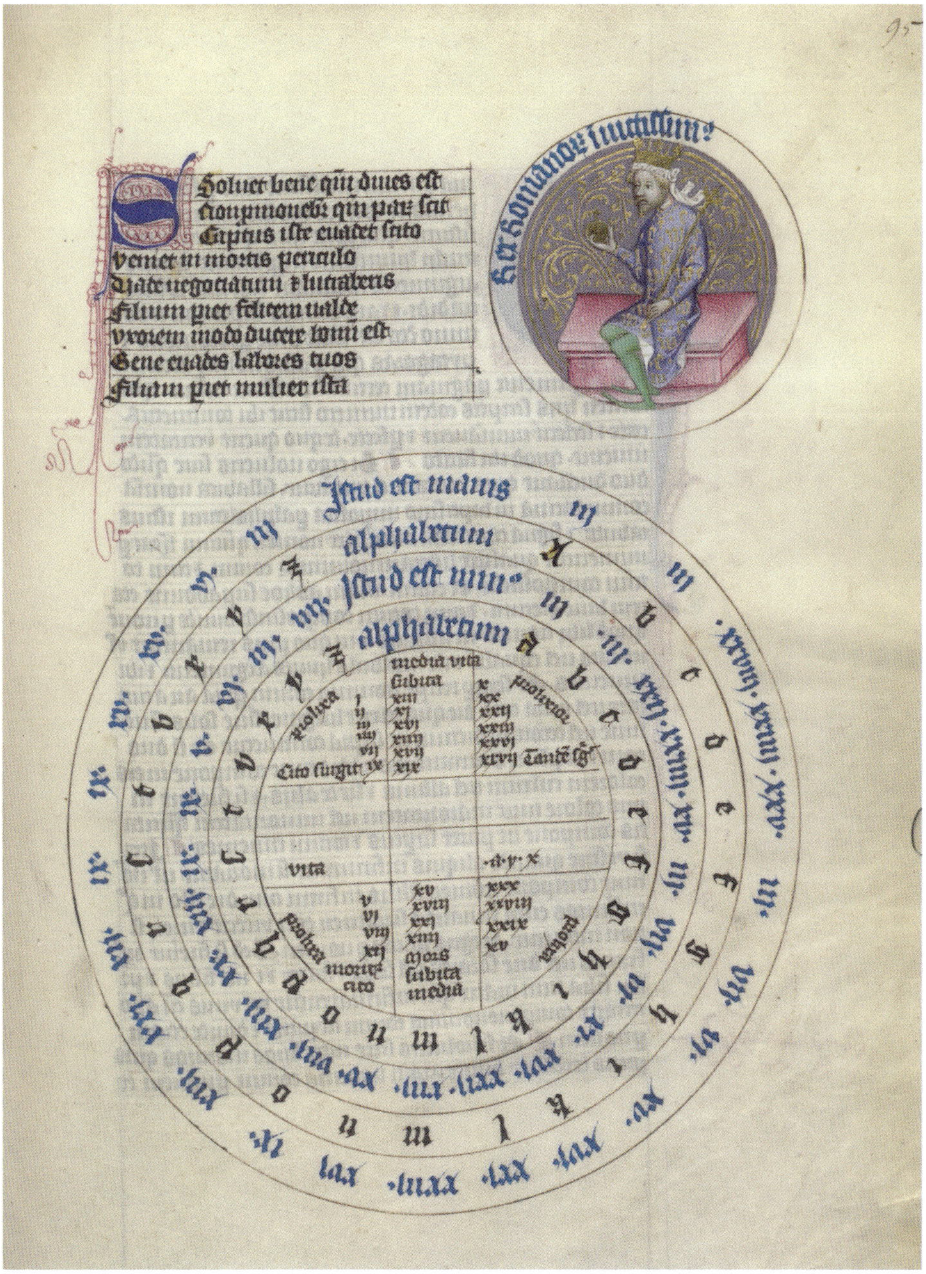

Figure 8. Wenceslas IV as 'Rex Romanorum invictissimus' (top), *Astrological and Astronomical Anthology with Alfonsine Planetary Tables*, from Prague, 1392/93. © Vienna, Österreichische Nationalbibliothek, Cod. 2352, folio 95^{r}, reproduced with permission.

---

ante secula ineffabilem et in magna sapiencia filium genuisti … dona michi suo tuo cor docile intelligendi acutam sublimitatem sanctorum tuorum secretorum. Qui cum deo patre et spiritu sancto vivis et regnas' (ff. 101^{v}–102^{r}). All translations are my own, unless otherwise indicated.

Figure 9. Wheel of Fortune, *Astrological and Astronomical Anthology with Alfonsine Planetary Tables*, from Prague, 1392/93. © Vienna, Österreichische Nationalbibliothek, Cod. 2352, folio 86ʳ, reproduced with permission.

This prayerful incantation is a coda that seems to harken back to the image of the Wheel of Fortune contained on folio 86ʳ (Fig. 9). Remarkably similar to the medallions of kings like Wenceslas IV in need of a well-augured future (Fig. 8), the personification of Fate within the Wheel of Fortune elevates one sovereign while deposing another. The kings required the fortune telling and divination tools from this section of the manuscript in order to rise above their foes (regnabo/I will rule!), to judiciously decree the fates of others (regno/I rule), or to recover should they stumble without humility (regnavi/Oops, I ruled!). If ever a king should find himself crushed beneath the wheel and the weight of his misfortune, having lost crown as well as kingdom, he must recognize his need to repent and regain that which he has lost in an endless cycle of redemption and failure (sum sine regno/I am without a kingdom, yikes).[75] This illustration makes it explicit: the courtly utility of the astronomical and astrological manuscripts created for Wenceslas IV is in the service of statecraft.

The third and final Wenceslas manuscript under consideration here (Vienna Cod. 2271), the *c.* 1400 Aegidius de Tebaldis translation of ʿAlī ibn Riḍwān's Commentary on Ptolemy's *Tetrabiblos* (Fig. 10), was intended to supply a conceptual and ideological bridge.[76] It linked Prague to earlier astrological, administrative centres of power, such as Toledo, and looked forward to a realized vision of eschatological Paradise inaugurated through spiritually apposite courtly activity in the present. In this context, yet another reference to Alfonso X appears. Aegidius de Tebaldis originally hailed from Parma but relocated to Alfonso X's court under whose patronage he became responsible, during the 1270s, for the Latin redaction of ʿAlī ibn Riḍwān's eleventh-century commentary[77] supplementing Plato of Tivoli's 1138 Latin edition of Ptolemy's *Tetrabiblos* and creating a set of scholarly pendants.[78] The historiated initial at the outset of the manuscript presents Wenceslas IV in the role of sage astronomically minded sovereign.[79] The Beautiful Style illuminator who painted this initial drew upon a standard iconography, celebrating learned kings undertaking rigorous study or interrogating philosophers such as Aristotle. This fourteenth-century form of portrait flourished at courts such as that of Charles V of France. In this context, the close working relationship between Nicole Oresme and the French king is reminiscent of the sort of situation that can be imagined between Terzysko, or any other putative otherworldly courtier in Prague, and Wenceslas IV.[80]

We must note, however, that Nicole Oresme did his best to encourage Charles V to avoid the deleterious interventions of otherworldly courtiers such as Terzysko, drafting two treatises warning French kings against overreliance upon court astrologers including Pèlerin de Prusse (discussed above) or Tommaso de Pizan (beginning 1364), Dominicus de Clavasio (from 1368), and (the aforementioned) Gervais Chrestien: the *Tractatus contra iudiciarios astronomos et principes in talibus se occupantes* (*c.* 1360) followed by the *Le livre de divinacions* (1361–65). Nicole Oresme argued that astrology could lead to precisely the sort of downfall seen in the Wheel of Fortune (Fig. 9) and experienced by Wenceslas IV, who perhaps should have heeded the sage counsel addressed to the '... princes and lords responsible for public governance'.[81] In any case, he did not, preferring to actively follow the example of Charles V. In a *c.* 1364 anthology containing texts by Nicole Oresme and Pèlerin de Prusse, Oxford, St John's College MS. 164 folio 1r, a miniature presents Charles V at his rotating reading table studiously ignoring Nicole Oresme's advice. *Mutatis mutandis*, this

75 Jenni and Theisen, *Mitteleuropäische Schulen*, IV, 109.

76 Jenni and Theisen, *Mitteleuropäische Schulen*, IV, 123–31.

77 Chabás and Goldstein, *The Alfonsine Tables of Toledo*, 229, 231–32; Legner, ed., III, 104; Krása, *Die Handschriften*, 52–56; Jenni and Theisen, *Mitteleuropäische Schulen*, IV, 128–29.

78 Olaf Pedersen, *A Survey of the Almagest*, rev. and ed. by Alexander Jones (New York: Springer, 2010), 16.

79 Jenni and Theisen, *Mitteleuropäische Schulen*, IV, 124.

80 Sherman, *Imaging Aristotle*, 13–22.

81 Laird and Fischer, *Pèlerin de Prusse on the Astrolabe*, 5–7 (pp. 6–7) for the citation and translation; Shore, 'A Case Study', 308–09; Sherman, *Imaging Aristotle*, 13–22.

Figure 10. Wenceslas IV studies and ciphers, 'Alī ibn Riḍwān's Commentary on Ptolemy's *Tetrabiblos/ Quadripartitus*, Latin by Aegidius de Tebaldis, from Prague, *c.* 1400. © Vienna, Österreichische Nationalbibliothek Cod. 2271, folio 1r, reproduced with permission.

glimpse into royal scholarly life supplies a telling record of the sort of situation imagined for Wenceslas IV with or without his otherworldly courtiers at his side (Fig. 11).[82]

Courtiers such as those employed by Charles V or Wenceslas IV reified presentations of their astrological science on their parchment folios. In Bohemia, court astrologers such as Terzysko performed a vital courtly role of pedagogical service, *inter alia*, cultivating Wenceslas IV's capabilities as an astute king who, by adopting the discerned models of Frederick II (d. 1250) or Alfonso X the Wise (d. 1284), could ultimately fulfil his destiny through study of the stars.[83] The three cycles of illustration, albeit limited in scope, underscore the courtly pressure placed upon a king who will rule and upon his courtier who will help him to navigate the tides of fate and ill-fortune via the tools recorded in these luxurious books. Although the exact role of Terzysko in the process cannot be ascertained, the *Astronomical Anthology* bearing his image and name reflects the range of sources and interests of other late medieval courts, especially that of Charles V in Paris.

3. The Case for a Curriculum

Wenceslas IV's individual astrological codices offer compelling insights when their respective contents are comparatively examined as wholes. On the other hand, it is similarly useful to query how the extant manuscripts, and it is vital to recall that the losses to Wenceslas IV's library are unfathomable, perform complementary functions and offer alternative texts. In summary, the compilation within the *Astrological and Astronomical Anthology with Alfonsine Tables* (Vienna Cod. 2352) brings together the Parisian Alfonsine Tables, Michael Scot's *Liber de signis*, and prophetic insights. An art historical link conjoins the terminal section on prognostication and the title page of John of Saxony's canons. In both cases, the blue royal robes of the wise, bearded rulers, with phylacteries about their heads, create an identification between Wenceslas IV (Fig. 8) and Alfonso X (Fig. 7).[84]

Although Alfonso did sponsor the compilation in 1272 of the original Castilian Alfonsine Tables, the later Middle Ages relied upon the Parisian Alfonsine Tables produced by figures such as John of Saxony's teacher and mentor, John of Lignères, and John of Murs.[85] The Parisian tables are the result of an efflorescence of astronomical activity in Paris during the 1320s, precisely when Charles IV, Wenceslas IV's father, was being reared at the court of his uncle, the final Capetian King Charles IV of France.[86]

---

82 Carey, *Courting Disaster*, 108; Sherman, *Imaging Aristotle*, 15–21, 341n41.

83 Blume, *Regenten des Himmels*, 47–51; Jenni and Theisen, *Mitteleuropäische Schulen*, IV, 9; H. Salvador Martínez, *Alfonso X, the Learned: A Biography* (Leiden: Brill, 2010), 45–87; Blume, Haffner, and Metzger, *Sternbilder des Mittelalters und der Renaissance*, II/1, 25–32, 42–46, 50–52, 69–79; Ramírez-Weaver, 'Reading the Heavens', 92.

84 Jenni and Theisen, *Mitteleuropäische Schulen*, IV, 91, 109; Burnett, 'Al-Qabīṣī's *Introduction to Astrology*', 53; Emmanuel Poulle, 'The Alfonsine Tables and Alfonso X of Castille', *Journal for the History of Astronomy*, 19 (1988), 97–105; Poulle, *Les Tables alphonsines*, 3–17. Also see Lynn Thorndike, *A History of Magic and Experimental Science*, 8 vols (New York: Columbia, 1923–58), III (1934), 257–58.

85 Burnett, 'Al-Qabīṣī's *Introduction to Astrology*', 53; Poulle, 'The Alfonsine Tables and Alfonso X of Castille', 102–05; Poulle, *Les Tables alphonsines*, 3–17; Thorndike, *A History of Magic and Experimental Science*, III, 254–58; Chabás and Goldstein, *The Alfonsine Tables of Toledo*, 1–2.

86 Poulle, 'The Alfonsine Tables and Alfonso X of Castille', 98–105; Boehm and Fajt, *Prague*, 4, 17n7.

Figure 11. Charles V of France studies his astrological and scientific manuscripts, compilation including Nicole Oresme, *Traitié de l'espere*, and Pèlerin de Prusse, *Astrological Treatises*, c. 1364. Oxford, Bodleian Library, St John's College MS 164, folio 1r. Photo: Bodleian Library, University of Oxford. Reproduced with permission of the Bodleian Library and the President and Fellows of St John's College, Oxford.

Charles IV was brought to Paris in April 1323; married a Valois princess, Blanche, in May; and nurtured in an astute cultural milieu cultivated by his godfather and uncle, Charles IV, the Fair, King of France. Charles IV, who in 1355 would become Holy Roman Emperor, developed an earnest interest and philological acumen for language study, remaining conversant over the course of his life in his native Czech, his adopted French, as well as other languages such as Latin, Italian, and the other Bohemian court language, German.[87] Charles IV's family became embroiled in the controversy over succession. John the Blind of Luxembourg, Bohemian Charles IV's father, advocated for the ascendancy of Phillip VI Valois to royal regent, while nail-biting Capetian courtiers awaited the birth of Jeanne d'Evreux's child in 1328, following the death of her husband King Charles IV of France.[88] Before the conclusion of the Capetian line and the ascendancy of the Valois dynasty resulted in Charles's ultimate return to Bohemia in 1330, he complemented his education in courtly affairs and the liberal arts with spiritual training at the hands of a future pontiff, Clement VI, otherwise known as Pierre Roger de Rosières. Charles described his boyhood fascination with the pious cleric, who '[caused] so much grace to flow over' him.[89] This profound spiritual sensitivity is one of the points of continuity across the reigns of Charles IV and his son, Wenceslas IV, who preferred instead the more scientific and occasionally occult pursuit of astrology. In any case, the sense of urgency manifested by some of Wenceslas IV's iconographic emblems or the illustrations in his books of astrology and astronomy suggests that one behaviour, modelled after his father, was a sincere belief that metaphysical assistance was vital to the welfare of Bohemia.[90] It is in this regard that representations of sage, astrologically minded kings performing Christian duty to their people must be further reconsidered. A juxtaposition of specific illustrations from Wenceslas IV's astronomical books helps to clarify how they worked together as a coherent set. In this context, the concepts of intervisuality discussed above—in which images borrow thematic and iconographic content from previous kinds of astronomical illustration or adapt alternative pictorial programmes—will help to clarify the symbolic content and latent ideologies contained within the imagery of Wenceslas IV's astronomical and astrological books.

As explained above, the title page of John of Saxony's canons, folio 34^{r} in the *Astrological and Astronomical Anthology with Alfonsine Planetary Tables*, supplies an *exemplum bonum* of a king who accurately augurs his kingdom's success through meditation upon the stars.[91] Knowledge in the service of power required first the purification and renewal of the royal intellect in order for the lessons learned to be processed suitably, as revealed by certain aspects of the typical Wenceslas iconography. A comparison between the medallion depicting the 'most invincible' prophesying King of the Romans—a title Wenceslas IV held from 1376 until 1400 (Fig. 8)—and the image of Alfonso studying the stars with the

87 František Šmahel, *The Parisian Summit, 1377–78: Emperor Charles IV and King Charles V of France*, trans. by Sean Mark Miller and Kateřina Millerová (Prague: Karolinum P, 2014), 20–32; Boehm and Fajt, *Prague*, 4, 17n7, 23.

88 Šmahel, *The Parisian Summit*, 32–33.

89 Šmahel, *The Parisian Summit*, 27–35; Boehm and Fajt, *Prague*, 4.

90 Ramírez-Weaver, 'Reading the Heavens', 73–94.

91 Jenni and Theisen, *Mitteleuropäische Schulen*, IV, 108–09.

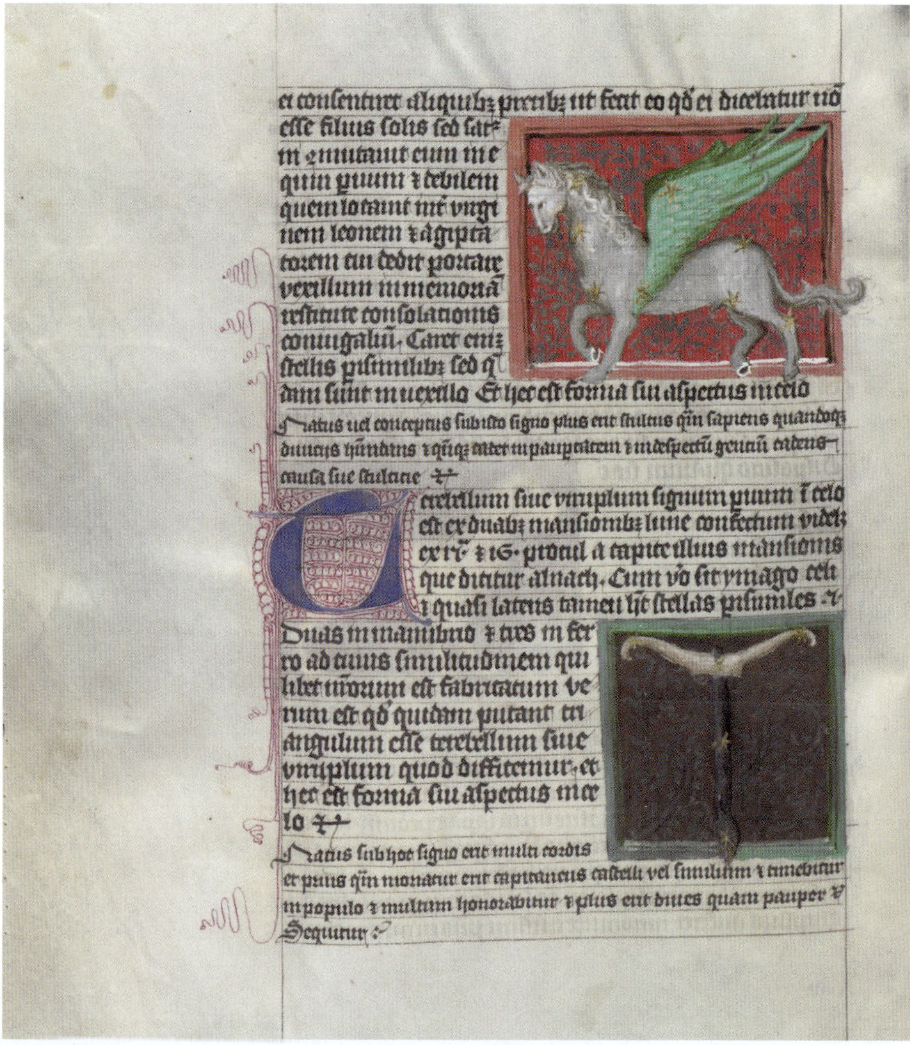

Figure 12. Equus secundus (above) and Terebellum (below), *Astrological and Astronomical Anthology with Alfonsine Planetary Tables*, 1392/93. © Vienna, Österreichische Nationalbibliothek, Cod. 2352, folio 25ᵛ, reproduced with permission.

aid of a quadrant is informative (Fig. 7).[92] Both kings have bound their foreheads with a knotted veil, or torse, akin to those routinely displayed as marginalia in manuscripts associated with the king. This symbol refers to the insignia of the chivalresque and royal Order of the King of Bohemia.[93] Of the sixteen kings depicted in the manuscript,

92 Jiří Spěváček, *Václav IV. 1361–1419: k předpokladům husitské revoluce* (Prague: Nakladatelství Svoboda, 1986), 314–21; Jenni and Theisen, *Mitteleuropäische Schulen*, IV, 109, supply inscriptions for all the medallions depicting prophesying kings.

93 Jenni and Theisen, *Mitteleuropäische Schulen*, IV, 7–9; Milada Studničková, 'Drehknoten und Drachen', 377–81; Milada Studničková, '*Gens Fera*', 214–39 (p. 222).

Wenceslas IV alone bears this mark of his dignity and duty. Theisen and Studničková have both argued that one recurring theme suggested by Wenceslas' emblems, including the knotted veil, is an emphasis upon the renewal of the royal body along with the body politic.[94] A courtier such as Terzysko would have considered meditation upon the spiritual Old Testament Law, and consideration of the statutes of the imperium, as important guarantors of proper royal conduct. The knotted veil, referred to as a *toczneigk* during the medieval period, as Studničková has explained, relates etymologically to ideas of rotation and revolution. In fact, the castle where Wenceslas IV summoned his astrological atlases was called Točník, at least in part because more likely than not it supplied a site for the rites of the Order of the King of Bohemia.[95] For an ideal Christian king, such as the Wenceslas IV perceived by his otherworldly courtiers, John of Saxony's opening citation from Aristotle resonates: *Tempus est mensura motus rerum motabilium et ut vult Aristotiles quarto Phisicorum* ('Time is the measure of motion of moving things, as Aristotle determines in Book IV of the Physics').[96] In the *Astrological and Astronomical Anthology with Alfonsine Planetary Tables*, the intervisual cross-reference identifying Wenceslas IV with Alfonso X creates an emphasis upon the eternal laws of the heavens, the perpetually rotating sphere of the fixed stars, and the beneficial value of its study through careful examination of the constellations. This suggests that the imagery within an individual codex was coordinated.[97]

Michael Scot's contributions in the *Liber de signis* likewise enable such meditation. One of the enduring features of the *Liber de signis* is its emphasis on astrological prediction and genethlialogy (or birth horoscopes) as applied to individual constellations.[98] Michael Scot offered an up-to-date, thirteenth-century revision of the standard Ptolemaic set of constellations in his *Liber de signis* and placed an integral, fundamental emphasis upon the pictorial information in the book.[99] After Gerard of Cremona supplied the definitive Latin translation of the Ptolemaic *Almagest* in *c.* 1175 using an intermediary Arabic text, scholars such as Michael Scot were confronted with forty-eight canonical constellations rather than the classical Aratean set of roughly forty-two star pictures contained within treatises such as the Carolingian *Libri computi* or later copies of Germanicus's first-century Latin

94 Jenni and Theisen, *Mitteleuropäische Schulen*, IV, 9; Studničková, '*Gens Fera*', 222.

95 Studničková, 'Drehknoten und Drachen', 381–83.

96 Charles Burnett offers a more complete translation of John of Saxony's opening remarks in 'Al-Qabīṣī's *Introduction to Astrology*', p. 53. Although used as the basis for my translation, mine is necessarily adapted. The Bohemian text emends the original in interesting ways. For the French translation and critical edition of the text that I also consulted, see Poulle, *Les Tables alphonsines*, with this passage reproduced at pp. 30–31. See also Thorndike, *A History of Magic and Experimental Science*, III, 257 (p. 257n10).

97 For basic information about the sphere of the fixed stars, see David C. Lindberg, *The Beginnings of Western Science: The European Scientific Tradition in Philosophical, Religious, and Institutional Context, Prehistory to A.D. 1450*, 2nd ed. (Chicago: University of Chicago Press, 2007), 41–42, and James Evans, *The History and Practice of Ancient Astronomy* (New York: Oxford University Press, 1998), 31–32, 75–77.

98 Jenni and Theisen, *Mitteleuropäische Schulen*, IV, 89–122; Wolfgang Metzger, 'Im Anfang war das Bild: Die Sternbilder in der Astrologie des Michael Scotus', *Transfert des savoirs au Moyen Âge/Wissenstransfer im Mittelalter: Actes de l'Atelier franco-allemand, Heidelberg, 15–18 janvier 2008*, ed. by Stephen Dörr and Raymund Wilhelm (Heidelberg: Universitätsverlag Winter, 2008), 149–51; Blume, *Regenten des Himmels*, 52–63; Josef Krása, *Die Handschriften*, 208–10; Ackermann, *Sternstunden am Kaiserhof*, 83–84.

99 Wolfgang Metzger, 'Im Anfang war das Bild', 150; Jenni and Theisen, *Mitteleuropäische Schulen*, IV, 82–84.

translation of Aratus's poem, the *Phaenomena*, as discussed above.[100] Michael seized this creative opportunity to offer new insights about the heavens. In an act of joint pictorial and textual scientific accomplishment, he compiled an alternative presentation of the heavens that appealed to the courtiers in Prague working for Wenceslas IV. The novelty of the *Liber de signis* made it an important complement to other star catalogues and a requisite component of what increasingly seems to have been an integrated astrological set of texts prepared for the king. The choices of codices and presentation of the materials are coherent and comprehensive enough to have even served as a curriculum, whether the books were in fact used in this way or not. As already stated, in the compilation of the *Liber de signis*, Michael routinely consulted the Madrid MS. 19 copy of the *Aratea*, available to him at Frederick II's court.[101] In the compilation of the *Liber de signis*, Ackermann has shown that Michael made errors while reading this precise resource, resulting in a constellation called the *Equus Secundus* (Fig. 12).[102] Similarly, a Bohemian painter could later offer his personal inflections, resulting in the Christological reinterpretation of a zodiacal constellation such as Aries recast as the *Agnus Dei* (Fig. 13).[103] Before the image of Aries and the description of its twenty stars, Michael Scot clarified its astrological significance: 'Whoever is born in this sign will be rich, comfortably satisfied with either secular or spiritual wealth, somewhat bold, strong, a vagabond, quarrelsome, and not very wise. He will fall into persecution or the confinement of persecutors, like such a beast; and it will be similar for him, and he will live in all aspects of his character and qualities. He will have it better in the first stage of life than in the second, since this is a moving sign.'[104]

---

100 Kunitzsch, 'Star Catalogues and Star Tables', 116–17; Paul Kunitzsch, *Der Almagest: Die Syntaxis Mathematica des Claudius Ptolemäus in arabisch-lateinischer Überlieferung* (Wiesbaden: Harassowitz, 1974), vii–viii; Pedersen, *A Survey of the Almagest*, 16–19; Ackermann, 'Habent sua fata libelli', 273–84; Ackermann, *Sternstunden am Kaiserhof*, 83–84; Metzger, 'Im Anfang war das Bild', 150; Eric Ramírez-Weaver, 'Classical constellations in Carolingian codices: Investigating the celestial imagery of Madrid, Biblioteca Nacional, MS 3307', in *Negotiating Secular and Sacred in Medieval Art: Christian, Islamic, and Buddhist*, ed. by Alicia Walker and Amanda Luyster (Farnham: Ashgate, 2009), 103–28. The best treatment of Aratean recensions remains Mechthild Haffner, *Ein antiker Sternbilderzyklus und seine Tradierung in Handschriften vom frühen Mittelalter bis zum Humanismus: Untersuchungen zu den Illustrationen der 'Aratea' des Germanicus* (Hildesheim: Georg Olms, 1997), 15–23, 75–79; Aratus, *Phaenomena*, trans. by Douglas Kidd (Cambridge: Cambridge University Press, 1997); Bernhard Bischoff and others, *Aratea: Kommentar zum Aratus des Germanicus MS. Voss. Lat. Q. 79, Bibliotheek der Rijksuniversiteit Leiden* (Luzern: Faksimile Verlag, 1989).

101 Ackermann, *Sternstunden am Kaiserhof*, 82–85; Jenni and Theisen, *Mitteleuropäische Schulen*, IV, 82–85. The manuscript in question is Madrid, Biblioteca Nacional MS 19 and contains the important clarificatory *Scholia Strozziana*. See Blume, *Regenten des Himmels*, 47–63; Haffner, *Ein antiker Sternbilderzyklus*, 125–29. Also see Ackermann, 'Habent sua fata libelli', 273–84; Metzger, 'Im Anfang war das Bild', 155.

102 Ackermann, 'Habent sua fata libelli', 273–76; Jenni and Theisen, *Mitteleuropäische Schulen*, IV, 98,105–6.

103 Jenni and Theisen, *Mitteleuropäische Schulen*, IV, 98,115–16.

104 Ackermann, *Sternstunden am Kaiserhof*, 148–49 supplies the critical edition of the text and an invaluable German translation that clarifies the significance of difficult passages. Natus in hoc signo erit dives, plenus convenienter diviciis temporalibus vel spiritualibus, aliquantulum audax, fortis, vagabundus, rixosus, non multum sapiens. Cadet in persecucionem vel custodiam persecutorum, ut tale animal, et huic erit similis in omni natura et proprietate ac vivet. Melius habebit in prima etate quam in secunda, quia est signum mobile.

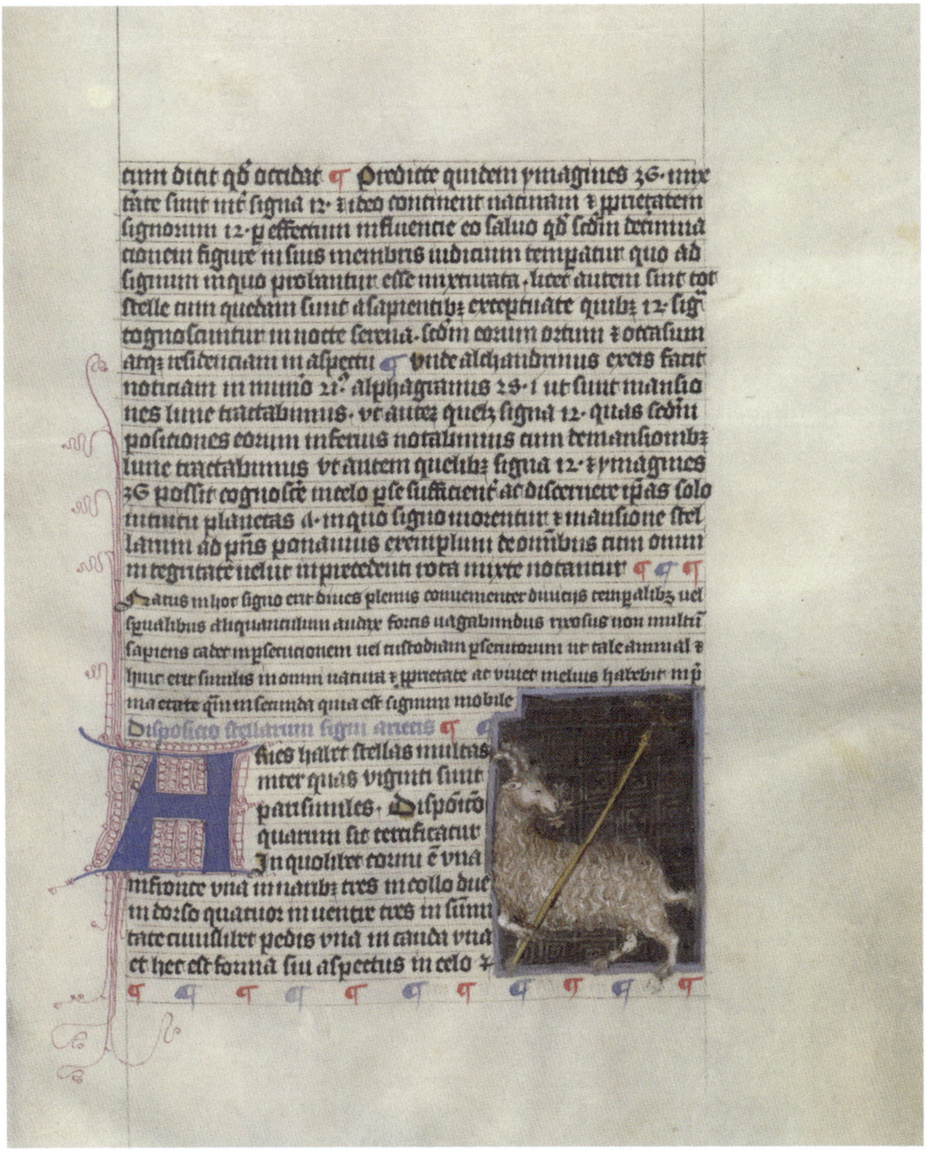

Figure 13. Aries, *Astrological and Astronomical Anthology with Alfonsine Planetary Tables*, 1392/93. © Vienna, Österreichische Nationalbibliothek, Cod. 2352, folio 7^{r}, reproduced with permission.

These remarks about personal proclivities betokened and augured by heavenly configurations encourage reflection in order to avoid misfortune and complement the material about nativities contained within Terzysko's diagram (Fig. 1).[105] Wenceslas IV's manuscript copies of the *Astronomical Anthology* in Munich and the *Libri de signis* in Vienna offered mutually reinforcing data regarding the heavens for the court astrologers and otherworldly

105 Ramírez-Weaver, 'Reading the Heavens', 83.

courtiers such as Terzysko in Prague. This fulfilled more than a bibliophilic desire for a comprehensive collection of astronomical books. The manuscripts worked in conjunction with one another, presenting Prague, Bohemia, and by extension Charles University, as sophisticated and stable centres within the Holy Roman Empire in which erudite astronomical investigation could thrive. The fact that this coherent codicological and bibliophilic index of planning and design remains discernible, despite the notorious losses to the library, makes the internal consistencies all the more, rather than less, compelling.

It is useful to pause and to reflect upon the significance of the Alfonsine corpus in this context. The intervisual links uniting Alfonso el Sabio (Fig. 7) and Wenceslas IV (Fig. 8) suggest that the identity of the ruler was as important as the text he engendered. Alfonso and the Alfonsine corpus offered one example of a late medieval scientific brand of kingship promulgated in Paris at the Court of Charles V Valois and in Prague at the court of Wenceslas IV. The Alfonsine Tables gained currency through revision. The intentional placement of John of Saxony's canons in Wenceslas IV's book provided another political connection between the two kingdoms and political centres that were the joint cities of Holy Roman Emperor Charles IV's boyhood. Such a sycophantic manoeuvre was beyond no savvy courtier. However, it is more important to underscore that the recast tables were included in a luxury codex in 1392/93 because they recorded Franco-Bohemian ties and emphasized Prague, via Charles University, as a new Central European intellectual seat of power on the Vltava. In other words, the association between Charles IV and Paris was paramount. Mastery of astronomical erudition was intended to be a kingly exercise. Terzysko's involvement in this process, as revealed in the illustration bearing his name, suggests that the courtiers in Prague were attempting to create an Alfonso, whereas Alfonso had made Toledo a centre for the creation of translation and astronomical science.

Conclusion

Recall once more the title page linking Alfonso and Wenceslas IV, the frontispiece to the Aegidius de Tebaldis translation of 'Alī ibn Riḍwān's Commentary on Ptolemy's *Tetrabiblos*, and the historiated initial introducing the *Introductorium quod dicitur principium sapientiae* from the *Astronomical Anthology for Wenceslas IV* in Munich. The intervisuality of these astronomical and astrological manuscripts produced in Prague has been emphasized in this iconographic analysis. The frontispiece to 'Alī ibn Riḍwān's commentary (Fig. 10) depicts Wenceslas IV undertaking astrological calculation and observation.[106] The text begins, *Scire et intelligere gloriosum et quia omnis sapiencia est a deo* ('To know and to understand is glorious, because all wisdom is from God').[107] The wisdom gleaned from study of the stars not only came from God. It could also renew the mind of a student such as King Wenceslas IV, who was reading the atlases prepared for him while he and his otherworldly courtiers were working and engaging in their various royal activities and duties including astronomical and astrological inquiry in the Točník Castle. This idea of

106 Jenni and Theisen, *Mitteleuropäische Schulen*, IV, 124.

107 Vienna, Cod. 2271, folio 1^{r}, reliant upon the reproduction in Jenni and Theisen, *Mitteleuropäische Schulen*, IV. 2, Abb. 71.

renewal is symbolized by the knotted veil on Alfonso's head, depicted in the title page for John of Saxony's canons, circling about his head like a phylactery or tefilin (Fig. 7).[108] Studničková recently argued that a wild man, such as the one across from Wenceslas holding the heraldic escutcheon of Bohemia and bearing a knotted veil about his waist, in the frontispiece to the commentary (Fig. 10) represents the spiritual law that transforms rebellious into disciplined servants of their God and king.[109] For Wenceslas IV, study of the three manuscripts discussed here performed the same function. The integrated astrological set of codices prepared by Terzysko, Bušek, and additional otherworldly courtiers supplied the king with the tools he needed to achieve intellectual renewal, following the examples of Alfonso X and Frederick II. In other words, contrary to first impressions, the astronomical books constituted a curricular pathway to renewal, proleptically aspiring to save Bohemia and Wenceslas IV in the process.

Across from ibn Ezra in the historiated initial introducing his *Introductorium* in the *Astronomical Anthology for Wenceslas IV* in Munich (Fig. 3), and opposite Alfonso X in the title page of John of Saxony's canon (Fig. 7), however, are bath maidens. Although Lenka Panušková has questioned the Bohemian provenance for the so-called *Celestial Atlas of the Bohemian Kings* (Cusanus 208, *c.* 1311, discussed above with regard to the *Astronomical Anthology for Wenceslas IV*), the significance of its glossed astrological readings of celestial conjunctions remain historically relevant for Central Europe at the outset of the fourteenth century. The book offers, *inter alia*, a glossed horoscope for the coronation of Wenceslas II in 1297 (see folio 88^{v}).[110] In the scant art historical literature devoted to the astronomical books of Wenceslas IV, scholars often mention a second gloss. On folio 85^{v}, as interpreted by Śnieżyńska-Stolot, a gloss indicates that the zodiacal constellation Leo (the Lion) 'signifies the King of Bohemia, Virgo signifies the people'.[111] The *Celestial Atlas* is replete with glosses. To highlight any individual gloss without context is therefore problematic. In Śnieżyńska-Stolot's view, nonetheless, the wooden buckets of bath maidens are linked to the iconography for the constellations (*qua* paranatellonta) that belong to the initial decan of the zodiacal sign said to be affiliated with the Bohemian people, Virgo.[112] There is no question that the bath maidens populating Wenceslas IV's astronomical and astrological books unite the ruler with his mission to serve.

Studničková has supplied a more nuanced clarification of the significance of bath maidens within the corpus of Wenceslas manuscripts. She argues that the bath maidens

108 Studničková, '*Gens Fera*', 222–24, 236n100.

109 Studničková, '*Gens Fera*', 220–24.

110 Benešovská, ed., *A Royal Marriage*, 350–57, 353 for the horoscope and textual gloss (catalogue entry by Lenka Panušková).

111 As reported in Benešovská, ed., *A Royal Marriage*, 356–57, and Milena Bartlová, 'The Magic of Image', in *The Role of Magic in the Past: Learned and Popular Magic, Popular Beliefs and Diversity of Attitudes*, ed. by Blanka Szeghyová (Bratislava: Pro Historia, 2005), 21, but confirmed by consultation of a digital reproduction supplied courtesy of the Hill Museum and Monastic Library, Saint John's Abbey and University, Collegeville, MN, and Bernkastel-Kues, Bibliothek des St Nikolaus-Hospitals. My translation is derived from that of Bartlová and was originally reported in Ramírez-Weaver, 'Reading the Heavens', 85. See Ewa Śnieżyńska-Stolot, 'Christian Interpretation of the Zodiac in Mediaeval Psalters', *Umění*, 37 (1989), 97–109 (p. 104). In addition, I wish to thank Alena Hadravová, Helena Avelar and Richard L. Kremer for their insightful remarks about the gloss on folio 85^{v}.

112 Śnieżyńska-Stolot, 'Christian Interpretation', 104; Bartlová, 'The Magic of Image', 21; Ramírez-Weaver, 'Reading the Heavens', 85–87; Studničková, 'Drehknoten und Drachen', 383.

supply a form of spiritual pendant for the wild men. The bath maidens according to Studničková are reformed and dutiful, obedient servants of their Lord. In the context of Wenceslas IV's astronomical codices, this idea finds special resonance. As the king worked through the judiciously selected celestial tractates, books such as the canons of John of Saxony and the Parisian Alfonsine Tables became more than scientific tools. They were politicized components of the king's astronomical study sessions which doubled as a devotional enterprise in service to Bohemia. Studničková also considers such bath maidens, when linked with Wenceslas IV, to personify the sort of virtues these astronomical books were supposed to cultivate in the mind of the ruler: love, mercy, wisdom, justice. In fact, the Alfonsine Tables were here ablutions for Wenceslas IV's mind, yielding pathways to wisdom after the model of Alfonso.[113]

In the guise of the bath maiden, according to the inscription and iconography just discussed, a vision of faithful Bohemian subjects appeared before the king. From within the initial 'T' of the title page to John of Saxony's canon (Fig. 7), Wenceslas literally embodies Alfonso's legacy and gazes at an image of his people, wearing a necklace dangling a golden 'W' about her neck—Bohemia's reminder of her sovereign and an invitation for Wenceslas IV to examine anew the astronomical texts contained within the manuscript. The compilers of these three astrological books knew that their curriculum was a model for earthly peace and that study of both divine and celestial law was of benefit to the king as well as the kingdom, in mutual reliance upon one another.[114]

The set of astronomical and astrological codices made for Wenceslas IV of Bohemia and reviewed here perform polyvalent forms of cultural labour when analysed semiotically as a set of books with both descriptive and prescriptive potential for their royal reader. The very act of codifying these source materials—of avowedly a somewhat dilettantish nature—for Wenceslas IV supplied a regulated and surveyed collection of resources culled presumably for the king's benefit and personal improvement. These prescribed texts, ranging from Michael Scot's *Liber de signis* to John of Saxony's canons for the Parisian Alfonsine Tables, offered the king a rudimentary descriptive apparatus validated and endorsed by their inclusion in luxury formats. In other words, in the collection of astronomical and astrological books associated with Wenceslas IV, the artistry and the astronomy participated equally in the attempted nurture of a wise king, resurrecting aspirations for a strong Bohemian sovereign in the image of Moses, Alfonso el Sabio, or Frederick II.

This was, however, not meant to be. Nevertheless, Wenceslas IV's astrological and astronomical manuscripts persist as vibrant records of the elevated cultural niveau and defiant erudition cultivated in late medieval Prague by a group of intellectual otherworldly courtiers such as Terzysko, Iohannes Šindel, and Křišt'an of Prachatice. In their efforts to illuminate Wenceslas IV, their talents eclipsed those of the king. And yet, the *de luxe* copies of astronomical and astrological manuscripts reviewed here reveal a careful curatorial process in the selections of texts slated for inclusion in the *Astronomical Anthology for*

113 Studničková, '*Gens Fera*', 224–32.

114 Jenni and Theisen, *Mitteleuropäische Schulen*, IV, 10–11; Śnieżyńska-Stolot, 'Christian Interpretation', 104; Ramírez-Weaver, 'Reading the Heavens', 87.

Wenceslas IV, the *Astrological and Astronomical Anthology with Alfonsine Planetary Tables*, and the Aegidius de Tebaldis translation of 'Alī ibn Riḍwān's Commentary on Ptolemy's *Tetrabiblos*.

Manuscript sources

Bernkastel-Kues, Bibliothek des St Nikolaus-Hospitals, Cus 208
Madrid, Biblioteca Nacional de España, 19
Munich, Bayerische Staatsbibliothek, Clm 826
Oxford, Bodleian Library, St John's College 164
Vienna, Österreichische Nationalbibliothek, Cod. 2271
Vienna, Österreichische Nationalbibliothek, Cod. 2352

Bibliography

Ackermann, Silke, 'Habent sua fata libelli—Michael Scot and the Transmission of Knowledge Between the Courts of Europe', in *Kulturtransfer und Hofgesellschaft im Mittelalter: Wissenskultur am sizilianischen und kastilischen Hof im 13. Jahrhundert*, ed. by Gundula Grebner and Johannes Fried (Berlin: Akademie Verlag, 2008), 273–84.

———, *Sternstunden am Kaiserhof: Michael Scotus und sein Buch von den Bildern und Zeichen des Himmels* (Frankfurt: Peter Lang, 2009).

Aratus, *Phaenomena*, trans. by Douglas Kidd, Cambridge Classical Texts and Commentaries, 34 (Cambridge: Cambridge University Press, 1997).

Bartlová, Milena, 'The Magic of Image', in *The Role of Magic in the Past: Learned and Popular Magic, Popular Beliefs and Diversity of Attitudes*, ed. by Blanka Szeghyová (Bratislava: Pro Historia, 2005), 19–28.

———, *Skutečná přítomnost: Středověký obraz mezi ikonou a virtuální realitou* (Prague: Argo, 2012).

Barton, Tamsyn, *Ancient Astrology* (New York: Routledge, 2006).

Bauer, Ulrike, *Der Liber Introductorius des Michael Scotus in der Abschrift Clm 10268 der Bayerischen Staatsbibliothek München: Ein illustrierter astronomish-astrologischer Codex aus Padua, 14. Jahrhundert* (Munich: Tuduv, 1983).

Beck, Roger, *A Brief History of Ancient Astrology* (Oxford: Blackwell, 2007).

Benešovská, Klára, ed., *A Royal Marriage: Elisabeth Premyslid and John of Luxembourg – 1310*, 2 vols (Prague: Muzeum hlavního města Prahy, 2011).

Bischoff, Bernhard, *Aratea: Kommentar zum Aratus des Germanicus MS. Voss. Lat. Q. 79, Bibliotheek der Rijksuniversiteit Leiden* (Luzern: Faksimile Verlag, 1989).

Blume, Dieter, *Regenten des Himmels: Astrologische Bilder in Mittelalter und Renaissance* (Berlin: Akademie Verlag, 2000).

———, 'Picturing the Stars: Astrological Imagery in the Latin West, 1100–1550', in *A Companion to Astrology in the Renaissance*, ed. by Brendan Dooley (Leiden: Brill, 2014), 333–98.

———, Mechthild Haffner, and Wolfgang Metzger, *Sternbilder des Mittelalters und der Renaissance: Der gemalte Himmel zwischen Wissenschaft und Phantasie*, 5 vols (Berlin: de Gruyter, 2012–16).

Bober, Harry, 'The Zodiacal Miniature of the *Très riches heures* of the Duke of Berry—Its Sources and Meaning', *Journal of the Warburg and Courtauld Institutes*, 11 (1948), 1–34.

Boehm, Barbara, and Jiří Fajt, eds, *Prague, The Crown of Bohemia, 1347–1437* (New York: Metropolitan Museum of Art, 2005).

Boll, Franz, *Sphaera: Neue griechische Texte und Untersuchungen zur Geschichte der Sternbilder* (Leipzig: Teubner, 1903; repr. 1967).

Boudet, Jean-Patrice, 'Charles V, Gervais Chrétien et les manuscrits scientifiques du collège de Maître Gervais', *Médiévales*, 52 (2007), 15–38.

———, 'A "College of Astrology and Medicine"? Charles V, Gervais Chrétien, and the Scientific Manuscripts of Maître Gervais's College', *Studies in History and Philosophy of Biological and Biomedical Sciences*, 41 (2010), 99–108.

Burnett, Charles, 'Michael Scot and the Transmission of Scientific Culture from Toledo to Bologna via the Court of Frederick II Hohenstaufen', *Micrologus*, 2 (1994), 101–26.

———, 'Al-Qabīṣī's *Introduction to Astrology*: From Courtly Entertainment to University Textbook', in *Studies in the history of culture and science: A tribute to Gad Freudenthal*, ed. by Resianne Fontaine, Ruth Glasner, Reimund Leicht, and Giuseppe Veltri, Studies in Jewish History and Culture, 30 (Leiden: Brill, 2011), 43–69.

———, 'Translation and Transmission of Greek and Islamic Science to Latin Christendom', in *The Cambridge History of Science*, vol. 2: *Medieval* Science, ed. by David Lindberg and Michael Shank (New York: Cambridge University Press, 2013), 341–64.

———, 'Teaching the Science of the Stars in Prague University in the Early Fifteenth Century: Master Johannes Borotin', *Aithér (AIΘHP)*, 2 (2014), 9–50.

———, Keiji Yamamoto, and Michio Yano, *Al-Qabīṣī (Alcabitius): The Introduction to Astrology* (London: Warburg Institute, 2004).

Camille, Michael, *Image on the Edge* (London: Reaktion, 1992).

Carboni, Stefano, *Following the Stars: Images of the Zodiac in Islamic Art* (New York: Metropolitan Museum of Art, 1997).

Carey, Hilary, 'Astrology at the English Court in the Later Middle Ages', in *Astrology, Science, and Society: Historical Essays*, ed. by Patrick Curry (Woodbridge: Boydell, 1987), 41–56.

———, *Courting Disaster: Astrology at the English Court and University in the Later Middle Ages* (London: Macmillan, 1992).

Carruthers, Mary, *The Book of Memory: A Study of Memory in Medieval Culture*, 2nd ed. (Cambridge: Cambridge University Press, 2008).

Caviness, Madeline, 'Hildegard as the Designer of the Illustrations to her Works', in *Hildegard of Bingen: The Context of her Thought and Art*, ed. by Charles Burnett and Peter Dronke (London: Warburg Institute, 1998), 29–62.

Cermann, Regina, 'Beschreibung einer problematischen Corvine: Cod. Guelf. 69.9 Aug. 2°', in *Corvina Augusta: Die Handschriften des Königs Matthias Corvinus in der Herzog August Bibliothek Wolfenbüttel*, ed. by Edina Zsupán (Budapest: Bibliotheca Nationalis Hungariae, 2014), 123–51.

Chabás, José, and Bernard Goldstein, *The Alfonsine Tables of Toledo* (Dordrecht: Kluwer, 2003).

Dokoupil, Vladislav, *Soupis Rukopisů Mikulovské Dietrichsteinské Knihovny*, Catalogi codicum manu scriptorum in Bibliotheca Universitatis Brunensis asservatorum, 2 (Brno: Státní pedagogické nakladatelství, 1958).

Evans, James, *The History and Practice of Ancient Astronomy* (New York: Oxford University Press, 1998).

Evans, Michael, 'The Geometry of the Mind', *Architectural Association Quarterly*, 12 (1980), 32–55.

Fajt, Jiří, 'Magister Theodoricus-Court Artist to Emperor Charles IV', in *Magister Theodoricus, Court Painter to Emperor Charles IV: The Pictorial Decoration of the Shrines at Karlštejn Castle*, ed. by Jiří Fajt (Prague: National Gallery, 1998), 217–77.

———, ed., *Karl IV.: Kaiser von Gottes Gnaden, Kunst und Repräsentation des Hauses Luxemburg 1310–1437* (Munich: Deutscher Kunstverlag, 2006).

Folta, Jaroslav, 'Clockmaking in Medieval Prague', *Antiquarian Horological Society*, 23 (1997), 405–17.

Hadravová, Alena, and Petr Hadrava, 'Astronomy in Prague: From the Past to the Present', *Highlights of Astronomy*, 14 (2006), 3–13.

———, and ———, *Sphaera octava IV, Mýty a věda o hvězdách, Katalogy hvězd a přemyslovský nebeský glóbus* (Prague: Artefactum–Academia, 2013).

Haffner, Mechthild, *Ein antiker Sternbilderzyklus und seine Tradierung in Handschriften vom frühen Mittelalter bis zum Humanismus: Untersuchungen zu den Illustrationen der 'Aratea' des Germanicus* (Hildesheim: Georg Olms, 1997).

Haskins, Charles Homer, *Studies in the History of Mediaeval Science*, 2nd ed. (New York: Frederick Ungar, 1960).

Jenni, Ulrike, and Maria Theisen, *Mitteleuropäische Schulen IV (ca. 1380–1400), Hofwerkstätten König Wenzels IV. und deren Umkreis* (Vienna: Verlag der Österreichischen Akademie der Wissenschaften, 2014).

Juste, David, ed., *Les Manuscrits astrologiques latins conservés à la Bayerische Staatsbibliothek de Munich* (Paris: CNRS, 2011).

Krása, Josef, *Die Handschriften König Wenzels IV.*, trans. by Herta Soswinski (Vienna: Forum Verlag, 1971).

———, 'Astrologické Rukopisy Václava IV', in *České Iluminované Rukopisy 13./16. Století*, ed. by Jiří Dvorský and Jiří Kropáček (Prague: Odeon, 1990), 180–203.

Krchňák, Alois, 'Die Herkunft der astronomischen Handschriften und Instrumente des Nikolaus von Kues', *Mitteilungen und Forschungsbeiträge der Cusanus-Gesellschaft*, 3 (1963), 109–80.

Kremer, Richard L., and Jerzy Dobrzycki, 'Alfonsine Meridians: Tradition Versus Experience in Astronomical Practice *c.* 1500', *Journal for the History of Astronomy*, 29 (1998), 187–99.

Křišťan z Prachatic, *Stavba a Užití Astrolábu*, ed. and trans. by Alena Hadravová and Petr Hadrava (Prague: Filosofia, 2001).

Kunitzsch, Paul, 'Ṣūfī-Latinus', *Zeitschrift der deutschen morgenländischen Gesellschaft*, 115 (1965), 65–74.

———, *Der Almagest: Die Syntaxis Mathematica des Claudius Ptolemäus in arabisch-lateinischer Überlieferung* (Wiesbaden: Harassowitz, 1974).

———, 'The Star Catalogue Commonly Appended to the Alfonsine Tables', *Journal of the History of Astronomy*, 17 (1986), 89–98.

———, 'Star Catalogues and Star Tables in Mediaeval Oriental and European Astronomy', in *The Arabs and the Stars: Texts and Traditions on the Fixed Stars, and their Influence in Medieval Europe* (Northampton: Variorum, 1989), 113–22.

Laird, Edgar, and Robert Fischer, *Pèlerin de Prusse on the Astrolabe: Text and Translation of his* Practique de astralabe, (Binghamton: SUNY Binghamton P, 1995).

Legner, Anton, ed., *Die Parler und der Schöne Stil 1350–1400*, 5 vols (Cologne: Schnütgen Museum, 1978–80).

Lejbowicz, Max, 'Guillaume Oresme, traducteur de la *Tetrabible* de Claude Ptolémée', *Pallas*, 30 (1983), 107–33.

Lemay, Richard, 'The Teaching of Astronomy in Medieval Universities, Principally at Paris in the Fourteenth Century', *Manuscripta*, 20 (1976), 197–217.

Levy, Raphael, *The Astrological Works of Abraham ibn Ezra: A Literary and Linguistic Study with Special Reference to the Old French Translation of Hagin* (Baltimore: Johns Hopkins P, 1927).

———, and Francisco Cantera, eds, *The Beginning of Wisdom: An Astrological Treatise by Abraham ibn Ezra* (Baltimore: Johns Hopkins P, 1939).

Lindberg, David C., *The Beginnings of Western Science: The European Scientific Tradition in Philosophical, Religious, and Institutional Context, Prehistory to A.D. 1450*, 2nd ed. (Chicago: University of Chicago Press, 2007),

Martínez, H. Salvador, *Alfonso X, the Learned: A Biography* (Leiden: Brill, 2010).

Metzger, Wolfgang, 'Im Anfang war das Bild: Die Sternbilder in der Astrologie des Michael Scotus', in *Transfert des savoirs au Moyen Âge/Wissenstransfer im Mittelalter: Actes de l'Atelier franco-allemand, Heidelberg, 15–18 janvier 2008*, ed. by Stephen Dörr and Raymund Wilhelm (Heidelberg: Universitätsverlag Winter, 2008), 149–61.

Novotná, Anna, *Die Prager Rathausuhr*, trans. by Artlingua (Prague: Práh, 2015).

Opačić, Zoë, 'Architecture and Religious Experience in 14th-Century Prague', in *Kunst als Herrschaftsinstrument: Böhmen und das Heilige Römische Reich unter den Luxemburgern im europäischen Kontext*, ed. Jiří Fajt and Andrea Langer (Berlin: Deutscher Kunstverlag, 2009), 136–49.

Pächt, Otto, 'Die Gotik der Zeit um 1400 als gesamteuropäische Kunstsprache', in *Europäische Kunst um 1400*, ed. by Vinzenz Oberhammer (Vienna: Kunsthistorisches Museum, 1962), 52–65.

Pedersen, Olaf, *A Survey of the Almagest*, ed. by Alexander Jones (New York: Springer, 2010).

Polívka, Miloslav, 'The Expansion of the Czech State during the Era of the Luxemburgs (1306–1419)', in *A History of the Czech Lands*, 2nd edition, ed. by Jaroslav Pánek and Oldřich Tůma, trans. by Justin Quinn, Petra Key, and Lea Bennis (Prague: Karolinum P, 2018), 123–56.

Poulle, Emmanuel, *Les Tables alphonsines avec les canons de Jean de Saxe* (Paris: Éditions du Centre National de la Recherche Scientifique, 1984).

———, 'Les Astronomes parisiens au XIVe siècle et l'astronomie alphonsine', *Histoire littéraire de la France*, 43 (2005), 1–54.

Ramírez-Weaver, Eric, 'Classical Constellations in Carolingian Codices: Investigating the Celestial Imagery of Madrid, Biblioteca Nacional, MS 3307', in *Negotiating Secular and Sacred in Medieval Art: Christian, Islamic, and Buddhist*, ed. by Alicia Walker and Amanda Luyster (Farnham: Ashgate, 2009), 103–28.

———, 'Creative Cosmologies in Late Gothic Bohemia: Illuminated Diagrams and Memory Tools for the Court of Wenceslas IV', *Manuscripta*, 54 (2010), 21–48.

———, 'Reading the Heavens: Revelation and Reification in the Astronomical Anthology for Wenceslas IV', *Gesta*, 53 (2014), 73–94.

———, *A Saving Science: Capturing the Heavens in Carolingian Manuscripts* (University Park: Pennsylvania State University Press, 2017).

Saxl, Fritz, *Verzeichnis astrologischer und mythologischer illustrierter Handschriften des lateinsichen Mittelalters: II. Die Handschriften der National-Bibliothek in Wien* (Heidelberg: Carl Winters Universitätsbuchhandlung, 1927).

Schmidt, Gerhard, 'Malerei bis 1450: Tafelmalerei—Wandmalerei—Buchmalerei', in *Gotik in Böhmen: Geschichte, Gesellschaftsgeschichte, Architektur, Plastik, und Malerei*, ed. by Karl Swoboda (Munich: Prestel Verlag, 1969), 167–321.

———, 'Kunsthistorischer Kommentar', in *Die Wenzelsbibel, vollständige Faksimile-Ausgabe der Codices Vindobonenses 2759–2764 der Österreichischen Nationalbibliothek Wien*, ed. by Hedwig Heger and others (Graz: Akademische Druck- und Verlaganstalt, 1998), 125–250.

Sherman, Claire Richter, *Imaging Aristotle: Verbal and Visual Representation in Fourteenth-Century France* (Berkeley: University of California Press, 1995).

Shore, Lys Ann, 'A Case Study in Medieval Nonliterary Translation: Scientific Texts from Latin to French', in *Medieval Translators and Their Craft*, ed. by Jeanette Beer, (Kalamazoo: Western Michigan University, Medieval Institute Press, 1989), 297–327.

Siraisi, Nancy, *Medieval and Early Renaissance Medicine: An Introduction to Knowledge and Practice* (Chicago: University of Chicago Press, 1990).

Spěváček, Jiří, *Václav IV, 1361–1419: k předpokladům husitské revoluce* (Prague: Nakladatelství Svoboda, 1986).

Stejskal, Karel, and Josef Krása, 'Astralvorstellungen in der mittelalterlichen Kunst Böhmens', *Sborník Prací Filosofické Fakulty Brněnske University*, 8 (1964), 61–85.

Studničková, Milada, 'Drehknoten und Drachen: Die Orden Wenzels IV. und Sigismunds von Luxemburg und die Polysemantik Ihrer Zeichen', in *Kunst als Herrschaftsinstrument: Böhmen und das Heilige Römische Reich unter den Luxemburgern im europäischen Kontext*, ed. by Jiří Fajt and Andrea Langer (Berlin: Deutscher Kunstverlag, 2009), 377–87.

———, '*Gens Fera*: The Wild Men in the System of Border Decoration of the Bible of Wenceslas IV', *Umění*, 62 (2014), 214–39.

Šíma, Zdislav, *Astronomy and Clementinum*, 2nd ed., trans. by Hana Vajnerová (Prague: Národní knihovna České republiky, 2006).

Šmahel, František, 'The Faculty of Liberal Arts 1348–1419', in *Die Präger Universität im Mittelalter: Gesammelte Aufsätze* (Leiden: Brill, 2007), 213–315.

———, *The Parisian Summit, 1377–78: Emperor Charles IV and King Charles V of France*, trans. by Sean Mark Miller and Kateřina Millerová (Prague: Karolinum P, 2014).

Śnieżyńska-Stolot, Ewa, 'Christian Interpretation of the Zodiac in Mediaeval Psalters', *Umění*, 37 (1989), 97–109.

———, *Ikonografia znaków zodiaku i gwiazdozbiorów w rękopisie monachijskim Abrahama ibn Ezry* (Cracow: Wydawnictwo Uniwersytetu Jagiellońskiego, 1998).

Theisen, Maria, *History Buech Reimenweisz: Geschichte, Bildprogramm und Illuminatoren des Willehalm-Codex König Wenzels IV. von Böhmen, Wien, Österreichische Nationalbibliothek ser. nov. 2643* (Vienna: Verlag der Österreichischen Akademie der Wissenschaften, 2010).

Thorndike, Lynn, *A History of Magic and Experimental Science*, 8 vols (New York: Columbia, 1923–58).

———, 'Albumasar in Sadan', *Isis*, 45 (1954), 22–32.

Vescovini, Graziella Federici, 'La Versio latina degli *Excerpta de Secretis Albumasar di Sadan*', *Archives d'histoire doctrinale et littéraire du Moyen Âge*, 65 (1998), 273–330.

Wallis, Faith, 'What a Medieval Diagram Shows: A Case Study of Computus', *Studies in Iconography*, 36 (2015), 1–40.

Part 2

Authors, Texts and their Receptions in Various Milieus

JOSÉ CHABÁS,
MARIE-MADELEINE SABY

Editing the Tables of 1322 by John of Lignères

1. Identification of the set

When the astronomy developed by the scholars at the court of King Alfonso X of Castile and León (reigned 1252–84) reached Paris, through channels not yet adequately explained, a group of astronomers working at or around the university assimilated this knowledge and integrated it into their scientific activity.[1] Among them were John Vimond, John of Murs, John of Lignères, and John of Saxony, who embraced the material coming from the Iberian Peninsula, to the detriment of astronomy based on the Toledan Tables, as practiced by their predecessors in Paris. The core of this material consisted of astronomical tables, based on the work of Muslim and Jewish astronomers mostly active in al-Andalus, the southern part of the Iberian Peninsula. They are now called the Castilian Alfonsine Tables, or simply 'Alfonsine Tables', as they were cited at the time in Paris.

A key role in this process of assimilation and adaptation of the Castilian Alfonsine Tables in Paris was played by John of Lignères, the author of several works on astronomy and mathematics. In particular, he compiled two sets of astronomical tables built on material from the Iberian Peninsula and he authored two texts or canons that have been associated with these tables.[2] One deals with daily rotation and addresses trigonometric problems of astronomical interest. It spans forty-four chapters beginning '*Cuiuslibet arcus propositi sinum rectum*'. The other, concerned with planetary motions and the computation

* Research presented in this chapter was supported by the ERC project ALFA: Shaping a European scientific scene, Alfonsine astronomy, CoG 723085, PI Matthieu Husson.
We thank Matthieu Husson and Richard L. Kremer for their useful comments on a draft of this paper, as well as the participants of the ALFA workshop held in Prague in October 2019, where a draft of this paper was presented, for their insightful remarks.

1 José Chabás and Bernard R. Goldstein, *The Alfonsine Tables of Toledo* (Dordrecht: Kluwer, 2003).

2 See Chabás and Goldstein, *Alfonsine Tables*, pp. 281–84; Emmanuel Poulle, 'John of Lignères', in *Dictionary of Scientific Biography*, ed. Charles Gillispie, 16 vols (New York: Charles Scribner's Sons, 1970–80), VII (1973), 122–28.

José Chabás • Universitat Pompeu Fabra, Barcelona

Marie-Madeleine Saby • Université Grenoble Alpes, Grenoble

Alfonsine Astronomy: The Written Record, ed. by Richard L. Kremer, Matthieu Husson, José Chabás, Turnhout, 2022 (*Alfonsine Astronomy: Studies and Sources*, 1), pp. 243-255
© BREPOLS PUBLISHERS 10.1484/M.ALFA.5.124928

of eclipses, is presented in forty-six chapters beginning '*Priores astrologi motus corporum celesti*'.[3] Although there is some overlap, the two canons are essentially complementary and one or often both at the same time can be found in astronomical manuscripts. These two texts saw remarkable success, if we are to judge from the great number of manuscripts in which they are preserved.

John of Lignères was also the author of two sets of tables. One is called the *Tabule magne*, dated 1325, and it goes along with a text, beginning '*Multiplicis philosophie*', describing the computational procedures that apply to these tables.[4] The text and the tables are extant in a reduced number of manuscripts. The *Tabule magne* provide the positions of the apogees of the Sun and the planets beginning in 1320, as well as the mean motions of the luminaries and the planets. We are also given tables for mean syzygies and the equations of the Sun, the Moon, and the planets. A salient feature of this set is the massive introduction of double-argument tables, a very convenient format rarely found in the previous literature in Latin, although not infrequent in earlier tables compiled by Arabic astronomers.

The other set has been named the Tables of 1322 by John of Lignères and has only recently been identified as such.[5] In contrast to the *Tabule magne*, this set is extant in a great number of manuscripts, a fact that facilitates the identification of the common tables belonging to it. As explained in Chabás 2019, thirteen manuscripts from the fourteenth and fifteenth centuries copied in a wide variety of places and milieus were used to define this set. It consists of thirty-two tables that make an altogether consistent set covering most problems in mathematical astronomy. The tables can be grouped in seven categories: trigonometry and spherical astronomy (sine, shadow, solar declination, ascensional difference, right and oblique ascensions), equation of time, latitudes (Moon and planets), planetary motions (unequal motion, retrogradation, stations, and phases), syzygies (mean syzygies, corrections, and velocities), parallax, and eclipses (solar and lunar eclipses, eclipsed parts, corrections, and proportions). Not all thirty-two tables are found together in any manuscript, but only a few are missing in each case. The order of presentation is mostly the same, and intercalated tables not belonging to the set are seldom found, which means that in the manuscripts we have reviewed, not much 'noise' surrounds the set of tables now identified. In addition, the set begins with the sine table in almost all manuscripts and it is often preceded by a general title indicating 'Here begin the tables of John of Lignères', as is the case shown in Figure 1.

The Tables of 1322 by John of Lignères have been associated with the canons beginning *Cuiuslibet* and *Priores*, which are mentioned above. It is an interesting exercise to examine the kind of correspondence between the tables mentioned in the two canons and those actually pertaining to this set.[6]

3 An edition of both texts can be found in Marie-Madeleine Saby, *Les Canons de Jean de Lignères sur les tables astronomiques de 1321* (unpublished thesis: École Nationale des Chartes, Paris, 1987); a summary appeared as 'Les canons de Jean de Lignères sur les tables astronomiques de 1321', *École Nationale des Chartes: Positions des thèses* (1987), pp. 183–90.

4 For an overview of this set and the manuscripts containing them, see José Chabás, *Computational Astronomy in the Middle Ages* (Madrid: Consejo Superior de Investigaciones Científicas, Madrid, 2019), pp. 199–206.

5 For the identification and description of the Tables of 1322 by John of Lignères, see Chabás, *Computational Astronomy*, pp. 175–98.

6 For a list of the tables mentioned in the canons, see Saby, 'Les Canons', pp. 418–29.

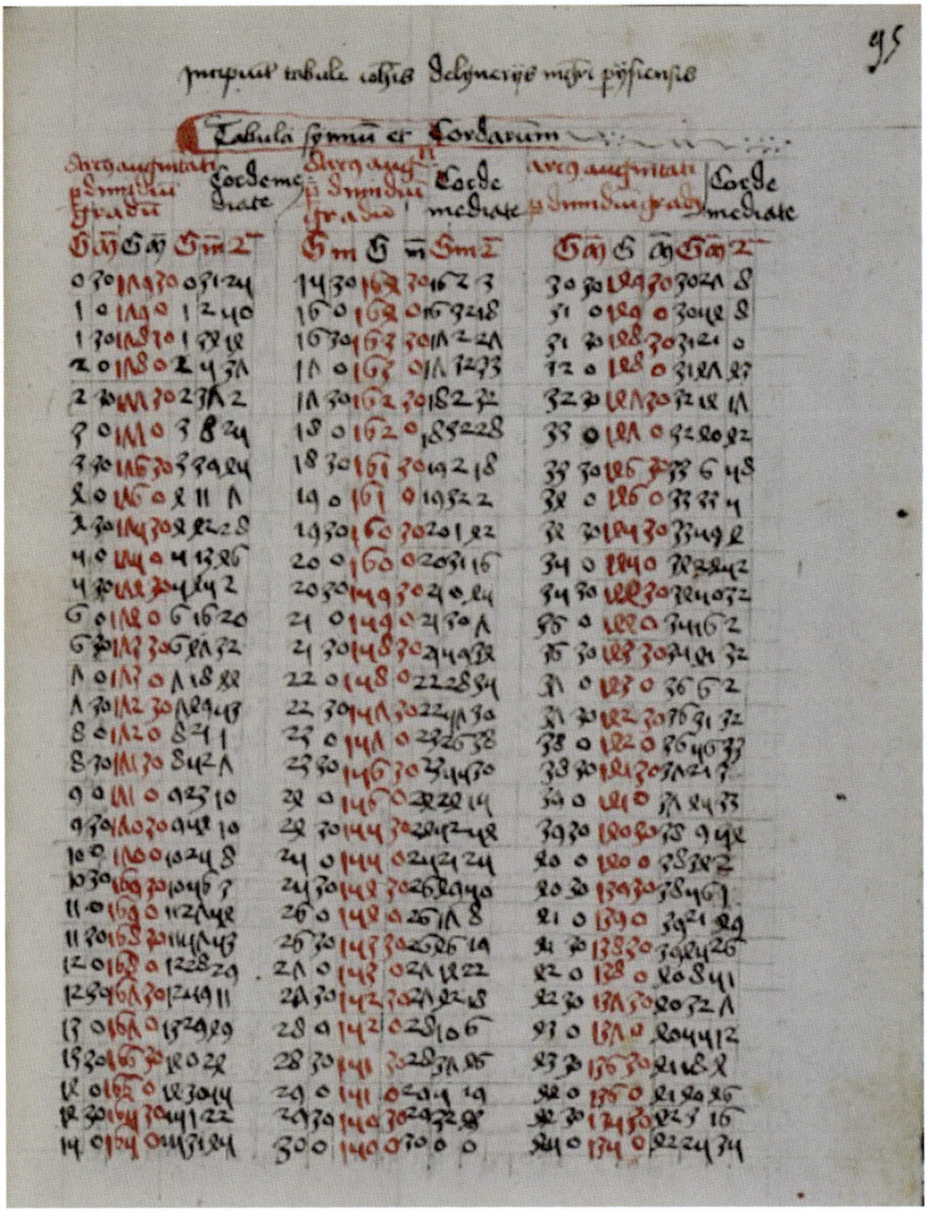

Figure 1. First half of the sine table in Vatican, Biblioteca Apostolica Vaticana, Pal. lat. 1412, 95[r]. Source digi.ub.uni-heidelberg.de/de/bpd/index.html.

The text beginning '*Cuiuslibet arcus propositi*' refers to six tables. The first five deal with trigonometry and spherical astronomy: sine, shadow, solar declination, ascensional difference, and right ascension. These are generally found as the first five tables in the set compiled by John of Lignères for 1322. The sixth table mentioned, for the equation of the astrological houses, is not among the tables identified in this set. This can be directly deduced from John of Lignères' own words in canon 37, where he suggests that [the practitioner] use one such table, 'if it exists for the considered region'; but he does not claim to have one for his own latitude.

The other text, beginning *Priores astrologi motus*, mentions many more tables. Canons 1–5 refer to several chronology tables for the various eras, aimed at reducing any period of time into a number of days expressed in sexagesimal form. These tables are not part of John of Lignères' Tables of 1322, but they were later integrated in the standard version of the Parisian Alfonsine Tables. Likewise, other canons allude to tables in collected years, months, days, and hours, for the mean motions of the Sun, the Moon, and the planets, as well as for the eighth sphere and the apogees. No such tables appear in this set for 1322. In contrast, the standard version of the Parisian Alfonsine Tables displays tables for mean motions, although not presented in collected years, and so on, but as sixty successive multiples of a basic parameter in degrees per day. An analogous situation occurs with the equations of the luminaries and the planets, mentioned repeatedly in the text, but with no corresponding tables in the set John of Lignères compiled for 1322. From canon 19 onwards, other tables are mentioned, most of them belonging to the set for 1322: solar declination, planetary latitudes, elongation, mean syzygies, solar and lunar velocities, equation of time, parallax, *attatium*, oblique ascension, proportions, solar and lunar eclipses, planetary phases, and unequal motion of the planets. Yet in the last two chapters, canons 45 and 46, two other tables are mentioned, one for the fixed stars and another for the excess of revolution of the years, which do not belong to the set compiled by John of Lignères for 1322.

In sum, many of the tables mentioned in the two canons by John of Lignères are not part of his set for 1322 and not all tables in the set are cited in the canons. It is therefore clear that, although similar astronomical problems are addressed, the canons beginning *Cuiuslibet* and *Priores* do not parallel this set of tables. Moreover, the two texts by John of Lignères, whether considered individually or jointly, do not fully describe or explain his set of tables for 1322. This situation is in sharp contrast with that of the other tabular work by John of Lignères, the *Tabule magne*, where the canons and the tables match nicely.

2. Methodology for the edition of astronomical tables

The identification of the set of thirty-two tables is only the first step in the process of edition. It turns out that the Tables of 1322 by John of Lignères cannot be defined only by means of their references in his canons. Another aspect to be considered in the edition of a set of tables is its wide dissemination throughout Europe in the fourteenth and fifteenth centuries and the present distribution in libraries. As far as we know, the set of tables by John of Lignères is partially or totally extant in at least forty-six manuscripts, indicating the importance Alfonsine astronomers attached to them. For the edition of this broadly diffused set of astronomical tables, we have only taken into account those manuscripts containing the most complete set.

As mentioned above, the tables and their associated canons are presented in a variety of ways in the manuscripts. In Paris, BnF, lat. 7286C, the canons surround the tables, in the wide margins of the manuscript. In London, BL, Egerton 889, the canons appear just after the tables. However, in most cases the canons precede the tables. Moreover, the tables can be separated from the canons by several other texts (see Basel, Universitätsbibliothek, F

Figure 2. Basel F II 7, 62^{r}. Reproduced with permission of the Universitätsbibliothek Basel.

Figure 3. Basel F II 7, 77^{v}. Reproduced with permission of the Universitätsbibliothek Basel.

II 7; Erfurt, Universitätsbibliothek CA 2° 377; Oxford, BL, Can. Misc. 27, among others).[7] Some manuscripts have only canons of John of Lignères without any tables. This is the case of Erfurt, CA 4° 366, in spite of what can be read in the explicit, 'Expliciunt canones magistri Johannis de Lineriis super tabulas eiusdem similiter et Alfonsi'. In Erfurt, CA 4° 376, only the canons beginning *Cuiuslibet arcus propositi…* are found. In contrast, BnF, lat. 7281 only contains texts; quite often manuscripts have only tables, for example, Vatican, BAV, Pal. lat. 1374.

Whereas in the Arabic astronomical tradition there is a close connection between tables and canons, the lose connection in the manuscripts of canons and tables for 1322 is in itself an issue. However, internal cross-references from text to tables are frequent in the 1322 canons; as evidence, canons 1, 2, 3, 4, 5, 6, and 10 of the first part of the treatise, *Cuiuslibet arcus propositi…*, call the *Tabula cordarum mediatarum* (also named *Tabula sinus*). To highlight the unity of the tables of 1322, many manuscripts include an incipit. In very few, we find both an incipit and an explicit, thus providing a frame to the tables.

7 In BnF, 7286C, the canons *Cuiuslibet arcus propositi…* (10^{r}–23^{r}) and *Priores astrologi…* (23^{v}–41^{r}) do not match at all with the tables they are surrounding; the Tables of 1322 are found on ff. 24^{v}–28^{v}, 48^{r}–52^{r}, 53^{v}–55^{r}, and 56^{r}, mixed with other tables, mainly by John of Lignères. In Basel, F II 7, the canons are found on ff. 38^{r}–57^{v} and the tables on ff. 62^{r}–77^{v}. Erfurt, CA 2° 377 also contains the canons (22^{r}–35^{r}) and the tables (41^{v}–47^{r}). See also Erfurt, CA 4° 366, 23^{v}.

Three out of thirty-six have both: Wolfenbüttel, Herzog August Bibliothek, Cod. Guelf. 36.21 Aug. (2401); Basel, F II 7 (see Figs 2 and 3); and BAV, Pal. lat. 1374.[8]

The numerous manuscripts also present some variability in the composition of the set of tables; in most cases the set begins with trigonometrical tables, but in the BAV, Pal. lat. 1374, the tables on the motion of the planets are placed at the very beginning. Moreover, the order of the tables within each set may differ; for example, in Erfurt, CA 2° 377, the copyist managed to include 11 tables on a single folio (f. 45^r). In a few manuscripts, some other tables are found mixed in with the tables of John of Lignères.[9] For the edition of the tables for 1322 we selected several among the forty-six manuscripts at our disposal, according to the following criteria.

The issue of milieu in the development of Alfonsine astronomy outside Castile played a central role. It can be characterized by the remarkable unity of space (Paris), time (the 1320s), and actors (John Vimond, John of Murs, John of Lignères, and John of Saxony). As for regional unity, Paris had a crucial role in the dissemination of Alfonsine astronomy through manuscripts copied or transferred to such important places in the history of medieval astronomy as Oxford, Erfurt, Cracow, and Prague. The time element is a part of the design of the milieu, and the dates of production of the copies are a central point: copies of the tables of John of Lignères were written from shortly after their compilation, as is the case of Erfurt, CA 2° 377, to the second half of the fifteenth century.[10] Additionally, most dated copies were made during the first half of the fifteenth century. The third element, describing the milieu, refers to the actors or the characters. The best way to address this point is to take into account the presence of other works of the above-mentioned astronomers or other Alfonsine works, that is to say, the inner coherence of the manuscript with Alfonsine astronomy.

Besides the criteria dealing with the milieu, several other features provide the editors with easy ways to select the manuscripts for the critical edition. The quality of the copy must not be ignored, that is, layout, paratext, annotations, corrections, incipits, or explicits. The clarity and readability of a manuscript are crucial for a reliable understanding of the numbers listed. The annotations, commentaries, or corrections in the margins relate to the use of the tables and they endow the editors with variants that either make sense — or that make no sense at all.

Finally, one critical factor is also to be taken into account in the selection process: the number of the tables of John of Lignères found in each manuscript. A review of the tables in fourteen manuscripts provided additional information about the manuscripts only lacking a few tables.[11] There are three manuscripts where only one table is missing,

---

8 The following manuscripts have an incipit: BAV, Pal. lat. 1412, 95^r; Paris, BnF, lat. 7295A; Paris, BnF, lat. 7282; and Madrid, Biblioteca Nacional, 10002. An example of a manuscript with both an incipit and an explicit is BAV, Pal. lat. 1374: *In nomine Jesu Christi. Incipiunt tabule magistri Johannis de Lineriis. Tabule medii motus lune…* (26^r) and *Expliciunt tabule reverendissimi magistri Johannis de Lineriis et finite Prage anno domini 1407 undecimo kalendas octobris* (46^v).

9 In BAV, Pal. lat. 1412, the Tables of 1322 are found between f. 95^r and f. 120^v, together with other tables taken from the *Tabule magne* (103^v–108^v).

10 The latest manuscripts containing the Tables of 1322 are BAV, Pal. lat. 1354, with only one table (74^v–75^v) and Rome, Biblioteca Casanatense, 653, with only a few tables (34^v–37^r, 40^r–64^r), copied in the late-fifteenth century.

11 Chabás, *Computational Astronomy*, 176–77, considered thirteen manuscripts; we have added Vatican, BAV, Pal. lat. 1374.

different in each case, one particular table is also missing in seven manuscripts (Table 20 for the velocities of the luminaries), and the number of missing tables in the fourteen manuscripts examined ranges from one to five[12]

3. Five manuscripts confronted with the table-editing criteria

3.1. Basel, Universitätsbibliothek, F II 7 (62^r–77^v): A base manuscript for the edition of the Tables of 1322

Date and chronology: This is a dated manuscript (1432, see f. 57^v)[13]. On the same folio, we are told that the copyist was Henricus Amici, a Doctor of Medicine in Montpellier. Amici collected or wrote several manuscripts, which he probably bequeathed to the Cartusian monastery of the city, including Basel, F II 15, that deals with astronomy and contains an astrological treatise by Peter d'Ailly (1350-1420).

Layout, legibility, paratext, and composition: The composition of Basel, F II 7, seems to be of a great coherence, for it contains canons by John of Saxony; the complete canons of John of Lignères for the *Primi mobilis* and thirty canons (out of forty-six) for the *Priores astrologi* as well as twenty-nine tables out of thirty-two; the table and canon by Nicholaus de Heybech of Erfurt;[14] and some other tables and canons attributed to John of Lignères. The layout also shows great coherence, as only one hand copied the entire manuscript and each work (except at the end of the manuscript, ff. 78^r–85^r) is bound by an incipit and an explicit. Moreover, the tables of John of Lignères seem to have been carefully corrected.

3.2. Cracow, BJ, 551 (74^r–95^r): An alternative base manuscript for the edition

This manuscript is described extensively in a modern catalog.[15]

Date and chronology: BJ 551 was probably copied for the most part in 1388. This date is mentioned several times in the manuscript (see folio 127^r, where 1388 is mentioned twice).

Place of production: According to the catalogue, the manuscript was copied in Prague.[16] On folio 32^v, the table for geographical coordinates begins with Prague.[17] About thirty years later, in 1420, this codex was in Cracow; in 1493 it was owned by a master of the University of Cracow, Johannes de Środa. It then belonged to Petrus Tomocki, Bishop of Cracow (d. 1535), who bequeathed it back to the college of theologians of the university.

12 A detailed list of the tables in each of the manuscripts can be found in José Chabás and Marie-Madeleine Saby, *The Tables of 1322 by John of Lignères: an edition with commentary,* forthcoming.

13 Albert Bruckner and others, *Katalog der datierten Handschriften in der Schweiz in lateinischer Schrift vom Anfang des Mittelalters bis 1550*, 6 vols (Dietikon-Zurich: Urs Graf, 1977–91), I (1977), 174n483.

14 See José Chabás and Bernard R. Goldstein, 'Nicolaus de Heybech and His Table for Finding True Syzygy', *Historia mathematica*, 19 (1992), 265–89.

15 Zofia Włodek and others, *Catalogus codicum manuscriptorum Medii Aevi Latinorum qui in Bibliotheca Jagellonica Cracoviae asservantur*, 11+ vols (Cracow: Ossolineum, 1980-), III (1984), 333–45.

16 Prague is mentioned repeatedly (ff. 5^r, 7^v, and 12^v) well as Paris, Magdeburg, Erfurt, Vienna, and Cracow.

17 In the bottom margin, we can also read additions (in another later hand) about the distance of several cities from Toledo and Paris (e.g. *Herfordia distat a Toleto*).

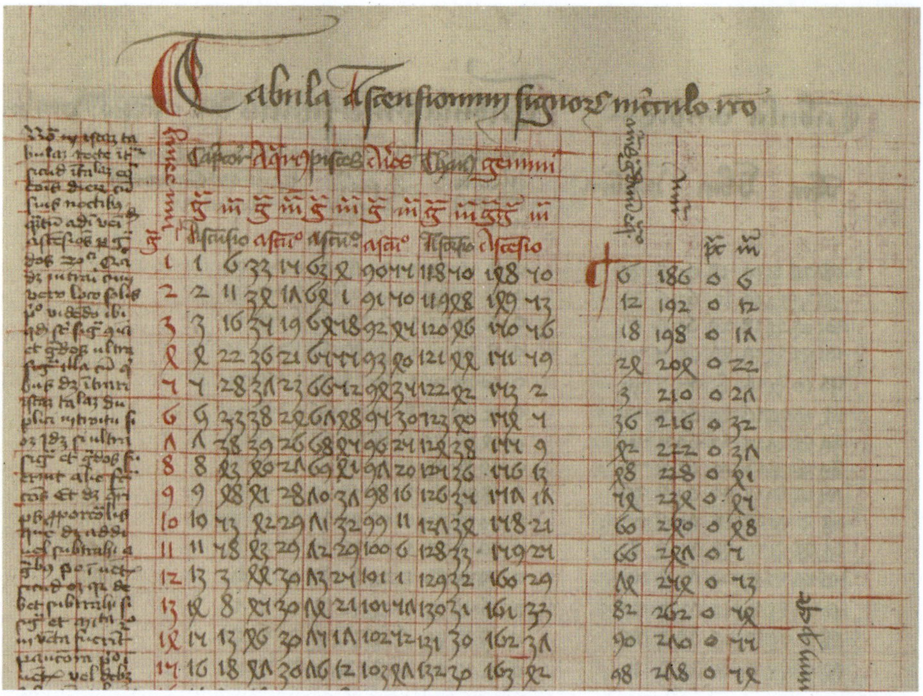

Figure 4. Cracow, BJ, 551, 76^{v}. Source jbc.bj.uj.edu.pl//dlibra/.

To sum up in terms of the geographical elements, the area of production and origin of the manuscript could be Prague, and then it reappears about three decades later in Cracow.

Layout, legibility, paratext, and composition: The contents of MS 551 are quite heterogeneous at first sight. Even if we can say that there is no accident in the composition of medieval manuscripts, coherence is not so evident in this case. In addition to various works on astronomy by John of Saxony and John of Lignères, there are others by Ptolemy (ff. 120^{v}-121^{r}), annotated medical works by Guillelmus Anglicus, *De urina non visa*, and treatises on astronomical instruments by Johannes Eligerus de Gunderslauen.

John of Lignères' works for 1322 are fully represented here: both sets of canons (one completed with forty-four canons, compared to thirty-eight out of forty-six canons in the other) and the set of tables (without any incipit or explicit). It may be worthwhile to point out that the set of tables is almost complete; it contains thirty-one tables out of thirty-two and hence has been used extensively for comparisons. The tables of John of Lignères are quite clearly copied and have no corrections, but some marginal notes in another more cursive hand are found. See Figure 4 for examples of the use of the tables or explanations about them.

3.3. Paris, BnF, lat. 7286C (24ᵛ–55ʳ): A manuscript for variants

This manuscript is very succinctly described in the eighteenth-century catalogue of the Royal Library.[18]

Date and chronology: This criterion does not hold for this manuscript because we have no information concerning its date. The handwriting can be dated to the late-fourteenth century.

Place of production: The manuscript could have been composed in Paris or in northern France, but no additional information is available.

Layout, legibility, paratext, and composition: The manuscript begins with a uniquely preserved work: John Vimond's tables with 1320 as the epoch.[19] Its composition and contents are also a matter of great interest. Besides Vimond's tables, most of the manuscript deals with works by John of Lignères, including his tables for 1322 and the canons *Cuiuslibet* and *Priores*.

At first glance, the layout is significant. The tables are surrounded by canons. Upon closer examination, it is clear that the tables do not match the canons on the same page. This fictious parallelism is an artifice due to the copyist. For example, he began copying the canons for the *Primi mobilis* in front of the tables for mean motions, instead of beginning on the page for the sine table. The result is a total disagreement, to the point that the canons for eclipses end on folio 41ʳ, but the tables continue until folio 55ʳ.

In a closer study of the tables, we notice that the manuscript contains twenty-nine tables of the set, which can be considered as a good 'score', given that the total number of tables in the set under examination is thirty-two. Many mistakes were also made in copying the tables; for example, in the table for ascensional difference (26ʳ), the scribe shifted a line in the column for the seconds. Finally, BnF, 7286C has never been annotated, corrected, or maybe even used by any late medieval scholar.

In sum, manuscript Paris, BnF, 7286C offers a quite valuable testimony regarding the works of 1322 of John of Lignères and the tables of John Vimond (*c.* 1320), two of the earliest sets of tables compiled in Paris by Alfonsine astronomers. While the layout suggests a correspondence between tables and canons, it is not actually the case. The tables were first copied by a careless scribe; then he or someone else copied the canons, leaving space for figures that were never drawn.

3.4. Vatican, BAV, Pal. lat. 1374 (26ʳ–46ᵛ): A manuscript for variants

This manuscript is used for collating purposes in the edition of the set of tables. A full description can be found in Schuba's catalogue.[20]

18 Guillaume de Villefroy and others, *Catalogus codicum manuscriptorum Bibliothecae Regiae*, 4 vols (Paris: Typographia regia, 1739–44), IV/3 (1744), p. 355 (the text only reads, '*Ibi continentur tabulae astronomicae: accedunt canones*').

19 José Chabás and Bernard R. Goldstein, 'John Vimond and the Alfonsine Trepidation Model', *Journal for the History of Astronomy*, 34 (2003), 163–70; Chabás and Goldstein, 'Early Alfonsine Astronomy in Paris: The Tables of John Vimond (1320)', *Suhayl*, 4 (2004), 207–94.

20 Ludwig Schuba, *Die Quadriviums-Handschriften der Codices Palatini Latini in der Vatikanischen Bibliothek* (Wiesbaden: Ludwig Reichert Verlag, 1992), 86–88.

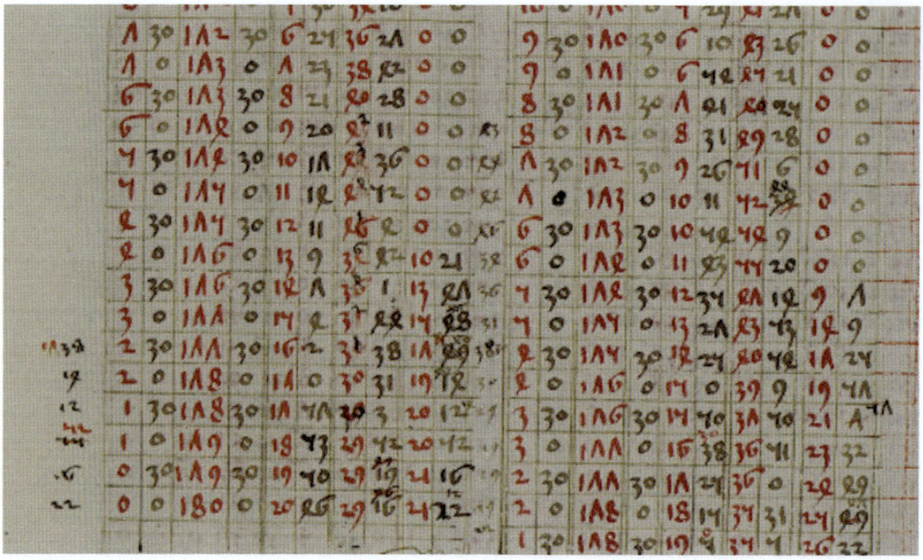

Figure 5. BAV, Pal. lat. 1374, 32r. Source digi.ub.uni-heidelberg.de/de/bpd/index.html.

Date and chronology: The date (1407) is expressly indicated twice in the first two major works: at the end a set of the Parisian Alfonsine Tables and at the end of John of Lignères' tables. Moreover, we are also given the name of the scribe, Reinhard von Nüremberg, Bachelor of Arts, who was perhaps studying to obtain a magister degree.

Place of production: It was copied in Prague and the copyist was from Nuremberg, according to his name.

Layout, legibility, paratext, and composition: Manuscript BAV 1374 only consists of tables, except for a short canon for the tables of Nicholaus Mülhus, a layman working in Zittau, modern-day Saxony. The corpus is quite original, for it includes a set of the Parisian Alfonsine Tables, tables of John of Lignères (1322), John of Genoa (1332), the above-mentioned Nicholaus Mülhus, and many other miscellaneous tables. As was the case of Basel, F II 7, an incipit and an explicit put a frame around the set of tables. The good quality layout of the tables, with rubrication and interchanging red and black between consecutive columns or titles, is also remarkable. Finally, the correction of the entries, as displayed in the table for lunar eclipses (see Fig. 5) are justified; the manuscript was used or collated by a much more attentive reader, who made corrections himself. The corrections seem to match mostly with the lessons in other manuscripts. Another example of the relevance of the corrections and annotations is the marginal note attributing the tables on folios 47–51v to John of Genoa.

3.5. Erfurt. Universitätsbibliothek, CA 2° 377 (41v–47r): A manuscript for variants

A detailed description of this manuscript is found in the library's nineteenth-century catalogue; a recent codicological study, in particular about the hands of the manuscript,

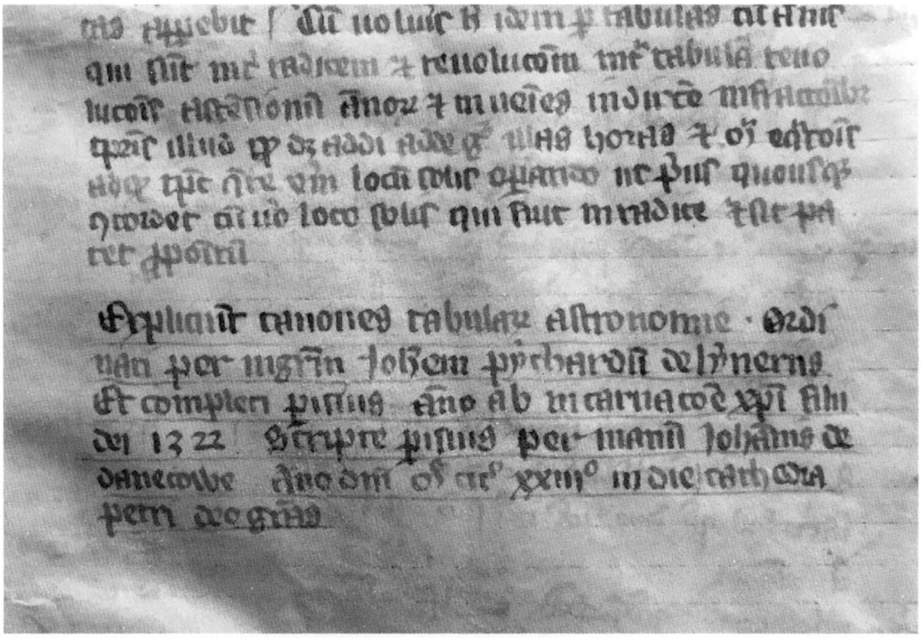

Figure 6. Erfurt, CA 2° 377, 35^{r}. Photo: Marie-Madeleine Saby.

is also available.[21] Erfurt, CA 2° 377 was selected as the base manuscript for the edition of the canons of John of Lignères' *Cuiuslibet* and *Priores*.[22]

Date and chronology: This composite manuscript consists of three parts bound together with a table of contents added in the fifteenth century. The second part deals mainly with Alfonsine astronomy and can be separated in two sections. The first consists of a text by John Vimond on an instrument he called *Planicelium* and the canons of 1322 by John of Lignères. The second is in a more cursive hand and addresses mathematical aspects of astronomy such as John of Lignères' *Algorismus* (38^{v}–41^{r}). This part was copied by 1323; the other works were copied between 1321 and 1324.

Place of production: No place is mentioned, but the dates, the contents, and the names of the scholars (John Vimond, John of Lignères, John of Murs, and John of Saxony, the copyist of John of Lignères' canons) suggest that the manuscript was produced in Paris.

Layout, legibility, paratext, and composition: The coherence of the central part of the manuscript is amazing: John Vimond, John of Lignères, and John of Murs are copied, at least for one part, by John of Saxony. Here, the four main astronomers associated with Alfonsine astronomy in Paris in the first decades of the fourteenth century are gathered together. Moreover, one copied his master's work, John of Lignères (see Fig. 6). The copyist of the second cursive part of the fourteenth century of this manuscript might also be

21 Wilhelm Schum, *Beschreibendes Verzechnis der Amplonianischen Handschriften-Sammlung zu Erfurt* (Berlin: Weidmannsche Buchhandlung, 1887), 262–64; see also Matthieu Husson and Marie-Madeleine Saby, 'Le Manuscrit Erfurt F.377 et l'astronomie alphonsine', *Micrologus* 27 (2019), 205–33.

22 See Saby, 'Les Canons'.

identified with John of Saxony, a scholar developing his own reflection on mathematical issues such as sexagesimal computation, among other questions.[23]

Figure 6 shows the explicit of John of Lignères's *Priores*, autograph by John of Saxony. It reads: 'Here end the canons to the astronomical tables by John of Lignères, from Picardie, completed in Paris in the year 1322 from Incarnation of Christ, the son of God. They are written in Paris by John de Danecowe [John of Saxony] in the year of our Lord 1323 on the day of the Chair of Peter [February 22]'.

In the manuscript, the astronomical tables (41^{v}–47^{r}) are not connected with the canons of John of Lignères and were copied in the cursive part of the fourteenth-century section of the manuscript. There is no incipit nor explicit. In the table of contents written in the fifteenth century, John of Lignères' set, with only twenty-seven of the thirty-two tables, is simply entitled *Item Tabule astronomice*. On each page, the presentation of the tables is very tight and the entries are often difficult to read. The tables are partly rubricated and the margins are very irregular. On folio 42, two tables are displayed vertically and a third one horizontally. The parchment is carelessly ruled with ink and the parchment is of very low quality. This is apparently a student copy. Nevertheless, it is a forefront manuscript for the history of Alfonsine astronomy and the set of tables it contains has to be considered as a very adequate witness for the edition.

4. Conclusion

The Tables of 1322 by John of Lignères, as defined above, are a highly coherent set, for they address the wide range of issues on mathematical astronomy faced by medieval scholars. This set saw enormous success as a complete repertoire to solve problems on astronomy by means of tables, and quickly spread throughout Europe. We note that, as was the case for the canons *Cuiuslibet* and *Priores*, it includes no tables directly associated with astrological matters. As indicated above, the Tables of 1322 by John of Lignères depend strongly on the Toledan Tables, and although he computed or adapted a few tables to the latitude of Paris, most of the tables he included were borrowed from the set developed on the Iberian Peninsula by Arabic astronomers more than two centuries before. When compared with later sets of astronomical tables, it turns out that many of the tables assembled by John of Lignères made their way into the standard version of the Parisian Alfonsine Tables, as first published in 1483 in Venice, which would indicate that the Tables of 1322 by John of Lignères presented here represent a transition from the Toledan Tables to the Parisian Alfonsine Tables.

The criteria described here for selecting the copies to be used in the edition of this table set focus both on external factors, such as the date of the manuscript copy and its place of production, and on internal factors, such as legibility, composition, and layout of the tables themselves. We are convinced that, together with the comments on each of the tables, they will provide a thorough and useful edition of the Tables of 1322 by John of Lignères, a work that has been waiting too long for scholarly attention.

---

23 Among the works copied in the second part of the fourteenth-century portion of this manuscript, in a more cursive hand, are some unusual texts: the *Prognosticationes* (36^{v}) attributed to John of Murs (the only known occurrence); the *De mare fusili* (36^{r}–36^{v}, also copied in Bruges, Public Library, 523); and the *Arbor Boecii* (35^{v}–36^{r}), both by John of Murs.

Manuscript sources

Basel, Universitätsbibliothek, F II 7
Basel, Universitätsbibliothek, F II 15
Bruges, Public Library, 523
Cracow, Bibliotheka Jagiellońska, 551
Erfurt, Universitätsbibliothek, CA 2° 377
Erfurt, Universitätsbibliothek, CA 4° 376
London, British Library, Egerton 889
Oxford, Bodleian Library, Can. Misc. 27
Paris, Bibliothèque nationale de France, lat. 7281
Paris, Bibliothèque nationale de France, lat. 7286C
Rome, Biblioteca Casanatense, 653
Vatican, Biblioteca Apostolica Vaticana, Pal. lat. 1354
Vatican, Biblioteca Apostolica Vaticana, Pal. lat. 1374
Vatican, Biblioteca Apostolica Vaticana, Pal. lat. 1412
Wolfenbüttel, Herzog August Bibliothek, Cod. Guelf. 36.21 Aug. (2401)

Bibliography

Bruckner, Albert and others, *Katalog der datierten Handschriften in der Schweiz in lateinischer Schrift vom Anfang des Mittelalters bis 1550*, 6 vols (Dietikon-Zurich: Urs Graf, 1977–91).

Chabás, José, *Computational Astronomy in the Middle Ages* (Madrid: Consejo Superior de Investigaciones Científicas, 2019).

———, and Goldstein, Bernard R., *The Alfonsine Tables of Toledo* (Dordrecht: Kluwer, 2003).

———, and ———, 'John Vimond and the Alfonsine Trepidation Model', *Journal for the History of Astronomy*, 34 (2003), 163–70.

———, and ———, 'Early Alfonsine Astronomy in Paris: The Tables of John Vimond (1320)', *Suhayl*, 4 (2004), 207–94.

Husson, Matthieu and Saby, Marie-Madeleine, 'Le Manuscrit Erfurt F.377 et l'astronomie alphonsine', in *Micrologus*, 27 (2019), 205–33.

Poulle, Emmanuel, 'John of Lignères', in *Dictionary of Scientific Biography*, ed. Charles Gillispie, 16 vols (New York: Charles Scribner's Sons, 1970–80), VII (1973), 122–28.

Saby, Marie-Madeleine, 'Les Canons de Jean de Lignères sur les tables astronomiques de 1321' (unpublished thesis, École Nationale des Chartes, Paris, 1987); abstract in *Positions des theses*, 1987, 183–90.

Schuba, Ludwig, *Die Quadriviums-Handschriften der Codices Palatini Latini in der Vatikanischen Bibliothek* (Wiesbaden: Ludwig Reichert Verlag, 1992).

Schum, Wilhelm, *Beschreibendes Verzechnis der Amplonianischen Handschriften-Sammlung zu Erfurt* (Berlin: Weidmannsche Buchhandlung, 1887).

Villefroy, Guillaume de, and others, *Catalogus codicum manuscriptorum Bibliothecae Regiae*, 4 vols (Paris: Typographia regia, 1739–44), IV/3 (1744).

Włodek, Zofia, and others, *Catalogus codicum manuscriptorum Medii Aevi Latinorum qui in Bibliotheca Jagellonica Cracoviae asservantur*, 11+ vols (Cracow: Ossolineum, 1980-).

ALENA HADRAVOVÁ,
PETR HADRAVA

John of Lignères's *Quia ad inveniendum loca planetarum*: An Edition and Translation

John of Lignères (Iohannes de Lineriis, Jean de Lignères; first half of the fourteenth century) was one of the first authors to have rendered the Alfonsine Tables and the accompanying canons into Latin shortly after the tables, which were compiled in Spain, became available in Paris. Together with John of Murs (Iohannes de Muris, Jean des Murs),[1] John Vimond (Iohannes Vimundi, Jean Vimond),[2] and John of Saxony (Iohannes de Saxonia, Jean de Saxe), he is one of the four most important authors who made the Latin adaptations of these tables and canons — as well as other related works of Alfonsine astronomy — available

* This work was supported by the ERC project ALFA: Shaping a European scientific scene, Alfonsine astronomy, CoG 723085, PI Matthieu Husson. It has also been supported by Projects RVO: 68378114 (Inst. Contemporary Hist.) and RVO: 67985815 (Astr. Inst.). We are extremely grateful to Marie-Madeleine Saby (Grenoble Alpes University) for her careful reviewing of the Latin edition, and Matthieu Husson (l'Observatoire de Paris) and Richard L. Kremer (Dartmouth College, New Hampshire) for their comments. Kremer also improved our English translation of the canons. Laure Miolo shared her new findings on Simon Bredon with us and provided photographs of Oxford, Hertford College MS 4. We are also indebted to José Chabás from whom we learned about copies of *Quia ad inveniendum* in the Vatican Library (Ott. lat. 1826 and Pal. lat. 1403). Anna Sobańska provided us with photographs of Cracow, BJ MS 548, and Philipp Nothaft supplied the photographs of MS Cusanus 212. Last but not least, we would like to thank Alexandre Tur and the staff at the Bibliothèque nationale de France for assisting our access to manuscripts in their library.

1 Beatriz Porres de Mateo and José Chabás, 'John of Murs's *Tabulae permanentes* for Finding True Syzygies', *Journal for the History of Astronomy*, 32 (2001), pp. 63–72. See Richard L. Kremer 'Cracking the *Tabulae permanentes* with Exploratory Data Analysis', in *Editing and Analyzing Numerical Tables: Towards a Digital Information System for the History of Astral Sciences*, ed. Matthieu Husson, Clemency Montelle and Benno van Dalen (Turnhout: Brepols, 2022), pp. 363-422.

2 A unique copy of the tables by John Vimond, designed for students at the University of Paris and elsewhere, is extant in Paris, BnF lat. 7286C (ff. 1^{r}–8^{v}), albeit without canons. The tables are calculated for the epoch of 1320. Vimond is also mentioned in Vatican, BAV Ott. lat. 1826, f. 153^{r}; the pertinent passage begins on f. 148ra ('Canon tabule sequentis, que intitulatur tabula motus diversi Solis et Lune in una hora'); see José Chabás and Bernard R. Goldstein, *The Alfonsine Tables of Toledo* (Dordrecht: Kluwer, 2003), pp. 267 and 269. For more information on John Vimond, see also José Chabás and Bernard R. Goldstein, 'Early Alfonsine Astronomy in Paris: The Tables of John Vimond (1320)', *Suhayl*, 4 (2004), 207–94.

Alena Hadravová • Institute of Contemporary History of the Czech Academy of Sciences

Petr Hadrava • Astronomical Institute of the Czech Academy of Sciences

Alfonsine Astronomy: The Written Record, ed. by Richard L. Kremer, Matthieu Husson, José Chabás, Turnhout, 2022 (*Alfonsine Astronomy: Studies and Sources*, 1), pp. 257-302

10.1484/M.ALFA.5.124929

in medieval Europe.[3] After the 1320s, these works began to spread, first to England, Italy,[4] Central Europe (Germany, Bohemia, Poland, Austria),[5] and then to other places, signalling the beginning of a long period marked by the influence and dominant position of Alfonsine astronomy in the whole of Europe. The tables were repeatedly copied and reused in a number of other works, adapted and adjusted to local meridians, while the canons were rewritten and reworked in multiple new editions to teach astronomers how to use the tables. The tables and canons were also published in incunables and early modern prints. The influence of Alfonsine astronomy only began to wane in the mid-sixteenth century.[6] The long period of dominance of Alfonsine astronomy in Europe is what makes the works of these first Parisian astronomers of this school so important.

Several recent publications focus on the works by John of Lignères[7], including an edition of his canons (*Cuiuslibet arcus propositi sinum rectum invenire*) and a forthcoming edition, with commentary, of his Tables of 1322.[8]

The *Quia ad inveniendum*[9] canons were probably created between 1322 and 1327.[10] Unfortunately, none of the copies consulted while preparing this edition of the canons yielded any explicit information that could lead to a more accurate dating or shed light on the circumstances in which the work was written. It is obvious that these canons were much less widespread than the canons written by John of Saxony in 1327 (*Tempus est mensura motus*); in a way, John of Saxony's canons marked the culmination of the first phase in the existence of the Parisian Alfonsine Tables in Latin and are extant in scores of manuscripts and several incunable editions (*editio princeps*: Ratdolt 1483).[11] The canons

3 Other authors, such as John of Montfort or John of Genoa, could also be said to belong to this group.

4 See, e.g., José Chabás and Bernard R. Goldstein, *The Astronomical Tables of Giovanni Bianchini* (Leiden: Brill, 2009).

5 See Beatriz Porres de Mateo, 'Les Tables astronomiques de Jean de Gmunden: Édition et étude comparative', unpublished PhD thesis, Paris, École Pratique des Hautes Études, 2003.

6 See José Chabás and Bernard R. Goldstein, *A Survey of European Astronomical Tables in the Late Middle Ages* (Leiden: Brill, 2012).

7 Emmanuel Poulle, 'John of Lignères', in *The Dictionary of Scientific Biography* (New York: Charles Scribner's Sons, 1970–1980), VII (1973), pp. 122–28, and Emmanuel Poulle, 'The Alfonsine Tables and Alfonso X of Castille', *Journal for the History of Astronomy*, 19 (1988), 97–113.

8 Marie-Madeleine Saby, 'Les Canons de Jean de Lignères sur les tables astronomiques de 1321. Édition critique, traduction et étude', unpublished thesis, Paris, École Nationale des Chartes, 1987; summary in *Positions des thèses soutenues par les élèves de la promotion 1987 pour obtenir le diplôme d'archiviste-paléographe* (Paris, École nationale de chartes 1987), pp. 183–90. José Chabás and Marie-Madeleine Saby's edition of the Tables of 1322 is in press. Other members of the ALFA team are currently preparing related editions: Laure Miolo of the *Opus astronomicum* by John of Genoa, likely a pupil of John of Lignères; and Matthieu Husson of John of Lignères's canons to the *Tabule magne* (*c.* 1325, inc. *Multiplicis philosophie variis radiis*).

9 Lynn Thorndike and Pearl Kibre, *A Catalogue of Incipits of Medieval Scientific Writings in Latin*, rev. and augmented ed. (Cambridge: Mediaeval Academy of America, 1963), col. 1213.

10 See Poulle, 'John of Lignères', p. 125: 'The date of these very succinct new canons (i.e. *Quia ad inveniendum*) cannot be determined from the text, but it is certainly later than that of the canons *Priores astrologi* (1322) and may perhaps be earlier than 1327, when John of Saxony produced a new version of the canons of the Alfonsine Tables'. John D. North, *Stars, Mind and Fate: Essays in Ancient and Mediaeval Cosmology* (London: The Hambledon Press, 1989), p. 330, writes: '*Quia ad inveniendum* was evidently written after 1322 and possibly even earlier than 1327, although John of Saxony in his work of that year abstracted eclipse canons from John of Lignères' 1322 work (*Cuiuslibet arcus*) and made no allusion to *Quia ad inveniendum*, so far as I can detect'. See also Chabás and Goldstein, *The Alfonsine Tables of Toledo*, p. 282.

11 Victor E. Thoren and Edward Grant, 'Extracts from the Alfonsine Tables and Rules for their Use (John of Saxony)', in *A Source Book in Medieval Science*, ed. Edward Grant (Cambridge: Harvard University Press, 1974), pp. 465–87; Emmanuel Poulle, *Les Tables alphonsines avec les canons de Jean de Saxe* (Paris: Centre national de la recherche

Quia ad inveniendum were certainly one of the sources used by John of Saxony, a student of John of Lignères who rated his work very highly, to formulate his canons.[12] John of Saxony's deference to his teacher is obvious from this note in Chapter 22 (*Tempus vere coniunctionis et oppositionis Solis et Lune*) of his canons:

> Et nota, quod est quedam tabula multum bona meo iudicio, in qua faciliter invenitur motus Solis et Lune in una hora, et licet ad hoc sint plures facte, tamen verior, quam vidi, est illa, que continetur in tabulis, quas magister Ioannes de Lineriis ordinavit, et intratur in eam cum argumento Solis et cum argumento Lune. (Take note: there is one table, which I think is very good and which makes it easy to find the motions of the Sun and the Moon in one hour; and even though many tables have been created for this purpose, the most accurate of those I have seen is the one included in the tables composed by John of Lignères, and it is entered with the argument of the Sun and the argument of the Moon'.)[13]

The text *Quia ad inveniendum* consists of a short introduction and sixteen chapters containing rules of different lengths. The introduction and the first seven rules deal with the counting of time according to a sexagesimal system of numeration with a unit of one day and with mutual conversions to dates in different epochs. The next three chapters (8–10) deal with the calculations of mean motion of the planets, mean conjunctions and oppositions and apogees. The remaining six chapters are devoted to calculations of the true positions of the planets (including the Sun and Moon) and to the search for possible eclipses of the Sun and Moon.

Preparing the edition of the canons *Quia ad inveniendum* was relatively complicated. The text is quite variable, likely because it was written at the beginning of the Latin tradition of Alfonsine astronomy and the astronomers of the time were searching for ways to interpret and formulate this type of text. This might be the reason why John of Saxony only a few years later created a new version of the canons (including *Tempus est mensura motus*), which, in addition to being more comprehensive, probably seemed easier to understand, so his version took hold and became much more widespread in manuscript copies.

This edition has been prepared using the following ten manuscripts:

A: Paris, Bibliothèque nationale de France (BnF) lat. 7286, ff. 1ra–3va (Paris, XIV).[14] The manuscript contains all the chapters (1–16), albeit without numbers, and leaves blank spaces for initials at the beginning of each chapter. It often contains very good readings. Moreover, the manuscript is written in calligraphy, which testifies that the text was

---

scientifique, 1984); Alena Hadravová and Petr Hadrava, 'John of Saxony', in *Medieval Science, Technology and Medicine: An Encyclopedia*, ed. by Thomas Glick, Steven J. Livesey, and Faith Wallis (New York: Routledge, 2005), p. 292. Even though the canons by John of Murs, *Prima tabula docet differenciam unius ere*, were published later (1339), they did not reach the widespread popularity of John of Saxony's earlier version.

12 See North, *Stars, Mind and Fate*, p. 330: 'Set of canons *Quia ad inveniendum* was designed to explain the use of what was to become the definitive version of the Alfonsine Tables on the European continent. To all intents and purposes, this is the version of the tables later printed together with the canons of John of Saxony (AD 1327), *Tempus est mensura motus*'.

13 Poulle, *Les Tables alphonsines*, pp. 7, 22, 82, and 83.

14 See a description of the manuscript: *Catalogus codicum manuscriptorum Bibliothecae regiae*, Pars tertia, Tomus tertius – quartus (Parisiis: ex typographia regia, 1744), p. 335a (https://archive.org/details/CatalogusCodicumManuscriptorumBibliotVol4/page/n. 341). This manuscript also contains a fifteenth-century copy of several sets of tables by Giovanni Bianchini (Iohannes Blanchinus).

considered to be a well-constituted version of the work. Since it was written in Paris and is among the oldest copies of the work, this manuscript was chosen as the basis for the preparation of this edition. It is probably quite close in time and space to the unknown autograph by John of Lignères.

B: Vatican, Biblioteca Apostolica Vaticana (BAV) Pal. lat. 1403, ff. 1^{v}–3^{v} (France, XIV^{1}).[15] The manuscript includes all the chapters (1–16); there are no chapter numbers, but each chapter starts with an initial written in a rubric.

C: Vatican, BAV Ott. lat. 1826, ff. 41ra–46vb (Italy, XIV^{mid}).[16] The manuscript contains all the chapters (1–16) and, unlike any manuscript except **G**, gives a title to each chapter (2–16). The titles along with their translations into English are included in this edition.

D: Oxford, Bodleian Library (Bodl.) Digby 168, ff. 145ra–146rb (XIV /1327–72/).[17] The manuscript contains all the chapters (1–16), albeit without numbers, and leaves blank space for initials at the beginning of each chapter. Although a good copy, it often represents a different version of the work. According to new findings by Laure Miolo, the copy of the *Quia ad inveniendum* in Digby 168 is of French origin; this part of the manuscript was acquired (perhaps already by 1333) by the Oxford astronomer Simon Bredon (*c.* 1300–72).[18]

E: Cracow, Biblioteka Jagiellońska (BJ) 548, ff. 30ra–33va (Prague, XIV).[19] Once it arrived in Cracow, the manuscript was owned by Caspar Franckenstein. It also contains a gloss written by Albertus de Brudzewo, a prominent Polish astronomer. It includes all the chapters (1–16), which are not numbered and have no titles, but mostly start with an initial written in a rubric. This Bohemian manuscript also includes other canons by John of Lignères, namely his *Canones Tabularum eclipsium* (inc. *Priores astrologi*) on ff. 56^{r}–73^{r}.[20]

15 See the digitized version at https://digi.vatlib.it/view/MSS_Pal.lat.1403 and the bibliographic references at https://digi.vatlib.it/mss/detail/Pal.lat.1403.

16 See the digitized version at https://digi.vatlib.it/view/MSS_Ott.lat.1826, the bibliographic references at https://digi.vatlib.it/mss/detail/Ott.lat.1826 and a handwritten catalogue: *Inventarii codicum manuscriptorum Latinorum Bibliothecae Vaticanae Ottobonianae* II, f. 92^{v}. See also David Juste, 'MS. Vatican, Biblioteca Apostolica Vaticana, Ottob. lat. 1826' (updated: 5 March 2018), *Ptolemaeus Arabus et Latinus. Manuscripts*, URL = http://ptolemaeus.badw.de/ms/203 and Miolo, 'Retracing the Tradition of John of Genoa's Opus astronomicum Through Extant Manuscripts', in *Alfonsine Astronomy: The Written Record*, ed. Richard L. Kremer, Matthieu Husson and José Chabás (Turnhout: Brepols, 2022), pp. 343-80.

17 See a description of the manuscript: William D. Macray, *Catalogi codicum manuscriptorum Bibliothecae Bodleianae, Pars nona: Codices a viro clarissimo Kenelm Digby anno 1634 donatos, complectens* (Oxford: Clarendon Press 1883), pp. 172–77 (https://archive.org/details/CatalogiCodicumManuscriptorumBibliothP9/page/n. 93). This manuscript also contains John of Saxony's canons of 1327 *Tempus est mensura motus* (ff. 131ra–135va), followed by the Parisian Alfonsine Tables (ff. 139^{r}–144^{v}, 146^{v}); see Jean-Patrice Boudet and Laure Miolo, 'Alfonsine Astronomy and Astrology in Fourteenth-Century Oxford: The Case of MS. Bodleian Library, Digby 176', in *Alfonsine Astronomy: The Written Record*, ed. by Richard L. Kremer, Matthieu Husson asnd José Chabás (Turnhout: Brepols, 2022), pp. 57-105.

18 Concerning Simon Bredon, see e.g. Keith V. Snedegar, 'The Works and Days of Simon Bredon, a Fourteenth-Century Astronomer and Physician', in *Between Demonstration and Imagination: Essays in the History of Science and Philosophy, Presented to John D. North*, ed. L. Nauta and A. Vanderjagt (Leiden and Boston: Brill, 1999), pp. 285–309, and more recently Boudet and Miolo, 'Alfonsine Astronomy'.

19 See *Catalogus codicum manuscriptorum medii aevi Latinorum, qui in Bibliotheca Jagellonica Cracoviae asservantur*, 11 vols (Wrocław: Ossolineum, 1980–2016), III (1984), pp. 319–23; Poulle, 'John of Lignères', pp. 122–28; Grażyna Rosińska, *Scientific Writings and Astronomical Tables in Cracow. A Census of Manuscript Sources (XIVth – XVIth Centuries)* (Wrocław: Ossolineum, 1984), p. 352.

20 There is another manuscript of Bohemian origin (dating to 1388), MS. Cracow, BJ 551, which includes a wide selection of works by John of Lignères: ff. 58^{r}–63^{r} contain his *Canones Tabularum primi mobilis* (inc. 'Cuiuslibet arcus propositi sinum rectum invenire'), ff. 63^{r}–70^{v} (*absque fine*) contain his *Canones tabularum eclipsium* ('Priores

F: Bernkastel-Kues, St Nikolaus Hospital, Stiftsbibliothek Cusanus (Cus.) 212, ff. 65ra–66vb (Paris?, Northern Italy?, XIV /or $XV^{in.}$?/).[21] The manuscript only contains Chapters 1–10, with the text ending abruptly in the middle of Chapter 10 (the writing continues until the bottom of the folio, but it ends in mid-sentence). The following folio (67^{r-v}) is blank, and another text starts on folio 68^{r}, which means that the last part of the work has been lost. The manuscript contains bold initials *in margine* at the beginning of each chapter and the chapters are numbered. Moreover, it is the only manuscript of those used for this edition that contains a rather lengthy discussion on tables of the differences between various eras (Nebuchadnezzar, Alexander the Great, *diluvium*, *Incarnacio Christi* etc.). We could not find any analogy to this text in the canons by John of Lignères (e.g. in his canons *Priores astrologi* contained in Cracow, BJ 551 and in other copies), and we do not know of any other parallel text by other authors. Since this additional text in the manuscript **F** is too long for the critical apparatus, it is included below:[22]

|*f.* 65ra:| Incipiunt *Canones super Tabulas magistri Alfoncii*, regis Castelle.

Quia ad inveniendum locum planetarum per *Tabulas Alfoncii*, Castelle regis illustrissimi, oportet reducere ad annos nobis notos, incipientes ab aliqua era ex eris in eisdem tabulis positis ad quarta, tercia, secunda et prima. Poscitis, quid sit era: era est… *(?)* temporis… *(?)* particulas continencia… *(?)*, sed tempus inter diluvium et nativitatem Domini nostri Iesu Christi. Si volueris habere eram alicuius regni seu regis, intra in *Tabulam differencie e<r>arum seu regum aut regionum*, que est prima tabula cuiuscumque regni aut regis volueris, et invenies in directo quarta, tercia, secunda et prima illi regno seu regi correspondencia. Que reduc ad annos per tabulas ad hoc factas et per modum tibi dictum in tercio canone ita tamen, quod cuiuscumque eram recipis per eiusdem ere tabulas, debes reducere quarta, tercia, secunda et prima ad annos Christi. Si recepis eram diluvii, tunc debes reducere predicta quarta, tercia et cetera ad annos per tabulas diluvii. Si autem recipias eram Arabum, tunc debes reducere dicta quarta et cetera ad annos per *Tabulas Arabum* et sic de aliis. Vel aliter: si volueris, multiplica quarta per 60 et tunc provenient tercia, quibus adde tercia prius habita, summa illa, iterum multiplica per 60 et habebis secunda, quibus adde secunda, que habes prius, hanc vero multiplica per 60 et habebis prima, quibus adde prima tua, que in tabula invenisti, et habebis prima, que sunt in illa era. Et illa prima dicuntur dies.

---

astrologi celestium corporum') and his tables can be found on ff. 74^{r}–96^{v}. See *Catalogus codicum…, qui in Bibliotheca Jagellonica Cracoviae asservantur*, vol. III (1984), pp. 333–45.

21 For a possible origin and date of the first part of manuscript, see Alois Krchňák, 'Die Herkunft der astronomischen Handschriften und Instrumente des Nicolaus von Kues', *Mitteilungen und Forschungsbeiträge der Cusanus-Gesellschaft*, 3 (1963), 109–81 (pp. 168–71); Jakob Marx, *Verzeichnis der Handschriften-Sammlung des Hospitals zu Cues bei Bernkastel a. Mosel* (Trier: Selbstverlag des Hospitals, 1905; repr. Frankfurt/Main, 1966), pp. 203–8; Elly Dekker, *Illustrating the Phaenomena. Celestial Carthography in Antiquity and the Middle Ages* (Oxford: Oxford University Press, 2013), pp. 181–82, 249, 356. The manuscript belongs to a collection of manuscripts purchased by Nicolaus Cusanus for his private use in Nuremberg in 1444, see e.g. Johannes Franz Hartmann, *Die astronomischen Instrumente des Kardinals Nikolaus Cusanus* (Berlin: Weidmannsche Buchhandlung 1919); Alena Hadravová and Petr Hadrava, *Sphaera octava. Mýty a věda o hvězdách IV. Katalogy hvězd a přemyslovský nebeský glóbus – Sphaera octava. Myths and Science on Stars IV. Catalogues of Stars and Premyslid Celestial Globe* (Prague: Artefactum – Academia, 2013), pp. 276–78. See also David Juste, 'MS. Bernkastel-Kues, Cusanusstiftsbibiothek, 212' (updated: 1 January 2019), *Ptolemaeus Arabus et Latinus. Manuscripts*, URL = http://ptolemaeus.badw.de/ms/18.

22 The question marks denote unreadable and uncertain text.

Sciendum est, quod quidam anni sunt solares, quidam sunt lunares et quidam sunt bisextiles, quidam vero non. Unde si anni fuerint solares et non bisextiles, tunc divide dies predictos per 365 et in numero quociens provenient anni, et quod post divisionem remanserit, vide, si potes, dividere per 30, tunc in numero quociens provenient menses. Et quod post divisionem remanserit, erunt dies mensis imperfecti. Si autem id, quod remanserit post divisionem per 365, non potest dividi per 30, tunc numerus ille erunt dies mensis imperfecti, qui sunt post annos completos. Si aut fuerint anni bisextiles, tunc certius est eos reducere ad annos per tabulas ad hoc factas, quod per divisionem quasi annus bisextilis facit diversitatem in tali operacione. Si aut per duas eras notas velis invenire terciam ignotam, subtrahe eram minorem ab era maiori nota et remanebit altera ignota… *(?)*. Si per eram diluvii et regis Alfoncii et per eram Nabuchodonosor et regis Alfoncii volueris invenire eram diluvii et Nabuchodonosor, subtrahe eram Nabuchodonosor et Alfoncii ab era diluvii et in *Tabulis Alfoncii* tibi remanebit era |*f. 65rb:*| diluvii et Nabuchodonosor. Et si subtraxeris eram diluvii et Nabuchodonosor eb era diluvii et regis Alfoncii, remanebit era Nabuchodonosor et regis Alfoncii. Et si subtraxeris ab era diluvii et regis Alfoncii eram diluvii et Incarnacionis Domini nostri Ihesu Christi, remanebit era Incarnacionis Ihesu Christi et regis Alfoncii et sic de ceteris.

Sciendum, quod omnis era maior continet minorem in se et hoc supposito, quod era sit tante quantitatis. Sed era, que est a diluvio usque ad diem presentem, continet omnes eras, que sunt inter diluvium et diem presentem, scilicet eram diluvii et Nabuchodonosor et eram diluvii et mortis Alexandri et sic de ceteris. Ideo quodcumque illarum subtraxeris ab era diluvii ad dies presentes, remanebit differencia illius usque ad diem presentem. Sic eciam potes per addicionem secundarum erarum annorum invenire terciam ignotam, sed si addideris eram Incarnacionis Ihesu Christi et regis Alfoncii super eram diluvii et Incarnacionis, proveniet era diluvii et regis Alfoncii. Et si addes eram Incarnacionis Ihesu Christi et Nabuchodonosor super eram diluvii et Nabuchodonosor, habebis eram diluvii et Incarnacionis Ihesu Christi. Et sic quociens sequenter sunt comprehense in… terci… *(?)*. Et si ille secunde ad id addantur, habentur tercia, que est maior, et hoc plane patet in operando. Item secundo sciendum, quod prius dictum, dies et quodlibet secundum usque 60 gradus et quodlibet tercium usque 60 secunda et quodlibet quarta usque 60 tercia. Et anni sunt reducti aput diversas eras, non sunt eiusdem quantitatis nec eciam incipiunt in eodem tempore, quia alii sunt bisextiles, alii vero non et alii sunt solares et alii lunares. Et de bisextilibus alii incipiunt in primo anno post bisextilem, alii in secundo, alii in tercio post bisextilem. Item alii incipiunt a Ianuario, alii ab aliis mensibus, quapropter plures oportuit componere *Tabulas tam in annis collectis, quam in expansis, quam in mensibus.*

G: Oxford, Hertford College (HC) 4, ff. 148^{v}–154^{r} (XV^{1} /1425–75/).[23] The manuscript contains all the chapters (1–16) and gives a title to each chapter (as does manuscript **C**). Folios are alternately superscribed by the words *Ihesus* and *Maria.*

H: Paris, BnF lat. 7405, ff. 1^{r}–4^{v} (Paris, XV^{1}).[24] — At the beginning of folio 1^{r}, the number 1349 has been entered by a different and more recent hand, but this is not the year of

23 See Paul Morgan, *Oxford Libraries Outside the Bodleian: A Guide* (Oxford: Oxford Bibliographical Society and Bodleian Library, 1973), pp. 44–47; Miolo, 'Retracing the Tradition', in this volume.

24 According to *Catalogus codicum manuscriptorum Bibliothecae regiae*, vol. 4, p. 352b, the manuscript was written in the fifteenth century (https://archive.org/details/CatalogusCodicumManuscriptorumBibliotVol4/page/n. 359).

the origin of this copy. The manuscript contains all the chapters (1–16), some of which are numbered *in margine*, and the initials are set apart from the rest of the text with a rubric. The manuscript is written in calligraphy (similar to MS. Paris, BnF 7286 / = **A**/); this means that it was considered a higher quality copy compared to regular copies such as those produced by university students. It belongs to the younger generation of manuscripts. Like manuscript **K**, it provides a partly improved and updated version of the text.

I: Vienna, Österreichische Nationalbibliothek (ÖNB) Cod. 5144, ff. 143^{v}–145^{v} (Vienna, XV^{1} /1440–60/).[25] The manuscript contains all the chapters (1–16), which are numbered *in margine*. It is one of the later manuscripts, and the text has been extended in many places, making it the longest of all the copies. It is also evident that it diverges significantly from the original version, although the added explanations are often highly pertinent. This manuscript does not contain any systematic overview of the tables.

K: Paris, BnF lat. 7281, ff. 175^{r}–178^{r} (Cambrai?, XV^{mid}.).[26] It contains all the chapters (1–16), including their numbers *in margine*. There are also blank spaces for larger initials, which were never added. In general, this manuscript is close to manuscript **H**, but contains slightly more errors. It was copied by the otherwise unknown 'Jo. B'. This manuscript contains many texts belonging to Alfonsine astronomy (notably those by John of Saxony, John of Lignères, John of Murs, and John of Genoa); from 1487 it was owned by Jean Avis and known to Simon de Phares.[27]

For the sake of completeness, the following manuscripts should, according to various sources and secondary literature, contain *Quia ad inveniendum*, but were mostly unavailable to us:[28]
Bruges, Openbare Bibliotheek (OB) 466, ff. 129^{r}–131^{r}
Cambridge, University Library (UL) Ii. 1. 27, ff. 54^{v}–56^{v} (from 1424)[29]
London, British Library (BL) Sloane 407, ff. 4^{r}-6^{v}
Milan, Biblioteca Ambrosiana (BA) N 217 Sup., ff. 7^{r}–8^{v}, 15^{v}, 27^{r}–28^{v} (fragments only; Cremona?, XIV)
Vatican, BAV Vat. lat. 3113, ff. 47^{r}–52^{v}.[30]

25 See *Tabulae codicum manu scriptorum praeter Graecos et orientales in Bibliotheca Palatina Vindobonensi asservatorum*, Band 4: Cod. 5001 – Cod. 6500, ed. Academia Caesarea Vindobonensis (Wien: Gerold, 1864–99), p. 39 (http://bilder.manuscripta-mediaevalia.de/hs//katalogseiten/HSK0751d_b0039_jpg.htm).

26 See the digitised version at https://gallica.bnf.fr/ark:/12148/btv1b525030045/f. 2.image.r = latin%207281 and a description of the manuscript: *Catalogus codicum manuscriptorum Bibliothecae regiae*, vol. 4, p. 334a (https://archive.org/details/CatalogusCodicumManuscriptorumBibliotVol4/page/n. 341).

27 See Jean-Patrice Boudet, *Lire dans le ciel. La bibliothèque de Simon de Phares, astrologue du xve siècle* (Brussels: Centre d'Études des Manuscrits, 1994), pp. 181–83. Boudet deals with the contexts of this manuscript also in the framework of the ALFA project; see also, Miolo, 'Retracing the Tradition', in this volume.

28 On the other hand, Paris, BnF lat. 7378A, which is often mentioned in this context, does not include *Quia ad inveniendum*; it only contains two other works by John of Lignères, namely *Canones primi mobilis* and *Canones minutiarum sive fraccionum*. And similarly, Milan, BA S 54 Sup., ff. 16^{r}–25^{r} ($XV^{mid.}$) contains only the incipit 'Quia ad inveniendum' on f. 16^{r}, but it is actually followed by a different text. See A. L. Gabriel, *A Summary Catalogue of Microfilms of One Thousand Scientific Manuscripts in The Ambrosiana Library, Milan* (Notre Dame: University of Notre Dame, 1968).

29 *A Catalogue of the Manuscripts preserved in the Library of the University of Cambridge*, ed. by Henry Richards Luard (Cambridge: Cambridge University Press, 1856–67), III (1858), p. 346.

30 Found by José Chabás in the time of proofreadings of this contribution.

The canons *Quia ad inveniendum* have never been issued in print.

The canons *Quia ad inveniendum* mention or pertain to the following tables found in the set known as the Parisian Alfonsine Tables:

Tabula in annis collectis, Tabula annorum collectorum ('Table of collected years')
Tabula in annis expansis, Tabula annorum expansorum ('Table of expanded years')
Tabula mensium ('Table of months')
Tabula mediorum motuum ('Table of mean motions')
Tabula radicum ('Table of radices')
Tabula ad sciendum horas et earum fractiones ('Table for finding the hours and fractions of hours')
Tabula differencie erarum ('Table of differences of eras')
Tabula annorum bisextilibus ('Table of leap years')
Tabula annorum non bisextilibus ('Table of non-leap years')
Tabula radicum notarum anni ('Table of radices of the days of the week in the year')
Tabula mediorum motuum ('Table of mean motions')
Tabula de medio centro Lune ('Table of the mean centre of the Moon')
Tabula de mediis argumentis trium superiorum ('Table of the mean argument of the three upper planets')
Tabula medie elongationis Solis et Lune ('Table of the mean elongation of the Sun and Moon')
Tabula coniunctionis et oppositionis in mensibus ('Table of conjunction and opposition in the months')
Tabula motus augium ('Table of motion of apogees')
Tabula equationis octave spere (*sive capitis Arietis*) ('Table of equation of the eighth sphere /or the head of Aries/')
Tabula equationis Solis ('Table of equation of the Sun')
Tabula equationis Lune ('Table of equation of the Moon')
Tabula equacionum planetarum ('Table of equations of the planets')

Below are several comments on the individual chapters of the canons. It is not our intention to offer a comprehensive commentary:

- All ten of our manuscripts explicitly state at the beginning, and several also at the end of the text, that the canons pertain to the Alfonsine Tables, but only manuscripts **C**, **H** and **K** mention that the canons were composed by John of Lignères.
- Chapter 5: The last word of the chapter is *diei* and is correctly recorded in all the manuscripts with the exception of **E**, which has the erroneous reading, *hore*. *Diei* is correct, since John of Lignères was counting sexagesimally and wrote about *minuta diei* (a minute of the day), which, in this context, equals twenty-four normal minutes. The scribe of **E**, who made the erroneous correction, was thinking about the counting of time in standard units, hence the wording *minuta hore* ('a minute of the hour').
- Chapter 7: Manuscripts **H** and **K** give a concrete title of the table to which this chapter refers, information missing in the other manuscripts. This could be an example of later additions (interpolations) to the text in these manuscripts.
- Manuscripts **H** and **K** also agree in using the word *consequenter* only four times in the whole text, while the other manuscripts use the word twice as often (seven or eight times).

- Chapter 11: A part of a sentence is added to **B**, **C**, **H**, and **K**. The readings in **H** and **K** are closest here; the added section, *equa tamen pro minutis, si sint minuta in argumento*, is the same in both, except that **H** corrects the dittography of *in argumento* with a dotted line under the repeated word while **K** leaves it uncorrected. Similarly, **B** and **C** agree in their readings, thus establishing two groups, or branches, of manuscripts. The additional part of a sentence is missing from **A**, **D**, **E**, **G**, and **I**. Manuscripts **H** and **K** are also the only ones that add the explanation *ab auge* in this chapter, albeit in the wrong place (it should have been included later in the text on six signs and not in the text on three signs).
- Chapter 13: Again, **H** and **K** are longer, both including an additional sentence at the end of the chapter. It is obvious that **H** and **K** belong to one distinct branch or family of the text. All manuscripts give the number of signs as twelve, with the exception of **A**, which gives the number as six. However, the readings of Manuscript **A** are often very good in other places, and in general it is a high-quality copy of the work.
- Chapter 15: Again, the last sentence is extended with *sicut sit in equacionibus in accipiendo partem proporcionalem* in **H** and **K**.
- A preliminary comparison of the manuscripts, when compiling the critical apparatus of the edition, indicates a particularly strong connection between the pairs **BC**, **DI** (especially noticeable in Chapter 14), and **HK**.

The above described observations of some tendencies and possible relations among the manuscripts can be supported using the statistical method of binary correlations[31] that yields the numbers of coincidences for each pair of copies, that is, how many times the two witnesses have the same variant reading. These data are presented in Table 1. The higher number in the cross of the row and column of two different sigla indicates a closer relation between the two corresponding manuscripts. On the diagonal of the table are given in bold the overall numbers showing how many times each manuscript appears in the variant readings. The bottom row gives the sums of the off-diagonal numbers in the columns.

The results of the method of binary correlations between the variant readings of the manuscripts confirm the estimates gathered from the linguistic treatment of the critical apparatus.

The highest overall numbers of coincidences with all other manuscripts (6600) appears in Manuscript **A**. In view of its highest coincidence and also because of the high quality of the writing, its age (fourteenth century), and its origin (Paris, where John of Lignères composed his rules), we chose it as the base manuscript for the edition. Just opposite, **F** exhibits the lowest number of coincidences (4433) with other manuscripts.

31 We have developed the method of binary correlations for a statistical treatment of variant readings in critical apparatus of editions and for a computer-aided stemmatology. It was used for the first time in our edition of treatises on astrolabes by Cristannus de Prachaticz (Alena Hadravová and Petr Hadrava, *Stavba a užití astrolábu*. Prague, Filosofia 2001). A recent description of the method can be found in Petr Hadrava and Alena Hadravová, 'Cristannus de Prachaticz's Treatises on the Astrolabe', in *Certissima signa. A Venice Conference on Greek and Latin Astronomical Texts*, ed. by Filippomaria Pontani (Venice: Edizioni Ca'Foscari, 2017), pp. 295–312.

Table 1. Binary correlations based on variant readings of the witnesses.

	A	B	C	D	E	F	G	H	I	K
A	**1274**	859	817	772	884	654	770	607	682	555
B	859	**1310**	946	755	731	551	697	554	636	507
C	817	946	**1322**	733	676	553	695	555	620	513
D	772	755	733	**1308**	653	483	691	620	735	570
E	884	731	676	653	**1285**	560	692	504	571	476
F	654	551	553	483	560	**889**	506	365	431	330
G	770	697	695	691	692	506	**1335**	537	644	483
H	607	554	555	620	504	365	537	**1281**	532	1097
I	682	636	620	735	571	431	644	532	**1308**	485
K	555	507	513	570	476	330	483	1097	485	**1282**
Σ	**6600**	**6236**	**6108**	**6012**	**5747**	**4433**	**5715**	**5371**	**5336**	**5016**

This is, however, mainly due to the fact that this witness is unfinished; it ends abruptly in the middle of Chapter 10. Its incompleteness is also reflected in the smallest number of the overall occurrence in the variant readings (889). Manuscript **F** coincides best with **A** (654 coincidences) and in renormalizing this to the mean number of the overall occurrences of the other manuscripts (about 1300), we would find a very high number (956) of coincidences with **A**.

The highest mutual coincidence is exerted by the pair of witnesses **HK** (1097). This shows that there is a quite close relation between **H** and **K** (both originated in Paris): **K** may be directly dependent on **H** or there could be a close interlinking. However, it does not mean that the witness **H** should be accepted as the basis because this manuscript was written in the fifteenth century; moreover, it is evident that its text improves older versions of the rules and brings many refinements, minute extensions, and explanations.[32] It was thus probably motivated by a need to consolidate and stabilize the then quite diversified text of the *Quia ad inveniendum*. This may also explain why it was written in calligraphy.

The next highest correlation of variant readings can be found for Manuscripts **B** and **C** (946 coincidences). This corresponds well with the fact that **B** was still written — as it is known — in France in the first half of the fourteenth century, while today it is saved in the Vatican Library. Manuscript **C** is directly or indirectly dependent on it and had already been written in the middle of the century in Italy. The third highest coincidence (884) can be found for the pair **AE**, which shows an influence of the Parisian witness **A** on

32 See e.g. cap. 7: facies … per tabulam ad hoc factam × facies … per tabulam ad hoc factam, *que dicitur Tabula ad sciendum horas et earum fracciones*; cap. 8: ex eris hic positis × ex eris hic *in tabulis* positis; cap. 10: quere medium motum octave spere × quere medium motum octave spere, *scilicet accessum et recessum*; cap. 11: a tribus signis × a tribus, *scilicet ab auge*, signis; cap. 13: et residuum erit verus locus capitis draconis in nona spera × et residuum erit verus locus capitis draconis in nona spera. *Habito vero loco Lune et capitis draconis subtrahe verum locum capitis a vero loco Lune et remanebit tibi latitudo Lune sive argumentum latitudinis Lune*; cap. 15: pro nichilo computentur × pro nichilo computentur, *sicut sit in equacionibus in accipiendo partem proporcionalem*.

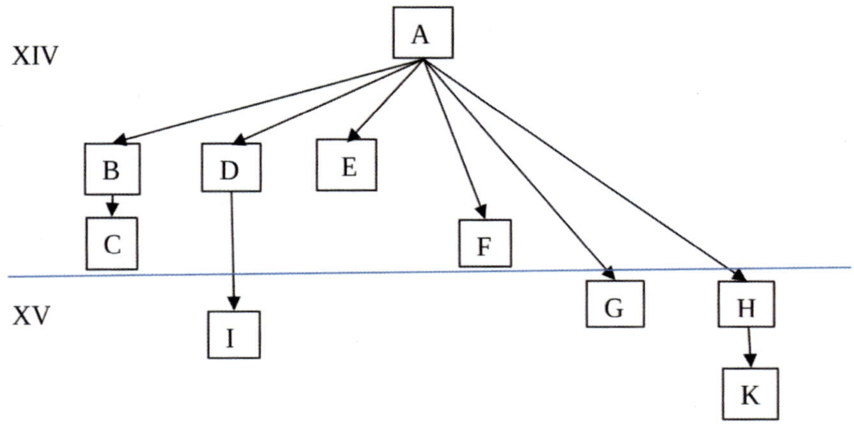

Figure 1. Computer-aided *stemma codicum*. The vertical displacement corresponds approximately to the chronology of the manuscripts.

E (of Bohemian provenance), saved today in Cracow. Another strong relation is visible between copies **AB** (859) and **AC** (817); it reveals a dependence of the Italian branch on the Parisian template. Other branches include **AD** (772) and **AG** (770), which trace the spreading of the treatise from France (Paris) to England (Oxford). The other smaller values of coincidences given in the table are already of a minor significance.

If the witnesses are ordered according to the decreasing overall coincidences with all other copies, we arrive at the following sequence:
A = 6600 (Paris, BnF, lat. 7286, XIV)
B = 6236 (Vatican, BAV, Pal. lat. 1403, XIV^{1})
C = 6108 (Vatican, BAV, Ott. lat. 1826, XIV^{mid})
D = 6012 (Oxford, Bodl., Digby 168, XIV /1327–72/)
E = 5747 (Cracow, BJ 548, XIV)
G = 5715 (Oxford, HC 4, XV /1425–75/)
H = 5371 (Paris, BnF, lat. 7405, XV)
I = 5336 (Vienna, ÖNB 5144, XV /1440–60/)
K = 5016 (Paris, BnF, lat. 7281, XV$^{mid.}$)
F = 4433 (Bernkastel-Kues, Cus 212, XIV /or XV$^{in.}$?/)

With the exception of the last witness, **F**, which is disadvantaged by its incompleteness as explained above, this sequence corresponds approximately to the age of the manuscripts.

The values from Table 1 can also be used as input data for the computer-aided stemmatology. Our suggested preliminary *stemma codicum* was created using the *Maximum binary tree* method.[33]

33 We have used the Jarník — Prim algorithm, cf. Hadrava and Hadravová, 'Cristannus de Prachaticz´s Treatises on the Astrolabe', p. 306.

Editorial note

The autograph of the work is not extant (or has not yet been identified) and it proved to be very difficult to reconstruct a reading that could be closest to the original version of the canons by John of Lignères. The text of Manuscript **A** was chosen as the basis of the edition because of its generally better quality and meticulous calligraphic writing when compared to other witnesses, which is attested by a lower number of scribal errors, and because of its dating to the fourteenth century. However, to offer readers better accessibility to the content of the treatise, we have decided to occasionally use variant readings from the other manuscripts in the edition when they contain a variant with which the text makes more sense. We focused our attention on preparing an edition that primarily aims to select factually correct readings that will help readers to understand the use of the corresponding set of tables. Whenever there was a choice of several factually correct readings, we favoured the readings of witness **A**; however, in some cases, we selected the readings found in a larger number of manuscripts, which means they were accepted by a majority of the scribes. The unusually large number of variant readings and errors in the copies (duplicate sections of the texts, omissions of large parts of sentences, etc.) can be attributed to the nature of the text with its rather stereotypical vocabulary and style. While copying the work, the scribes would often skip to other parts of the text due to similar wording.

The complete apparatus on which the statistical analysis is based was too extensive, so we decided to shorten it for the present publication. We thus mostly chose variant readings that document the preferred readings from other manuscripts rather than those from the basic one. In our opinion, a text like this should be processed using the modern editing method XML-TEI,[34] which makes it possible to document all the copies of a given text and makes them electronically searchable, so that readers can look up specific queries. Our edition should be viewed as a first attempt at determining the reading of John of Lignères' *Quia ad inveniendum*, rather than a closure to the topic.

The basic Manuscript **A** mostly uses the spelling of *ti-* against of *ci-* (e.g. *coniunctio* × *coniunccio*, *fractio* × *fraccio*, *etiam* × *eciam*, *elongatio* × *elongacio*, etc.), which we follow in the whole edition for consistency. The spelling of other Latin words is left unchanged even if it varies throughout the text. This concerns, for example, the inconsistent use of *i* and *y* (e.g. *diameter* × *dyameter*), double versus single letters (*agrego* × *aggrego*), and other phenomena (*nihil* × *nichil*, *era* × *hera*, *Toletum* × *Tolletum* × *Tholetum*); their specific usage in manuscripts may implicate relations between individual witnesses (especially **HK**).

Common paleographic abbreviations are transcribed in full, as is customary. The spelling is chosen so as to match the full forms in the surrounding text (*numquam* × *nunquam*).

Upper-case letters are used regardless of the inconsistent medieval usage to mark the beginning of a sentence, proper names, adjectives and adverbs derived from proper names, the first word of a work of literature, etc.

Insufficient or old rhetorical punctuation is substituted by logical punctuation, as used in Czech.

34 For a recent description of this method, see e.g. Massimiliano Carloni, 'Towards a Digital Edition of the Aratean Tradition', in *The Stars in the Classical and Medieval Traditions*, ed. Alena Hadravová, Petr Hadrava, and Kristen Lippincott (Prague: Scriptorium 2019), pp. 171–88.

Abbreviations (2^{a}, 3^{a}, 4^{a}; 2^{m}, 3^{m}, 4^{m}, etc.) are written out in full (*secunda, tertia, quarta; secundum, tertium, quartum*).

The question mark *(?)* denotes unreadable places of the manuscripts.

Where longer sections of texts are omitted, the notes are located at the beginning of these sections, after the first omitted word (as opposed to all other variant reading notes in the critical apparatus, which are always located at the end of the respective section).

<> Angle brackets are used to mark chapter numbers and titles added for ease of reading.
{ } Curly brackets are used to athetize redundant letters or words, for example, *angulus* {angulus}.
[…] = lacuna

Abbreviations

add. = addidit (added)
cap. = caput, capitulum (chapter)
cf. = confer (compare)
corr. = correxit (corrected)
corr. ex = correxit ex (corrected from)
corr. in = correxit in (corrected to)
del. = delevit (deleted, removed, stricken out)
expl. = explicit (end, ends)
f., ff. = folio, folia
in mg. = in margine (in the margin of the folio)
inc. = incipit (beginning, begins)
ms. = codex manu scriptus (manuscript)
om. = omisit (omitted)
p. = pagina (page)
sic = sic (thus)
suprascr. = suprascripsit, suprascriptum (wrote above, written above)

Conspectus siglorum

A: Paris, BnF, lat. 7286, ff. 1ra–3va (Paris, XIV)
B: Vatican, BAV, Pal. lat. 1403, ff. 1^{v}–3^{v} (France, XIV^{1})
C: Vatican, BAV, Ott. lat. 1826, ff. 41ra–46vb (Italy, $XIV^{mid.}$)
D: Oxford, Bodl., Digby 168, ff. 145ra–146rb (XIV /1327–72/)
E: Cracow, BJ 548, ff. 30ra–33va (Prague, XIV)
F: Bernkastel-Kues, Cus. 212, ff. 65ra–66vb (Paris?, Northern Italy?, XIV /or $XV^{in.}$?/)
G: Oxford, Hertford Coll., 4, ff. 148^{v}–154^{r} (XV^{1} /1425–75/)
H: Paris, BnF, lat. 7405, ff. 1^{r}–4^{v} (Paris, XV^{1})
I: Vienna, ÖNB, Cod. 5144, ff. 143^{v}–145^{v} (Vienna, XV^{1} /1440–60/)
K: Paris, BnF, lat. 7281, ff. 175^{r}–178^{r} (Cambrai?, $XV^{mid.}$)

Canons *Quia ad inveniendum* by John of Lignères (post 1322 – ante 1327)

Incipiunt *Canones Tabularum Alfoncii,* ordinati per magistrum Ioannem de Lineriis.

Quia ad inveniendum loca planetarum per *Tabulas Alfonsi,* regis Castelle illustris, oportet reducere annos nobis notos incipientes ab aliqua era ex eris hic positis in tabulis istis ad quarta, tertia, secunda et prima. Ideo primo sciendum, quod prima dicuntur dies et quodlibet secundum valet 60 prima et quodlibet tertium valet 60 secunda et quodlibet quartum valet 60 tertia. Et quia anni reducendi apud diversas eras non sunt eiusdem quantitatis nec etiam incipiunt in eodem tempore, quia alii sunt bisextiles, alii non bisextiles, alii lunares, alii solares. Et de bisextilibus alii incipiunt primo anno post bisextum, alii secundo, alii tertio, alii a Ianuario, alii ab aliis mensibus, quapropter plures oportuit componere *Tabulas tam in annis collectis, quam in expansis, quam in mensibus.*

Capitulum primum

Cum ergo volueris reducere ad quarta, tertia, secunda, prima aliquem numerum annorum incipientium a principio illius ere tibi note, intra cum illis annis in tabulis illi here deservientibus, secundum quod potes videre per titulos tabularum. Et si precise numerum illorum annorum poteris invenire, invenies in directo quarta, tertia, secunda et prima illis annis equivalentia. Si autem non inveneris precise, accipe minorem numerum propinquiorem et quarta, tertia, secunda et prima, que inveneris in directo, scribe extra in tabula eo ordine, quo sunt. Deinde residuum illorum annorum vel numerum minorem propinquiorem quere, ut prius, in eisdem tabulis et quarta, tertia, secunda et prima in directo inventa extra sub aliis scribe, ita scilicet pro regula generali, quod quod-

---

1 Incipiunt – Lineriis] *om. ADBEI*: Incipiunt Canones super Tabulas illustris regis Alfunsi Castelle, bone memorie. Rubrica *C*: Incipiunt Canones super Tabulas magistri Alfoncii, regis Castelle *F*: Ihesus. Canon Tabularum Alfontis. Capitulum primum de reductione annorum notorum alicuius ere ad quarta, tercia, secunda et prima *G*: Canones tabularum Alfonsii, regis Castelle, per Iohannem de Lineriis *K* **2** Alfonsi – illustris] Alfonsii *B*: Alfunsi *C*: Alfonci *D*: Alfoncii *EHK*: Alfoncii, Castelle regis illustrissimi *F*: Alfontis *G*: Alphoncii *I* **3** reducere annos] annos *A*: reducere ad annos *F*: nos reducere annos *GI* era – eris] ex eris *B*: hera ex heris *HK* **3/4** hic – istis] in eisdem tabulis positis *ADEFGI*: hic positis in eisdem tabulis *C*: inter positis in eisdem tabulis *B* *an addition of the manuscript F is transcribed in the Foreword of the edition, see pp. 261-62* **5** prima] dies *ABCE* **6** valet] *om. AE* reducendi] reducturi *A* **7** alii²] et alii *AE* non bisextiles] non sunt bisextiles *EK*: *om. AD* **8** alii¹ – solares] solares, alii lunares *A*: lunares et alii solares *D*: alii lunares *E*: et alii lunares et alii solares *G* **9/10** quapropter – in mensibus] *om. HK* **11** Capitulum primum *(mg.)*] *om. ABCDEFGI*: 1 *K* **12** ad quarta – prima] *om. ABCDEFG*: ad quarta *I* **13** annis] *om. A* **13/14** illi – deservientibus] ad hoc factis propriis et ad hoc deservientibus, scilicet illi ere *A*: proprias deservientes illi (huic *DGI*) ere *BCDGI*: propriis deservientibus illi ere *E*: ad hoc factas et proprias illi ere *F* **14/15** precise – invenire] precise potes numerum illorum invenire *A*: precise invenire poteris numerum annorum illorum *B*: precise poteris numerum illum annorum invenire *C*: poteris precise numerum annorum illorum (illorum annorum *F*) invenire *DF*: prescise poteris illum numerum invenire annorum *E* **16** illis annis equivalentia] equivalentia *ABCEF*: predictis annis equivalencia *D*: illis annis equipollencia *HK* **17** inveneris – directo] in eius directo inveneris *A*: inveneris e directo eius *B* **18** illorum] *om. ABCDEFGI* **18/19** numerum – quere] minus propinquius quere *A*: minorem propinquiorem ei (ei *om. CK*) quere *BCK*: numerus propinquus quere *EF*: numerum annorum propinquiorum *GI*

Translation of canons *Quia ad inveniendum* *Rules for the Tables of Alfonso* composed by Master Iohannes de Lineriis.

To find the positions of planets using the Tables of Alfonso, the illustrious King of Castille, it is necessary to convert the known years since the beginning of any era[1] of those given in the tables to fourths, thirds, seconds, and units.[2] It is thus necessary to know that units denote the days and that each second has an amount of sixty units, that each third has an amount of sixty seconds, and each fourth has an amount of sixty thirds. The years to be converted do not have equal lengths in different eras and neither do they start at the same times, because some of them are leap, others are non-leap, some are lunar, and others are solar. And for leap years, some start in the first year after the leap year, others in the second year, and others in the third. Some years start in January, others in different months; this is why it is necessary to compile many *Tables both in collected as well as in expanded years, and also in months.*

Chapter one

If you wish to convert any number of years, starting from the beginning of any known era, to fourths, thirds, seconds, and units, enter with these years into the appropriate tables that correspond to that era, as you can see from the titles of those tables. And if you can find exactly that number of years, you will directly obtain the fourths, thirds, seconds, and units that correspond to the given number of years. If you cannot find them exactly, take the closest smaller number and write beside the table, in the same order they are given, the fourths, thirds, seconds, and units that you find in the table. Then, search for the remaining years, or the closest smaller number in the same tables as before, and write the fourths, thirds, seconds, and units below the others, following the

1 The explanation relates to Alfonsine Table No. 1 (*Tabula differentiarum unius regni ad aliud et nomina regum atque cuiuslibet ere cognite*), cf. Poulle, *Les Tables alphonsines*, p. 107.

2 The Latin terms *quarta, tercia, secunda,* and *prima* (translated here as 'fourth', 'third', 'second', and 'unit') denote the orders in the sexagesimal system of numeration, which was widely used in Alfonsine astronomy. The primes are units (in particular one day, 24 hours), a second has 60 primes, a tercia has 60 seconds, a quarta has 60 tercias. The sixtieth part of prime is named minute (*minutum*), and its sixtieth part is named second and has 60 tercias or 3600 quartas. The terms seconds, thirds, fourths, etc. are thus multivalent and may denote either multiples or fractions of the units.

libet scribatur sub suo genere: scilicet quarta sub quartis, tertia sub tertiis, secunda sub secundis, prima sub primis. Eodem modo intra cum residuo, si sit residuum, intrando tam *Tabulas annorum collectorum quam expansorum*, quociens oportuerit. Et similiter cum mensibus perfectis *Tabulas mensium*, ita scilicet, quod in annis non bisextilibus intrabis tabulam superiorem mensium et in bisextilibus tabulam inferiorem, semper subscribendo, quod inveneris in directo, quodlibet sub suo genere extra in tabula quousque invenias totum numerum annorum et mensium. Si autem restant aliqui dies de mense aliquo imperfecto, quia sunt prima, scribe sub primis. Postea vero agrega omnia adinvicem incipiendo a primis, et si ex agregatione primorum excrescant 60, pro illis 60 adde unitatem in ordine secundorum. Eodem modo fac de secundis respectu tertiorum et sic de aliis. Et alia, que non possunt complere 60, remaneant in locis suis. Quo facto provenient quarta, tertia, secunda et prima, que in toto numero annorum, mensium et dierum continebantur, cum quibus intrare debes *Tabulam mediorum motuum*.

Capitulum secundum

Item de reductione annorum cuiuslibet here ad quartas, tertias, secundas et primas

Si habueris quarta, tertia, secunda et prima, que sunt a principio alicuius ere tibi note, et velis scire hoc idem a principio alicuius alterius ere, vide, utrum illa era precedat eram tuam vel sequitur, quod potes scire per *Tabulas differentie erarum*. Quod si precedat, adde differentiam illius ere et ere tibi note, quam in eisdem tabulis invenies supra illud, quod habes. Si autem illa era sequatur, subtrahe differentiam illarum et invenies, quod queris.

---

23 tam Tabulas] in Tabulis tam *AEF*: tam in Tabulis *BCI*: tam in Tabulas *G* **24** Tabulas] in Tabula *AEF*: in Tabulis *BC*: in Tabulas *D*: intra Tabulas *G* **24/25** ita – inferiorem] *om. AEF* **25** tabulam superiorem] in tabula superiorum *B*: in tabula superiori *CI*: in tabulam superiorem *D*: in tabula superiorem *G* et in] in annis vero *H*: in annis non *K* tabulam inferiorem] in inferiori *BC*: in inferiorem earum *DI*: in tabula inferiorem eorum *G* **30/31** fac – aliis] si ex agregatione secundorum adinvicem proveniant 60, adde similiter pro illis unitatem in ordine tertiorum et sic de tertiis et quartis *AEF*: fac (fac *om. B*, fit *HK*) de secundis respectu tertiorum et sic de aliis (ceteris *I*) *BHIK* **31** Et alia – 60] residua autem *AEF*: et alia prima, si non possunt complere *B*: et alia (aliqua prima *C*, ea *G*, aliam *K*)… 60 *CGK* remaneant] sint *AEF*: dimittantur *C*: remanent *HK* **33** Tabulam] in Tabulis *AEFI*: in Tabula *B*: in Tabulam *C*: Tabulas *D*: in Tabulas *G* **34** Capitulum secundum] *om. ABCDEF*: secundum *mg. H*: *om. et in mg. add. I*: 2 *mg. K* **35/36** Item – primas] *om. ABDEFHIK*: docens invenire eram ignotam per notam *G* **37** habueris] autem habitis *ABCEF*: autem habueris *GI* quarta – prima] quartis, tertiis, secundis et primis *ABCEF* **38** et velis – idem] vel scire hoc idem *A*: scire volueris hoc idem *B*: velis scire *E*: et (et *om. CF*) velis (vel *D*) scire hoc idem *CDF* **39** sequitur] sequatur *AHK* **40** tabulis] *om. ABCDE*

general rule that each value is written with its kind: fourths below fourths, thirds below thirds, seconds below seconds, and units below units. Then enter in the same way with the remainder, if there is one, into the *Tables of either collected or expanded years* as many times as needed. And similarly, with the completed months, enter the *Table of months* in such a way that in non-leap years you enter the upper table of months and in leap years the lower table of months. Always write beside the table what you find there directly, each value under its kind, the total number of years and months found. If, however, some days of an uncompleted month remain, because they are units, write them below the units. Then, add everything together starting with the units. And if the sum of units exceeds 60, then for these 60 add one in the order of seconds. In the same way, treat the seconds in respect to thirds, and so on. The other remainders, which cannot complete 60, will remain in their positions. In this way, the fourths, thirds, seconds, and units will result which were contained in the total number of years, months and days; with this result, enter the *Table of mean motions*.

Chapter two

On converting the years of any era to fourths, thirds, seconds, and units[3]

If you have fourths, thirds, seconds, and units from the beginning of some era known to you and you would like to know the same from the beginning of another era, see if that era precedes your era or follows after it. You can recognize it using the *Tables of differences of eras*. If it precedes it, add the difference between that era and that known to you, which you will find in the same tables, to the number you have. If that era follows, subtract their difference and you will find the required result.

3 This title, written according to Manuscript C, does not correspond to the content of the chapter, which does not deal with the converting of years of a certain era into fourths etc., but with the converting of the number of days (given in fourths, thirds, seconds, and units) from one era to another.

Capitulum tertium

De sciencia annorum cuiuslibet here per quartas et cetera

Si autem propositis quartis, tertiis, secundis et primis, que sunt a principio alicuius ere usque ad aliquod tempus propositum, velles scire, quot anni, menses et dies in eisdem contineantur et est conversum capituli primi, sic facies: quere illa in tabulis propriis illius ere, si precise poteris invenire, et annos, quos inveneris scriptos in directo, sunt anni, quos queris. Si autem illa non inveneris precise, quere in eisdem tabulis numerum minorem propinquiorem et numerum annorum in directo inventum extra scribe. Et postea cum residuo intra et fac, ut prius, et hoc tociens fac tam in annis collectis, quam expansis, quousque totus numerus quartorum, tertiorum, secundorum et primorum consumatur vel saltem tam parvus remaneat, quod non possit complere mensem sequentem illos menses, quos per operationem habuisti, si aliquos habuisti. Et tunc illud residuum erunt dies mensis imperfecti sequentis menses, quos per operationem habuisti, si aliquos invenisti, vel erunt dies primi mensis nondum perfecti, si nullum mensem habuisti. Et sic per istud capitulum et per capitulum precedens poteris scire omnes annos, menses et dies, qui sunt a principio alicuius ere, cuiusvis ex eris hic positis, dum tamen una sit nota usque ad quodcumque tempus vis determinatum. Verbi gracia: quarta, tertia, secunda et prima, que sunt a principio ere Christi usque ad aliquod tempus, sunt tibi nota per primum capitulum, quia anni Christi sunt tibi noti, sed quarta, tertia, secunda et prima, que sunt a principio ere Arabum, sunt etiam tibi nota per secundum capitulum. Et per istud tertium capitulum erunt tibi noti anni, menses et dies Arabum operando per *Tabulas Arabum*, secundum quod dictum est. Cavendum tamen est in hoc et in primo capitulo, ne annus imperfectus sit bisextilis. Unde sciendum, quod *Tabula mensium* replicatur bis. Unde tabula suprascripta mensium deservit annis non bisextilibus et tabula subscripta deservit annis bisextilibus. Unde si annus inperfectus sit bisexti-

---

43 Capitulum tertium] *om. ABCDEF*: tercium *mg. H*: *om. et in mg. add. I*: 3 *mg. K* **44** De – cetera] *om. ABDEFHIK*: de reductione quartarum, terciarum ad alicuius ere note ad annos *G* **49** tabulis] *om. ABCDE* **49/50** numerum minorem] minorem *AEF*: numerum *BC*: minorem numerum *I* **50** propinquiorem] propinquiorem tamen (tam *E*) *AE*: et propinquiorem *H*: et propinquiorem scilicet quartis *K* Et] *om. ABCDEFG*: subtrahe ergo quarta, tercia, secunda et prima, que ibi invenisti, ab illis quartis, terciis, secundis et primis, que habuisti *G*: prius *I*: et que subtrahe minorem de maiore et *K* **51** intra] vel minori propinquiori tamen intra in eisdem tabulis et numerum annorum in directo inventum extra sub aliis scribe. Deinde cum residuo, si sit residuum, tociens intra *A*: vel minori propinquiori tam et numerum annorum in directo inventorum extra sub aliis scribe. Deinde cum residuo, si sit residuum tociens, intra *E*: illi minori propinquiori intra in eisdem tabulis et numerum annorum in directo inventum extra sub aliis scribe. Deinde cum residuo, si sit […] tociens intra *F*: intra vel cum minori proporcioni tamen et numerus annorum in directo inventum extra sub aliis scribe subtrahendo semper quarta cum terciis, et cetera a tuis quartis et terciis et cetera. Deinde cum residuo, si sit residuum, tociens intra *G* et[1] – prius] *om. AEF*: et fiat ut prius *C* hoc – fac[2]] *om. AEF*: hoc facies (facias *C*) tociens *BC*: hoc fac tociens *D*: sic facies tociens *I* in – collectis] in tabulis annorum collectorum *AEF*: annorum expansorum *G* **52/53** expansis – consumatur] expansorum quam etiam mensium, quousque nichil sit residumm de propositis quartis, tertiis, secundis et primis *AEF*: collectorum quam eciam mensium, quousque totus numerus quartorum, terciorum, secundorum et primorum consumatur *G* **53** tam – remaneat] si sit residuum, sic (si *E*) ita parvum *AEF*: tam parvum remaneat *BDG*: pars remaneat *C*: tam parum remaneat *I* **54** illos menses] menses *AF*: mensem *mg. E* **55** erunt] erit *A*: erit vel erunt *G* **55/56** imperfecti – mensis] *om. B* **55** sequentis] sequentes *ACEFI* **59** quodcumque – vis] quod vis tempus *AF*: quodcumque vis tempus *BC*: quodcumque tempus *DGI*: quamvis tempus *E* determinatum] deteriatum *ABEFH*: determinatum vis *DG*: determinatum volueris *I* **61/62** sed – nota] *om. E* **65** Tabula] tabule *ABCDEFG*: *om. I* **66** replicatur bis] deservientes annis bisextilibus, quelibet (tabula *add. G*) replicatur bis (scilicet sub et supra *add. G*) *ABCDEFG*: replicatur, sed si sub et supra *I*

Chapter three

On finding the years of any era from fourths, et cetera

If, with given fourths, thirds, seconds, and units since the beginning of any era up to a given time, you wish to know how many years, months and days are contained in this [period], which is an inverse of the first chapter, do as follows: search for them in the tables appropriate for that era, and if you can find them precisely, then the years which you find directly written are the years you are seeking. If you cannot find them precisely, search in the same tables for the closest smaller number and record and set aside that directly found number of years. Then, enter with the remainder and work as before. Enter the table repeatedly, in both the collected and expanded years, until the whole number of fourths, thirds, seconds and units are exhausted or at least until so little time remains that it cannot fill up the month following those months which you filled by the previous operations. Then, this remainder will be days of the yet incomplete month after those months that you found by the operations if you have some months, or it will be the days of the first incomplete month if you have none [completed months]. By using this and the previous chapter, you can find all years, months, and days from the beginning of an arbitrary era of those that are placed [in the table], namely, if one is given for the time you want. For example, using the first chapter you can find the fourths, thirds, seconds, and units from the beginning of the Christian era; according to the second chapter, you can find fourths, thirds, seconds and units from the beginning of the era of the Arabs. And from this third chapter, you will know the years, months, and days of the Arabs, using the *Tables of the era of Arabs* according to what has been said. It is only necessary to consider, in this and in the first chapter, whether the unfinished year is not a leap year. However, remember that the *Table of months* is repeated twice, so that the upper serves in non-leap years and the lower serves in leap years.

lis et oporteat intrare *Tabulam mensium*, intra tabulam inferiorem, et si non est bisextilis, intra tabulam superiorem.

Capitulum quartum

De nota cuiuslibet anni vel mensis vel diei invenienda

Si autem velis scire notam alicuius anni vel mensis vel etiam cuiusvis diei, totum tempus precedens istam diem, de qua vis scire, que feria est, incipiens a principio alicuius ere, reduc ad quarta, tertia, secunda et prima per doctrinam primi capituli. Postea cum numero quartorum intra tabulam ad hoc factam, quam invenies per titulum in linea numeri crescente usque ad 60 et dicitur linea communis, quia in eam intramus cum quartis, tertiis, secundis et primis et numerum inventum in directo sub titulo quartorum extra scribe. Deinde eodem modo quere numerum tertiorum in eadem linea communi et numerum in directo inventum sub titulo tertiorum sub alio eodem modo extra scribe. Eodem modo fac de secundis et primis. Deinde ad huc subscribe numerum feriarum, per quem intravit illa era, ad quam operatus es, quam habes in quadam tabula ad hoc facta, que intitulatur *Tabula radicum notarum anni*. Postea totum adde insimul, et quod proveniet, divide per septem et tunc aut aliquod erit residuum, aut nichil. Si nichil, dies sequens intrabit feria septima, et si aliquod fuerit residuum, intrabit per talem feriam, qualis est numerus infra 7 remanens.

Capitulum quintum

De reductione horarum ad fractiones dierum

Cum volueris reducere horas ad fractiones dierum, quod oportet ad hoc, quod habeamus medios motus planetarum per istas tabulas, intra cum horis in tabulam ad hoc

---

68 Tabulam mensium] in tabula mensium (mensium *om. et in mg. add. E*) *ABEF*: *om. C* tabulam inferiorem] in tabula inferiori *AEF* **69** tabulam superiorem] in superiori *A*: in superioribus *E*: in superiori tabula *F* **70** Capitulum quartum] *om. ABCDEF*: quartum *mg. H*: *om. et in mg. add. I*: 4 *mg. K* **71** De – invenienda] *om. ABDEFHIK*: docens invenire notam quamlibet *G* **72** alicuius anni] anni alicuius *AE*: alicuius ere *D*: anni *FHK* **73** istam diem] diem illam (illum *D*) *ABDEI*: die illa et non computes diem illam *C*: diem istam *F*: illam *G* **74** tertia – prima] tertia et cetera *AEF* **75** tabulam] in tabulam *ABF* **76** et – communis] que dicitur Tabula radicum notarum anni vel mensis et cetera et dicitur communis *HK*: *om. I* in – intramus] in ea intramus *AEF*: per eam intramus *B*: in eam intravimus *D*: intramus in eam *G*: cum ea intramus *HK*: interius cum ea tabula *I* **77** tertiis – primis] tertiis et ceteris *AHK*: terciis et secundis et primis *D*: terciis et primis *E*: terciis *F* **78** eadem – communi] eadem linea *ABDEG*: ea linea *F*: linea communi *HK* **79** sub[2] – scribe] extra sub aliis (alio *D*) scribe *AD*: extra scribe sub alio *C*: sub alio (aliis *F*) extra scribe *BEFGI* **80** ad – subscribe] subscribe ad huc *A*: scribe ad huc *E*: subtrahe ad hoc *F*: ad huc adde *HK* **81** habes] habebis *AEG* **82** que – anni] posita in fine istorum canonum (canonum istorum *EG*) *ADEG*: in fine illorum canonum *F*: in hiis tabulis posita *HK*: posita ante quartam tabulam *I* **84** fuerit – feriam] intrabit per talem feriam *AEF*: per talem feriam intrabit *BC*: intrabit, intrabit per talem feriam *HK* **86** Capitulum quintum] *om. ABCDEF*: quintum *mg. H*: *om. et in mg. add. I*: 5 *mg. K* **87** De – dierum] *om. ABDEFHIK* **89** in tabulam] in tabula *AEF*: ad tabulam *G* **89/90** ad – factam] ad hoc facta *AEF*: ad sciendum minuta dierum et cetera ad hoc factam *H*: minuta dierum et quod ad hoc factam *K*

Hence, if the unfinished year is a leap year and one must enter a *Table of months*, enter into the lower table; if the year is non-leap, enter into the upper table.

Chapter four

On finding 'nota' of any year, month, and day[4]

If you wish to know the 'nota'[5] (or day of the week) of any year, month, and also any day, then convert the total time, from the beginning of any era to the day you desire, into fourths, thirds, seconds, and units according to the instructions of the first chapter. Then, enter with the number of fourths into the table given for it. You will find it using the inscription in the column of numbers rising up to sixty which is called the column of common values[6] because we enter it with fourths, thirds, seconds, and units and set aside the number found below the inscription of fourths. Then, search in the same way for the number of thirds in the same column of common values and set aside, in the same way below the previous value, the number that you will find directly below the inscription of thirds. In the same manner, treat the seconds and units. Then, write there also the sequence number of the day of the week that starts the era in which you have worked, a number given in the table provided for this which is entitled *Table of known radices of a year*. Then sum everything together, divide the sum by seven, and you will either have some remainder or not. If there is no remainder, then the following day will be the seventh day of the week. And if something remains, the day of the week will be given by the remaining number which is smaller than seven.[7]

Chapter five

On converting the hours to fractions of days[8]

If you wish to convert hours to fractions of days, which is needed to find the mean motions of planets using these tables, enter with the hours — namely with the number

4 The subject of Chapter (rule) 4 was later explained in Chapter (canon) 6 by John of Saxony, cf. Poulle, *Les Tables alphonsines*, p. 40. It refers to Alfonsine Table No. 7 (*Tabula notarum anni vel mensis cuiuscumque*), cf. Poulle, *Les Tables alphonsines*, p. 122.

5 The term *nota* is used here as a synonym of *feria*, 'day of the week', similarly to John of Saxony's Canon 5: *intelligendum est primo, quod nota secundum quod hic accipitur, est idem quod feria* (Poulle, *Les Tables alphonsines*, p. 40) — 'it is needed to know first that *nota* as it is used here is the same as *feria* (a day of the week)'.

6 Latin expression *linea* means column (in a table), cf. Poulle, *Les Tables alphonsines*, p. 27; the term *linea communis* means column of common values (usually the leftmost column of independent variables); *linea numeri* means column of units.

7 The calculation is described by Poulle, *Les Tables alphonsines*, pp. 191–92 for an example of Friday 3 July 1327, where the remainder is 6. The remainder 0 is thus Saturday, 1 is Sunday, etc.

8 The subject of Chapter 5 was later explained in Chapter 7 by John of Saxony, cf. Poulle, *Les Tables alphonsines*, p. 42. It relates to the Alfonsine table No. 8 (*Tabula conversionis horarum in minuta et secunda dierum*), cf. Poulle, *Les Tables alphonsines*, p. 123.

factam, scilicet cum numero horarum, et invenies in directo minuta et secunda dierum propositis horis correspondentia. Dicitur autem minutum diei 60 pars diei unius.

Capitulum sextum

De reductione minutorum et ceterarum fractionum horarum ad fractiones dierum

Si autem velis reducere minuta horarum et alias fractiones ad fractiones dierum, cum numero illorum minutorum, seu aliquarum aliarum fractionum, intra in tabulam ad hoc factam in linea communi, que crescit usque ad 60 et dicitur communis, quia tenetur pro quavis fractione horarum. Si igitur intras cum minutis, invenies in directo in secunda linea minuta et in tertia linea secunda et in quarta linea tertia illis minutis correspondentia. Si tamen intres cum secundis, quod invenies in secunda linea, erunt secunda, et quod invenies in tertia, tertia et sic de ceteris. Et si etiam intrares cum tertiis, quod invenies in secunda linea, erunt tertia et deinde quarta. Unde sicut prima linea dicebatur linea communis, ita quelibet linearum sequentium potest dici communis, quia quelibet accipitur pro diversis fractionibus, quod denotatur per quandam parvam tabulam subscriptam, que potest dici titulus inferior ipsius tabule.

Capitulum septimum

De reductione fractionum dierum ad fractiones horarum

Si autem velis reducere fractiones dierum ad fractiones horarum, facies eodem modo, quo dictum est in capitulo precedenti, per tabulam ad hoc factam, intrando in linea communi crescente usque ad 60. Et similiter alie due linee sequentes illam lineam dicuntur communes, quia si intres cum minutis dierum, quod invenies in secunda linea, erunt hore, et quod in tertia, minuta. Et si intres cum secundis, quod invenies in eadem

90 minuta] minuta dierum *AFGI*: prima dierum *E* secunda dierum] secunda *AEFGHIK* **91** Dicitur – diei[1]] dicuntur autem minuta dierum *ABEFGI* diei unius] unius diei *AFG*: diei *BC*: unius hore *E* **92** Capitulum sextum] *om. ABCDEF*: sextum *mg. H*: *om. et in mg. add. I*: 6 *mg. K* **93/94** De – dierum] *om. ABDEFHIK*: ad sciendum minuta dierum et eius fractiones per minuta horarum et alias fractiones *G* **99** minuta – linea[2]] minuta et in tertia *ADF*: erunt minuta et in secunda erunt *C*: communi et in communi *E*: minuta dierum et in tercia linea *GI* **101** et quod – ceteris] deinde tercia et (et *om. FG*) post (post *om. B*) quarta *ABFG*: et sic deinceps *C*: deinde tercia *DE*: deinde quarta *HK* **101/102** Et – quarta] *om. HK* **103** linearum] trium linearum *AEFG* communis[2]] linea communis *AEFG* **104** tabulam] *om. A* subscriptam] sub illa scriptam *AEF* **106** Capitulum septimum] *om. ABCDEF*: septimum *mg. H*: *om. et in mg. add. I*: 7 *mg. K* **107** De – horarum] *om. ABDEFHIK*: ad sciendum horas et suas fractiones per minuta dierum *G* **109** tabulam – factam] tabulam ad hoc factam, que dicitur Tabula ad sciendum horas et earum fracciones et cetera (et cetera *om. K*) *HK*: tabulas ad hoc factas *I* **112** Et] *om. A*

of hours — into the table which is prepared for this, and you will find directly the minutes and seconds of days which correspond to the given hours. One sixtieth of a day is denoted as one minute of a day.[9]

Chapter six

On converting minutes and other fractions of hours to fractions of days[10]

If you wish to convert minutes of hours and other fractions to fractions of a day, enter with the number of these minutes or other fractions into the table prepared for this, namely into the column of common values, which rises up to sixty and is called common because it is suitable for any division of hours. If you thus enter into the table with minutes of hours, then you will directly find minutes in the second column, seconds in the third, and thirds in the fourth column, which correspond to these minutes. If you enter with seconds, you will find seconds in the second column, thirds in the third column, and so on. And similarly, if you enter with thirds, you will find thirds in the second column and then fourths. As the first column was named common, similarly each of the following columns can be named common because each can be considered for different fractions, which is indicated by means of a small table inscribed below it, which can be named the lower inscription of this table.[11]

Chapter seven

On converting fractions of days to fractions of hours[12]

If you wish to convert fractions of days to fractions of hours, work in the same way as described in the previous chapter, using the table that is made for this, namely, so that you enter the column with the common value increasing up to sixty. And similarly, the next two columns that follow after this common column, are also named common; because if you enter with the minutes of days, what you will find in the second column will be the hours, and in the third column will be the minutes.[13] And if you enter with seconds,[14] then what you will find in the same second column will be minutes of hours,

9 It is twenty-four minutes of an hour.

10 The subject of Chapter 6 was later explained in Chapter 7 by John of Saxony, cf. Poulle, *Les Tables alphonsines*, p. 44. It relates to the first sub-table of the Alfonsine table No. 8 (*Tabula ad sciendum minuta dierum et eorum fractiones per minuta horarum et eorum fractiones*), cf. Poulle, *Les Tables alphonsines*, p. 123.

11 The 'lower inscription of this table' refers to the small triangular table at the end of the first sub-table in the Alfonsine Table No. 8 (Poulle, *Les Tables alphonsines*, p. 123), which has a similar purpose to the inscriptions at the top of the table.

12 The subject of Chapter 7 was later explained in Chapter 8 by John of Saxony, cf. Poulle, *Les Tables alphonsines*, p. 44. It relates to the second sub-table of the Alfonsine table No. 8 (*Tabula ad sciendum horas et horarum fractiones per minuta dierum et eorum fractiones*), cf. Poulle, *Les Tables alphonsines*, p. 123.

13 I.e., minutes of hours.

14 I.e., seconds of days.

secunda linea, erunt minuta horarum, et quod in tertia, secunda, secundum quod potes videre per titulum inferiorem, id est per parvam tabulam subscriptam.

Capitulum octavum

De medio motu planetarum inveniendo

Cum volueris scire medios motus planetarum in aliquo tempore tibi noto incipiente a principio alicuius ere ex eris hic positis, oportet, sicut dictum est, totum illud tempus reducere ad quarta, tertia, secunda et prima. Si autem cum diebus integris haberes horas et minuta horarum, oportet reducere ad fractiones dierum, secundum quod superius dictum est. Unde sciendum est, quod in *Tabulis de mediis motibus* linea prima, que crescit continue usque ad 60, dicitur linea communis, quia tenetur pro omnibus differentiis temporum, sicut pro quartis, tertiis, secundis et primis. Et etiam tenetur pro omnibus fractionibus dierum, ut in ea omnis differentia temporis, et in directo medius motus illius differentie inveniatur, et ideo superius intitulatur prima, secunda, tertia, quarta et inferius minuta, secunda, tertia, quarta. Preterea sciendum est, quod ad minus post lineam communem sunt 8 linee et intitulantur quatuor prime prima, secunda, tertia, quarta et iterum replicantur isti tituli super quatuor postremas lineas, secundum quod ibi apparet. Sed sciendum, quod ista prima, secunda, tertia et quarta non representant idem, quod representant in linea communi et etiam in capitulis precedentibus, quia ibi representant fractiones temporis, sed hic significant signa et gradus et fractiones graduum. Tempore igitur proposito, reducto ad quarta, tertia, secunda et prima, intrabimus primo cum numero quartorum in linea communi, et quod invenimus in directo sub titulo quartorum, quod est in quinta linea versus dextram, cum eo, quod sequitur procedendo directe versus dextram, scribemus extra eo ordine, quo est in tabula. Et tunc illud, quod invenisti sub titulo quartorum, scilicet in quinta linea, sunt signa, et illud, quod invenisti in sexta linea, sunt gradus et in septima sunt minuta et sic consequenter. Postea vero eodem modo intrabis cum numero tertiorum in eadem linea communi et accipias, quod invenies in directo sub titulo tertiorum cum aliis etiam consequenter procedendo versus dextram sicut prius et illud, quod modo invenis sub titulo tertiorum, sunt signa, deinde sequuntur gradus, deinde minuta et cetera secundum ordinem. Scribas igitur, que hic accepta sunt sub tertiis, sub aliis extra scriptis, sub quartis scilicet, quia sunt signa. Deinde alia consequenter scribantur secundum ordinem. Eodem modo intrabimus cum secundis in linea communi accipiendo, quod erit in directo sub linea secundo-

---

113 horarum] *om. ABCDEFI* **115** Capitulum octavum] *om. ABCDEF*: octavum *mg. H*: *om. et in mg. add. I*: 8 *mg. K* **116** De – inveniendo] *om. ABDEFHIK*: ad inveniendum medios motus planetarum *G* **117** in] ab *BCDEFHIK* **126** Preterea – est] postea sciendum *AFG*: propterea sciendum est (est *om. CD*) *BCD*: sciendum *E*: et sciendum *I* **126/129** ad minus – ibi apparet *om. I* **128** postremas lineas] postremas *ACDEF*: lineas postremas *BG* **133** communi] *om. A* **138** eodem – intrabis] eodem modo intrabimus *ABCDFI*: intrabimus eodem modo *G* **141** sequuntur – ordinem] sequentia gradus, deinde minuta pre [*sic*] ordinem (*del.*: eodem) *A*: gradus et tibi *B*: gradus et cetera *CDHK*: sequuntur gradus et deinde minuta secundum ordinem *F*: sequencia gradus, deinde minuta secundum ordinem *G* **143** scribantur] subscribantur *A*: scribas *B* **144** communi] numeri *A*

and in the third column will be the seconds,[15] which you can see in the lower title written as a small table.[16]

Chapter eight

On finding the mean motion of planets[17]

If you wish to know mean motion of planets at some time known to you from the beginning of any era that is given in the tables, it is necessary, as noted, to convert that whole time to fourths, thirds, seconds, and units. If you have, together with the whole days, the hours and minutes of hours, convert them to fractions of days, as described above. Note that the first column in the *Tables of mean motions*, which gradually increases up to sixty, is called the common values because it is valid for all differences of time, that is, for fourths, thirds, seconds, and units. It is also valid for all fractions of days because these are found in all differences of time and also directly as the mean motion corresponding to this difference. That is why it is labelled unit, second, third, and fourth above, and then below as minute, second, third, and fourth. Furthermore, note that after the column of common values there follow at least eight other columns. The first four of them are entitled unit, second, third, and fourth, and these titles are repeated again for the four final columns, as can be seen. And know that these units, seconds, thirds, and fourths do not represent the same quantities that they represent in the column of common values in the previous chapters because there they represent parts of time, but here they represent signs,[18] degrees, and fractions of degrees.[19]

Once we have the given time converted to fourths, thirds, seconds, and units, we first enter the column of common values with the number of fourths and what we find directly under the fourths heading, which is in the fifth column to the right, together with what follows if we proceed directly to the right, we write down in the same order given in the table. And what you have found under the fourths heading, namely in the fifth column, are the signs. And what you have found in the sixth column are degrees, in the seventh are minutes, and so on. Then, enter in the same way the column of common values with the number of the thirds and find directly under the heading of thirds, and continuing to the right as before, the signs, followed by degrees, then minutes and others, according to their order. And write down beside the table what you have found below the thirds, which means below the fourths, because these are also the signs. Then, the other values will be written sequentially according their order. Similarly, we enter into the column of common values with seconds, taking what will be directly in the

15 I.e., seconds of an hour.

16 'Table placed below' refers to the small triangular table; such a table is placed, for example, at the end of the second sub-table of Alfonsine Table No. 8 (cf. Poulle, *Les Tables alphonsines*, p. 123).

17 The subject of Chapter 8 was later explained also in Chapter 12 by John of Saxony, cf. Poulle, *Les Tables alphonsines*, p. 50. It relates to Alfonsine Tables 15, 16, 21, 22, and 23 (*Tabula medii motus Solis, Veneris et Mercurii*; *Tabula medii motus Lune*; *Tabula medii motus Saturni*; *Tabula medii motus Iovis*; *Tabula medii motus Martis*), cf. Poulle, *Les Tables alphonsines*, pp. 134, 135, and 140–42.

18 Here and in the following text, the 'sign' means the so-called 'physical sign', which has sixty degrees to keep the strictly sexagesimal counting of angles. The whole circle of the zodiac thus contains six physical signs only, each one consisting of two thirty-degree 'natural signs'.

19 The dual type of variables, on the temporal and angular variables, is pointed out here.

rum et aliarum fractionum consequenter procedendo versus dextram, et quod nunc accepisti sub secundis, sunt signa. Subscribantur igitur cum aliis extra scriptis, signa scilicet sub signis et alia consequenter secundum ordinem. Eodem modo intrabimus cum primis faciendo sicut prius et hoc, quod accepisti sub titulo primorum, erunt signa. Subscribantur igitur aliis extra scriptis, scilicet signa sub signis et alia consequenter. Si iterum habueris minuta et secunda dierum et cetera, cum quibus velis intrare, cum numero illorum intra in linea communi, et quod inveneris in directo in prima linea sequenti lineam communem, erunt gradus et in sequenti minuta et sic consequenter, et subscribantur aliis gradus, scilicet sub gradibus, et cetera consequenter. Si autem intrares cum secundis, quod invenies in directo in linea sequenti lineam communem, erunt minuta et sic consequenter. Subscribantur ergo sub aliis, minuta scilicet sub minutis, secunda sub secundis et tertia sub tertiis et sic de aliis consequenter, quodlibet sub suo genere. Secundum ergo varietatem fractionum dierum variantur similiter fractiones signorum et graduum in linea sequenti lineam communem et in aliis etiam consequentibus, et que fractio graduum correspondeat fractioni diei, habetur per quandam parvam tabulam existentem sub qualibet *Tabula mediorum motuum*, que potest dici titulus inferior. Hiis igitur extra scriptis eo ordine, quo dictum est, aggrega omnia adinvicem, incipiendo ab ultimo genere fractionum graduum, et sint prope tertia. Verbi gracia: tertia aggregabis tertiis adinvicem et si ex agregatione tertiorum proveniant 60 vel plus, pro illis 60 ponemus unitatem in secundis et residuum, si sit, stet in loco suo. Eodem modo faciemus de secundis et minutis et etiam de gradibus, quia in istis tabulis 60 gradus accipiuntur pro uno signo et 6 signa pro 12 signis, scilicet pro toto circulo. Et si ex agregatione signorum proveniant plura signa quam 6 signa, proiciamus inde 6 signa, quociens poterimus, et residuum sit loco suo, vel 0, si nichil remaneat. Quo facto proveniet medius motus in toto tempore proposito sine radice. Addatur ergo radix illius ere, a qua anni tui sumpserunt inicium, que invenietur in *Tabulis radicum*. Iste est modus inveniendi medium motum Solis et medium motum omnium planetarum et medium motum capitis draconis et medium argumentum Lune, Veneris et Mercurii et argumentum latitudinis Lune et medium motum accessus et recessus octave spere et etiam motum augium penitus sine aliqua diversitate et medium motum elongationis Solis et Lune.

---

145 fractionum] *om. ABCF* nunc] nunc etiam *AG*: nunc ibi quos *F*: tunc *I* **146** signa[1]] gradus 18 *A*: signa erunt signa *G* Subscribantur] scribantur *AI* cum] *om. ABCFG* **147** secundum ordinem] *om. ABCFG* **148** et hoc] quia etiam hoc *A*: *om.* *C*: quia et hoc *F*: et hoc eciam *G*: et *I* **148/149** quod – consequenter] *om.* *C* **148** accepisti] (hic *add. G*) acceptum est *AG*: accipiamus *I* **149** aliis – sub] aliis signis, scilicet sub *A*: sub aliis signis *BG*: aliis scilicet signis sub *F* **150** habueris] haberes *AEFG*: habes *CI*: habeas *D* minuta – dierum] minuta dierum et secunda *AEFG*: numerum dierum *I* **151** inveneris] invenies *AEFGHK* prima linea] linea prima *AF*: secunda linea *I* **153/155** Si – consequenter] *om. I* **154** quod invenies] quod invenires *A*: quod inveneris *DH*: quod *E*: tunc quod invenies *G*: quod invenis *K* **155** ergo – aliis] igitur aliis *ABFI*: igitur sub aliis, scilicet *D*: aliis *E*: eciam aliis *G*: ergo *H*: *om. K* **156** secunda – secundis] *om. ABCEFGI*: et secunda sub secundis *HK* **156/157** et tertia – genere] et cetera *ADE*: et cetera et sic de aliis consequenter *BC*: et cetera consequenter *F*: et sic consequenter *G*: et cetera de aliis consequenter *I* **158** in[2]] *om. AF* **162** sint – tertia[1]] sint propter *A*: sicut propter *BE*: sint ipse *C*: sic prope tercia *D*: et si sint *F*: sint *G* **162/163** tertia[2] – tertiis] tertia (tertia *om. F*) agregabimus igitur tercia *AF*: tercia agregabis tercia *B*: aggregabis tercia *D*: tercia agrega ergo tercia *E*: signa, gradus, minuta, secunda, tercia aggrega ergo tercia *G*: aggregabis omnia (omnia *om. I*) tercia *HIK* **164** in[1]] sub *AEFG* **168** et] et si *A* **171** medium[1] – planetarum] medios motus planetarum *AFG*: medium motum omnium planetarum *BCD*: medios motus *E*: motum medium omnium planetarum *I* **173** Lune] *om. ABCDEI* octave spere] circuli octavi *B*: octavi circuli *C*

column of seconds and other fractions continuing to the right, and what you now have below the seconds are the signs. They will thus be written below the others, that is, signs below the signs and others continuing according to the order. In the same way, enter with the units as before, and what you will find below the heading of units will be the signs. These will also be written below the other values, that is, signs below signs and the others continuing.

Also, if you have minutes and seconds of days and so on that you wish to enter, enter the column of common values with this number and what you find directly in the first column, which follows after the column of common values, will be the degrees, and in the next column will be the minutes, and so on. And write these down below the other degrees and so on, subsequently. And if you enter the column of common values with seconds, what you find directly in the column which follows the column of common values will be the minutes and so on. They will be written below the others, minutes below minutes, seconds below seconds, thirds below thirds and so on, each below its kind. Thus, according to the change of fractions of days, similarly also the fractions of signs and degrees are changing in the column which follows after the column of common values and also in the subsequent columns, and the fraction of degrees which corresponds to the fraction of a day will be taken according to the small table placed below any *Table of mean motions,* which can also be called the lower inscription. After having written down everything beside the table in the order here described, then sum everything together, starting from the last order of fractions of degrees, that is, near the thirds. For example, add thirds with thirds, and if the sum is sixty or more, then for these sixty a unit will be given to seconds, and if anything remains it will stay in its place. And treat similarly the seconds, minutes, and also degrees, because in these tables sixty degrees has a value of one sign and six signs the value of twelve [thirty-degree] signs, that is, of a whole circle. And if the addition of signs results in more than six signs, we shall thus remove six signs as many times as possible, and there will be a remainder, or zero if nothing remains.

In this way, the mean motion over the entire interval of time, without the radix, will be found. The radix of the era from which your time interval began must be added and can be found in *Tables of radices*. Such is the procedure to find the mean motions of the Sun, all planets, and of the head of dragon; the mean argument of the Moon, Venus, and Mercury; the argument of lunar latitude; the mean motion of access and recess of the eighth sphere; and also the inner motion of the apogees without any correction and the

Notandum tamen, quod omnia ista habebis ad meridianum Toleti. Si autem vis habere hoc ad aliam civitatem, que sit orientalior vel occidentalior, vide, in quanto tempore sit orientalior vel occidentalior, et in tanto tempore quere omnia ista nunc enumerata et subtrahe pro civitate orientaliori et adde pro occidentali. Notandum etiam pro capitulo sequenti, quod tempore medie coniunctionis invento per *Tabulam medie elongationis Solis et Lune* additur tempus illud, scilicet existens inter duos meridianos pro civitate orientali, et removetur pro occidentali. Tamen medio motui Solis et Lune et medio argumento Lune et argumento latitudinis invento hora medie coniunctionis non fit addicio nec etiam subtractio.

Capitulum nonum

De media coniunctione et oppositione Solis et Lune invenienda

Cum volueris scire coniunctionem Solis et Lune mediam et oppositionem, intra cum quocumque tempore volueris in *Tabulam medie elongationis Solis et Lune* per modum dictum in capitulo precedenti penitus et si operatione facta remaneant 6 signa precise, tunc in illa hora est media coniunctio Solis et Lune. Et si remaneant 3 signa precise, tunc in illa hora est media oppositio eorum. Si autem non remaneant nec 6 signa nec 3 precise, tunc elongationem inter Solem et Lunam inventam in illa hora subtrahe a 6 signis, si vis habere coniunctionem. Vel si ista elongatio fuerit minor 3 signis et vis scire oppositionem, subtrahe a tribus signis, et cum residuo intra in eandem *Tabulam medie elongationis*, si ipsum precise poteris invenire. Si autem non, intra cum numero minori propinquiori tamen et fac eodem modo penitus, sicut fecisti in capitulo tertio, scilicet in reducendo quarta, tertia, secunda et prima ad annos, menses et dies, scilicet accipiendo de prima linea in directo residui dies, minuta, secunda et cetera, que adhuc restant ultra tempus propositum usque ad coniunctionem vel oppositionem istius temporis, ad quam operatus es. Unde si in residuo illo contineantur aliqua signa vel unum signum, illud, quod invenies in directo in linea communi, erunt dies, et si non essent nisi gradus cum aliis fractionibus et possent inveniri in prima linea sequenti lineam communem,

---

175 omnia – habebis] omnia habebis *A*: illa (illud *D*) habebis *BD*: hoc est *C*: ista habebis *I* Toleti] Toletum *A*: Tholeti. Vide, in quanto tempore sit orientalior vel occidentalior *B*: Tolletanum *C*: Tholeti *EG*: Tolleti *HK* **175/176** vis – hoc] velles hoc *AF*: hoc volueris *B*: hoc velles *C*: volueris hoc *D*: veles hec *E*: velles omnia ista habere *G*: velles habere *I* **178** Notandum etiam] nota etiam *AHK*: notandum, quod meridies Toleti sequitur meridiem Parisiensem 46 minutis et quod longitudo meridianus Parisiensis est ab occidente vero 40 gradus 30 minutis et eius latitudo est 48 gradus 50 minutis et eciam *D*: notandum autem *I* **179** quod] scilicet quod *AG* tempore] tempori *AHK* **181** Tamen] cum *BEI*: tantum *DG* **184** Capitulum nonum] *om. ABCDEF*: nonum *mg. H*: *om. et in mg. add. I*: 9 *mg. K* **185** De – invenienda] *om. ABDEFHIK*: ad inveniendum medium coniunctionem Solis et Lune et eorum oppositionem *G* **186** cum] in *ABCF* **189** tunc[1] – est] tunc est illa hora *C*: illa est hora *D*: tunc in ista hora est *E*: illa hora est (erit *K*) *HK* **190** nec 3 precise] precise nec 3 precise *A*: precise nec tria signa precise *B*: precise nec precise tria *C*: precise neque tria signa precise *D*: prescise (*sic*) nec tria signa *E*: precise nec tria *F*: vel tria precise *G*: precise nec 3 signa precise *I* **192/193** si – signis] si illa elongatio fit (fuerit *D*) minus 3 signis et velis (volueris *D*) scire oppositionem (opposicionem habere *D*), subtrahe a 3 signis *AD*: a tribus signis, si volueris opposicionem *B*: a tribus, (sed *add. F*) si vis opposicionem *CF*: si sit elongacio (elongacio est *I*) minus tribus signis et velis (vis *I*) scire oppositum (opposicionem *I*), subtrahe a tribus signis *EI*: si sit minus 3 signis *G* **194** Si – non] sin autem *ABCDEFG*: si autem *I* intra] *om. AEI* numero] *om. ABCEFGHIK* **195** fac – penitus] facias penitus *ADEFHIK*: fac penitus *G* capitulo tertio] tertio capitulo *AEFGI* **197** minuta – cetera] minuta et secunda et cetera *AF*: mensis *B*: minuta et dies, minuta (secunda *add. G*) et cetera *DG*: in minuta, secunda et cetera *E* **198** istius temporis] illius scilicet *ACEFHK*: illius autem *B*: *om. D*: istius *G* **201** inveniri] invenire *A*

mean motion of elongation of the Sun and Moon. And note that all radices are given for the meridian of Toledo. If you want to know radices for another city, to the west or east, determine how much time it is east or west of Toledo, and for that time search all the radices listed here and for an eastern city subtract and for a western city add. In the next chapter on finding the time of mean conjunction using the *Table of mean elongations of the Sun and Moon*, the time difference is added for an eastern city and subtracted for a western one. However, to find the mean motion of the Sun and Moon, the mean argument of the Moon, and the argument of the latitude at the hour of the mean conjunction, nothing is added or subtracted.

Chapter nine

On finding the mean conjunction and opposition of the Sun and Moon[20]

If you wish to know the mean conjunction of the Sun and Moon and their opposition, enter, with the time you want, the *Table of mean elongations of the Sun and Moon*, as discussed in the previous chapter. And if, after the completing this step, precisely six signs remain, then in that hour the mean conjunction of the Sun and Moon will occur. And if precisely three signs remain, then their mean opposition will occur in that hour. If, however, it is not precisely six signs or three that remain, then subtract the elongation of the Sun and Moon found for that hour from six signs if you want the conjunction. And if this elongation is less than three signs and you would like to know the opposition, subtract it from the three signs and enter with the remainder into the same *Table of mean elongations*, if you can find it precisely. If not, enter with the closest smaller number and proceed as you did in the third chapter when you converted the fourths, thirds, seconds, and units to years, months, and days. Take from the first column directly the number of days, minutes, seconds, and so on that remain after the given time of the conjunction or opposition. If the remainder contains several signs or one sign, then what you find directly in the column of common values will be the days. And if you have only degrees with some fractions and they can be found in the first column after the column with common values, then what is found directly in the column of common values are

20 The subject of Chapter 9 was later explained in Chapter 13 by John of Saxony, cf. Poulle, *Les Tables alphonsines*, p. 54.

tunc quod in directo continetur in linea communi, erunt minuta. Si autem non posssent inveniri, sed oporteret querere in alia linea sequenti, tunc illud, quod continetur in directo in linea communi, sunt dies. Unde secundum quod illud residuum sit, gradus vel minuta vel secunda, invenitur in prima linea sequenti lineam communem vel in alia sequente, quando non potest inveniri in prima. Secundum hoc etiam variantur fractiones temporis invente in directo in linea prima. Et hoc invenitur per quandam tabulam parvam positam sub titulo inferiori ipsius *Tabule medie elongationis.* Unde notandum, quod si isto modo velles querere coniunctionem et oppositionem primam, que est in principio anni cuiuslibet, et ad illud tempus etiam querere medium motum Solis et Lune et medium argumentum Lune et argumentum latitudinis, similiter postea habebis omnia illa eadem in toto illo anno per *Tabulam coniunctionis et oppositionis in mensibus,* addendo scilicet dies, horas et minuta, que sunt a principio Ianuarii usque ad tempus illius coniunctionis prime vel oppositionis cum diebus, horis et minutis et cetera unius coniunctionis invente in directo primi mensis. Si vis primam coniunctionem post vel duarum coniunctionum si vis coniunctionem secundam post et sic de aliis sequentibus illius anni, de quacumque illarum volueris et habebis postea dies, horas et minuta, que sunt a principio Ianuarii usque ad illam coniunctionem. Si autem annus bisextilis sit tunc et transivit locus bisexti, tunc illa coniunctio erit in die precedenti in simili fractione. Sed nota, quod minuta et secunda prime coniunctionis, que sunt inventa per *Tabulam medie elongationis,* oportet reducere, quia sunt fractiones diei ad horas et ad fractiones horarum, quia fractiones contente in *Tabula coniunctionis et oppositionis in mensibus* sunt hore et fractiones horarum. Similiter etiam cum medio motu et argumento medio et argumento latitudinis prime coniunctionis anni addas medium motum et medium argumentum et argumentum latitudinis illius coniunctionis, quam queris, inventos in *Tabula mensium,* et habebis illam coniunctionem propositam. Sed nota, quod ista non diversificantur in anno bisextili.

Capitulum decimum

De loco augis cuiuslibet planete inveniendo

Cum volueris invenire locum augis cuiuslibet planete, quod oportet ad hoc, quod inveniantur vera loca omnium planetarum, queras primo medium motum augium in *Tabula motus augium,* secundum quod dictum est, et serva. Deinde eodem modo quere medium motum octave spere, scilicet accessum et recessum, addita radice eius. Quo

203 illud] *om. ABCEFG* **205/206** vel[2] – sequente] vel in (in *om. C)* alia secunda *ABCFG*: vel minuta, secunda *E*: et in (in *om. HK)* alia sequente *HIK* **207** invenitur] habetur (habentur *F) ABCDFI*: videtur *EG* **210** etiam querere] querere (quere *G)* eciam *AEFG*: eciam velles querere *I* **220** coniunctionis que] fraccionis coniunccionis, que *E*: coniunctionis anni *AG* **221/222** diei – horarum] dierum ad fractiones horarum et ad horas *A*: dierum (et *add. I)* ad horas et fracciones horarum *BCFGI*: *om. DE* **222/223** quia – fractiones horarum *om. I* **228** Capitulum decimum] *om. ABCDEF*: decimum *mg. H*: *om. et in mg. add. I*: 10 *mg. K* **229** De – inveniendo] *om. ABDEFHIK*: ad inveniendum loca augium cuiuslibet planete *G* **230** Cum volueris] Si velis *ACEF*: Si volueris *B*: cum velis *G* **231** vera loca] loca vera *AEF*: loca *BC* medium] *om. ABCDEFGI* **233** spere] scilicet accessum et recessum *suprascr. H*: scilicet accessus et recessus *K*

minutes. If, however, degrees cannot be found, then it is necessary to search in the following column and what is found directly in the column of common values will be the days. Thus, according to what the remainder is, either degrees, minutes, or seconds and others will be found in the first column which follows after the column of common values or in some subsequent column if it cannot be found in the first one. And according to the remainder, the fractions of time vary; these can be found directly in the first column and they can be found using the small table below the lower title of the *Table of mean elongations*.

And note that if you wish to search in this way for the first conjunction and opposition at the beginning of some year, and if you wish at the same time to search for the mean motion of the Sun and Moon, the mean argument of the Moon and the argument of latitude, you will find these all similarly for the whole year using the *Table of conjunctions and oppositions in the months*, namely if you add the days, hours, and minutes, which are from the beginning of January up to the time of the first conjunction or opposition, to the days, hours, minutes, and so on of the already directly found conjunction in the first month. If you want to know the next conjunction after this one,[21] or the second one, and so on, for all other conjunctions of that year, then you will have the days, hours and minutes, which are from the beginning of January until that conjunction. If, however, it is a leap year and the leap day has passed, then that conjunction would be in the previous day in the same part of the day. But note that it is necessary to convert the minutes and seconds of the first conjunction, which were found using the *Table of mean elongations*, which are fractions of the day, into hours and fractions of hours because the parts contained in the *Table of conjunction and opposition in months* are hours and fractions of hours. Similarly, add also to the mean motion, to the mean argument and to the argument of the latitude of the first conjunction of the year the mean motion, mean argument and argument of the latitude of the sought conjunction which you have found in the *Table of months* and you will have that presumed conjunction. And note that this does not differ in a leap year.

Chapter ten

On finding the position of apogee of any planet[22]

If you wish to find the position of apogee of any planet, which is necessary for determining the true positions of all planets, search first for the mean motion of apogees in the *Table of motion of apogees*, according to what has previously been said, and set it aside. Then, in the same way, search for the mean motion of the eighth sphere, that is,

21 The Latin text is confusing here.

22 The subject of Chapter 10 was later explained in Chapter 16 by John of Saxony, cf. Poulle, *Les Tables alphonsines*, p. 62.

invento resolve ad gradus, si ibi contineantur signa. Deinde cum numero illorum graduum intra *Tabulam equationis motus octave spere sive capitis Arietis*, que crescit usque ad 90 gradus, quos gradus ibi invenies, si sint pauciores 90. Si autem sint a 90 usque ad 180, computandum est econverso ita, quod 89 sint 91 et 88 sint 92 et sic usque ad principium computando ita, quod gradus 180 nichil habent de equatione. Si autem sint a 180 usque ad 270, invenies illos taliter sicut prius, quod primus gradus sit 181, secundus 182 et sic usque ad 90, qui erit etiam 270. Si autem sint a 270 usque ad 360, incipies computare ordine econverso ita scilicet, quod 89 sint 271 et 88 sint 272 et sic usque ad 360, qui etiam nichil habent de equatione. Cum numero igitur graduum intra *Tabulam equationis octavi circuli* et in directo eius invenies equationem motus octave spere. Si autem cum gradibus contineantur minuta, intra tunc iterum cum illis gradibus eandem *Tabulam equationis* addito uno gradu et suscipe equationem in directo existentem et tunc vide differentiam istius equationis secunde et prime et illius differentie accipe partem proportionalem secundum proportionem minutorum, que erant cum gradibus ad 60. Que pars proportionalis debet addi equationi prime, si prima equatio fuerit minor secunda, vel ab ea removeri, si fuerit maior. Et quod postea provenerit, erit equatio motus octave spere illis gradibus et minutis correspondens. Que quidem equatio erit addenda, si gradus et minuta, mediantibus quibus eam invenisti, fuerint ab uno gradu usque ad 180. Si vero fuerint a 180 usque ad 360, est minuenda. Si igitur est addenda, adde eam cum motu augium iam servato, quod totum adde radici augium cuiuslibet planete invente in sua tabula illius ere, scilicet ad quam operatus es. Si autem sit minuenda, tunc adde motum augium prius servatum eidem radici et ab illo toto subtrahe equationem motus octave spere et proveniet tibi locus augis cuiuslibet planete preterquam de Luna, quia alio modo invenitur. Notandum tamen est hic, quod modus accipiendi partem proportionalem alicuius numeri secundum proportionem alterius ad aliquem alium potest esse duplex, scilicet aut per denominationem, aut per multiplicationem. Per denominationem scilicet, quod si alter sit alterius tertia pars vel quarta et cetera, quod accipitur illius numeri tertia pars vel quarta. Per multiplicationem vero sic scilicet, quod multiplicabis secundum per tertium et divide per primum et proveniet tibi quesitus ignotus numerus, quem querebas. Unde primus numerus est ille, qui est notus et etiam sua pars est nota. Secundus numerus est illa pars nota. Tertius numerus est ille, qui licet sit notus, tamen pars est ignota.

---

235 Tabulam] in Tabulam (Tabulas *D*, Tabula *E*) *ADE* spere – Arietis] spere *ABDEFGHIK* **236** gradus[1]] *om.* *ABCDEFGHK* **237** sint[2]] *om.* *AEFG*: sit *C* **239** invenies – prius] invenies etiam ibi ita (ista *B*, secunda *E*) scilicet (si *E*) *ABE*: tunc invenies eos taliter scilicet *C*: invenies illud eciam ita *D*: invenies eciam ibi ita *F*: invenies eciam eos ibi ita scilicet *G*: invenies et ita *I* secundus 182] et secundus sit 182 *DI*: *om.* *BCG*: et (*eciam E*) secundus gradus sit 182 *AE*: 182 *F* **242** Tabulam] in tabulam *ABCF*: in tabula *G* **242/243** octavi circuli] *om.* *ACDEFHIK*: accessus et recessus argumenti capitis Arietis *G* **244** eandem] in eandem *ABCEF*: in eadem *G* Tabulam equationis] tabulam *ACEF*: lineam *B*: tabula *G* **246** istius – et prime] secunde et prime *A*: istius secunde equacionis et prime *C*: secunde equacionis et prime *DHK*: illius equacionis secunde et prime *F*: istius equacionis secunde et prime subtrahendo minorem a maiori *G* **248** prima – fuerit] fuerit *ABCEF*: fuerit ipsa prima equacio *D*: ipsa fuerit *G*: fuerit prima equacio *HK* **253** iam – augium[2]] *om.* *E* **258/259** potest – duplex] (alium *add.* *AG*) est duplex modus (modus *om.* *AEG*) *ABCEFG* **259** per[2] [...] *manuscript F omitted the rest of the treatise* **260** et – accipitur] quod etiam accipitur (accipiatur *C*) *ABC*: quod accipiatur eciam *D*: *om.* *EG*: et eciam accipiatur *I* **262** quesitus – numerus] eius ignotus *A*: quartus ignotus *BCEGI*: quartus ignotum *D*

the access and recess,[23] adding this to its radix. When you find it [mean motion of the eighth sphere], convert it into degrees if it contains signs. Then, enter with the number of these degrees into the *Table of equation of the motion of the eighth sphere or the head of Aries*, which increases up to ninety degrees and find there these degrees if they are less than ninety. If they are from ninety up to 180, then calculate backwards so that eighty-nine is ninety-one and eighty-eight is ninety-two, and calculate to the beginning, whereby 180 degrees has no equation. If, however, they are from 180 up to 270, then you will find it as before, namely, so the first degree is 181, the second 182 and so on up to ninety, which will be 270. If they are from 270 up to 360, start to calculate in the inverse order so that eighty-nine is 271, eighty-eight is 272 and so on up to 360, which also has no equation. Then enter with the number of degrees into the *Table of equations of the eighth sphere* and you directly find the equation of the motion of the eighth sphere. However, if there are also minutes contained with the degrees, then enter with these degrees again in the same *Table of equation*, add one degree and directly find the equation there. And then consider the difference of this second and the first equation and find the proportional part of this difference according to the proportion of minutes that were with the degrees up to sixty. This proportional part must be added to the first equation if the first equation is less than the second, or subtracted from it if it is greater. The result will be equation of the motion of the eighth sphere corresponding to these degrees and minutes. This equation must be added, if the degrees and minutes for which it was found are between one and 180 degrees. If they are from 180 up to 360 degrees, subtract the equation. If it is to be added, add it with the motion of the apogees, which is already set aside, and to all this add the radix of apogee of any planet that you find in its table of the era, namely, of that on which you have worked. If, however, it is to be subtracted, then add the previously set aside motion of the apogee with its radix and from this total subtract the equation of the motion of the eighth sphere and you will arrive at the position of apogee for any planet except the Moon, because its apogee is sought in another way.

It is necessary to realize here that there are two methods to find the proportional part of some number in its ratio to another number, either by division or multiplication. Using division, if some number is a third or a quarter of the other number, take the third or the quarter of that number. Using multiplication, multiply the second number by the third and divide by the first and the result will be the required unknown number that you were seeking. Then the first number is that which is known and also its part is known; the second number is the known part. And the third number is the one that is known, but its part is unknown.[24]

23 The 'eighth sphere' (or the 'eighth circle' in some manuscripts) is the sphere of the stars. Its motion is the precession, which appears as a rotation of the sphere of the stars with respect to the vernal point in the ecliptic. The apogees (auges) were believed to share this motion. Precession was supposed to be periodic in the pre-Alfonsine Toledan Tables, that is, subjected to the 'accession and recession'. The authors of the Parisian Alfonsine Tables modified the idea of precession to include both a periodic and a (Ptolemaic) uniform motion. For both the Toledan and the Parisian Alfonsine Tables, it is necessary to calculate the equation of the accession and recession.

24 The procedure using the multiplication is nowadays the standard method of linear interpolation: the first number is the step of the table (one degree in this case), the second number is the number of minutes – it means an increment of the independent variable – the third number is the difference of the dependent variable – in this case of equations – and the fourth number is the required increment of the dependent variable. The procedure using the division is described in more detail, for example, in the canons belonging to the *Tabulae resolutae* with incipit *Mirabilis in altis Dominus* (second rule, an example of the year 1430); our edition is in preparation.

Capitulum undecimum

De vero loco Solis inveniendo

Cum volueris invenire verum locum Solis, primo medium motum eius quere et ipsum serva. Deinde augem ipsius quere, quam subtrahe de medio motu ipsius, et remanebit tibi argumentum, cum quo intrabis *Tabulam equationis Solis* et equatio in directo existens est equatio Solis, si precise poterit inveniri. Si autem sint ibi minuta, equa pro eis, sicut in equatione motus octave spere. Que quidem equatio debet addi medio motui Solis, si fuerit argumentum a tribus signis usque ad sex signa, et debet subtrahi, si fuerit idem argumentum ab uno gradu usque ad tria signa.

Capitulum duodecimum

De vero loco Lune inveniendo

Cum volueris invenire verum locum Lune, quere primo eius medium motum et medium argumentum. Deinde medium motum Solis subtrahe a medio motu Lune et residuum duplica, et quod provenerit, pro centro Lune tene. Cum quo centro *Tabulam equationis Lune* ingredere et eius simile in lineis numeri quere et accipe, quod inveneris in directo de equatione centri et minutis proportionalibus, et unumquodque seorsum per se scribe. Si autem cum gradibus in centro Lune essent minuta, equa pro eis, sicut fecisti in equatione motus octave spere. Tunc si centrum Lune fuerit plus tribus signis, equationem centri subtrahe a medio argumento Lune. Si fuerit minus, eandem equationem adde, et sic idem argumentum remanebit equatum. Cum quo argumento equato intra easdem tabulas iterum et eius simile quere in lineis numeri et accipe, quod in directo eius inveneris de diversitate dyametri circuli brevis et de equatione argumenti. Et unumquodque per se et seorsum extra scribe, equando pro minutis, si sint minuta primo examinata in argumento vero vel equato. Postea vero accipe de diversitate dyametri circuli brevis partem proportionalem secundum proportionem minutorum proportionalium ad 60, quam partem proportionalem adde equationi argumenti et resultans vocabitur equatio examinata. Que quidem equatio argumenti sic examinata addatur medio motui Lune, si fuerit argumentum equatum plus tribus signis, vel ab eo minuatur, si fuerit minus tribus signis, et habebis verum locum Lune in nona spera.

---

266 Capitulum undecimum] *om. ABCDE*: undecimum *mg. H*: *om. et in mg. add. I*: 11 *mg. K* **267** De – inveniendo] *om. ABDEHIK*: ad inveniendum verum locum Solis *G* **269** de] a *AEG* ipsius[2]] Solis *ABCE*: Solis, si potes, quod si non potes, adde 6 signa, postea subtrahe *G* **270** Tabulam] in Tabulam *AEI*: in Tabula *G* **271/272** si – spere] *om. ADEGI*: si precise inveniri potest… equacio… spere *C*: equa tamen pro minutis, si sint minuta in argumento *HK* **272** quidem equatio] *om. ADEGI: equacio HK* **273** a – signis] a tribus, scilicet ab auge (scilicet ab auge *suprascr. H*), signis *HK* ad – signa] ad 6 *ADEGHI*: a 6 *K* **274** usque – signa] usque ad tercium signum vel ad tria signa *B*: usque ad tria signa, et habebis verum locum Solis *D*: in tria signa *HK* **275** Capitulum duodecimum] *om. ABCDE*: duodecimum *mg. H*: *om. et in mg. add. I*: 12 *mg. K* **276** De – inveniendo] *om. ABDEHIK*: ad inveniendum verum locum Lune *G* **277** Cum – Lune] *om. G* **284** argumento] motu (*corr. in:* argumento) *A* Lune] *om. A* **288** Et] *om. ABCDEG* et seorsum] *om. ABCDEGI* **288/289** minuta – equato] ibi minuta *ABCEG*: minuta *DI* **290** circuli – proportionalem] partem proportionalem *ABCDG*: partem proporcionalem circuli brevis *K* **291/292** et – examinata[1]] *om. ABCDGI* **293** ab – minuatur] ab eo minue *ABCG*: ab eo minime *E*: vel minuatur *HK* **294** tribus signis] *om. ABCDEI*

Chapter eleven

On finding the true position of the Sun[25]

If you wish to find the true position of the Sun, search first for its mean motion and set it aside. Then, search for its apogee, subtract it from the mean motion of the Sun, and the argument remains, with which you enter the *Table of equation of the Sun* to directly find the equation of the Sun if you can find it precisely. If there are some minutes, treat them as in the equation of the motion of the eighth sphere. This equation must be added to the mean motion of the Sun if the argument is from three up to six signs.[26] If the argument is from one degree up to three signs, it must be subtracted.

Chapter twelve

On finding the true position of the Moon[27]

If you wish to find the true position of the Moon, search first for its mean motion and mean argument. Then, subtract the mean motion of the Sun from the mean motion of the Moon and multiply the remainder by two; what results is the centre of the Moon, which you set aside. With this centre, enter the *Table of equation of the Moon* and search similarly in the column with units and find there directly the equation of the centre and the minutes of proportion and write both down separately. However, if there are also some minutes with the degrees in the centre of the Moon, treat them as in the equation of motion of the eighth sphere. If the centre of the Moon is greater than three signs, subtract the equation of the centre from the mean argument of the Moon. If it is less, add the equation, and the resulting argument will be corrected. Enter with this corrected argument into the same tables, search for what corresponds to it in the column of common units, and write down what you find directly for the variation of the diameter of the epicycle and the equation of the argument. Write down both separately, correcting for the minutes if there are some minutes found earlier in the true or corrected argument. Then get from the variation of the diameter the proportional part according to the ratio of the proportional minutes to sixty. Add this proportional part to the equation of the argument and the result will be called the examined equation. This equation of argument, which has been determined in this way, is added to the mean motion of the Moon if the corrected argument is more than three signs[28] or it will be subtracted if it is less than three signs. This gives the true position of the Moon on the ninth sphere.

25 The subject of Chapter 11 was later explained in Chapter 17 by John of Saxony, cf. Poulle, *Les Tables alphonsines*, p. 64.

26 The signs are calculated according to the sexagesimal system, meaning that one sign encompasses sixty degrees, that is, two usual signs of thirty degrees each. Manuscripts H and K add to the text on three signs 'calculated from the apogee' (*ab auge*); it should be understood that the signs are counted from the apogee, which is at zero, or equivalently at six (sexagesimal) signs (i.e. sixty degrees long), not that the apogee is at the three signs.

27 The subject of the Chapter 12 was also later explained in the Chapter 18 by John of Saxony, cf. Poulle, *Les Tables alphonsines*, p. 70.

28 This again deals with the sixty-degree signs.

Capitulum tredecimum

De loco capitis draconis

Cum volueris invenire verum locum capitis draconis, quere medium motum eius, quem subtrahe de 6 signis, et residuum erit verus locus eius.

Capitulum quattuordecimum

De veris locis trium superiorum planetarum inveniendis

Cum volueris invenire vera loca trium superiorum planetarum, scilicet Saturni, Iovis et Martis, querantur primo medii motus eorum, et tunc medius motus cuiusvis ipsorum subtrahatur a medio motu Solis et residuum servetur, quia illud est medium argumentum. Deinde augem ipsius subtrahe a medio motu eorum, et quod relinquitur, est centrum medium, quod etiam serva sub argumento. Cum quo centro *Tabulam equationis* illius ingredere querendo simile in lineis numeri, et quod in directo eius inveneris de equatione centri, exterius per se nota, equatum pro minutis, si sint ibi minuta. Si igitur centrum, cum quo tabulas intrasti, fuerit plus tribus signis, scribe super equationem 'addatur' et adde eam centro et minue eam ab argumento. Et si fuerit minus tribus signis, scribe super eam 'minuatur' et minue eam a centro et adde eam argumento et habebis utrumque equatum, centrum scilicet et argumentum. Cum centro igitur equato eandem tabulam intra, querendo simile in lineis numeri, et minuta proportionalia in directo inventa exterius per se nota. Eodem modo intra cum argumento equato, et quod in directo eius inveneris de diversitate dyametri circuli brevis et hoc in altera longitudinum et equatione argumenti unumquodque seorsum et per se extra scribe, equando pro minutis, si sint in argumento vero minuta. Accipies enim de longitudine longiori, si minuta proportionalia sint accepta in parte decrescente, procedendo a principio tabule

---

295 Capitulum tredecimum] *om. ABCDE*: tredecimum *mg. H*: *om. et in mg. add. I*: 13 *mg. K* **296** De – draconis] *om. ABDEHIK*: ad inveniendum verum locum capitis draconis *G* **298** 6] 12 *BCDEHIK* eius] eius. *(in mg. add.:* Canones ostendentis practicam in tabulis Alfoncii regis vide residuum retro*) B*: capitis draconis in nona *(*nona *corr. ex:* octava *H)* spera. Habito *(*dico *K)* vero loco Lune et capitis draconis subtrahe verum locum capitis a vero loco Lune et remanebit tibi latitudo Lune sive argumentum latitudinis Lune *HK* **299** Capitulum quattuordecimum] *om. ABCDE*: quattuordecimum *mg. H*: *om. et in mg. add. I*: 4 *mg. K* **300** De – inveniendis] *om. ABDEHIK*: ad inveniendum vera loca trium superiorum *G* **302** eorum et tunc] ipsorum tunc *ABCDEI*: eorum et *G* ipsorum] eorum *ABCG* **303** quia – argumentum] quia est illius *(*eius *B*, istius *C)* argumentum *ABCDEI*: pro eius argumento *G* **304** eorum] *om. ADEI*: suo *BC*: subtrahe *G* **305/306** quod – ingredere] *om. K* **305** Tabulam] in tabula *(*tabulam *BCE) ABCE* **305/306** equationis illius] equationis illius planete *A*: *om. BC*: illius equacionis *G* **307** equatum – minuta] equa tamen pro minutis, si sint ibi *(*ibi *om. I)* minuta *ABEI*: quia est equacio centri *HK* **309/310** tribus signis] *om. ABCE*: et habebis utrumque equatum, scilicet centrum et argumentum. Et si centrum fuerit minus *G* **310** scribe – eam[1]] superscribe *ABCE*: scribe super equacionem *D*: subscribe *G*: supra equacionem scribe *I* adde eam] eandem adde *AEG*: adde *D*: adde ipsam *I*: adde eam de *K* **311** habebis – equatum] sic utrumque habebis equatum *AEG*: sic unumquodque habebis equatum *BC*: iterum utrumque erit equatum *DI* centrum – argumentum] scilicet centrum et argumentum *ABCE*: *om. DI*: et centrum, et argumentum *G* **314** directo eius] eius *(*in *add. E)* directo *ACEG* **315** seorsum – scribe] per se scribe *AEG*: per se scribe extra *B*: per se extra scribe *(*nota *D) CD*: extra nota *I* **316** in – minuta] *om. AE*: ibi minuta *BCG*: minuta, et nota, quod si minuta proportionalia sint accepta in parte tabule decrescente procedendo, scilicet a parte superiori tabule versus inferiorem *DI* **316/318** Accipies – versus] *om. D*

Chapter thirteen

On the position of the head of dragon[29]

If you wish to find the true position of the head of dragon, search first for its mean motion and subtract from the six signs. The remainder will be the true position of the head of dragon.[30]

Chapter fourteen

On finding the true positions of three upper planets[31]

If you wish to find the true positions of the three upper planets, that is, of Saturn, Jupiter, and Mars, it is necessary to first find their mean motion. Then, the mean motion of the chosen planet is subtracted from the mean motion of the Sun, and the remainder is set aside as the mean argument. Then, subtract its apogee from its mean motion, and what remains is the mean centre, which is also to be saved below the argument. With this centre, enter the *Table of equation* in the column of common values and directly find there the equation of the centre, corrected for minutes, if there are some minutes, and then set aside. If the centre with which you entered into the tables is greater than three signs, inscribe above the equation 'to be added' and add it to the centre and subtract it from the argument. If it is fewer than three signs, inscribe above it 'to be subtracted' and subtract it from the centre and add it to the argument and you will have corrected both the centre and the argument.

With this corrected centre, enter the same table and search similarly in the column of numbers for the proportional minutes, which you will find directly and set aside. In the same way, enter with the corrected argument and directly find the variation of diameter of the epicycle, and that in the other distance and the equation of argument, and set aside everything separately, corrected always for minutes if there are some minutes in the argument. And you will thus obtain the farther distance,[32] if the proportional minutes are in the decreasing part proceeding from the beginning of the table to its end.

29 The subject of the Chapter 13 was later explained in Chapter 19 by John of Saxony, cf. Poulle, *Les Tables alphonsines*, p. 72.

30 The motion of the head of the dragon is retrograde: this is why it is to be subtracted from 360 degrees. This value is expressed as six sexagesimal signs in manuscripts A and G while in all other manuscripts it is in contradiction to the previous chapters expressed as twelve zodiacal signs, each sign per thirty degrees. Manuscripts H f. 3^v and K f. 17^v supplement here the following text which explains next usage of the position of the head of dragon: 'If you have the true position of the Moon and the head of dragon, subtract the true position of the head from the true position of the Moon and there will remain to you the latitude of the Moon, or argument of lunar latitude'.

31 The subject of Chapter 14 was later explained in Chapter 20 by John of Saxony, cf. Poulle, *Les Tables alphonsines*, p. 74.

32 The farther distance (*longitudo longior*) is a distance of the apogee of any planet from the Earth.

versus finem. Si vero sint accepta in parte crescente, accipe de propinquiori. De qua diversitate accipe partem proportionalem secundum proportionem minutorum proportionalium ad 60, quam partem proportionalem adde equationi argumenti, si fuerit diversitas dyametri accepta de longitudine propinquiori. Si vero fuerit accepta de longitudine longiori, eandem partem subtrahe ab equatione argumenti et habebis tunc equationem argumenti primo examinatam, supra quam scribe 'addatur', si fuerit argumentum equatum minus tribus signis. Si vero fuerit plus, scribe super eandem 'minuatur'. Postea vero equationem centri et equationem argumenti considera. Si super utramque scribatur 'addatur', adde eas simul et totum adde medio motui illius planete. Et si super utrasque scribatur 'minuatur', iunge etiam illas simul et totum minue a medio motu illius planete. Si autem super unam scribatur 'addatur' et super aliam 'minuatur', tunc minue minorem a maiori, et si super maiorem scribatur 'addatur', residuum adde medio motui, et si super maiorem scribatur 'minuatur', residuum minue de medio motu, et quod post augmentum et diminucionem provenerit, erit verus locus illius planete in nona spera.

Capitulum quindecimum

De vero loco Veneris et Mercurii inveniendo

Si velis invenire verum locum Veneris et Mercurii, fac eodem modo, quo dictum est in tribus superioribus, nisi quod argumenta eorum extrahuntur ex tabulis et medius motus eorum est idem cum medio motu Solis. Et similiter diversificatur motus in Mercurio, quia in addendo partem proportionalem diversitatis dyametri circuli brevis non respicimus, utrum illa longitudo sit diversitas accepta a longitudine longiori vel breviori, sed respicimus titulum minutorum proportionalium. Nam si super minuta proportionalia scriptum sit 'addatur', debemus illam addere, et si 'minuatur', minuere. Hic notandum, quod qui vellet sepe equare planetas ad hoc, quod faciliter inveniret loca ipsorum, expediret, quod quereret in complemento cuiuslibet anni medios motus eorum et media argumenta et media centra, et reduceret ad meridianum suum et servaret pro radice in toto illo anno. Tunc mediantibus *Tabulis de medio centro Lune et de mediis argumentis trium superiorum* faceret sic ita scilicet, quod si expectaret postea ad equandum planetas per spatium 20 primorum, scilicet dierum, quod idem est, quereret medium motum Solis in sua tabula in 20 primis et adderet radici Solis, Veneris et Mercurii servate. Eodem modo quereret medium centrum Lune et medium eius motum et medium

---

319 accipe] accipias *AE*: accipis *B*: accipere *C*: accipies *G* **319/320** secundum – proportionalem] *om. GI* **322/323** eandem – examinatam] *om. D* **322** subtrahe – argumenti] proportionalem ab equacione argumenti (argumenti *om. I*) subtrahe *AEGI* **327** illas simul] utrasque simul (insimul *D*) *ABCD*: utramque simul *E*: utrumque simul *I* **328** illius] *om. ADEI*: ipsius *G* **331** et[2]] vel *ACDEG* illius planete] planete *ADI*: planete illius *E* **333** Capitulum quindecimum] *om. ABCDE*: quindecimum *mg. H*: *om. et in mg. add. I*: 15 *mg. K* **334** De – inveniendo] *om. ABDEHIK*: ad inveniendum verum locum Veneris et Mercurii *G* **335** Si – Mercurii] *om. G* velis invenire] autem velis scire *A*: volueris invenire *B*: autem volueris invenire *E*: velis scire *I* **336** ex] in *ABE* **337** motus[2]] modus *ABEHK* **339/340** utrum – respicimus] *om. HK* **342** ad hoc – ipsorum] *om. BC* quod[2]] ut *AEG* loca ipsorum] loca eorum *AE*: vera loca ipsorum *D*: vera loca planetarum *I*: loca ipsarum *K* **348** servate] servato *ABE*: servatur C: servatis (*suprascr.* radicibus *I*) *DI* **349/350** et[1] – superiorum] *om. I*

If, however, they are found in the increasing part, you will obtain the nearer distance.[33] Find from this variation of diameter the proportional part according to the ratio of proportional minutes to sixty. Add this proportional part to the equation of argument, if the found variation of diameter relates the nearer distance. If, however, it relates to the farther distance, then subtract this part from the equation of argument and you will have the examined equation of argument. Inscribe above it 'to be added' if the corrected argument is fewer than three signs. If it is greater, inscribe above it 'to be subtracted'. After this, consider the equation of centre and equation of argument. If they are both inscribed 'to be added', add them together and add the sum to the mean motion of that planet. If, however, they are both inscribed 'to be subtracted', then again add them together and subtract the sum from the mean motion of that planet. If 'to be added' is inscribed above one and above the other you have 'to be subtracted', then subtract the smaller from the greater, and if 'to be added' is inscribed above the greater, add the remainder to the mean motion. And if 'to be subtracted' is inscribed above the greater, subtract the remainder from the mean motion and the result, after the adding or subtracting, will be the true position of that planet on the ninth sphere.

Chapter fifteen

On finding the true position of Venus and Mercury[34]

If you wish to find the true position of Venus and Mercury, work in the same way as described for the upper three planets, except their arguments are obtained from tables and their mean motion is the same as the mean motion of the Sun. Likewise, the motion of Mercury differs because in adding the proportional part of the variation of diameter of epicycle we do not consider whether this difference is taken from the farther or nearer distance, but we do consider the inscription of the proportional minutes. Namely, if 'to be added' is inscribed above the proportional minutes, we must add it, and if 'to be subtracted' is inscribed, we must subtract it.

It should be noted here that whoever wishes to repeatedly correct the planets in order to easily find their positions[35] must find in the complement of any year the mean motion of the planets, mean arguments and mean centres, and convert them to his meridian and set aside the radices of that whole year. Then, use the *Table of the mean centre of the Moon* and the *Table of the mean arguments of the three upper planets*, correcting the planets for steps of twenty units or days, which are the same. In the table for the mean motion of the Sun, search for twenty units and add it to the set-aside radices of the Sun, Venus, and Mercury. In the same way, find the mean centre of the Moon, its mean motion, mean argument, and the mean motion of some of the upper three plan-

33 The nearer distance (*longitudo propinquior /brevis/*) is a distance of the perigee of any planet from the Earth.

34 The subject of Chapter 15 was later explained in Chapter 21 by John of Saxony, cf. Poulle, *Les Tables alphonsines*, p. 78.

35 This second paragraph introduces a new topic, viz., preparation of almanacs or ephemerides giving planetary longitudes at regular intervals of every day, every fifth day, and every tenth day.

argumentum et medios motus cuiuslibet trium superiorum et media argumenta eorum et argumenta Veneris et Mercurii et adderet quodlibet ad suum genus radicis servate in principio anni. Medium autem motum Solis in illis 20 diebus seu primis adderet medio argumento Solis prius servato et mediis centris Veneris et Mercurii et habebis ad illam diem medium argumentum Solis et medium centrum Veneris et Mercurii. Medium similiter motum Saturni, Iovis et Martis in eodem tempore addet medio centro eorum servato et habebis media centra eorum. Et per istum modum posset aliquis facere almanach in quolibet anno, scilicet querendo per easdem tabulas in quolibet die pro Sole et Luna, de 5 diebus in 5 diebus pro Venere et Mercurio, de decem diebus in decem diebus pro tribus superioribus et postea equare planetas, secundum quod dictum est mediantibus tabulis de minutis proportionalibus, et ita faciliter possunt inveniri loca omnium planetarum. Nota tamen, quod potes tantum expectare, quod oportet intrare cum secundis et primis. Et intrare etiam potes cum minutis et secundis, si velis. Nota etiam, quod sufficit extrahere de tabulis signa, gradus, minuta et secunda, et tunc, si tertia sequentia secunda sunt plura 30, accipe ea pro uno secundo, et si sint pauciora, pro nichilo computentur.

Capitulum sedecimum

De possibilitate eclipsis Solis vel Lune invenienda

Cum volueris scire, utrum in aliquo anno eclipsis Solis vel Lune sit possibilis, primo quere mediam coniunctionem primam in illo anno et argumentum latitudinis ad illud tempus et quere eodem modo primam mediam oppositionem et argumentum latitudinis ad illud tempus. Et si volueris scire de eclipsi Solis, argumentum latitudinis ad primam coniunctionem in 12 locis scribe et ei adde argumentum latitudinis cuiuslibet coniunctionis sequentis inventum in *Tabula coniunctionis et oppositionis in mensibus*, et sic habebis argumentum latitudinis ad quemlibet mensem illius anni. Et in quo mense inveneris argumentum latitudinis a 0 in signis et gradibus usque ad 12 gradus vel a duobus signis et 48 gradibus usque ad tria signa integra, in illo mense est eclipsis Solis possibilis. Eodem modo penitus queras argumentum latitudinis ad oppositionem Solis et

350/351 eorum – Veneris] eorum et argumentum Veneris *ADK*: eorum Veneris argumentum *I* **351** radicis servate] radici servate *ABEHK*: *om. G*: radice servata *I* **352** adderet] adde *AHK*: addito *BC*: addet *E* medio] *om. A* **353/354** et[3] – Mercurii] *om. BC* **354** medium[1]] *om. A* medium centrum] centrum medium *AD* **355** addet] adde *AHK*: adderet *I* **356** media – eorum] eorum media centra (medium centrum *D*) *ACDEG* **357** quolibet die] qualibet die *AGI*: quamlibet diem *E* **358** diebus[1]] *om. A* diebus[2]] *om. ABCDEGI* de[2] – diebus[4]] de decem in decem (diebus *add. GI*) *AGI*: et (et *om. BC*) de (in *E*) 10 diebus in 10 *BCDE* **360** omnium] *om. A* **361** oportet] oporteret *AHK*: oportebit *D* **361/363** cum – sufficit] *om. K* **362** etiam potes] potes etiam *AEG*: potes *C*: eciam posses *I* **363** extrahere] tibi extrahere *AI*: subtrahere *BC*: extractione *D*: eum extrahere *I* de – signa] de signis *ABCEG*: de tabulis *D*: signa *HK* et[1]] *om. A* **364** ea] *om. ABCE*: illa *D*: ipsa *G* **365** computentur] computantur *EI*: computentur, sicut sit in equacionibus in accipiendo partem proporcionalem *HK* **366** Capitulum sedecimum] *om. ABCDE*: sedecimum *mg. H*: *om. et in mg. add. I*: 16 *mg. K* **367** De – invenienda] *om. ABDEHIK*: ostendens possibilitatem eclipsis Solis et Lune in quolibet anno *G* **368** Cum – possibilis] *om. G* vel] et *AEI* primo] *om. ABCDEGI* **369** illud] idem *ABCE* **371/372** Et – latitudinis] *om. E* **371** scire] *om. ABCDGI* **375** inveneris] invenies *ADEI* a 0 – gradibus] a (ab *E*) 0 *ABCE*: a 0 in signis *G*: a 0 in signis et gradus *HK* **376** integra] *om. ABCEG*: et 12 gradus et iterum a 5 signis et 48 gradus (gradibus *I*) usque ad 6 signa *DI* eclipsis Solis possibilis] eclipsis possibilis *ABCDEG*: possibilis eclipsis Solis *HK*

ets, their mean arguments, and the arguments of Venus and Mercury. And all of these must be added, in order, to the radices for the beginning of the year that you set aside. Add the mean motion of the Sun in these twenty days, that is, units, to the mean argument of the Sun that you set aside, and to the mean centres of Venus and Mercury, and you will get for this day the mean argument of the Sun and mean centres of Venus and Mercury. Similarly, in the same period add the mean motion of Saturn, Jupiter and Mars to their set-aside mean centres, and you will have their mean centres. And in this way, anyone can create an almanac for any year, namely, by using these tables to search for positions of the Sun and Moon for every day, positions of Venus and Mercury for every fifth day, and for the upper three planets for every tenth day. And then you can correct the planets according to what has already been said, using tables of the proportional minutes, and thus you can easily find positions of all planets. Note, however, that you enter only with seconds and units, as you may expect.[36] And if you wish, you can enter with minutes and seconds. Note also that it is sufficient to extract from the tables the signs, degrees, minutes and seconds and then, if the thirds after the seconds are greater than thirty, take them for one second; if they are fewer, they are counted for nil.

Chapter sixteen

On finding the possibility of an eclipse of the Sun and Moon[37]

If you wish to know whether an eclipse of the Sun and Moon is possible in any year, search initially for the first mean conjunction in that year and for the argument of latitude at the same time. In the same way, search for the first mean opposition and the argument of latitude at this time. And if you wish to know an eclipse of the Sun, write down the argument of latitude for the first conjunction at twelve places[38] and add to it the argument of latitude at any subsequent conjunction found in the *Table of conjunction and opposition in months*, and so you will have the argument of latitude for any month of that year. And in whichever month you find the argument of the latitude from zero in the signs and degrees up to twelve degrees or from two signs and forty-eight degrees up to the whole three signs, in this month an eclipse of the Sun is possible. In the same way, search for the argument of latitude at the opposition of the Sun and Moon, and in

36 The number of days in one year is at maximum 365, that is, six seconds and five units (= days).

37 The subject of Chapter 16 was later explained in Chapter 14 by John of Saxony, cf. Poulle, *Les Tables alphonsines*, p. 60.

38 'At twelve places' because this value will be used for calculations in all twelve months of the year.

Lune, et in quo mense inveneris argumentum latitudinis a o in signis et gradibus usque ad 12 gradus ante et retro, vel si argumentum latitudinis sit tria signa completa et per 12 gradus ante vel retro, dic eclipsim Lune possibilem in illo mense.

Expliciunt *Canones* super *Tabulas illustris regis Alfunsi*, regis Castelle, compillati per magistrum Iohannem de Lineriis, diocesis Ambianensis.

378 inveneris] invenies *ABE* **378/379** a o – ad] a (ab *E*) o usque ad *ABCEI*: et cum fuerit a o usque ad *D*: a o in signis usque ad *G*: sic tria signa complete et per *K* **379** si] quod *ABEI*: *om.* *C* **379/380** et per – retro] vel (si sit *add.* *G*) per (per *om.* *D*) 12 gradus ante vel post *ADG*: vel quod per 12 gradus ante vel post *B*: *om.* *C*: et (vel *I*) 12 gradus ante vel post *EI* **381/382** Expliciunt – Ambianensis] *om.* *AEK*: Explicit *BD*: Explicit. Deo gracias. — Tabula, de qua dictum est supra. / Tabula radicum notarum anni / Radix diluvii 5 / Radix Nabugodonosor 4 / Radix mortis Alexandri 1 / Radix Alexandri Magni 2 / Radix Cesaris 1 / Radix Incarnacionis 7 / Radix Dyocleciani 6 / Radix Arabum 5 / Radix zezdarert, id est Persarum 3 / Radix Alfontis 7 *G*: Expliciunt Canones magistri Iohannis de Lineriis super Tabulas Alfoncii. Deo gracias *H*: Expliciunt Canones super Tabulas regis Alphoncii *I*

whichever month you find the argument of latitude from zero in signs and degrees up to twelve degrees forwards or backwards, or if the argument of latitude is three whole signs and up to twelve degrees forwards or backwards, then an eclipse of the Moon is possible in that month.

The *Rules to the Tables of the illustrious King Alfonso,* the King of Castille, composed by Master John of Lignères, from the diocese of Amiens, the end.[39]

39 John of Lignères originated from the bishopric of Amiens (*Dioecesis Ambianensis,* Diocèse d'Amiens), which had already been founded in the third century.

Manuscript sources

Bernkastel-Kues, Bibliothek des St Nikolaus-Hospitals, Cus 212
Bruges, Openbare Bibliotheek, 466
Cambridge, University Library, Ii. 1. 27
Cracow, Biblioteka Jagiellońska, 548
Cracow, Biblioteka Jagiellońska, 551
London, British Library, Sloane 407
Milan, Biblioteca Ambrosiana, N 217 Sup
Milan, Biblioteca Ambrosiana, S 54 Sup
Oxford, Bodleian Library, Digby 168
Oxford, Hertford College, 4
Paris, Bibliothèque nationale de France, lat. 7281
Paris, Bibliothèque nationale de France, lat. 7286
Paris, Bibliothèque nationale de France, lat. 7286C
Paris, Bibliothèque nationale de France, lat. 7378A
Paris, Bibliothèque nationale de France, lat. 7405
Vatican, Biblioteca Apostolica Vaticana, Ott. lat. 1826
Vatican, Biblioteca Apostolica Vaticana, Pal. lat. 1403
Vienna, Österreichische Nationalbibliothek, Cod. 5144

Bibliography

Boudet, Jean-Patrice, *Lire dans le ciel. La bibliothèque de Simon de Phares, astrologue du* xv*e* *siècle* (Brussels: Centre d'Études des Manuscrits, 1994).

Carloni, Massimiliano, 'Towards a Digital Edition of the Aratean Tradition', in *The Stars in the Classical and Medieval Traditions*, ed. Alena Hadravová, Petr Hadrava and Kristen Lippincott (Praha: Scriptorium 2019), pp. 171–88.

A Catalogue of the Manuscripts preserved in the Library of the University of Cambridge, ed. Henry Richards Luard, 5 vols (Cambridge: Cambridge University Press, 1856–67), III (1858).

Catalogus codicum manuscriptorum Bibliothecae regiae, Pars tertia, Tomus tertius — quartus (Parisiis: ex typographia regia, 1744).

Catalogus codicum manuscriptorum medii aevi Latinorum, qui in Bibliotheca Jagellonica Cracoviae asservantur, 11 vols (Wrocław: Ossolineum, 1980–2016).

Chabás, José, and Bernard R. Goldstein, *The Alfonsine Tables of Toledo* (Dordrecht: Kluwer, 2003).

———, and ———, 'Early Alfonsine Astronomy in Paris: The Tables of John Vimond (1320)', *Suhayl*, 4 (2004), 207–94.

———, and ———, *The Astronomical Tables of Giovanni Bianchini* (Leiden: Brill, 2009).

———, and ———, *A Survey of European Astronomical Tables in the Late Middle Ages* (Leiden: Brill, 2012).

Dekker, Elly, *Illustrating the Phaenomena. Celestial Carthography in Antiquity and the Middle Ages* (Oxford: Oxford University Press, 2013).

Gabriel, A. L., *A Summary Catalogue of Microfilms of One Thousand Scientific Manuscripts in The Ambrosiana Library, Milan* (Notre Dame: University of Notre Dame 1968).

Hadrava, Petr, and Alena Hadravová, 'Cristannus de Prachaticz´s Treatises on the Astrolabe', in *Certissima signa. A Venice Conference on Greek and Latin Astronomical Texts*, ed. Filippomaria Pontani (Venice: Edizioni Ca´Foscari, 2017), pp. 295–312.

Hadravová, Alena, and Petr Hadrava, *Stavba a užití astrolábu*. Prague, Filosofia 2001.

———, and ———, 'John of Saxony', in *Medieval Science, Technology and Medicine: An Encyclopedia*, ed. Thomas Glick, Steven J. Livesey and Faith Wallis (New York: Routledge, 2005), p. 292.

———, and ———, *Sphaera octava. Mýty a věda o hvězdách IV. Katalogy hvězd a přemyslovský nebeský glóbus — Sphaera octava. Myths and Science on Stars IV. Catalogues of Stars and Premyslid Celestial Globe* (Prague: Artefactum — Academia, 2013).

Hartmann, Johannes Franz, *Die astronomischen Instrumente des Kardinals Nikolaus Cusanus* (Berlin: Weidmannsche Buchhandlung 1919).

Juste, David, 'MS. Vatican, Biblioteca Apostolica Vaticana, Ottob. lat. 1826' (updated: 5 March 2018), *Ptolemaeus Arabus et Latinus. Manuscripts*, URL: http://ptolemaeus.badw.de/ms/203.

———, 'MS. Bernkastel-Kues, Cusanusstiftsbibiothek, 212' (updated: 1 January 2019), *Ptolemaeus Arabus et Latinus. Manuscripts*, URL: http://ptolemaeus.badw.de/ms/18.

Krchňák, Alois, 'Die Herkunft der astronomischen Handschriften und Instrumente des Nicolaus von Kues', *Mitteilungen und Forschungsbeiträge der Cusanus-Gesellschaft*, 3 (1963), 109–81.

Kremer, Richard L. 'Cracking the *Tabulae permanentes* with Exploratory Data Analysis', in *Editing and Analyzing Numerical Tables: Towards a Digital Information System for the History of Astral Sciences*, ed. Matthieu Husson, Clemency Montelle and Benno van Dalen (Turnhout: Brepols, 2022), pp. 363-422

Macray, William D., *Catalogi codicum manuscriptorum Bibliothecae Bodleianae, Pars nona: Codices a viro clarissimo Kenelm Digby anno 1634 donatos, complectens* (Oxford: Clarendon Press 1883).

Marx, Jakob, *Verzeichnis der Handschriften-Sammlung des Hospitals zu Cues bei Bernkastel a. Mosel* (Trier: Selbstverlag des Hospitals, 1905; repr. Frankfurt/Main, 1966).

Morgan, Paul, *Oxford Libraries Outside the Bodleian: A Guide* (Oxford: Oxford Bibliographical Society and Bodleian Library, 1973).

North, John D., *Stars, Mind and Fate: Essays in Ancient and Mediaeval Cosmology* (London: The Hambledon Press, 1989).

Porres de Mateo, Beatriz, 'Les Tables astronomiques de Jean de Gmunden: Édition et étude comparative', unpublished PhD thesis, Paris, École Pratique des Hautes Études, 2003.

———, and José Chabás, 'John of Murs's *Tabulae permanentes* for Finding True Syzygies', *Journal for the History of Astronomy*, 32 (2001), 63–72.

Poulle, Emmanuel, 'John of Lignères', in *The Dictionary of Scientific Biography*, 16 vols (New York: Charles Scribner´s Sons, 1970–1980), VII (1973), pp. 122–28.

———, *Les Tables alphonsines avec les canons de Jean de Saxe* (Paris: Centre national de la recherche scientifique, 1984).

———'The Alfonsine Tables and Alfonso X of Castille', *Journal for the History of Astronomy*, 19 (1988), 97–113.

Rosińska, Grażyna, *Scientific Writings and Astronomical Tables in Cracow. A Census of Manuscript Sources (XIVth – XVIth Centuries)* (Wrocław: Ossolineum, 1984).

Saby, Marie-Madeleine, 'Les Canons de Jean de Lignères sur les tables astronomiques de 1321. Édition critique, traduction et étude', unpublished thesis, Paris, École Nationale des Chartes, 1987.

Snedegar, Keith V., 'The Works and Days of Simon Bredon, a Fourteenth-Century Astronomer and Physician', in *Between Demonstration and Imagination: Essays in the History of Science and philosophy, presented to John D. North*, ed. by L. Nauta and A. Vanderjagt (Leiden: Brill, 1999), pp. 285–309.

Tabulae codicum manu scriptorum praeter Graecos et orientales in Bibliotheca Palatina Vindobonensi asservatorum, ed. Academia Caesarea Vindobonensis, 10 vols (Wien: Gerold, 1864–99).

Thoren, Victor E., and Edward Grant, 'Extracts from the Alfonsine Tables and Rules for their Use (John of Saxony)', in *A Source Book in Medieval Science*, ed. by Edward Grant (Cambridge: Harvard University Press, 1974), pp. 465–87.

Thorndike, Lynn, and Pearl Kibre, *A Catalogue of Incipits of Medieval Scientific Writings in Latin*, rev. and augmented ed. (Cambridge: Mediaeval Academy of America, 1963).

JOSÉ CHABÁS

New Texts and Tables Attributed to John of Lignères: Context and Analysis

Introduction

John of Lignères was a prolific and authoritative scholar in astronomy and mathematics.[1] His name is frequently found in miscellaneous manuscripts on astronomy, in particular those including a set of tables for 1322, which saw remarkable success and laid the foundations for the standard Parisian Alfonsine Tables.

This gave him widespread renown, and it is not uncommon to find works attributed to him in late medieval astronomical manuscripts, in particular those dealing with astronomical tables. Among these are two short texts and their associated tables, both on mean conjunctions. The first text and its corresponding table are uniquely extant in a manuscript in Madrid, Biblioteca Nacional, and the second one is found in two manuscripts at the Vatican, Biblioteca Apostolica.

1. *Madrid, Biblioteca Nacional, 9288*

This is an exceptional fourteenth-century manuscript of Franciscan origin (Order of Friars Minor), almost entirely devoted to calendars and almanacs, including a text on an

* This work was supported by the ERC project ALFA: Shaping a European scientific scene, Alfonsine astronomy, CoG 723085, PI Matthieu Husson.
I wish to thank Bernard R. Goldstein for his valuable comments on a draft of this paper and Matthieu Husson, Richard L. Kremer and the participants of the ALFA workshop held in Prague in October 2019, where this paper was presented, for their insightful remarks.

1 For a review of John of Lignères' works, see Emanuel Poulle, 'John of Lignères', in *Dictionary of Scientific Biography*, ed. Charles Gillispie, 16 vols. (New York: Charles Scribner's Sons, 1970–80), VII (1973), pp. 122–28; Marie-Madeleine Saby, 'Les Canons de Jean de Lignères sur les tables astronomiques de 1321' (unpublished thesis, Paris, École Nationale des Chartes, 1987); José Chabás and Bernard R. Goldstein, *The Alfonsine Tables of Toledo* (Dordrecht-Boston: Kluwer, 2003), pp. 281–84; José Chabás, *Computational Astronomy in the Middle Ages* (Madrid: Consejo Superior de Investigaciones Científicas, 2019), pp. 175–206.

José Chabás • Universitat Pompeu Fabra, Barcelona

Alfonsine Astronomy: The Written Record, ed. by Richard L. Kremer, Matthieu Husson, José Chabás, Turnhout, 2022 (*Alfonsine Astronomy: Studies and Sources*, 1), pp. 303-316
 10.1484/M.ALFA.5.125139

astronomical instrument. The five items in this manuscript were composed in the first decades of the fourteenth century in southern France and in Paris.[2]

1^{r}–9^{v}: *Kalendarium* by Geoffrey of Meaux

The work of this Parisian scholar consists of a short text and a few tables, including twelve monthly tables. The *Kalendarium* by Geoffrey of Meaux is still waiting for an edition and/or a thorough examination. Besides the Madrid codex, it is extant in the following manuscripts:

Erfurt, Universitätsbibliothek, CA 4° 369, 6^{r}–14^{v}
Los Angeles, J. Paul Getty Museum, Ludwig XII.6, 1^{r}–11^{r}
Paris, Bibliothèque nationale de France, lat. 15118, 67^{r}–75^{v}
Toledo, Biblioteca de la Catedral, 99–5, 13^{v}–25^{v}
Uppsala, Universitetbibliotek, C 653, 189^{r}–197^{r}

The text (1^{r}–2^{v}) begins, '*Cunctis solis et lune scire desiderantibus vera loca,* and ends, *Quo habito alia festa mobilia de facili possunt sciri*'. It consists of an introduction and four chapters on the positions of the Sun and the Moon, mean conjunctions, and movable feast days.

Among the tables (3^{r}–9^{v}), the first one displays the yearly radices, from 1320 to 1340, of four quantities: solar longitude, lunar longitude, lunar anomaly, and argument of lunar anomaly. These are the four quantities appearing in all standard tables for computing mean syzygies, indicating that this is a purpose of the calendar. In addition to the radices, we are given the values for the progress of these four quantities in a year. This table usually precedes the monthly tables, but it is found after them in other astronomical manuscripts.

At the beginning of the text, Geoffrey of Meaux indicates that the tables begin at noon of the last day of December 1320 completed, and mentions the Alfonsine Tables. He explicitly rejects *radices Alphonci* and prefers *radices Arzachelis*, meaning those used in the Toledan Tables. Despite the date of the radix, 1 January 1321, we do not know the precise date of Geoffrey's text, and it is thus not possible to compare it with those of the works done by other Parisian astronomers using the Alfonsine Tables, originally compiled in Toledo, and already in use in Paris towards 1320. The value of the yearly progress of the Sun, 11s 29;44,50°, in the first table clearly shows that we are dealing with some derivative of the Toledan Tables, usually ascribed to Azarquiel, for this is the standard value used in sidereal astronomy based on the Toledan Tables, in contrast to the corresponding value used in tropical Alfonsine astronomy, 11s 29;45,40°.[3]

It is possible that Geoffrey's explicit preference for the Toledan Tables over the Alfonsine Tables is a sign of a debate being held among Parisian astronomers in the early fourteenth century concerning the use of an old set of tables, the Toledan Tables, referring to the

2 Summary descriptions can be found in Manuel de Castro, *Manuscritos franciscanos de la Biblioteca Nacional de Madrid* (Madrid: Ministerio de Educación y Ciencia, 1973), pp. 403–4; and Biblioteca Nacional, *Inventario general de manuscritos de la Biblioteca Nacional*, 15 vols. (Madrid: Dirección General de Archivos y Bibliotecas, Servicio de Publicaciones, 1953–2001), XIII (1995), p. 273. We note that both catalogues enlarge Geoffrey of Meaux's *Kalendarium* to 98 folios, do not mention the canon and the tables attributed to John of Lignères, ignore the Almanac of Jacob ben Makhir, reduce Bancal's work to two folios (99^{r-v}), and give notice of the astronomical instrument attributed to John of Lignères.

3 See Chabás, *Computational Astronomy*, pp. 104–14.

eighth sphere (sidereal coordinates), or a new one, the Alfonsine Tables, based on the ninth sphere (tropical coordinates). The twelve monthly tables list various calendrical features, such as the golden number, the Sunday letter, and names of saints, as well as daily entries for the mean motions of the Sun, the Moon, and the lunar node, beginning in 1 January 1321. The entries of all quantities are given to minutes and start at 0s 0;0°. The last table, just after that of December, displays values for the mean motions of the four quantities in the table containing the radices, for each hour in a day and for the minutes in an hour, at two-minute intervals.

10^{v}–14^{v}: *Kalendarium* attributed to John of Lignères

After a blank folio follows a short text headed *Canon supra kalendarium magistri Johannes de Lineriis* (10^{v}) and a table (11^{r}–14^{v}) for all mean conjunctions in the period from 1321 to 1396. This *Kalendarium* is not among the works hitherto ascribed to John of Lignères, and, as far as we know, it is uniquely preserved in the present manuscript. For a close examination of the text and the table see below.

15^{r}–92^{v}: *Almanac* by Jacob ben Makhir

This almanac consists of a number of tables (15^{r}–88^{v}) that determine the true positions of the planets and the two luminaries and the circumstances of eclipses, followed by a text that includes a prologue and canons (89^{r}–92^{v}). It was originally written in Hebrew by Jacob ben Makhir Ibn Tibbon (*c.* 1236-*c.* 1305), also known by his Provençal name, Profeit Tibbon, later rendered in Latin as Profatius. The almanac was mainly diffused in Latin and it is preserved in about thirty manuscripts in a mixture of both languages.[4]

To compute the entries in his almanac, Jacob depended on the Toledan Tables and used tropical coordinates, in contrast to that set of tables, where sidereal coordinates are used, and thus applied a correction for precession. In Jacob's *Almanac*, the year begins in March and the year radix is 1301 in the Hebrew manuscripts and 1300 in the Latin manuscripts. Some of the tables in the *Almanac* are really ingenious and unprecedented in the west, especially the large table for the true lunar anomaly, given on a daily basis for a period of about twenty-four years, and the double argument table for the complete lunar equation. Further details are provided in Chabás and Goldstein.

93^{r}–99^{v}: *Kalendarium* by Raymond Bancal

Of this calendar compiled by the Franciscan minister for Aragon in 1326, only one other copy is known: Paris, Bibliothèque nationale de France, lat. 7420A, 57^{v}–63^{r} and 71^{r}.[5] It is

4 For a detailed study of its mathematical content, see José Chabás and Bernard R. Goldstein, 'The *Almanac* of Jacob ben Makhir', in *Editing and Analyzing Numerical Tables: Towards a Digital Information System for the History of Astral Sciences*, ed. Matthieu Husson, Clemency Montelle and Benno van Dalen (Turnhout: Brepols, 2022), pp. 53-78.; see also Gerald J. Toomer, 'Prophatius Judaeus and the Toledan Tables', *Isis*, 64 (1973), 351–55.

5 See Philipp Nothaft, 'Medieval Astronomy in Catalonia and the South of France: The 'Improved' Lunar *Kalendarium* of Friar Raymond (Ramon) Bancal (*c.* 1311) and its Predecessors', *Llull*, 38 (2015), 101–25.

an improved calendar based on the Toledan Tables giving the time of mean conjunctions for a period of seventy-six years, from 1311 to 1386, that is, for four nineteen-year cycles. The information on mean conjunctions is presented on a monthly basis and fills in six columns for each month: one displays the *numerus cicli*, which is the number of the year in a cycle of nineteen years; another is for the day of the month; and the other four list the times of mean conjunction, in hours and minutes, in each of the four nineteen-year cycles considered.

An additional table to extend the calendar to dates prior to 1311 or after 1386 follows the monthly tables. The procedure is based on a value of about 5;48h, resulting from the difference between the times of two mean conjunctions exactly separated by the 940 mean synodic months elapsed in the period of recurrence of seventy-six years. A list of sub-multiples of the mean synodic month, 29d 12;44,3,20h, used in the Toledan Tables, is also provided. They are said to correspond to the lunar aspects, of little relevance in a calendar, but of interest in astrological medicine.[6] A short canon (99^{r-v}) explains how to identify the mean conjunctions in the twelve monthly tables, the use of the additional table, and the meaning of the lunar aspects.

100^{r}–105^{v}: *Saphea* by John of Lignères

The manuscript closes with some canons attributed to John of Lignères, hitherto not correctly identified. It is indeed a text by John of Lignères on an astronomical instrument, the saphea, based on an analogous universal instrument developed by Azarquiel, *al-safiha*. Azarquiel's canons were translated from Arabic into Castilian and into Latin by the astronomers at the court of King Alfonso in Toledo in the second half of the thirteenth century, and into Hebrew by Jacob ben Makhir.[7] John of Lignères' text has no diagrams and consists of thirty-four chapters, beginning, '*Descriptiones que sunt in facie instrumenti notificare. Limbus seu circulus exterior*'. The colophon reads, '*Expliciunt canons magistri Iohannis de lineriis supra quoddam instrumentum mirabile, cuius anima cum Christo in eternum possideat sempiterna. Amen*'. This text is preserved in three other manuscripts:

Erfurt, Universitätsbibliothek, CA 4° 355, 73^{r}–81^{v};
Erfurt, Universitätsbibliothek, CA 4° 366, 40^{r}–49^{r};
Paris, Bibliothèque nationale de France, lat. 7295, 2^{r}–14^{r}.

We note that in Paris, Bibliothèque de France, lat. 7295, 14^{r}, the text in the colophon reads, '*quod vocatur saphea*' after '*instrumentum mirabile*'.

To sum up, in this exceptional Madrid manuscript gathering 'old' sidereal astronomy together with 'new' tropical astronomy, in addition to one text on an instrument by an Alfonsine astronomer, John of Lignères, there are two calendars based on the Toledan Tables enabling the computation of mean syzygies, one almanac also based on the Toledan Tables, and a text and a table for mean conjunctions attributed to John of Lignères. We now turn to a more specific analysis of the latter.

6 See Nothaft, *Kalendarium*, p. 114.

7 Roser Puig, *Los Tratados de construcción y uso de la azafea de Azarquiel* (Madrid: Instituto Hispano-Árabe de Cultura, 1987).

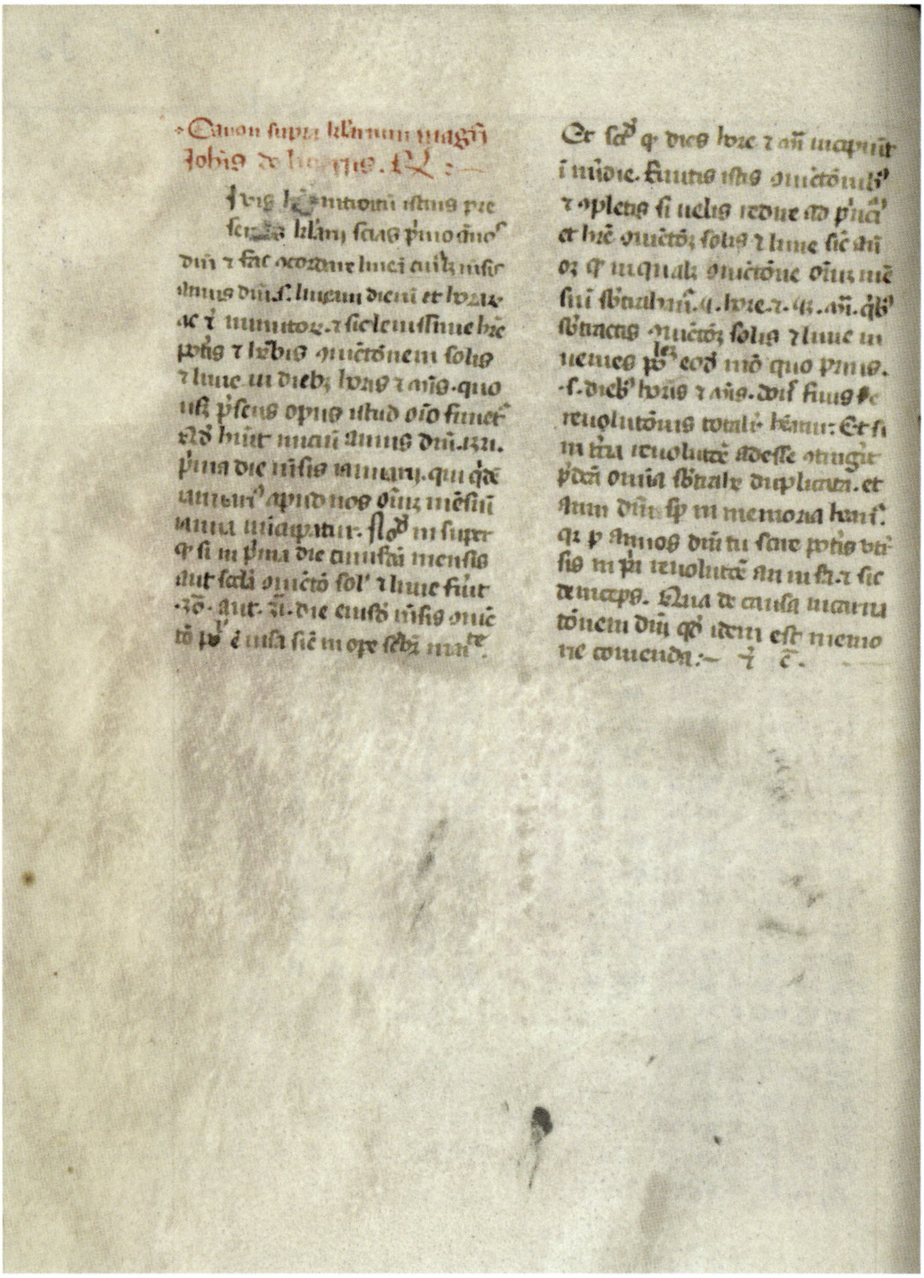

Figure 1. Text on mean conjunctions attributed to John of Lignères in Madrid, Biblioteca Nacional, 9288, 10ᵛ. Image taken from the holdings of the Biblioteca Nacional.

As pointed out above, the text in the Madrid manuscript (10ᵛ) is titled *Canon supra kalendarium magistri* Johannes *de Lineriis,* and it begins, '*Si vis* [...] *introitum presentis kalendarii scias primo*'. The text is presented in a double column for a total of 36 lines (see Fig. 1). The information provided is scarce: the entries represent conjunctions, given in months, days, hours, and minutes, beginning in January 1321; the day begins at noon. To find the time of conjunctions after the base period of 1321–1396, 5;53h have to be subtracted from the time of those in the base period, and an analogous rule is given for conjunctions occurring after the base period. We note that no specific locality is mentioned.

In the table, for each year from 1321 to 1396, we are given the time in days, hours, and minutes of the mean conjunction of each month of the year, beginning in January. Although not specified, we are dealing here with mean, not true, conjunctions, because there is a constant difference of 29d 12;44h or 29d 12;45h between successive entries. See, for example, the difference between the time of the conjunctions of 27 April 1321 (8;14h) and 26 May 1321 (20;58h) is 29d 12;44h. We note that already in the first line displayed in Fig. 2, corresponding to 1321, two entries were erroneously copied, for 27 February and 28 March (see Table 1). Quite a few such errors are spread throughout the table.

The value mentioned in the text, 5;53h, allows for an estimation of the length of the mean synodic month underlying this table. This amount corresponds to the time between two mean conjunctions 940 lunations apart. In fact, when considering a mean synodic month to thirds, 29d 12;44,3,3h, we obtain 5;52,13h = 76 · 365.25d – 29d 12;44,3,3h · 940. When using a rounded value of the mean synodic month to seconds, one obtains 5;53,0h = 76 · 365.25d – 29d 12;44,3h · 940. Thus, in this table, the embedded value is 29d 12;44,3h, a value often used in Alfonsine astronomy.

In itself, this table is straightforward. However, it is possible to go further by analysing the astronomical context in which this table was produced. Various Parisian astronomers in the early 1320s, including John of Lignères, addressed the issue of computing syzygies and compiled tables for mean and/or true conjunctions and oppositions.

In his *Patefit*, completed before 1335, John of Murs compiled a table, 'for the meridian of Toledo and according to the radices of Alfonso, King of Castile', displaying the time of the mean conjunction (to seconds), and the corresponding values of three other quantities: the mean motion of both luminaries, lunar anomaly, and the argument of lunar latitude. The table ranges from 1321 to 1396, as in the present table attributed to John of Lignères'.[8] However, John of Murs only tabulated one conjunction for each year, the last one in December. Therefore, the table attributed to John of Lignères could not depend on that by John of Murs because many conjunctions are missing and, on the contrary, John of Murs could not have borrowed the information from this table, because of an insufficient precision.

John of Murs's *Pateft* contains another table for mean conjunctions and oppositions for the period 1321–1396. For most of this seventy-six-year period, there are entries for the four quantities mentioned above, but only for time (given to minutes here) is the full interval covered.[9] Although not specified in the title of the table, all entries were again computed for Toledo, following the radices set up by King Alfonso's astronomers. Table 1 shows the comparison of selected entries in the table of Madrid and John of Murs's table, the latter being taken from the Biblioteca Apostolica Vaticana, Vat. lat. 3116, 11ʳ–23ᵛ.

8 See Chabás, *Computational Astronomy*, p. 154.

9 See Chabás, *Computational Astronomy*, pp. 155–56.

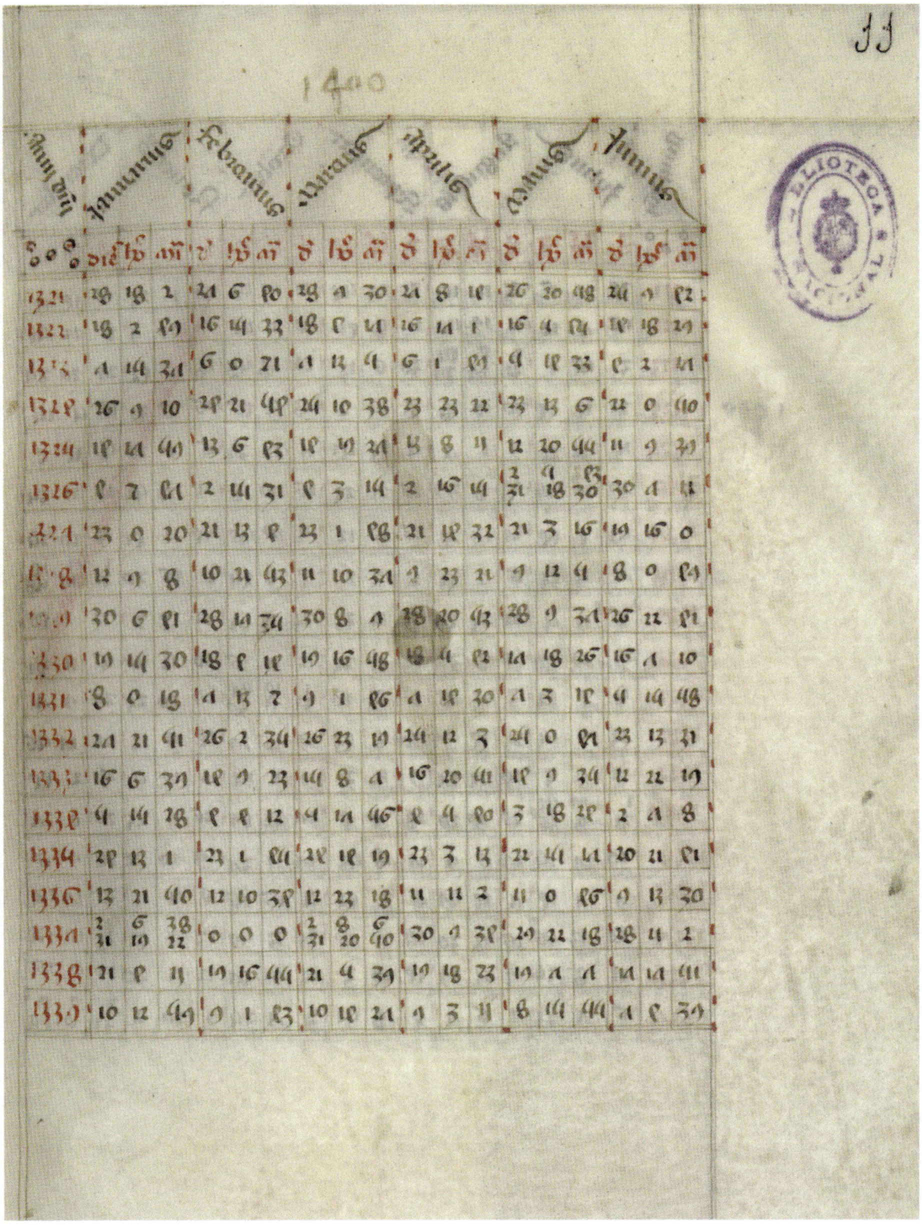

Figure 2. Excerpt of the table for mean conjunctions attributed to John of Lignères in Madrid, Biblioteca Nacional, 9288, 11[r]. Image taken from the holdings of the Biblioteca Nacional.

Apart from scribal errors, the entries in both tables differ in 0;48h, corresponding to the time difference between Toledo (John of Murs) and Paris, thus indicating that the table attributed to John of Lignères was compiled for Paris, and that it is possibly an adaptation of that of John of Murs.

Table 1. Time of mean conjunctions in Madrid 9288 and John of Murs (Vatican 3116).

Date	Madrid (h)	Vatican (h)	Difference (h)
28 January 1321	18; 2	17;14	0;48
27 February 1321	6;40 [a]	5;58	–
28 March 1321	7;30 [b]	18;42	–
27 April 1321	8;14	7;26	0;48
26 May 1321	20;58	20;10	0;48
25 June 1321	9;42	8;54	0;48
…			
5 July 1396	19;1	18;13	0;48
4 August 1396	7;45	6;57	0;48
2 September 1396	20;29	19;41	0;48
2 October 1396	9;13	8;25	0;48
31 October 1396	21;58	21; 9	0;49
30 November 1396	10;42	9;53	0;49
29 December 1396	23;26	22;38	0;48

[a] Madrid reads 6;40h instead of 6;46h
[b] Madrid reads 7;30h instead of 19;30h

It is worth noting that the *Patefit* contains yet another table for syzygies for the period 1321–1396, also computed for the meridian of Toledo. In this case, we are dealing with true conjunctions and oppositions.[10] Now, computing a mean syzygy does not require much effort, but computing a true syzygy does. And computing 1880 different true syzygies implies an outstanding effort. All of them were computed for Toledo. In this sense, John of Murs is no doubt the most 'Toledan' table-maker among the Parisian astronomers in the early fourteenth century. This was not the first time John of Murs compiled tables for syzygies. In his *Kalendarium solis et lune* for 1321, he listed the times and the positions of the Sun and the Moon of 235 consecutive mean conjunctions in nineteen yearly tables.[11] Here again, John of Murs computed the entries for Toledo; they also agree with those later found in his *Patefit*. In any case, as the number of conjunctions is limited here to nineteen years, this particular table by John of Murs cannot be the direct source of the table attributed to John of Lignères for seventy-six years.

In his Tables for 1322, John of Lignères gave tables for mean syzygies computed for the meridian of Paris and beginning in 1321.[12] However, these tables do not list the times of successive mean syzygies, in contrast to the table in the Madrid manuscript and in all the tables by John of Murs mentioned above. Instead, John of Lignères gave tables for expanded and collected years as well as for the months in a year to compute mean syzygies, which he later included in his *Tabule magne*. The sub-table for expanded years gives the time of the

10 See Chabás, *Computational Astronomy*, p. 156.

11 José Chabás and Bernard R. Goldstein, 'John of Murs Revisited: The *Kalendarium solis et lune* for 1321', *Journal for the History of Astronomy*, 43 (2012), 411–37 (pp. 423–24).

12 See Chabás, *Computational Astronomy*, pp. 184–85.

first conjunction of any year from 1321 to 1609 at intervals of twenty-four years, and thus has only four entries in common with the table in the Madrid manuscript (for years 1321, 1345, 1369, and 1393). The first entry, corresponding to January 1321, is 18d 17;14,6h. The precision here is given to the second. It is clear that the table attributed to John of Lignères in the Madrid manuscript is not an adaptation of his tables for mean conjunctions, but it is a natural complement that can be easily deduced from them.

Everything points to the conclusion that the table for mean conjunctions attributed to John of Lignères in the Madrid manuscript was indeed computed by him, or derived from tables compiled by him, building on those of John of Murs by applying a correction of 0;48h to account for the time difference between Toledo and Paris.

2. Two manuscripts at the Biblioteca Apostolica Vaticana: MSS Pal. lat. 1390 and 1445

The contents of the two miscellaneous manuscripts are quite similar: astrological texts and tables for computing the positions of the celestial bodies.[13]

Vatican, Biblioteca Apostolica Vaticana, Pal. lat. 1390 (henceforth MS A) is a late fourteenth- or early fifteenth-century manuscript originating from Frankfurt. Among other items, it contains a star catalogue of more than 1000 stars, lacking their names, said to be compiled for the year 1300, with an increment in precession since Ptolemy's catalogue of 16;12° (1^r–7^v); various astrological treatises by Ptolemy, Māsh^{c}āllāh, Alī ibn Ridwān, and Abū Ma^{c}shar; and a miscellaneous series of texts and tables in the framework of Alfonsine astronomy beginning on 84^r. This series includes a text and several tables attributed to John of Lignères (86^r–92^r), which are examined below; an additional chapter of the canons by John of Saxony to the Parisian Alfonsine Tables on lunar eclipses (93^r–v); the canons to the Parisian Alfonsine Tables beginning, '*Tempus est mensura*' (96^r–116^v); and two copies of the Parisian Alfonsine Tables (119^v, 120^v, 121^v–151^r and 165^r–182^v, 190^r–194^v).

The other manuscript, Vatican, Biblioteca Apostolica Vaticana, Pal. lat. 1445 (henceforth MS B), is a late fifteenth-century manuscript copied in Southern Germany. It also contains several astrological texts by Leopold of Austria, Abū Ma^{c}shar, Ptolemy, Sahl ibn Bishr (Zael), Johannes de Wachenheim, and Albertus Magnus. Then come the text and tables attributed to John of Lignères (219^r–221^v), which are examined here, followed by other astronomical texts and, as was the case with MS A, a long list of stars, dated 1429, with the names of the associated planets in each case (224^r–245^v).

Both manuscripts share the same text attributed to John of Lignères (MS A, 86^{r}–^{v}; MS B, 221^{r}–^{v}), beginning, '*Volens invenire medios motus planetarum*' and ending, '*Explicit canon Iohannis de Lineriis in tabulas sequentes*' (MS A), or, '*Explicit canon Iohannis de Lineriis in tabulis precedentibus brevis et utile valde*' (MS B), an excerpt of which is shown in Fig. 3.

---

13 Descriptions of both manuscripts are found in Ludwig Schuba, *Die Quadriviums-Handschriften der Codices Palatini Latini in der Vatikanischen Bibliothek* (Wiesbaden: Reichert, 1992), pp. 148–53 and 251–53.

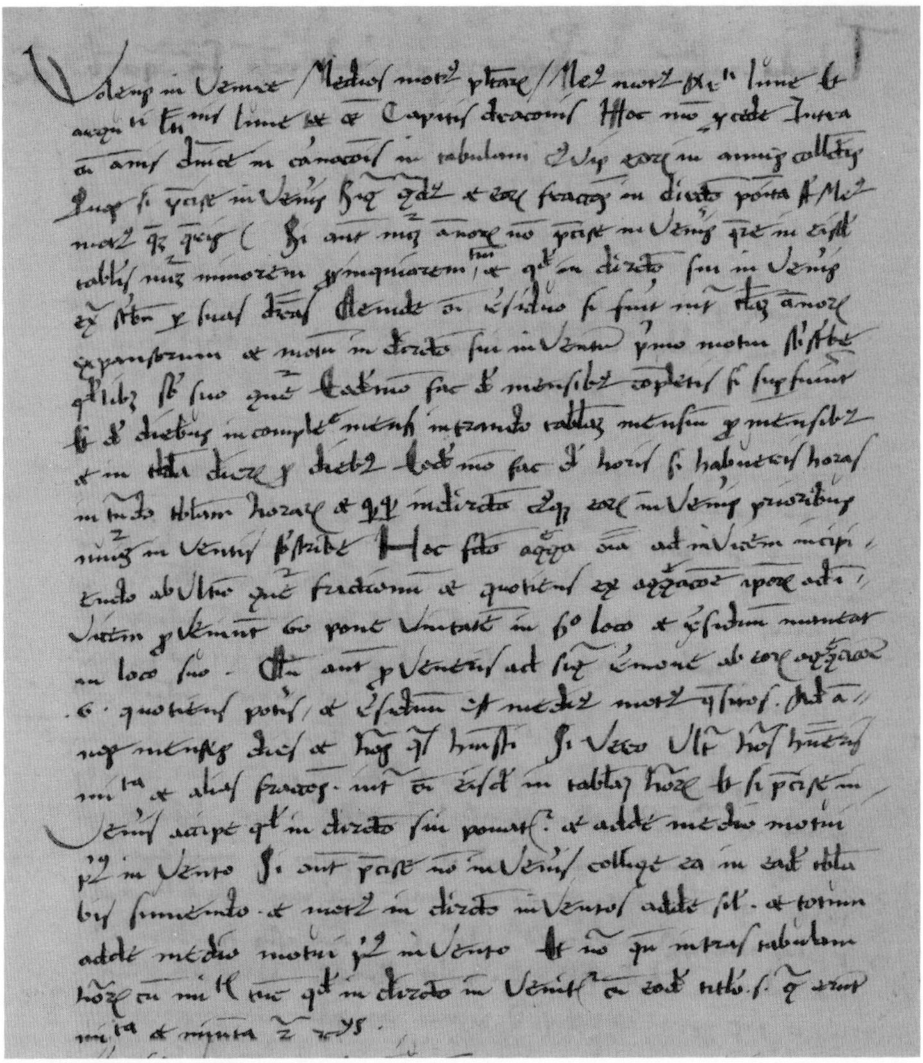

Figure 3. Beginning of the text in Vatican, BAV, Pal. lat. 1445, 221r. Source digi.ub.uni-heidelberg.de/de/bpd/index.html.

The text is divided into three paragraphs on mean motions, mean conjunctions, and apogees of the planets. We are given simple rules on how to enter the tables, and some relevant information: the year begins in March and complete years are used. This is already not a good sign for attributing the following tables to John of Lignères. We can safely say this because in all of his known tables, the year always begins in January. The canons also explain that signs of sixty degrees are used and that computations were done '*ad meridiam Tholeti sicut in tabulis Alfonsij*'. The second clue is that this is not a practice followed by John of Lignères, whose tables are always computed for Paris.

The associated tables with these canons are as follows.

1. *Radices and mean motions: MS A, 86ᵛ–90ʳ; MS B, 219ʳ–220ʳ*

There are twelve small tables displaying cyclical radices of twelve quantities in collected years, from 1363 to 1559, at intervals of twenty-eight years, for the Sun, the Moon, the five planets, the fixed stars, and the eighth sphere. The entries are given to thirds. Recomputation shows that all entries were indeed computed for Toledo.

MS A, 86ᵛ, displays a similar table for the radices of the Sun, the Moon in longitude, lunar anomaly, argument of lunar latitude, and node, in collected years, from 1251 to 1335, also at intervals of twenty-eight years. In this case, all quantities are given to seconds.

Then follow the tables for mean motions of ten quantities, all previous ones except for those for the fixed stars and the eighth sphere, in expanded years (from 1 to 28), months in a year (beginning in March), days (from 1 to 31), hours (from 1 to 30 for the Sun and the Moon in longitude and anomaly, and from 1 to 24 for the rest), with entries given to thirds. All parameters belong to the Alfonsine tradition.

The interval of twenty-eight years for mean motions is not the one used by John of Lignères or any of the early Alfonsine astronomers for this purpose. Instead, it was used previously in several adaptations of the Toledan Tables.

2. *Apogees of the Sun and the planets: MS A, 91ᵛ–92ʳ; MS B, 220ʳ*

This table displays the values of the planetary apogees at twenty-year intervals, from 1380 to 1520. In MS A, the entries for 1540, 1560, and 1580 were left blank. All tabulated values are given to seconds. Those for 1380 are as follows:

Sun and Venus	1s 30;1,29°
Saturn	4s 11;59,48°
Jupiter	2s 52;13,6°
Mars	2s 13;48,19°
Mercury	3s 29;15,39°

The twenty-year interval for the planetary apogees is precisely that used by John of Lignères in his *Tabule magne* for the period from 1320 to 1520.[14] As a matter of fact, the entries in MSS A and B fully agree, but for scribal errors, with those in the *Tabule magne*. We now understand why the entries for 1540, 1560, and 1580 were left blank: they just did not appear in the original table. Also included in MS A, 92ʳ, are two additional columns for the difference between the positions of the apogees twenty years apart and for the motion of the apogees in a year. In sum, this table was taken from a larger one belonging to a set compiled by John of Lignères.

MS A, 92ʳ, also displays part of the table for planetary apogees, also derived from the *Tabule magne*, together with several unfinished tables and working notes.

14 See Chabás, *Computational Astronomy*, pp. 199–201.

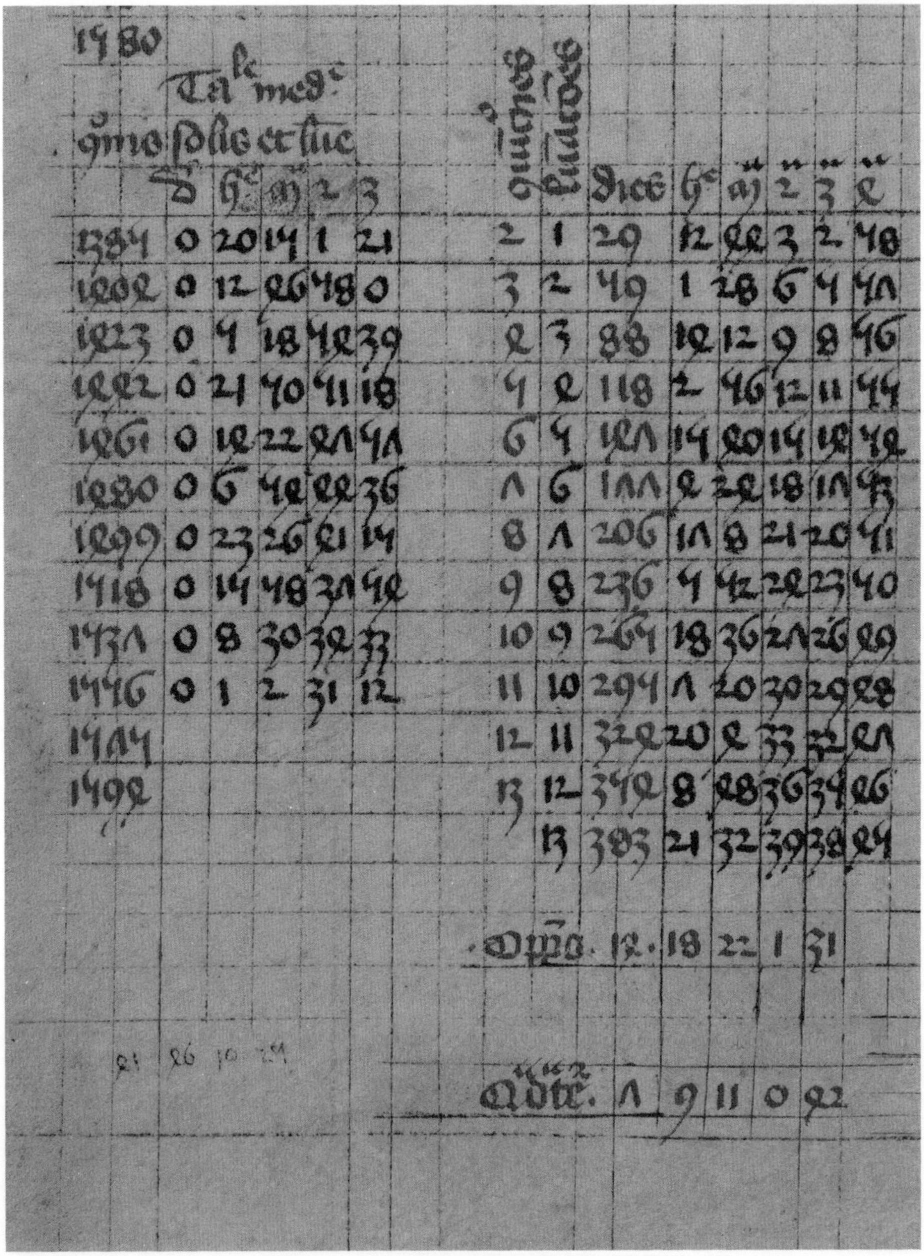

Figure 4. Two sub-tables for mean conjunctions in Vatican, BAV, Pal. lat. 1390, 91v. Reproduced with permission. Source digi.ub.uni-heidelberg.de/de/bpd/index.html.

3. *Mean conjunctions: MS A, 91v; MS B, 220r*

There are three sub-tables for mean conjunctions. The first one lists the time, in days and hours, to thirds, of the mean conjunctions from 1385 to 1556, at intervals of nineteen years (see Fig. 4). Only one entry is given for each of these years. The entry for 1385 is od 20;15,1,21h, and this is indeed the time of the first mean conjunction occurring in March 1386 (that is, 1385 completed) in Toledo. As is the case for the rest of the entries, all computations were done on the meridian of Toledo. Again, this seems to discard the direct intervention of John of Lignères.

The second sub-table gives the duration of thirteen successive lunations. The first entry, 29d 12;44,3,2,58°, is a precise value for the length of the mean synodic month in Alfonsine astronomy.

The third sub-table gives the time associated with the golden number, from 1 to 19. The entry corresponding to 19 is od 16;31,56,39h, which is the amount to be added to an entry in the first table to obtain the following one (mod. 24h).

The conclusion seems straightforward. The set of tables associated with the canons beginning, '*Volens invenire*' cannot have been compiled by John of Lignères, even though it contains a table for the apogees of the Sun and the planets derived from his *Tabule magne*. It does, however, seem a hectic combination of Alfonsine tabular material partially computed for Toledo, which was assembled in the late fourteenth century.

To sum up, of the two texts and tables examined here, one (the *Kalendarium*, as it is called in the Madrid manuscript) was most likely produced by John of Lignères. On the other hand, as uncomfortable as it is to contradict what is explicitly said in a manuscript, it is not plausible to attach the name of John of Lignères as the author of the text and tables in the two Vatican manuscripts.

One wonders why, in this second case, the text and tables were attributed to John of Lignères. As mentioned at the very beginning of this paper, John of Lignères was a prolific and authoritative voice in astronomy, with an extensive and appreciated body of work, especially his sets of tables, if we are to judge from the number of copies of his Tables for 1322 and the *Tabule magne*. Therefore, it would not be surprising that his name was used as *auctoritas* – one could maybe call it a trademark – in order to provide reliability and authority to a text or a table to add value to them.

Manuscript sources

Erfurt, Universitätsbibliothek, CA 4° 355
Erfurt, Universitätsbibliothek, CA 4° 366
Erfurt, Universitätsbibliothek, CA 4° 369
Los Angeles, J. Paul Getty Museum, Ludwig XII.6
Madrid, Biblioteca Nacional, 9288
Paris, Bibliothèque nationale de France, lat. 7295
Paris, Bibliothèque nationale de France, lat. 15118
Toledo, Biblioteca de la Catedral, 99–5
Uppsala, Universitetbibliotek, C 653
Vatican, Biblioteca Apostolica Vaticana, Pal. lat. 1390
Vatican, Biblioteca Apostolica Vaticana, Pal. lat. 1445
Vatican, Biblioteca Apostolica Vaticana, Vat. lat. 3116

Bibliography

Biblioteca Nacional, *Inventario general de manuscritos de la Biblioteca Nacional*, 15 vols (Madrid: Dirección General de Archivos y Bibliotecas, Servicio de Publicaciones, 1953–2001).

Castro, Manuel de, *Manuscritos franciscanos de la Biblioteca Nacional de Madrid* (Madrid: Servicio de Publicaciones del Ministerio de Educación y Ciencia, 1973).

Chabás, José, *Computational Astronomy in the Middle Ages* (Madrid: Consejo Superior de Investigaciones Cientificas, 2019).

———, and Goldstein, Bernard R., *The Alfonsine Tables of* Toledo (Dordrecht: Kluwer, 2003).

———, and ———, 'John of Murs Revisited: The *Kalendarium solis et lune* for 1321', *Journal for the History of Astronomy*, 43 (2012), 411–37.

———, and ———, 'The *Almanac* of Jacob ben Makhir', in *Editing and Analyzing Numerical Tables: Towards a Digital Information System for the History of Astral Sciences*, ed. Matthieu Husson, Clemency Montelle and Benno van Dalen (Turnhout: Brepols, 2022), pp. 53-78.

Nothaft, Philipp, 'Medieval Astronomy in Catalonia and the South of France: the 'Improved' Lunar *Kalendarium* of Friar Raymond (Ramon) Bancal (*c.* 1311) and its Predecessors', *Llull: Revista de la Sociedad Española de Historia de las Ciencias y de las Técnicas*, 38 (2015), 101–25.

Poulle, Emmanuel, 'John of Lignères', in *Dictionary of Scientific Biography*, ed. by Charles Gillispie, 16 vols (New-York: Charles Scribner's Sons, 1970–80), VII (1973), pp. 122–28.

Puig, Roser, *Los Tratados de construcción y uso de la azafea de Azarquiel* (Madrid: Instituto Hispano-Árabe de Cultura, 1987).

Saby, Marie-Madeleine, 'Les Canons de Jean de Lignères sur les tables astronomiques de 1321' (unpublished thesis, Paris, École Nationale des Chartes, 1987); abstract in *Positions des thèses* (1987), pp. 183–90.

Schuba, Ludwig, *Die Quadriviums-Handschriften der Codices Palatini Latini in der Vatikanischen Bibliothek* (Wiesbaden: Reichert, 1992).

Toomer, G. J., 'Profatius Judaeus and the Toledan Tables', *Isis*, 64 (1973), 351–55.

MATTHIEU HUSSON

Work Cohesion as a Test of Manuscript Transmission: The Case of John of Lignères' *Tabule magne*

Introduction

The identification and analysis of the discrete elements or 'works' of mathematical astronomy in medieval manuscripts are complex. In the Latin sources, the corpus of mathematical astronomy is generally formed of multiple-text manuscripts. And inside a given codex, texts and table sets are often not copied one after the other; instead, parts of these elements are mixed according to different ordering methods. Hence, the identification and classification of the medieval Latin works of mathematical astronomy require expert knowledge and the ability to analyse, in some cases by means of statistical tools, elements deeply embedded in the content of the documents. The material aspects of manuscript transmission are thus finely intertwined with the intellectual aspects of mathematical astronomy. The usual approach to this issue has been to seek reliable identification tools for works, such as an incipit for a text or the morphology of quantitative material with regard to tabular content.[1] Within the framework of ALFA, and especially in the context of surveying the Alfonsine text corpus, we frequently rely on these well-established methods.

In this contribution, my aim is to take a different approach, to ask whether it is now possible to consider the complexity of manuscript transmission not only as a problem to solve in order to identify works but also as a resource to better understand the astro-

* Research presented in this chapter was supported by the ERC project ALFA: Shaping a European scientific scene, Alfonsine astronomy, CoG 723085, PI Matthieu Husson. I am in debt to all members of the ALFA team for the many exchanges of ideas, and in particular to Richard L. Kremer, José Chabás, and Nick Jacobson for their precious comments on a draft of this chapter.

1 Such methods have been developed over the last seventy years. Seminal works include E. S. Kennedy, 'A Survey of Islamic Astronomical Tables', *Transactions of the American Philosophical Society*, N. S. 46 (1956), 123–77, and for the Latin sources G. J. Toomer, 'A Survey of the Toledan Tables', *Osiris*, 15 (1968), 5–174. Several other works were completed in this style, the most recent and pertinent in the context of this chapter being José Chabás and Bernard R. Goldstein, *A Survey of European Astronomical Tables in the Late Middle Ages*, (Leiden: Brill, 2012) and José Chabás, *Computational Astronomy in the Middle Ages: Sets of Astronomical Tables in Latin*, (Madrid: Consejo Superior de Investigaciones Científicas 2019).

Matthieu Husson • SYRTE-Observatoire de Paris-PSL, CNRS

Alfonsine Astronomy: The Written Record, ed. by Richard L. Kremer, Matthieu Husson, José Chabás, Turnhout, 2022 (*Alfonsine Astronomy: Studies and Sources*, 1), pp. 317-342
© BREPOLS PUBLISHERS 10.1484/M.ALFA.5.124930

nomical practices of historical actors. The complexity of the manuscript transmission of mathematical astronomical material in the Alfonsine milieus is, at least partially, the result of the choices of historical figures, namely, the ways in which they considered the texts they were copying, compiling, or composing.[2] Understanding these choices and attitudes could shed new light on the intellectual practice of mathematical astronomy in the Alfonsine context. The complexity of manuscript transmission might also convey information about the ways historical actors would or could work with the manuscripts they shaped in various contexts. Why did a particular scribe select a specific combination of texts in a manuscript? In what ways would this particular combination be useful later on? Which astronomical quantities can be computed, to what level of accuracy, and according to what method? What are the various astronomical phenomena that can be analysed and how can this be done?

Although these general questions are the drivers of this contribution, I do not address them directly or in their full generality. Instead, I shall consider the case of an individual work, and then allow the manuscript tradition of this work to inform me of its parts, how they are held together, and how they are reconfigured in different situations. This, in turn, allows me to raise the issue of the level of cohesion (or lack thereof) between the different parts of the work as completed by different historical actors. Together, this question and this approach clarify the aforementioned more general issues concerning the attitudes of historical actors towards written mathematical astronomical material and the consequences of their choices on the way they were able to interact with the manuscripts they produced.

John of Lignères' *Tabule magne* will be the subject of this analysis. This work, likely composed sometime between 1320 and 1325, includes a set of canons and tables.[3] It was integrated by its author in a larger collection of his works, which incorporated two treatises on instruments (the *saphea* and the *equatorium*).[4] Thus, the initial composition of the

2 Most of the Alfonsine manuscripts were copied by fourteenth- and fifteenth-century hands and binding may, in some cases, have occurred much later (even in the nineteenth century); but in those cases, a material analysis of the codex often provides strong indications about the medieval organization of the codices.

3 Emannuel Poulle, 'John of Lignères', in *Dictionary of Scientific Biography*, ed. Charles Gillispie, 16 vols (New York: Scribners, 1970–80), VII (1973), pp. 122–28; John D. North, 'The Alfonsine Tables in England', in *Prismata, Naturwissenschaftsgeschichtliche Studien: Festschrift für Willy Hartner*, ed. Y. Maeyama and W.G. Satzer (Wiesbaden: Steiner, 1977), pp. 269–301; José Chabás and Bernard R. Goldstein, *The Alfonsine Tables of Toledo* (Dordrecht, Kluwer Academic Publishers, 2003), pp. 282-3; Chabás and Goldstein, *Survey*; José Chabás and Bernard R. Goldstein, 'The Moon in the Oxford Tables of 1348', *Journal for the History of Astronomy*, 47 (2016), 159–67; Chabás, *Computational Astronomy*, pp. 199–206.

4 'Feci unicum instrumentum modici sumptus levi ponderis, quantitate parvum, virtute et continentia magnum, quod et planetarum equatorium nuncupatur eo quia in eo faciliter eorum equationes habentur. Sed quia talium instrumentorum non sunt compositores ubique et etiam ubi sunt pauci reperiuntur perfecti, quia etiam per talia non ita precise veritas sicut per tabulas haberi potest, idcirco composui has permanentes tabulas in veritate perpetuas et omnium radicum susceptivas in quibus omnium tabularum labor excluditur et instrumentorum inprecisio amovetur. Post hoc igitur videns quod in opere tabularum instrumenta primi nobilis antecedunt. *Tria* illorum precipua adinvicem comparavi: speram scilicet solidam, astrolabium et sapheam Azerchelis [...] *unum* composui instrumetum omnium instrumentorum predictorum vires ac etiam excellentias continens, quod etmerito universale astrolabium nuncupatur. [...]' Paris, BnF lat. 10263, f. 71^r (I have made a unique, lightweight and low-cost instrument, which although small in size, is rich in virtues and content. It is called the equatorium of the planets, because it is easy to obtain their equations with it. But the designers of such instruments are not to be found everywhere, and of those that exist, very few are perfect, and thus with such instruments the truth cannot be obtained as precisely as it

text by John of Lignères is already interesting with respect to the issue of cohesion, firstly because the work is composed of two distinctive types of content (tables and canons), and secondly because this composite work was apparently integrated into a larger intellectual unit by its author.[5] The *Tabule magne* were important in the transmission of Alfonsine material in England,[6] but the text also circulated in other parts of Europe in the fourteenth and fifteenth centuries. Thus, different attitudes towards the *Tabule magne* are witnessed in the corpus, which might enlighten us on the subject of cohesion.

This chapter is divided in two parts: canons, and tables. In both parts, the same three steps are followed. Firstly, I briefly describe the manuscripts (date, place, content) in order to understand the intellectual profile of the scribes who produced them. Secondly, I detail the content of the work to see the topic addressed, the originality, and level of technicity of each table or canon included in it. Thirdly, I study how this content is organized in the different witnesses (omission, addition, ordering) in order to assess how each manuscript handles the canons, tables, and the relationship between them.

We see that the scribes transmitting the *Tabule magne* canons or tables have largely relied on the modularity of those writings to build different sets. Their art of compilation reveals the main components of the work, which seems to be expressed with more liberty in the more technical parts (canons on complex topics, or even more tables). Compilation thus can have different effects on canons and tables, suggesting that a straightforward procedural relation between both types of writings was not central for these scribes. These results also raise methodological issues surrounding the techniques used to identify texts and tables, respectively, and concerning the enterprise of preparing critical editions that I also discuss in the conclusion.

1. Canons to the *Tabule magne*

1.1. *Manuscript witnesses*

Among the seven manuscripts witnessing the canons to the *Tabule magne*, all but one concentrate on mathematical astronomy and astrology with highly technical content. On the other hand, John of Lignères emphasized in his dedication to Robert Bardis, Dean of Glasgow, that the purpose of the association of the *Tabule magne* with an *equatorium* and a *saphea* was to offer an autonomous and simple set of tools related to spherical and planetary astronomy.[7] The text of the canons (e.g., the arithmetical canons on addition

can with tables. Thus, I have composed permanent tables, truly perpetual because all the roots are written on them, in which the work of all the tables is excluded, and the imprecision of the instruments is avoided. After this we see thus that, in the operation with tables, an instrument for the first mobile precedes. [...] I composed an instrument .... deserving the name of universal astrolabe).

5 The first characteristic is very common in Alfonsine astronomy, the latter is more original. However, no manuscripts propose these three works together.

6 North, 'Alfonsine Tables in England', pp 273-4. See also, Boudet and Miolo, in this volume.

7 Two equatoria texts are attributed to John of Lignères; see Emmanuel Poulle, *Les Instruments de la théorie des planètes selon Ptolémée: Équatoires et horlogerie planétaire du xiiie au xvie siècle*, 2 vols (Geneva: Droz, 1980), I, pp. 363–74. Nothing in the text of the *Tabule magne* introduction indicates which of those two texts was intended to be associated with the saphea or the *Tabule magne*.

and subtraction) as well as some aspects of the transmitted tables (e.g., the double-argument tables for planetary equations or the time from mean to true syzygy) confirm this initial statement. Thus, a text designed by its author for an educated but non-technical audience was transmitted to us through manuscripts giving the perspective of a highly technical group of scribes and users. Many of the transmission phenomena attested in this manuscript tradition can be explained by this tension.

Erfurt CA 4° 349 is a manuscript, mostly on paper (except for ff. 78–97 and 160–61), of 171 folios.[8] It is composed of eight parts (ff. 1–77, 78–97, 98–120, 121–32, 133–45, 146–59, 160–61, and 162–71), likely assembled during the second half of the fourteenth century. The manuscript was copied by several French or English hands, including, for the first codicological part, that of Johannes de Wasia who is also probably responsible for a table of contents on the manuscript's fly-leaves.[9] It is thus likely that Johannes de Wasia assembled this manuscript during his presence at the University of Paris from 1369 to 1383. Johannes de Wasia's interests then turned towards theology and he finished his career as the first dean of the theology faculty at Cologne University, where he died in 1395. Johannes de Wasia authored several astronomical and mathematical works, from which some notes on the *Almagest* and a *Tractatus de proportionionibus* are copied in Erfurt CA 4° 349. He also composed a series of mean motions tables for the meridian of Paris and some *Questiones de spera*.[10] Johannes de Wasia owned an important collection of scientific manuscripts, which was later purchased by Amplonius Rating de Berka (1363–1435). Now preserved in Erfurt, they are an essential source for the study of early Alfonsine astronomy. Although he was perhaps not a first-rank Alfonsine astronomer, Johannes de Wasia was certainly competent in mathematical astronomy. The technical content of Erfurt CA 4° 349 reflects this. Mathematical astronomy texts from the pre-Alfonsine and Parisian Alfonsine milieus form the core of the manuscript's first codicological unit, with works from John of Lignères, Henry of Langenstein, and Petrus de Dacia. Other sections of the manuscript also reflect the Parisian milieu, with works by Geoffrey of Meaux and Franco of Polonia, for example. The larger influence of texts circulating in the Parisian arts faculty can be seen with the works of Oresme, or earlier authors like Grosseteste, Jordanus de Nemore, Alexander of Villedieu, or pseudo-Albert the Great. These texts concern mathematical astronomy, mathematics, computus, astrology, and natural philosophy (Grosseteste *De coloribus*). In this manuscript, the canons to the *Tabule magne* (ff. 11^r–17^v) are not identified by a title where they begin but by an explicit on folio 17^v: *Expliciunt canones magistri Johannes de Lineriis super tabulas magnas*.

8 Wilhelm Schum, *Beschreibendes Verzeichnis der Amplonianischen Handschriften Sammlung zu Erfurt* (Berlin: Weidmannsche Buchhandlung 1887), pp. 583–87; David Juste, 'MS Erfurt, Universitäts- und Forschungsbibliothek, CA 4° 349' (updated: 14.11.2019), *Ptolemaeus Arabus et Latinus. Manuscripts*, URL: http://ptolemaeus.badw.de/ms/31.

9 For a short note on Johannes de Wasia, see David Juste, 'Johannes de Wasia, *Notes on the Almagest*' (updated: 29.10.2019), *Ptolemaeus Arabus et Latinus. Works*, URL: http://ptolemaeus.badw.de/work/75.

10 'Tabella radicum mediorum motuum subscriptorum ad annum domini 1369 completa ad meridianum Parisiensem per Iohannem de Wasia calculata', Erfurt 4° 362, ff. 13^v–14^r (table for the subscribed radices of the mean motion for the complete year 1369 at Paris meridian, calculated by Johannes de Wasia). The questions on the sphere can be found in Erfurt CA 4° 298, ff. 31^r–58^r.

Erfurt CA 4° 366 shares a similar profile.[11] Like 4° 349, 4° 366 is a student manuscript reflecting the advanced teaching of mathematical astronomy in the Parisian arts faculty in the mid-fourteenth century. The first part of the manuscript witnesses many works from the Parisian Alfonsine milieu. Interestingly, the second part of the manuscript presents texts from the earlier Toledan astronomy with, for instance, a fragment of the most common version of the canons to the Toledan Tables (f. 80). The canons to the *Tabule magne* are distinguished on folio 28r with the following title: *Canones tabulas magne magister Johannes de Lineriis*. Erfurt 4° 366 is also important in the transmission of the *Tabule magne* because the text of those canons is followed by a commentary by John of Spira.

Cambridge Gonville & Caius MS 110 is a parchment manuscript of 368 pages.[12] It is formed of multiple codicological parts (1–19; 20–40; 41–100; 101–98; 199–294; 295–342; 343–62; 363–68), most of them from the fourteenth century. A note from R. Marchall along with a table of contents for the full manuscript found on page 20 indicates that the document was probably assembled in England during the fifteenth century. Although essentially technical, the manuscript seems to reflect a practitioner's collection rather than a university student's workbook. This practical rather than pedagogical orientation of the codex is indicated by the fact that the manuscript is mainly filled with 'almanacs' and features a full section on astrological computations (houses, aspects). The term 'almanac' was used by historical participants to refer to what appears to us today as distinct material. For instance, we find in the same manuscript an ephemerides by John of Saxony that is called 'almanac'.[13] Although assembled in England, the manuscript's most represented authors are John of Lignères and John of Saxony, thus showing a connection between the Parisian and English milieus in the practice of mathematical astronomy. The manuscript also transmits a version of the standard canons to the Toledan Tables. The canons to the *Tabule magne* are the opening piece of the document. They are introduced with the following title on page 1: *Incipiunt canones super magnum almanac omnes planetarum magistro Iohanne de Lineriis Ambianencis dyocesis composui super meridianum Parisiensem*. In coherence with the content of the rest of the manuscript, the *Tabule magne* are presented as an 'almanac'. In this context, we might understand this title as pointing to a practically-oriented set of tables and canons. The manuscript also transmits both the canons and the tables of the *Tabule magne*.

Paris BnF lat. Colbert 60 is a parchment manuscript of 182 folios.[14] Like the Cambridge manuscript, it is also a practitioner's collection assembled probably near the end of the fifteenth century from different codicological units produced in the fourteenth and fifteenth centuries. The practical orientation of the document is attested to, for instance, by the appearance of different types of arithmetical tables throughout the manuscript, some of which remarkably indicate an interest in decimal numbers. Although the manuscript

11 Schum, *Verzeichnis*, pp. 612–14; Fritz S. Pedersen, *The Toledan Tables: A Review of the Manuscripts and the Textual Versions with an Edition* (Copenhagen: C. A. Reitzels Forlag, 2002), pp. 109–10. I was also able to consult a full digitalization of the manuscript.

12 Pedersen, *Toledan Tables*, p. 96. I was also able to consult a partial digitalization of the manuscript.

13 José Chabás and Bernard R. Goldstein, 'The Master and the Disciple: The Almanac of John of Lignères and the Ephemerides of John of Saxony', *Journal for the History of Astronomy*, 50 (2019), 82–96.

14 See https://archivesetmanuscrits.bnf.fr/ark:/12148/cc95559t (last consuled: 18.02.20). I also thank Alexandre Tur and José Chabás for information on this manuscript.

contains different almanacs, the central elements of the codex, in terms of content, are different versions of the Oxford Tables and the *Tabule magne*. Thus, Col. 60 also attests to a connection between Parisian and English milieus in the transmission of the canons to the *Tabule magne*. In this case, the *Tabule magne* somehow returned to the continent accompanied by new sets of tables produced in England. The canons to the *Tabule magne* are introduced on folio 34^r by an incipit almost identical to that of the Cambridge manuscript: *Incipiunt canones super magnum almanac omnes planetarum magistro Iohanne de Lineriis Picardi Ambianencis dyocesis composui super meridianum Parisiensem*. Col. 60, like the Cambridge manuscript, is also a rare witness transmitting the canons and tables of the *Tabule magne* in the same codex.

Prague KMK N.VIII is a parchment manuscript from the end of fourteenth century or the early fifteenth century, composed of 118 folios conserved in an early binding.[15] It is a composite manuscript with two distinctive parts in terms of content. The first contains the canons to the *Tabule magne* along with the initial part of John of Lignères' *Priores astrologi*. The remaining sections of the manuscript are purely theological. Although the example of Johannes de Wasia, encountered in Erfurt 4° 349, may confirm the existence of disparate textual configurations, it is difficult to see any kind of specific intellectual project that the composite codex might serve. The canons to the *Tabule magne* are not distinguished by any explicit or title in this manuscript.

Paris BnF lat. 7281 is a paper manuscript of 279 folios produced in the middle of the fifteenth century around Paris and northern France by the unidentified scribe, Jo. B.[16] While several hands add different kinds of marginal notes up until the sixteenth century, the main interest of the manuscript lies in its very peculiar collection and organisation of texts. The twenty-seven astronomical treatises constitute a sort of anthology of mathematical astronomy as perceived by an author towards the middle of the fifteenth century. In this very specific context, the canons to the *Tabule magne* are introduced on folio 201^v under the following title: *Canones super tabulas magnas per Johannes de Lineriis compilatas ex tabulis Alfonsi*. Thus, the connection of the set of tables to the Alfonsine Tables is quite clear. The table set is not described as an almanac, but Jo. B. uses the title attested in the two Erfurt manuscripts. Paris, BnF, lat. 7281 is neither a student nor a practitioner produced manuscript, but it is certainly a manuscript assembled by a skilled and very well-informed astronomer.

Paris, BnF lat. 10263 is a paper manuscript of 172 folios from the second half of the fifteenth century, copied in Naples by Arnaud de Bruxelles.[17] Written in a humanistic hand, it contains mainly texts. It belongs to a larger collection of BnF manuscripts attributed to Arnaud de Bruxelles,[18] an important humanist in the latter part of the fifteenth century. The large collection of scientific manuscripts associated with him demonstrates that he was

15 I thank Alena Hadravova for the information she provided on this manuscript.

16 Jean-Patrice Boudet, *Lire dans le ciel: La bibliothèque de Simon de Phares, astrologue du xve siècle* (Brussels: Centre d'Etudes des Manuscrits, 1994), pp. 175–89; see https://archivesetmanuscrits.bnf.fr/ark:/12148/cc66474 p. (last consulted 18 February 2020).

17 E. Poulle, *La Bibliothèque scientifique d'un imprimeur humaniste au xve siècle: Catalogue des manuscrits d'Arnaud de Bruxelles à la Bibliothèque nationale de Paris* (Geneva: Librairie Droz, 1963), pp. 45–53, See https://archivesetmanuscrits.bnf.fr/ark:/12148/cc71998b (last consulted: 18.02.20).

18 Paris, BnF lat. 10252, Paris, BnF lat. 10253, and Paris, BnF lat. 10263 to BnF lat. 10271.

also very well-informed in mathematical astronomy. The works of Arnaud de Bruxelles and Jo. B. demonstrate that the astronomical and intrinsic interest of the *Tabule magne* became associated with a more 'patrimonial' approach to the domain. In lat. 10263, the canons to the *Tabule magne* are introduced by the following incipit on folio 70^r: *Canones clarissimi mathematici Magistri Johannes de Lineriis Ambianensis diocesis amatoris scientie astrorum super tabulas per eum compositas ad inventionem motuum et locorum planetarum necessariorum ad iudicia astronomica.* Somehow, this title also indicates the practical orientation of the canons to the *Tabule magne*, and especially their astrological utility. These canons are immediately followed by John of Spira's commentary. The manuscript was intended to be coupled with Paris, BnF lat. 10264, which contains the table set of the *Tabule magne*.

This brief introduction to the seven witnesses shows that the text originated in Paris and circulated in England, Italy, Germany, and Central Europe. In these different regions, the text was transmitted in highly specialized manuscripts. These manuscripts suggest different contexts and historical participants in the transmission of the canons: students, practitioners, and scientific humanists. When present, the incipit or explicit to the canons also indicates different aspects of the work: its practical orientation as an 'almanac' and its link to astrological computation or to Alfonsine astronomy.

1.2. Content of the set of canons

From this point, it is possible to investigate more precisely how these different manuscripts organize the content of the *Tabule magne*. The texts seem to be composed of eleven separate canons or chapters.[19] Apart from the introduction, these canons can be gathered, for ease of description, into three different thematic groups. In summarizing the content of these different parts, I also signal possible parallels in canons that could have been known to John of Lignères.

The first thematic group addresses sexagesimal arithmetic with canons on addition, subtraction, and the computation of proportional parts.[20] It is common to find texts dedicated to the computation of proportional parts in astronomical canons, especially when this computation is made with the support of a specific proportional table, as is the case here. For instance, canon 9 of John of Lignères' *Priores astrologi* presents such a case.[21] It is less frequent to find separate canons specifically on the subjects of addition and subtraction. When these topics are addressed in canons, it is usually done in paragraphs inside another canon dedicated to a specific astronomical issue. For instance, a note on addition is inserted in the canons dedicated to mean motion (canon 5) of the *Priores astrologi*. A note on subtraction is incorporated into the canons concerning the true place of the sun in John of Saxony's *Tempus est mensura motus*. When composing the canons to the

19 The canons feature a little more than 6000 words.

20 The manuscript Erfurt CA 4° 366 uniquely adds a fourth canon to this section on arithmetic with an alternative and more general handling of proportional parts. It is probably an interpolation by Johannes de Wasia. Because it is uniquely attested in this manuscript, I do not consider this fourth canon in the following analysis. This shows that, in some contexts, historical participants had specific interests in these fundamental arithmetical questions. One may also think of the many arithmetical tables found in Paris, BnF Col. 60.

21 Marie-Madeleine Saby, 'Les Canons de Jean de Lignères sur les tables astronomiques de 1321: Edition critique, traduction et étude' (unpublished thesis, Paris, École Nationale des Chartes, 1987), p. 194.

Tabule magne, John of Lignères had probably already completed his *Algorismus minutiarum*, a specific arithmetical treatise on computation with usual and sexagesimal fractions. John of Lignères' choice to dedicate specific canons to these fundamental arithmetical issues echoes his explicit intention to provide, with the *Tabule magne*, an *equatorium*, and a *saphea*, that is, autonomous and simple mathematical tools for studying and computing spherical and planetary astronomy.

The second thematic group addresses questions of planetary astronomy in five canons. The central piece of this section is the canons on planetary equations that explain how to use the double-argument tables John of Lignères prepared for this specific purpose. This canon insists on the symmetries of the table layout and the way it can be read. It also precisely addresses the algebraic 'sign' of the equation obtained as the table entry with respect to the way the table is read. The interpolation procedures, unusual in this double-argument context, are presented in detail.[22] The mathematical approach to the computation of planetary equations is very original, but the canons do not pursue this aspect in depth. The computation of the true place of the Moon, although not unprecedented in the Latin sources, is also different from the conventional Toledan, Alfonsine, or *Priores Astrologi* approach.[23] The procedure of the *Tabule magne* relies on mean syzygy times and a double-argument table. This double-argument table gives the increment in longitude of the Moon which is to be added to its mean longitude at the preceding syzygy to obtain the true lunar place from its age and the mean lunar anomaly. The lunar canon insists on the same practical aspects of the computation, interpolation, and reading of the table as does the planetary equation canon. A canon concerning the computation of the true place of the Sun is also presented. On this topic, the *Tabule magne* follow a quite common procedure and the canon is very similar, for instance, to canon 13 of the *Priores astrologi*.[24] Finally, this planetary motion section also contains two canons that address planetary apogees and mean motions. These canons insist that computations be made for the meridian of Paris and explain in detail how they can be adjusted for a different meridian. In this respect, John of Lignères followed an option he had already adopted in his *Priores astrologi* canons.

The third and final group consists of two canons dealing with the computation of syzygies. The first explains how to find the first mean conjunction or opposition of a given year, and thus of all mean syzygies for that year. This is a very typical canon, of which equivalent versions can be found in the canons to the Toledan Tables or the *Priores astrologi*.[25] The second canon of this last group addresses the issue of computing the time from mean to true syzygy. It describes an algorithm relying on a double-argument table

22 Matthieu Husson, 'Ways to Read a Table: Reading and Interpolation Techniques in Canons of Early Fourteenth-Century Double-Argument Tables', *Journal for the History of Astronomy*, 43 (2012), 299–319.

23 For a similar approach to this issue, see the tables of John Vimond and John of Murs: José Chabás and Bernard R. Goldstein, 'Early Alfonsine astronomy in Paris: The Tables of John Vimond (1320)', *Suhayl*, 4 (2004), 207–94; José Chabás, and Bernard R. Goldstein, 'John of Murs's Tables for 1321', *Journal for the History of Astronomy*, 40 (2009), 297–320.

24 *Priores astrologi's* set of canons is a little longer and more complex, because it precisely cross-references all the preceding canons on which the computation of the Sun's true place rely: for example, the computation of mean motion, apogees, mean argument, and proportional parts (Saby, 'Les Canons de Jean de Lignères', p. 199).

25 Saby, 'Les Canons de Jean de Lignères', pp. 260–64; Pedersen, *Toledan Tables*, pp. 446–49.

that gives the result of the division of the true elongation by the difference of the solar and lunar velocities. The described algorithm also relies on a table providing the equations and velocities of the two luminaries. This approach to the computation of true syzygies has precedent; for instance, it is described, among other possibilities, in canon 32 of the *Priores astrologi*.[26] However, in this latter case, the final division does not rely on a dedicated table but must be performed by the user.[27]

Hence, John of Lignères appears to have composed the canons to the *Tabule magne* by compiling different types of materials. Firstly, there are a few canons that rely on original or seldom used approaches to specific computations (e.g., planetary equations, true position of the Moon, or true syzygies). These canons appear to favour double-argument tables. Secondly, a set of much more common canons and procedures are provided, allowing the user to produce the initial data required for the more original canons. These more common canons often have very close parallels in the *Priores astrologi* or, for arithmetical canons, in the *Algorismus minutiarum*. There, John of Lignères adapted and re-used some of his previous work. Lastly, a general introduction is written, which unifies the set and ties it to two texts on astronomical instruments. Thus, the canons to the *Tabule magne* appear to be a somehow heterogeneous compilation of materials that might have been produced or borrowed by John of Lignères at different moments. We know nothing about the relationship between John of Lignères and Robert Bardis, but it could be that a specific opportunity presented itself and prompted the compilation of the work.

1.3. Organization of the manuscript witness content

Information regarding the various ways in which different manuscripts organise this content is presented in Figure 1. On the left of the table, the different canons are organized according to the thematic grouping I used to describe them above. At the top of the table, the shelf-mark of the manuscript is given. They are roughly chronologically ordered, left to right. At the intersection of a line and a column, one finds an ordinal number giving the rank of that particular set of canons in the manuscript. A cell with no ordinal number denotes the absence of the canon from the manuscript. Finally, when the absence of a given canon causes a procedural gap in the flow of the text, as presented in the witness, the concerned cell is shaded and an arrow points to the canon that would, in principle, not be applicable without the information of the missing one.[28]

Different manuscript witnesses present different parts of the work. One of the earliest witnesses, Erfurt 4° 349, proposes the most compact version of the text with only six canons: three on arithmetic and three on planetary motion. It completely ignores the issue of syzygies.[29] Furthermore, BnF lat. 10263, probably the latest witness, proposes the most comprehensive compilation of canons. This contrast between the two manuscripts

26 Saby, 'Les Canons de Jean de Lignères', p. 221.

27 For an earlier zij where a table is provided for division, see Ibn al-Kammād in José Chabás, and Bernard R. Goldstein, 'Computational Astronomy: Five Centuries of Finding True Syzygy', *Journal for the History of Astronomy*, 28 (1997), 93–105.

28 See the discussion in the next section for a definition of 'procedural gap'.

29 The scribe of the canons concluded them with an explicit showing that this shortened version was intentional.

		post 1300			circa 1400			ante 1500
		Erfurt Q 349	Erfurt Q 366	Cambridge G&G Ms 110	BnF, Mel. Col. 60	Prague Met. Chap. N VIII	BnF lat. 7281	BnF lat. 10263
Introduction	Multiplicis philosophie	1	1	1	1	1	1	1
Artihmetic	Modum additionis	2	2	2	2	2	2	2
	modum subtrahere	3	3	3	3	3	3	3
	veram partem proportionalem		4	4	4	6	4	6
Planetary	Locum augis	5	5	5	5			5
	medium motum	4	6	10	10	4 ↓		4
	dato centro medio et argumento vero…verum locum per tabulas	6 ↓	7	6	6	5	5 ↓	7
	Dato numerorum dierum…usque at 15 dies		8	7	7	9	6	9
	dato medium motus solis et verum locum eius invenire		9					10
Syzygies	tempus prime conjunctionis			9	9	7		8
	Tempus vere conjunctionis		10 ↓	8 ↓	8 ↓	8 ↓	7 ↓	11

Figure 1. Organization of the canons to the *Tabule magne* in the different manuscript witnesses.

cannot be directly correlated to the profile of the documents. Erfurt 4° 366, quite similar in style to Erfurt 4° 349, has only one canon missing. BnF lat. 7281, like BnF lat. 10263 (insofar as both present some sort of scientific-humanist view on the text), drastically offers only seven of the eleven canons. The most often omitted canon is that on the computation of the true place of the sun. It is interesting to note that its content is also among the least original of the work. The second most omitted canon is also a common one that deals with the computation of mean syzygies. On the other hand, only four canons are present in every manuscript witness: the introduction, the arithmetical canons on addition and subtraction, and the canon on planetary equation. The systematic presence of the canons on planetary equations is clearly understood as being central in the *Tabule magne,* since their most original contribution is the double-argument planetary equation tables. The case of the introduction canon is probably an artefact of the method of identifying works in manuscripts by their incipit. A manuscript giving the canons to the *Tabule magne* but omitting this first canon would not have been identified. Given the fluidity of the manuscript transmission of these kinds of texts, it is highly likely that many more manuscript witnesses of the canons will be found, once new methods for text identification are available.[30] It is more difficult to identify a reason for the systematic presence of the two rather elementary canons regarding addition and subtraction in a manuscript tradition that seems to have been produced by experts.

These variations in the collection of canons found in a given witness create a procedural gap in almost every manuscript, except for BnF lat. 10263, which is complete. All the other manuscripts present at least one procedural gap, that is, a canon that in principle would

30 The constantly progressing availability of digital surrogates for manuscripts and the improving performance of image processing and hand-written text recognition will certainly induce new methodologies and discoveries for this domain in the near future.

not be applicable because it requires information or a method from a missing canon. For instance, the computation of planetary positions begins with the computation of the planetary mean argument. This quantity is obtained as the difference between the mean longitude and the apogee of the planet. Manuscripts that lack the canons allowing the user to compute the apogee of a planet present a procedural gap with respect to the planetary equation canons. Similarly, the absence of canons on the true solar position or on the mean syzygies in some manuscripts yields a procedural gap with respect to the true syzygy canon.

The order of the canons is also an interesting indication of the procedural structure of the text. Let us consider the case of the canons for computation of proportional parts. In principle, this canon is to be used for the computation of interpolation in an equation (or any kind of non-linear) table. For a manuscript like Erfurt 4° 366, all the arithmetical canons are grouped together, followed by the equation canon. This manner of separating and isolating the arithmetical canons at the beginning of the text is a peculiar aspect of the canons to the *Tabule magne*. In this case and for this particular issue, the canons follow a procedural order. Three other manuscripts adopt the same organization for this canon on proportional parts.

Paris, BnF lat. 10263 does not break the procedural order, but adopts an additional ordering method. It presents the canon on proportional parts immediately before the canon on the equation, but the former is disconnected from the other arithmetical canons. The adopted ordering conforms to a frequent organisational method for canons, in which a specific mathematical set of instructions is given 'on the spot' where required, even at the cost of being repeated several times and in slightly different ways in the canons.[31] Prague Met. Chap. N VIII presents an ordering similar to that of BnF lat. 10263, but breaches the procedural order as the canon for proportional parts is presented immediately after the first equation computation canon.

Additional variations appear even on a more refined level of text analysis. For instance, BnF lat. 10263 has, at the end of the mean motion canon, a sentence relating it to the canon on apogees; at the same time, Erfurt, 4° 349, which shows the same succession for these two canons, lacks that sentence.[32] Of course, this sentence is also missing from Erfurt 4° 366. Going deeper, I have shown elsewhere that the interpolation procedures presented in these two manuscripts for double-argument tables are very different.[33] In fact, BnF lat. 10263 significantly simplifies the procedure and renders it independent of the astronomical context. Erfurt 4° 366 presents more complex procedures, which are sensitive to the astronomical nature of the arguments or to the astronomical object concerned (e.g. interpolation procedures for the Moon differ from those employed for the planets). This is a major modification to the text and might even change the planetary positions calculateed withthe tables.

31 For instance, the computation of proportional parts is presented twice by John of Saxony in *Tempus est mensura motus*: Emmanuel Poulle, *Les Tables alphonsines avec les canons de Jean de Saxe: Édition, traduction et commentaire* (Paris: Éditions du Centre national de la recherche scientifique, 1984), pp. 64, 66–68.

32 '4us canon de locis augis cuiuslibet et argumentis solis et centro aliorum 5e planetarum invenendio', Paris, BnF lat. 10263, 74^{r} (Fourth canon regarding the position of apogee, used to find the argument of the Sun and the five planets).

33 Husson, 'Ways to Read a Table'.

A more general view of the ordering of the canons in the different witnesses raises other issues. BnF lat. 10263 begins with the two elementary arithmetical canons concerning addition and subtraction. Following this, a group regarding planetary true positions is presented with canons on mean motions, radices, proportional parts, and planetary equations. A group on syzygies is later presented with canons on mean syzygies, computation of the true lunar and solar places, and computation of true syzygies. Erfurt 4° 366 organizes and delimits these three series differently. The arithmetical series is more comprehensive and even includes the canons on proportional parts. The planetary equation group is, in fact, more of a true position group and also includes the canons on the Sun and Moon. The group on syzygies is reduced to only one canon and presents a procedural gap. In some cases, the ordering of the canons suggests that a given procedural gap was identified by the scribe, who chose to remedy the issue by adding one of the missing chapters to the end. This may be the case of the canon *dato numerorum dierum* found in Prague, Met. Chap. N VIII. A similar hypothesis can be proposed for the case of the canon *medium motum* in Cambridge, C&G MS 110 and BnF Col. 60.

The situation of the different canons is also diverse. For example, the beginning of the text is very stable; all manuscripts present the first three canons in the same order. By contrast, the canons on the computation of the true location of the Moon (when present) are always in a different position, ranging from rank 6 to rank 9.[34] It would appear as if canons copied later and presenting more technical content were more likely to be reordered, omitted, and modified.

The expert scribes who transmitted the canons to the *Tabule magne* have amplified the compilation aspect of the work by omitting different canons and variously ordering them. Apparently, some of the working habits adopted during the production of the canons are also those of the scribes who transmitted them. Even the content of the text can be modified, even in strategic places such as interpolation procedures. Hence, the manuscript tradition of the canons to the *Tabule magne* does not show a great concern for procedural coherence. This may suggest that the interests of the scribes of those witnesses were not confined by the procedural content of the canons. Perhaps they were able to supply the 'missing information' by other means, for example, from texts in a different part of the codex, from a different codex available to them, or simply from their own personal expertise. This kind of distant and open copying of a set of canons, initially created for a non-expert but educated audience, raises interesting research questions about the motivation of expert readers to copy such works. What does this tell us about the astronomical and computational practices of these experts?

The issue of the relation between canons and tables being one of the points of this chapter, it is important to briefly summarize how the canons refer to tables before we turn to the analysis of the table set in the manuscript tradition. We first note that not all canons are related to a specific table, including the introduction and the first two arithmetical canons. Strikingly, these are also the only canons that directly manipulate numerical content regarding specific quantities. All the other canons avoid discussion of any specific numerical examples but do refer to tables. They often do this by using a

34 Except, of course, for the two manuscripts presenting exactly the same ordering.

title for the table. They also describe the reading process for the table. These descriptions might refer to the layout of the table (including the step of the argument) and some of its non-numerical content, such as headings, units of measurement, or abbreviations, e.g., providing information on the 'additive' or 'subtractive' aspect of an entry. In these cases, thescanons, to a certain extent, constrain the type of table to which they can be directly applied. However, such textual constraints put on the table set by the canons are rather loose; by avoiding mention of the numerical content of the tables, the canons do not commit themselves to any specific values for astronomical parameters.

2. Tables of the *Tabule magne*

2.1. Manuscript witnesses of the tables

The twenty tables manuscripts, like those with the canons, confirm the presence of the work in Paris, England, Germany, Central Europe, and Italy. They would have circulated from the mid-fourteenth century to the end of the fifteenth century. Like the canons manuscripts, those transmitting the tables can generally be described as expert practitioners' codices, sometimes more carefully executed than others. The tables are not found, for instance, in a courtly presentation manuscript. Thus, the manuscript tradition suggests that the audience for the table set differs from the educated yet non-specialized users who were explicitly targeted by John of Lignères in the introduction to the canons. I distinguish three manuscript groups according to the relative proportions of text and tables they contain, as well as to the intellectual profiles of the codices.

Seven manuscripts contain only, or mainly, numerical tables.[35] They span a wide range of dates across the fourteenth and fifteenth centuries. Four of them have an identified geographical origin, either in Paris, Prague, Vienna, or Italy. Paris, BnF lat. 7286C is among the earliest manuscript of the corpus. It is a small (30x22 cm) parchment codex from the fourteenth century, featuring fifty-eight folios. It contains material produced by John of Lignères and John Vimond in the early 1320, when the Parisian Alfonsine Tables were most likely compiled. The tales by John Vimond are uniquely preserved in this manuscript. It is the first set attesting, in tabular format, parameters and theories specific to Parisian Alfonsine astronomy.[36] The section attributed to John of Lignères extends from folios 9^{r} to 56^{v} and essentially contains tables with the canons *Priores astrologi* and *Cuiuslibet arcus* copied in their margin as commentary. Most of the tables in this set are related to John of Lignères' Tables of 1322. Only two sub-sets, related to mean motions in years (ff. 10^{v}–11^{r}) and in hours or minutes of hours (ff. 23^{v}–24^{r}), can be connected to the *Tabule magne* thanks to their specific layout. Moreover, John of Lignères' Tables of 1322 derive from the standard version of the Toledan Tables.[37] The compilation of John of Lignères' Tables of

35 Erfurt CA 2° 388; Ajuda MS 52-XII-35; British Library Add Ms 24070; BnF lat. 7286C; BnF lat. 7300A; Vatican MS Pal. lat. 1374; Venice MS lat. VI 29.

36 José Chabás and Bernard R. Goldstein, 'John Vimond and the Alfonsine Trepidation Model', *Journal of the History of Astronomy*, 34 (2003), 163–70.

37 Chabás, *Computational Astronomy*, p. 193.

1322 with tables from the *Tabule magne* is a recurrent pattern in the manuscript tradition of the latter. Such. a compilation effect might occur because John Vimond and John of Lignères themselves were working in the same astronomical mileu at the same time.

The most important group of witnesses in the table manuscript tradition contains eleven elements. They are quite similar in style to those transmitting the canons.[38] These eleven manuscripts are evenly distributed over the fourteenth and fifteenth centuries and attest the presence of the *Tabule magne* in Paris, England, Germany, Central Europe, and Italy. They are astral science manuscripts dedicated mainly to mathematical astronomy and astrology. Two manuscripts of this group are described above in the section on canons: Cambridge, G&C Col. Ms 110 and Paris, BnF Col. 60. Erfurt 2° 376 is a typical witness of this group.[39] This is a mid-fourteenth-century manuscript of 102 parchment folios with an early leather and wood binding. It opens with works on the sphere (Sacrobosco, Peckham) and continues with the *Tabule magne*'s tables. The codex concludes with treatises on arithmetic, the astrolabe, and planetary theory. This collection of texts is an instance of what O. Perdesen called the *corpus astronomicum*.[40] In contrast to Paris, BnF lat. 7286C, this manuscript's table set is solely composed of tables that can be related to the *Tabule magne*. The first four folios (30^{v}–34^{v}) contain mean motion and mean syzygy tables. Folios 35^{r} to 53^{v} contain the planetary double argument equation tables.

Finally, two manuscripts transmitting the table include content that extends beyond astral sciences in the direction of the *scientiae mediae*, with, for instance, works on optics, music, or even natural philosophy.[41] They both date to the fifteenth century and come from Italy and Central Europe. Given the very low number of codices in this group, it is difficult to make any definitive statements about a potential meaning of this geographical and temporal distribution. On the other hand, the fact that only two out of the twenty table manuscripts have a scope going beyond the astral sciences supports our suggestion that the manuscript tradition of this text is predominantly technical and specialized. Bernkastel-Kues, Cusa MS 212 belonged to Nicolaus Cusanus. It is a 407-folio parchment manuscript mostly featuring works related to the astral sciences, but also Johannes Peckham's *Perspectiva communis* (239^{r}–250^{v}). In this manuscript, similarly to Paris, BnF lat. 7286C, what circulated under the incipit *Tabule Parisienses cum canonibus* is mainly composed of John of Lignères' Tables of 1322 with their canons in marginal notes. Two 'clandestine' elements from the *Tabule magne* are, however, incorporated: a set of apogees from 1320 to 1520 on folio 93^{r}, and a set of mean syzygy tables on folios 91^{v}–92^{r}. Paris, BnF lat. 10264 is especially important in the context of this paper because it is closely linked to the manuscript Paris, BnF lat. 10263 (described in the section on canons). Paris, BnF lat. 10264 is a large 286-folio paper manuscript related to Arnaud de Bruxelles.[42] Thus, many

38 Bernkastel-Kues, MS 210; Oxford, Hertford College, MS E.4; Cambridge, Gonville and Caius College, MS 110; Erfurt CA 2° 376; Erfurt CA 2° 384; Milan, MS N217 sup; BnF Col. 60; Vatican, MS Pal lat. 1367; Vatican MS Pal lat. 1376; Vatican, MS Pal lat. 1412; Venice, MS Cic 2309.

39 See https://ptolemaeus.badw.de/jordanus/ms/3356 (last consulted: 18.02.20).

40 Olaf Pedersen, 'The corpus astronomicum and the Tradition of Medieval Astronomy', *Studia Copernicana*, 13 (1975), 57–96.

41 Bernkastel-Kues, MS 212; BnF lat. 10264.

42 https://ptolemaeus.badw.de/jordanus/ms/4755 (last consulted: 18.02.20).

of the remarks I have made concerning Paris, BnF lat. 10263 also hold for this manuscript. The document begins (ff. 1^r–36^r) with a miscellany of astronomical tables, including most of the *Tabule magne*. Different arithmetical and proportion tables accompany them. These arithmetical tables cannot easily be linked to the *Tabule magne*, although they could be very useful in computing with the *Tabule magne*. Furthermore, folios 36^v to 56^v are left blank. Some of these folios are ruled with ink to leave templates for tables (e.g. 45^r). Arnaud de Bruxelles thus probably intended to complete the table set with some more material and left ample room to do so. The rest of the manuscript features texts on completely different topics, such as optics with Alhacen's *De perspectiva*, natural philosophy with Roger Bacon's *Opus tertiums*, and Albert the Great's *Cosmographia*. This kind of manuscript could be interpreted as showing a transmission of the *Tabule magne* in a less specialized context, perhaps targeted to the kind of users for which John of Lignères had compiled his work. However, in this case, Arnaud de Bruxelles and Nicolaus Cusanus were certainly first-rank practitioners of astronomy, not beginners.

One particularly interesting transmission phenomenon attested in this manuscript tradition, which is highly likely to be more common in table transmission, is the silent compilation of tables belonging to different sets. This phenomenon is almost never witnessed for the canon transmission of the *Tabule magne*,[43] where the transmitting scribes may omit canons, present them in different orders, and add, suppress, or modify sentences or paragraphs; but they do not blend multiple texts.[44] Perhaps when text identification methods more easily allow the identification of parts of texts rather than relying on the incipits, more complex phenomena of textual transmission will be identified. But already in our current set of twenty manusripts the existence of this compilation effect for tables shows a certain attitude of scribes and table compilers with respect to a table set. The *Tabule magne* are often mixed with John of Lignères' Tables of 1322.[45] Thus, they form a composite set of the two works that began to circulate very early in the mid-fourteenth and well into the fifteenth century, often as a complement to the Parisian Alfonsine Tables.[46] This early compilation of tables might be the consequence of a choice made by John of Lignères himself or at least in a milieu closely connected to him.[47] John of Lignères composed both sets during the same period of his life, roughly between 1320 and 1325. The impression I have is that compiling new tables and combining them into sets are two distinct processes. This leads to situations in which the same table is integrated into different table sets, some of which are dignified by the redaction of canons while others are not. The combination of tables from the *Tabule magne* and the Tables of 1322 lacks any set of canons. When incorporated into this larger set, the *Tabule magne* lose much of their practical, user-friendly qualities.

43 Only one canon on proportional parts from Erfurt 4° 366 was discarded from the canons because it was present only in this witness.

44 Prague, MS N VIII, for instance, first features the canons *Multiplicis philosophie* and then the *Priores astrologi*, but the two texts are not blended together.

45 In some situations, John of Genoa's tables are also aggregated to this collection (see Laure Miolo, in this book).

46 Vatican MS Pal. lat. 1367 also offers an example of this phenomenon in the second group of manuscripts.

47 BnF lat. 7286C, or Erfurt 2° 388 and Erfurt 2° 376.

2.3. Table set content

Following the organization of the canons, the content of the table set can be described in three different thematic groups. The first group concerns planetary positions, the second syzygies, and the third arithmetical tables. In a similar fashion to the canons, some elements of these groups are original and have a strong identity, while others are more common and are often shared by multiple table sets.[48]

The group of tables dedicated to the computation of planetary position is the largest. It can be further divided into two sub-groups: mean motions and equations. For both, John of Lignères composed tables with specific layouts, ensuring a strong identity for the *Tabule magne*. Mean motion tables are given for the meridian of Paris, similar to the Tables of 1322. The two main tables for mean motions concern (i) the position of the true apogees of the Sun and the planets and (ii) the mean motions of the Sun, Moon, and planets. Both cover intervals of twenty years, but in different ways.[49] The apogee tables list the true apogees with a precision of seconds from 1320 to 1520 (complete years). The mean motion table displays, in the same grid, twelve useful mean quantities.[50] One grid is given for each of the main calendar units: years, months, days, hours, and minutes with a precision of seconds or thirds. These quantities are needed to operate an *equatorium* of the type John of Lignères associated with the *Tabule magne*. They are also required to employ the equation tables of the *Tabule magne*. The expanded and collected years are organized in one grid within the table for the the years. It begins by giving entries for each year from one to twenty, then for every twenty years up to 100, then for every hundred years up to 1000. It ends with a value for 2000 years. A set of radices for Paris in 1320 is associated to these mean motion tables. John of Lignères also uses this peculiar format for mean elongations in the Tables of 1322.[51]

The most voluminous of this thematic group on planetary motion are the tables for lunar and planetary equations. They are double-argument tables that provide the total equation and share a common layout; the head of the table (and its bottom line) display the mean centre, and the left and right columns display the mean anomaly at intervals of six degrees. This choice of layout and computational organization is very original and gives a strong identity to the *Tabule magne*. The format would continue in table making until the modern period and would spread globally, in both astronomical and trigonometrical contexts. John of Lignères' Parisian contemporaries, John of Murs and John Vimond, carried out further experiments with double-argument tables.[52] An echo of these experiments is found in an alternative double-argument table given to find the true lunar position. The first argument of the table is a number of days, from one to fifteen, representing the time elapsed since the preceding mean conjunction. The second argument, to the left of the table, is given

48 Descriptions here largely follow Chabás, *Computational Astronomy*, pp. 199–206.

49 The use of this period is, for instance, attested in the Castillan Alfonsine canons; cf. Chabás and Goldstein, *The Alfonsine Tables*, 2003.

50 Mean motion of the Sun, Venus, and Mercury; mean motion of the Moon, lunar centre; lunar anomaly, ascending lunar node, mean motion of Saturn; argument of Saturn, mean motion of Jupiter, argument of Jupiter, mean motion of Mars, argument of Mars, argument of Venus, and argument of Mercury.

51 Chabás, *Computational Astronomy*, p. 202.

52 Chabás and Goldstein, 'John of Murs's Tables'; Chabás and Goldstein, 'Early Alfonsine Astronomy'.

at intervals of six degrees from zero to 180 and represents the mean lunar anomaly. The entries of the table give the increment in longitude of the Moon to be added to its mean longitude at the preceding conjunction. Interestingly, the canons for the Moon describe this particular table and do not mention the first double-argument tables addressing the Moon's true position as that of any regular planet.[53] The solar equation table is entirely standard, with a maximum value of 2;10° reached at 92°-94°. It cannot be distinguished from that found in John of Lignères' Tables of 1322 or the Parisian Alfonsine Tables.

The second thematic group of tables concerns syzygy computation. For this group, the canons describe two subjects. The first concerns mean syzygies and refers to tables also found in the Tables of 1322. The entries provide mean motions of the two luminaries, the mean lunar anomaly, and the mean argument of lunar latitude. These values, for the meridian of Paris, are given for the first syzygy of January, at intervals of twenty-four years from 1321 until 1609. A table of expanded years gives the values for years from one to twenty-three; and a table of months gives them for the first day of each month from February through December.

There are distinct sub-tables for conjunctions and oppositions. The time of syzygy is given to seconds and the mean motions to thirds. A second group of tables concerns the computation of the interval between mean and true syzygy. The canons describe a double-argument table providing the result in hours and minutes of the division of the true elongation by the *superatio* or difference between the true velocity of the Moon and the Sun. When attested in the manuscript tradition, this table is often associated with, and in some case replaced by, another table that gives the equations and velocities of the Sun and Moon in one single grid at one-degree intervals. These velocity and equation tables are not described in the *Tabule magne* canons but in the *Priores astrologi* (although with an interval of six degrees). It is also found circulating in the Tables of 1322.

Finally, the canons also describe a proportion table that can be used to compute proportional parts in the context of interpolation. Such tables are found in the manuscript tradition of the *Tabule magne*, but they are common in many astronomical manuscripts and have very few distinctive characteristics. I have thus not yet attempted to describe them or associate them with our table set.

Like the compilation of the canons in the *Tabule magne*, the compilation of the table set appears to be organized around a few very specific tables that give the set its own identity. In the case of the *Tabule magne*, the most original tables are undoubtedly those of the planetary equations. However, the mean motion tables, organized as if for an instrument, and the division table for the computation of true syzygies, are also quite specific. Other more common tables are also found in the manuscripts, such as the solar equation, the table for mean syzygies, or the equation and velocity tables. They are borrowed from or shared with other table sets, compiled (or not) by John of Lignères, and are not necessarily described in the canons.

53 José Chabás, and Bernard R. Goldstein, 'The Medieval Moon in a Matrix: Double Argument Tables for Lunar Motion', *Archive for History of Exact Sciences*, 73 (2019), 335–59 (pp. 352–56).

2.3. Manuscript content organization

It is now possible to address the issue of the ways the different manuscript witnesses of the table set arrange and organize this tabular content. I summarize the information in Figures 2 and 3. Figures 1 and 2 present their content similarly. In Figure 2, we see two different arrays one above the other, each with ten manuscripts divided broadly over the fourteenth and fifteenth centuries. When tables from a given group are present in a specific manuscript, the cell is numbered. Empty cells mark the absence of a group of tables. The number in the cell corresponds to the order in which the tables appear in the manuscript. In contrast to Figure 1, here I have not attempted, for several reasons, to define or even mark what could be a procedural gap in a table set. Firstly, tables from the *Tabule magne* are copied with many other tables from different sets. Thus, if the *Tabule magne* set were missing a table in a given manuscript, it is highly likely that the missing information would be provided by another table from a different set. Secondly, and more fundamentally, tables are not intrinsically procedural or even discursive objects. It is thus difficult to define a procedural gap for a table set without thereby projecting onto it some sort of procedure. More than two thirds of the cells in Figure 2 are empty. It is thus difficult to assess, from Figure 2, the relative frequency of the different table groups in the manuscript tradition. Figure 3 provides this information.

More than two thirds of the cells in Figure 2 are empty, in stark contrast to Figure 1 where only one sixth of the cells are empty. The same phenomenon can be seen from another comparison of numbers; four canons out of eleven (i.e. more than a third) are present in every manuscript witness of the canons, while the most represented table is attested in only half of the witnesses forming the table manuscript tradition. This information is important when comparing the manuscript traditions for the canons and the tables and points to a contrast between the types of transmissions that table and canon sets undergo in the manuscript milieus of these expert astronomers. Table sets are blended (that is, have fluid boundaries), whereas in the witnesses so far identified, canon sets are not. Figure 2, however, shows striking contrasts among the various manuscripts. Three manuscripts boast nearly half of the table appearances, while fifteen offer three or fewer. This manuscript distribution of tables from the *Tabule magne* might thus reveal different situations in which the codices originated.

The first group consists of three manuscripts that present an almost complete set from the *Tabule magne*: Paris, BnF Col. 60; Paris, BnF lat. 10264; and Cambridge, G&C Col. MS 110. These three witnesses are all related to the canons. The Colbert and Cambridge manuscripts contain a version of the canons; Paris, BnF lat. 10264 is closely related to Paris, BnF lat. 10263. The scribes of these manuscripts held together both sides of the work and seemingly proposed their 'edition' of it. Colbert 60 is the most complete witness. It orders the tables broadly in three groups. The first group of tables is related to the computation of mean motion, the second to the computation of syzygies, and the third to the computation of planetary positions. Manuscript 10264 also addresses these three topics, but in a different order; mean motions come first, followed by planetary positions, and, finally, syzygy computations. This order is more congruent with the general organization of the canons than is the one adopted by Colbert 60. The Cambridge manuscript, generally

		Post 1300									circa 1400
		Paris, BnF, Lat 7286C	Erfurt, UB, F 376	Cambridge, G&C Col. Ms 110	Milano, Ambrosiano, Ms N217 sup	Bernkastel-Kues, Cusa, Ms 210	Efurt, UB, F 384	Venice, Bib Naz Marciana, Ms lat VI 29	Vatican, Pal lat. 1374	Bernkastel-Kues, Cusa, Ms 212	Erfurt, UB, F 388
Planetary	radice for Paris 1320			6		2					
	Apogees			1	1	1	1	1		2	
	Mean motions	1	1	3		3					
	Solar equation										2
	True lunar longitude double argument										3
	Lunar equation double argument										
	Planetary equation double argument		3								1
Syzygies	Mean syzygies		2	2			2		1	1	4
	Solar and lunar Equations and velocities			4			3		2		
	True zysygies			5					3		
Arithmetic	Proportion tables			7							

		circa 1400									ante 1500
		Venice, Museo Civico correr, Ms Cic 2309	Paris, BnF, lat 7300A	Paris, BnF, Colbert 60	Vatican, Pal lat. 1412	Vatican, Pal lat. 1376	Oxford, Heford Col., Ms 4	Vatican, Pal lat. 1367	Lisbon, Ajuda, MS 52-XII-35	London, BL, Ms add 24070	Paris, BnF, lat 10264
Planetary	radice for Paris 1320			6	1						1
	Apogees	1		1			1				
	Mean motions			3	2						2
	Solar equation									1	
	True lunar longitude double argument			10		2		2			5
	Lunar equation double argument		1	8						2	
	Planetary equation double argument		2	9		1			1	3	4
Syzygies	Mean syzygies			2	3			1			6
	Solar and lunar Equations and velocities			5				3			7
	True zysygies			7							8
Arithmetic	Proportion tables			4				4			3

Figure 2. Distribution of the different tables in the manuscript tradition of the *Tabule magne*.

Number of witness per table (total number of witness is 20)	
Mean syzygies	10
Apogees	9
Planetary equation double argument	8
Mean motions	7
Solar and lunar Equations and velocities	6
radice for Paris 1320	5
True lunar longitude double argument	5
True zysygies	4
Proportion tables	4
Solar equation	2
Lunar equation double argument	2

Figure 3. Number of witnesses for the different tables in the manuscript tradition of the *Tabule magne*.

speaking, follows the same order as Colbert 60, to which it is closely linked, except that it entirely omits the topic of planetary positions.

Manuscripts from the second group place tables from the *Tabule magne* within a larger collection of tables, often related to John of Lignères' Tables of 1322. They do not present any kind of edition of the table set. They do, however, keep some elements of its identity and inform us about the core elements of the set. Comprised of twelve manuscripts, this group is the most important and suggests that expert users can blend table sets in ways that reveal their interests. An initial sub-group of manuscripts transmits the planetary equations. This is the case for six manuscripts.[54] Among them, only Erfurt 2° 376 and Erfurt 2° 388 also provide mean motions from the *Tabule magne* (the later only for mean syzygies). All the others have inserted equations from the *Tabule magne* into a different context. A manuscript such as Lisbon, Ajuda 52-XII-35 has extracted only the planetary equations and omitted solar and lunar equations. Another group of tables closely linked to the identity of the *Tabule magne* concerns mean motions. Six manuscripts have extracted this thematic set from the *Tabule magne* and enclosed it in another tabular context.[55] The extreme case in this group is Paris, BnF lat. 7286C, which keeps only the mean motion tables. Some manuscripts, such as Bernkastel-kues, Cusa MS 210, extend a little beyond this core and include the tables for apogees and radices. Others, like Erfurt 2° 384 or Vatican Pal. lat. 1367, extend the core in the direction of syzygy computations.

The final manuscript group is constituted by those manuscripts attesting only one small table from the *Tabule magne*.[56] In these manuscripts, the link to the *Tabule magne* has almost faded completely. These manuscripts are more difficult to identify than those of the first three groups. It is thus likely that many more manuscripts of this kind have not

54 Erfurt 2° 376; Erfurt 2° 388; Ajuda MS 52-XII-35; British Library Add Ms 24070; BnF lat. 7300A; Vatican MS Pal, lat. 1367.

55 Bernkastel-Kues, MS 210; Bernkastel-Kues, MS 212; Erfurt 2° 384; BnF lat. 7286C; Vatican MS Pal, lat. 1367; Vatican MS Pal, lat. 1412.

56 Oxford, Hertford College, MS E.4; Milan, MS N217 sup; Venice, MS Cic 2309; Venice, MS lat. VI 29.

yet been identified. All of the manuscripts in this final group attest the apogee table. This is probably because this table provided direct positions of apogees over a long period of time; hence, the table was simple and would have remained useful long after its creation.

Consequently, apogee tables are the second most frequent in the table manuscript tradition. The most frequent are mean syzygy tables. There was a rather constant interest in those tables over the two centuries of the manuscript tradition. This was not the case for the planetary equation tables, which rank third in terms of frequency; seven witnesses (of eight) are from the fifteenth century. The relative absence of this central group from the *Tabule magne* during the first century of its transmission is striking, as is the attestation only in the fifteenth century of the lunar equation double argument table.

Our analysis of canons showed that the one dedicated to these double argument tables was a favourite locus for editorial intervention by the scribe, especially with respect to interpolation patterns. Similarly, planetary equation tables are also a locus for editorial intervention. Lisbon, Ajuda, MS 52-XII-35 uses sixty-degree signs, two ink colours, and displays differences only from one column to the next. Paris, BnF lat. 10264 and Erfurt 2° 376 use sixty-degree signs and display differences horizontally and vertically. Erfurt 2° 388 uses thirthy-degree signs and does not display differences, and so on. These interventions on the layout of the table have some consequences for the type of scribal mistakes likely to arise, on these tables' relation to the canon and on the way they can be manipulated for computation. They might then change, to some extent, the computational results one gets when using a different table witness.

The *Tabule magne* table set, like its related canons, is organized around two core table sub-groups with strong identities. The most important is the planetary equation group. The mean motions group (including apogee, radices and mean syzygies) closely follows it. Much less original satellite tables are associated with these two cores. The paradigmatic case here is the solar equation table. The lunar double-equation table is also interesting because its rarity and late appearance in the manuscript tradition might imply that it is a latter addition. However, the type of manuscript transmission occurring for table sets is different from that appearing in the canons (at least given what can be known with the current identification methods of texts and tables). Table sets are blended together in pairs, while canon sets are not. This blending of table sets is not random; the breaks occur primarily around the core group of tables that constitute the *Tabule magne's* identity as a table set.

A group of manuscripts proposes in the same, or closely related codices, both canons and tables. For these manuscripts, the relation between tables and canons is not organized around a procedural concern. For instance, the Colbert 60 and Cambridge manuscripts have the same set of canons in the same order, but their tabular sets are different. For the other types of manuscripts witnessing the tables, the blending of the *Tabule magne* with other table sets results in different kinds of editorial interventions and makes the relation between tables and canons more complicated than a simple procedural reading of tables and table sets might suggest. Expert users, apparently, did not always rely on canons to understand how tables were to be manipulated; they were perfectly at ease in constituting hybrid table sets that actually relate to no specific canons. Hence, we might surmise that from the expert astronomer's perspective, the procedural relation of canons to tables became secondary and that some other dimensions of these two types of mathematical descriptions concerning celestial phenomena became more important. What these could be and how they could be recovered from the extant sources remains an open research question.

Conclusion

In this survey of the *Tabule magne* manuscript tradition, I have attempted to identify the elements composing the works, their cohesion, the possible genealogy of the project, and its reception. This effort leads to different results concerning the *Tabule magne* themselves and the editing practices of different kinds of expert astronomers in the fourteenth and fifteenth centuries. It also raises more general methodological and historical questions.

The *Tabule magne* appear to have been composed in between 1320 and 1325. John of Lignères was, at the time, working towards composing various tables and related canons. Early in the period, he fixed the larger part of this work in his Tables of 1322 and their associated *Cuiuslibet arcus* and *Priores astrologi* canons. Later, in 1325, an opportunity was perhaps offered to him in relation to Robert Bardis and he compiled the *Tabule magne* with less, yet more original, tabular material, the canons *Multiplicis philosophie* and two instrument texts. In this context, it is important to note that the process of composing tables and canons and the process of compiling them in different works are distinct. Both the canons and tables sets are composed from the material at hand. They are structured around core elements with a strong intellectual identity. In the case of the *Tabule magne*, this core consists of tools for planetary equations and mean motions. It is complemented by certain satellite tables and canons, which are usually less original and possibly borrowed. The reception of the canons and tables in their manuscript witnesses shows this in the omission of the canons and tables or in the ways they were arranged in different orders according to patterns that respect the core elements of the work.

Tables were blended with tables from other works into sets that acquired a certain level of stability but to which no canons correspond. In these mixed table sets, the identity of the *Tabule magne* as a distinctive work may entirely disappear. Tables from the *Tabule magne* are often found mixed with tables from John of Lignères' Tables of 1322 in a set that complements the Parisian Alfonsine Tables. This mixed set might reflect the more extended tabular material from which John of Lignères composed, respectively, the *Tabule magne* and the Tables of 1322. The *Tabule magne* were probably intended for an educated but non-expert audience. However, the manuscript tradition known to us suggests the reading and editorial interventions by expert practitioners of astronomy. Their editorial interventions reveal much about the cohesion of the work and its possible genealogy, because they underline the core elements around which the *Tabule magne* are organized. These interventions modify, in some cases, very intricate details of the canons and tables by changing the layout, types of sexagesimal numbers, or even interpolation procedures. Overall, the editorial intervention of the expert practitioners who transmitted the *Tabule magne* to us indicate that they might have been interested in more than a procedural reading of the canons and uses of the table set. The relation between the canons and the tables is too distant. A reader with no knowledge of their procedural relation would have many difficulties in inferring how table sets and canons can be used together from these manuscripts.

In this work, I have suggested that historians of astronomy address the variability of the manuscript transmission of mathematical astronomical texts not as a problem to overcome in the identification and critical edition of texts but as a resource to better understand how historical actors worked with their textual and tabular material. This proposition raises two separate questions. The first concerns the method of text and table identification, in particular, the incipit method for identifying texts and the morphological approach for identifying tables. My conclusion depends on these two identification methods, yet they also reveal that compilation, in various forms, was a distinctive feature of the intellectual habits developed in late medieval scholarly milieus. Recent research on canon texts in Alfonsine astronomy shows that redactors of canons also actively compiled materials. These compilations can gain cohesion, over time, and become authoritative texts attributed to named authors, such as John of Saxony's *Tempus est mensura motus* or John of Lignères' *Priores astrologi*.[57] As digital humanities progress, it may soon be possible that methods of text identification, relying on the analysis of digital surrogates of manuscripts, will be more sensitive to textual parts than is today's incipit method. In this new context, it may be possible to further fine-tune the conclusions derived here. For instance, one might distinguish, on a larger scale and more meticulously, the canon texts built as compilations from those that are compositions of another kind or are even dignified as authorities.

A second question emerging from my approach deals with the nature of 'critical editions'. Many critical editions, over centuries stretching back to the Renaissance, have been built around the notion of the 'author' composing a no longer accessible *Urquelle*. In the case explored in this chapter, however, 'authorial agency' was largely counteracted by the editorial interventions of the expert practitioners who transmitted the textual and tabular material from the milieus where the work first emerged. A critical edition, recognizing these dynamics, should attempt to reveal the different voices of the *Tabule magne's* manuscript tradition.

Manuscripts sources of the *Tabule magne*

Canons

Cambridge, Gonville & Caius College, 110, pp. 1–5 (England, mid-fourteenth century)
Erfurt, Universitätsbibliothek, CA 4° 349, 11^r–17^v (Paris, second half of the fourteenth century)
Erfurt, Universitätsbibliothek, CA 4° 366, 28^r–32^v (Paris, mid-fourteenth century)
Paris, Bibliothèque nationale de France, lat. 7281, 201^v–205^v (Paris, mid-fifteenth century)
Paris, Bibliothèque nationale de France, lat. 10263, 70^r–78^r (southern Italy, second half of the fifteenth century)
Paris, Bibliothèque nationale de France, Col. 60, 34^r–36^r (Paris, fifteenth century)
Prague, Knihovna Metropolitní kapituly, N VIII, ff. 1^r–10^v (Prague, fifteenth century)

57 Nicholas A. Jacobson, *Ordering Language to Order the Heavens: On Alfonsine Astronomical Canons (1350–1500)* (Turnhout: Brepols, forthcoming).

Tables

Bernkastel-Kues, Bibliothek des St Nikolaus-Hospitals, Cus 210, 89^{v}, 103^{r}–117^{r} (mid-fifteenth century)
Bernkastel-Kues, Bibliothek des St Nikolaus-Hospitals, Cus 212, 91^{v}–93^{r} (Italy, 1415–21)
Oxford, Hertford College, 4, 57^{v} (England, mid-fifteenth century)
Cambridge, Gonville & Caius College, 110, pp. 7–18 (England, mid-fourteenth century)
Erfurt, Universitätsbibliothek, CA 2° 376, 30^{v}–53^{v} (Paris, mid-fourteenth century)
Erfurt, Universitätsbibliothek, CA 2° 384, 26^{r}–28^{v} (1346–55)
Erfurt, Universitätsbibliothek, CA 2° 388, 1^{r}–42^{v} (fifteenth century)
Lisbon, Biblioteca da Ajuda, 52-XII-35, ff. 67^{r}–92^{v} (fifteenth century)
London, British Library, Add Ms 24070, ff. 24^{v}–42^{v} (fifteenth century)
Milano, Biblioteca Ambrosiana, N217 sup, f. 26^{v} (Cremona, mid-fifteenth century)
Paris, Bibliothèque nationale de France, Col. 60, ff. 36^{v}–62^{v} (fifteenth century)
Paris, Bibliothèque nationale de France, lat. 7286C, ff. 10^{v}–11^{r}; 23^{v}–24^{r} (Paris, mid-fourteenth century)
Paris, Bibliothèque nationale de France, lat. 7300A, ff. 94^{v}–112^{r} (fifteenth century)
Paris, Bibliothèque nationale de France, lat. 10264, ff. 1^{r}–28^{v} (southern Italy, second half of the fifteenth century
Vatican, Biblioteca Apostolica Vaticana, Pal. lat. 1367, ff. 60^{v}–62^{r} (Bavaria, mid-fifteenth century)
Vatican, Biblioteca Apostolica Vaticana, Pal. lat. 1374, ff. 26^{r}–27^{v}, 51^{v} (Prague, 1407)
Vatican, Biblioteca Apostolica Vaticana, Pal. lat. 1376, ff. 46^{r}, 102^{r}–130^{r} (Regensburg, 1447–58)
Vatican, Biblioteca Apostolica Vaticana, Pal. lat. 1412, ff. 95^{r}–123^{r} (Paris, 1453–54)
Venice, Museo Civico Correr, 2309, ff. 71^{r} (Italy, fifteenth century)
Venice, Biblioteca Nazionale Marciana, lat. VI 29, f. 77^{r} (Italy, second half of the fifteenth century)

Other manuscript sources

Erfurt, Universitätsbibliothek, CA 4° 298
Erfurt, Universitätsbibliothek, CA 4° 362
Paris, Bibliothèque nationale de France, lat. 10252
Paris, Bibliothèque nationale de France, lat. 10253

Bibliography

Boudet, Jean-Patrice, *Lire dans le ciel: La bibliothèque de Simon de Phares, astrologue du XVe siècle* (Brussels: Centre d'Etudes des Manuscrits, 1994).

Chabás, José, *Computational Astronomy in the Middle Ages: Sets of Astronomical Tables in Latin* (Madrid: Consejo Superior de Investigaciones Científicas 2019).

———, and Goldstein, B. R., 'Computational Astronomy: Five Centuries of Finding True Syzygy', *Journal for the History of Astronomy*, 28 (1997), 93–105.

———, and ———, *The Alfonsine Tables of Toledo* (Dordrecht, Kluwer Academic Publishers, 2003).

———, and ———, 'John Vimond and the Alfonsine Trepidation Model', *Journal of the History of Astronomy*, 34 (2003), 163–70.

———. and ———, 'Early Alfonsine Astronomy in Paris: The Tables of John Vimond (1320)', *Suhayl*, 4 (2004), 207–94.

———, and ———, 'John of Murs's Tables of 1321', *Journal for the History of Astronomy*, 40 (2009), 297–320.

———, and ———, *A Survey of European Astronomical Tables in the Late Middle Ages* (Leiden: Brill, 2012).

———, and ———, 'The Moon in the Oxford Tables of 1348', *Journal for the History of Astronomy*, 47 (2016), 159–67.

———, and ———, 'The Medieval Moon in a Matrix: Double Argument Tables for Lunar Motion', *Archive for History of Exact Sciences*, 73 (2019), 335–59.

———, and ———, 'The Master and the Disciple: The Almanac of John of Lignères and the Ephemerides of John of Saxony', *Journal for the History of Astronomy*, 50 (2019), 82–96.

David Juste, 'Johannes de Wasia, *Notes on the Almagest*' (updated: 29.10.2019), *Ptolemaeus Arabus et Latinus. Works*, URL: http://ptolemaeus.badw.de/work/75

Husson, Matthieu, 'Ways to Read a Table: Reading and Interpolation Techniques in Canons of Early Fourteenth-Century Double-Argument Tables', *Journal for the History of Astronomy*, 43 (2012), 299–319.

Jacobson, Nicholas A., *Ordering Language to Order the Heavens: On Alfonsine Astronomical Canons (1350–1500)* (Turnhout: Brepols, forthcoming).

Kennedy, E. S., 'A survey of Islamic Astronomical Tables', *Transactions of the American Philosophical Society*, N. S. 46 (1956), 123–77.

North, J. D., 'The Alfonsine Tables in England', in *Prismata, Naturwissenschaftsgeschichtliche Studien: Festschrift für Willy Hartner*, ed. by Y. Maeyama and W. G. Satzer (Wiesbaden: Steiner, 1977), pp. 269–301.

Pedersen, Fritz S. *The Toledan Tables: A Review of the Manuscripts and the Textual Versions with an Edition*, 4 vols (Copenhagen: C. A. Reitzels Forlag, 2002).

Pedersen, Olaf, 'The *corpus astronomicum* and the Tradition of Medieval Astronomy', *Studia copernicana*, 13 (1975), 57–96.

Poulle, Emmanuel, *La Bibliothèque scientifique un imprimeur humaniste au* XV*e siècle: Catalogue des manuscrits d'Arnaud de Bruxelles à la Bibliothèque nationale de Paris* (Geneva: Librairie Droz, 1963).

———, 'John of Lignères', in *Dictionary of Scientific Biography*, ed. by C. Gillispie, 16 vols (New York: Scribners, 1970–80), VII (1973), pp. 122–28.

———, *Les Instruments de la théorie des planètes selon Ptolémée: Équatoires et horlogerie planétaire du* XIII*e au* XVI*e siècle*, 2 vols (Geneva: Droz, 1980).

———, *Les Tables alphonsines avec les canons de Jean de Saxe: Édition, traduction et commentaire* (Paris: Éditions du Centre national de la recherche scientifique, 1984).

Saby, Marie-Madeleine, 'Les Canons de Jean de Lignères sur les tables astronomiques de 1321: Edition critique, traduction et étude' (unpublished thesis, Paris, Ecole Nationale des Chartes, 1987).

Schum, Wilhelm, *Beschreibendes Verzeichnis der Amplonianischen Handschriften Sammlung zu Erfurt* (Berlin: Weidmannsche Buchhandlung 1887).
Toomer, G. J., 'A Survey of the Toledan Tables', *Osiris*, 15 (1968), 5–174.
Wilhelm Schum, *Beschreibendes Verzeichnis der Amplonianischen Handschriften Sammlung zu Erfurt* (Berlin: Weidmannsche Buchhandlung 1887), pp. 583–87

LAURE MIOLO

Retracing the Tradition of John of Genoa's *Opus astronomicum* Through Extant Manuscripts

Introduction

The prediction and calculation of true syzygies and eclipses were among the common topics in medieval astronomy. In the context of fourteenth-century Paris, Alfonsine astronomers could refer to several sources in order to compute eclipses, including the Toledan Tables and their canons, late-thirteenth-century commentaries such as John of Sicily's *Scriptum super canones Azarchelis*, or the *Sicut dicit Hermes* by an anonymous Parisian astronomer of 1290.[1] In the framework of the Parisian Alfonsine tradition, several canons and tables were devoted to syzygies and eclipses. Major sources for eclipse computations were John of Lignères' tables and canons, especially the canons *Priores astrologi*, which devote six chapters to eclipse computations, and John of Saxony's canons of 1327 which offer a method for finding the time of true conjunction or opposition in Chapter 22.[2] Two anonymous eclipse canons, frequently associated with John of Saxony's *canones* and sometimes attributed to him, were composed in 1330 and are found in the *editio*

* Research presented in this chapter was supported by the ERC project ALFA: Shaping a European scientific scene, Alfonsine astronomy, CoG 723085, PI Matthieu Husson. I am very grateful to Jean-Patrice Boudet, José Chabás, Matthieu Husson, and Richard L. Kremer for their helpful comments and suggestions on earlier drafts.

1 Fritz S. Pedersen, *The Toledan Tables, A Review of the Manuscripts and the Textual Versions with an Edition* (Copenhagen: C.A. Reitzels, 2002), pp. 331–499 (see especially the eclipse canons derived from version Cb); Fritz S. Pedersen, 'Johannes de Sicilia: Scriptum super canones Azarchelis', *Cahiers de l'Institut du Moyen-Âge grec et latin*, 51–52 (1986), 2–268; Fritz S. Pedersen, 'Anonymous Parisian Astronomer of 1290', *Cahiers de l'Institut du Moyen-Âge grec et latin*, 72 (2001) 169–269, 73 (2002), 61–166.

2 Emmanuel Poulle, *Les Tables Alphonsines avec les canons de Jean de Saxe* (Paris: Éditions du CNRS, 1984), pp. 80–87; José Chabás and Bernard R. Goldstein, 'Nicholaus de Heybech and His Table for Finding True Syzygy', *Historia mathematica*, 19 (1992), 265–89; Richard L. Kremer, 'Thoughts on John of Saxony's Method for Finding Times of True Syzygy', *Historia mathematica*, 30 (2003), 263–77.

Laure Miolo • University of Cambridge

Alfonsine Astronomy: The Written Record, ed. by Richard L. Kremer, Matthieu Husson, José Chabás, Turnhout, 2022 (*Alfonsine Astronomy: Studies and Sources*, 1), pp. 343-380
© BREPOLS PUBLISHERS 10.1484/M.ALFA.5.124932

princeps of 1483.[3] These two short texts were quite popular according to the number of extant manuscripts.

A part of John of Murs's work was also devoted to syzygies and eclipses.[4] He wrote the *Patefit* (1329–35), which includes true syzygy tables for years from 1321 to 1395 as well as instructions on how to directly find the time of true conjunction.[5] Later he also developed the *Tabule permanentes*, which provide the time difference between mean syzygy and true syzygy in a double-argument table.[6] John of Murs also authored the *Canones tabularum Alfonsii*, completed in 1339, which is shorter than John of Lignères' *Priores astrologi* or John of Saxony's canons. In his canons, he dedicated chapters to the method of finding the times of true syzygies and eclipses.[7] The aforementioned Parisian masters were fully devoted to establishing a more precise method of true syzygy and eclipse calculations. They also intended to provide precise rules and instructions on those astronomical practices in order to make them more accessible.

It is in this context that John of Genoa's *Opus astronomicum* should be considered. This little-known figure of the Alfonsine scene produced three known works and a table, all dedicated to eclipse calculations. Except his lunar true velocity table and a small part of the *Canones eclipsium*, John of Genoa and his works have received little attention.[8] This article aims to shed some light on his treatises and table by examining the extant manuscripts. After a *mise au point* on this quite unknown astronomer, his background, and his place in the Parisian milieu, I examine each of his four works, considering, respectively, their manuscript transmission as well as their structure and content.

1. Who was John of Genoa?

The prominence of some figures of the Parisian Alfonsine milieu, such as John of Lignères, John of Murs, or John of Saxony, has led to the eclipsing of other individuals, although their contributions can be now considered as significant. John of Genoa (*Johannes de*

3 *Tabule astronomice illustrissimi Alfontii regis Castelle* (Venice: Erhard Ratdolt, 1483), ff. b4r-b6r and b6v-b7v. Cf. the *membra adiuncta* (Poulle, *Les Tables Alphonsines*, pp. 23–26). The first canon begins: 'Eclypsim solis quantitatem et durationem per tabulas invenire. Quere coniunctionem mediam solis et lune; the second: Eclypsis lune quantitatem et durationem invenire. Quere primo mediam oppositionem illam.'

4 José Chabás and Bernard R. Goldstein, *The Alfonsine Tables of Toledo* (Dordrecht: Kluwer, 2003), pp. 277–81.

5 Richard L. Kremer, 'John of Murs, Wenzel Faber and the Computation of True Syzygy in the Fourteenth and Fifteenth Centuries', in *Mathematics Celestial and Terrestrial: Festschrift für Menso Folkerts zum 65. Geburtstag*, ed. by Joseph W. Dauben and others (Stuttgart: Wissenschaftliche Verlagsgesellschaft, 2008), pp. 147–60.

6 See the critical edition in Richard L. Kremer, 'Cracking the *Tabulae permanentes* of John of Murs and Firmin of Beauval with Exploratory Data Analysis', in *Editing and Analysing Astronomical Tables: Towards a Digital Information System for the History of Astral Sciences*, ed. Matthieu Husson, Clemency Montelle and Benno van Dalen (Turnhout: Brepols, 2022), pp. 363-422.

7 C. Philipp E. Nothaft, 'Jean des Murs's *Canones tabularum Alfonsii* of 1339', *Erudition and the Republic of Letters*, 4 (2019), 98–122.

8 Bernard R. Goldstein, 'Lunar Velocity in the Ptolemaic Tradition', in *The Investigation of Difficult Things: Essays on Newton and the History of the Exact Sciences*, ed. by P. M. Harman and Alan E. Shapiro (Cambridge: Cambridge University Press, 1992), pp. 3–18; Bernard R. Goldstein, 'Lunar Velocity in the Middle Ages: A Comparative Study', in *From Baghdad to Barcelona: Studies in the Islamic Exact Sciences in Honour of Prof. Juan Vernet*, ed. by J. Casulleras and J. Samsó (Barcelona: Instituto 'Millás Vallicrosa, 1996), pp. 181–94; on a part of the *Canones eclipsium*, see Nothaft, 'Jean des Murs's *Canones tabularum Alfonsii*', pp. 117–22.

Janua) is one such little-known figure, who dedicated his intellectual production to eclipse calculations. Beyond his extant astronomical works, little evidence remains of his life. Although John of Genoa's *Opus astronomicum* can be placed within the large framework of the Parisian Alfonsine tradition, as we shall see shortly, it is also important to place him in a precise context.

John of Genoa's astronomical works provide precious hints concerning the place where he composed them, as well as his university grade. He is said to be a master (*magister*) of arts, so we can assume that he had already obtained his arts degree at the time he elaborated his works.[9] Nevertheless, the place where he was active is rarely specified in the incipits or explicits of his works. One of these rare examples is written at the end of the *Canones eclipsium*, contained in Prosdocimo de' Beldomandis's student manuscript (Padua, *c.* 1405), Florence, BML, Ashburnham 206, folio 76^{r}: *Expliciunt canones de eclipsi sollis* [*sic*] *et lune ordinati a magistro Johanne de Ianua Parisius anno domini .1332. incompleto die 22 Januarii. Amen.* However, another piece of evidence can be found within the *Investigatio eclipsis solis 1337*, where John himself refers twice to the meridian of Paris. In the first part of his calculation for finding the time of the true conjunction, he says:

> Deinde intravi cum isto tempore tabulam argumenti latitudinis lune, et quesivi argumentum latitudinis, addita sua radice reducta ad meridianum Parisiensis [...].
>
> Then, I entered this time [the time of the first conjunction] in the table of the argument of the lunar latitude, and I sought the argument of the latitude, having added its radix reduced to the meridian of Paris.[10]

Later in the same text, after finding the time of the true conjunction, he pursues:

> Deinde intravi cum horis distancie coniunctionis a meridie, scilicet cum .3. horis, tabulam diversitatis aspectus factam ad Parisius [...].
>
> Then, I entered the hours of the distance of the conjunction at noon, that is with three hours [after noon], in the parallax table made for Paris [...].[11]

The connection of John of Genoa to Paris can also be inferred from references to his table. His canons, *Verum motum solis et lune*, usually associated with his table of lunar and solar velocities (including the table of radii of the Sun, the Moon, and the shadow of the Earth) specify that this table was made from the equation of the Alfonsine Tables.[12] This passage suggests that John of Genoa likely based his table on an equation table extracted from the

---

9 These are some examples found in manuscript witnesses of John of Genoa's *Opus astronomicum*, introducing the astronomer as a master. However, many other witnesses present the same assessment. London, British Library Royal MS 12.C.XVII, f. 213ra (*c.* 1344): *Incipiunt canones eclipsium magistri Johannis de Janua* [...]; Paris, BNF lat. 7282, f. 129^{v} (1468): *Canon tabule precedentis quam composuit magister Johannes de Janua;* Paris, BNF lat. 7281, f. 210^{v} (s. XV^{med}): *Explicit doctrina ad inveniendum eclipsis solis anno domini, 2* (corr. 3) *die martii data a magistro Johanne de Janua.*

10 Cambridge, UL Ee.III.61, f. 75^{r} and Douai, BM 715, f. 37^{r}.

11 Cambridge, UL Ee.III.61, f. 77^{r} and Douai, BM 715, f. 39^{r}.

12 See, for instance, Paris, BNF lat. 7282, f. 129^{v}, Vatican, BAV Reg. lat. 1241, f. 153^{r}: *Nota quod hec tabula est facta super equationes tabularum Alfonsii per magistrum Johannem de Janua.* Pierre Duhem had already noticed this mention in *Le Système du monde. Histoire des doctrines cosmologiques de Platon à Copernic*, 10 vols (Paris: A. Hermann et fils, 1913–53), IV (1916), p. 75; see also, Goldstein, 'Lunar Velocity in the Ptolemaic Tradition', pp. 3–18.

Parisian adaptation of the Alfonsine Tables. As we shall see, his links to — at least — two Parisian astronomers show that he was involved in this milieu. Thus, in two of his works, he makes explicit and glowing references to John of Lignères (*Johannes de Lineriis*).[13] The latter is the only contemporary that John of Genoa quotes in his works.

External evidence of John of Genoa's affiliation to Paris can be found in University records. In the *Computus* of 1329–30, a certain John of Genoa (*Johannes de Janua*) is mentioned with two *socii* in the district of *rue de Judas* (*in vico Jude*).[14] This source corresponds to a financial account recording monies collected from the students and masters of the University of Paris for a specific and extraordinary purpose, which is not mentioned in the document.[15] This *computus* is now the last quire of the oldest register of the proctors of the Anglo-German nation.[16] It was first edited in 1891 in the second volume of the *Chartularium universitatis Parisiensis*, and recently edited again, dated and analysed in detail by William J. Courtenay.[17] The University assessment begins with a street-by-street financial collection and ends with a list of names followed by payments. The amount paid for this irregular taxation depended on the *burse* — that is the weekly expenditures — of each scholar or an individual and his *familia*. The *bursa* was a unit of measurement for determining a person's ability to pay, which often depended on the scholar's grade, meaning whether they were students or masters. This ability was also based on people's oath.

In this source, John of Genoa is mentioned paying nine *sous* — for him and his two *socii* — for the extraordinary taxation of 1329–30.[18]

> [...]
> In vico Jude:
> [...]
> Johannes de Janua cum 2 sociis, 9 solid. solv[antur].
> In Judas street:
> [...]
> John of Genoa with two fellows, nine *sous* were paid.

The district of *rue de Judas* is the last area concerned by this collection; it is the last one recorded in the *computus*. This street is located to the east of the University area, close to

13 These references can be found in his *Canones eclipsium* and in his *Investigatio eclipsis solis 1337*. See *infra* his sources.

14 William J. Courtenay, *Parisian Scholars in the Early Fourteenth Century: A Social Portrait* (Cambridge: Cambridge University Press, 1999), pp. 176, 230; Rotuli Parisienses: *Supplications to the Pope from the University of Paris, I: 1316–1349*, ed. by William J. Courtenay (Leiden: Brill, 2002), p. 34.

15 These irregular collections were often related to diplomatic missions on behalf of the university to the papal Curia. Sending a nuncio was a concern for the whole community of students and masters, hence this taxation. See Courtenay, *Supplications*, p. 29.

16 Paris, BIU Sorbonne, Reg. 2. 1, ff. 58r–65v.

17 *Chartularium Universitatis Parisiensis*, II, ed. by Henri Denifle and Émile Châtelain (Paris: Delalain, 1891), p. 661–71 (henceforth *CUP*). Denifle and Châtelain gave the date range of 1329–36; this dating was revised by Courtenay, who also provided a detailed study of the computus, in Courtenay, *Parisian Scholars*, Chapter 1.

18 Courtenay, *Parisian Scholars*, p. 230. Paris, BIU Sorbonne, Reg. 2. 1, ff. 58r–65v. It is noteworthy that a *Johannes de Saxonia* is mentioned with his *socius* in the second quire of the *Computus*, but the street is not named: 'Johannes de Saxonia cum socio suo, iiii sol. simul'. Of course, there is not enough evidence to conclude that this John of Saxony is the same individual as the astronomer and author of the canons of 1327.

the Collège de Navarre.[19] John of Genoa was living in this area with two fellows, for whom he paid the taxation. The term *socius* can have various meanings in the university context. It can designate: 1) a colleague; 2) a junior fellow or a student; 3) a fellow of a college, or more rarely a member of a nation.[20] In the *computus*, the named person — here John of Genoa — represents the group of *socii* with whom he lived. This representative could be a wealthy student in charge of the others, but it is more likely that every time a named individual is mentioned with *socii*, he is a master of arts with his students, although the University members' titles are never specified in the *computus*.[21] Thus, the *magister* was living with his *scholares*, and so was responsible for his group with regard to taxation.[22] In the case of John of Genoa, the amount of nine *sous* corresponds to a master's payment, which is consistent with the status of *magister* that the astronomer John of Genoa held when he wrote his *Canones eclipsium*.[23] Other evidence can be found in the *Chartularium* of the University.

William Courtenay's prosopography of the individuals mentioned in the *computus* also shows that the *vicus Jude* included a certain number of medical students and doctors.[24] This assumption is quite consistent with two records related to the John of Genoa mentioned in the *computus*. By 20 January 1332, as a bachelor in medicine, he took — with other bachelors of medical science — the usual oath before receiving the authorization to begin their lectures.[25] Only two months later (30 March) he received his *licentia* in medicine.[26] It is likely that this same John of Genoa ten years later became the surgeon of Pope Clement VI until 1348.[27] At the curia, he financially supported two clerks from Metz from 1342 to 1344.[28] According to Guy de Chauliac, the pope's surgeon was also the nephew of the surgeon Anselm of Genoa[29] and the author of an ointment recipe called

19 On the Collège de Navarre see Nathalie Gorochov, *Le Collège de Navarre, de sa fondation (1305) au début du xve siècle (1418). Histoire de l'institution, de sa vie intellectuelle et de son recrutement* (Paris: Honoré Champion, 1997).

20 Mariken Teeuwen, *The Vocabulary of Intellectual Life in the Middle Ages* (Turnhout: Brepols, 2003), pp. 135–36.

21 On the role of the master within the university, see Jacques Verger, 'Teachers', in *A History of the University in Europe, I: Universities in the Middle Ages*, ed. Walter Füegg (Cambridge: Cambridge University Press, 1992), pp. 144–69.

22 Courtenay, *Parisian Scholars*, pp. 85–88.

23 Courtenay, *Parisian Scholars*, p. 176.

24 Thomas de Carliolo (a surgeon), Jacobus de Cantarana (doctor in medicine and regent master by 1328) and Johannes de Dia (doctor in medicine and regent master by 1328 until 1332). See their respective entries in Ernest Wickersheimer, *Dictionnaire biographique des médecins en France au Moyen Âge* (Geneva: Droz, 1979), pp. 758, 323, 398; see also Courtenay, *Parisian Scholars*, pp. 214, 165, 173.

25 *CUP*, II, 940. If this 'John of Genoa' is the same individual as the astronomer, it is important to note that the date of 20 January 1332 preceded the completion of the *Canones eclipsium* by two days.

26 It is unlikely that *Johannes de Janua*, master of medicine by October 1328 and canon of Genoa, is the same as the 'John of Genoa' mentioned in the *computus* and in the faculty of medicine in 1332. On John of Genoa, canon of Genoa, see *Jean XXII (1316–1334). Lettres Communes. Analysées d'après les registres dits d'Avignon et du Vatican*, VIII, ed. by Guillaume Mollat (Paris: E. De Boccard, 1924), p. 41, n. 43089.

27 John of Genoa is quoted in Pope Clement VI's expenditures as a master and surgeon. He is often mentioned with Petrus Augerii, also a surgeon at the Curia from 1339–48. See K. H. Schäfer, *Die Ausgaben der apostolischen Kammer unter Benedikt XII, Klemens VI, und Innocenz VI* (Paderborn: F. Schöningh, 1914), pp. 202, 234, 324, 390; Wickersheimer, *Dictionnaire biographique*, p. 424; Pierre Pansier, 'Les Médecins des Papes d'Avignon (1308–1403)', *Janus*, 14 (1909), 422–23.

28 Bernard Guillemain, *La Cour pontificale d'Avignon (1309–1376): Étude d'une société* (Paris: De Boccard, 1962), p. 377–78.

29 On Anselm of Genoa, also known as Anserin de la Porte, see Guillemain, *La Cour pontificale*, p. 29. Lanfranc de Milan, Henri de Mondeville and Guy de Chauliac were refering to ointment recipes created by Anselm and offered by him to Philippe le Bel and Bonifacius VIII.

unguent de Gloire.[30] Although a family link between John and Anselm, mentioned in the *Chirurgia magna* (1363), is questionable, it seems likely that the pope's surgeon *Johannes de Janua* was the same person as the author of works on eclipses. Accordingly, if we assume that the aforementioned John of Genoa of the *computus* and the faculty of medicine was the same as the master at the papal curia, it may be possible to obtain a clearer view of the astronomer's background.

A final piece of evidence supports the hypothesis that John of Genoa, the Parisian astronomer, was the member of the papal curia. The earliest known copy of the *Canones eclipsium*, completed 22 January 1332, was owned by John of Murs. The second part of what is now the composite manuscript, London, British Library, Royal 12.C.XVII (now *L*) was indeed commissioned by John of Murs in the 1340s, perhaps to present to the Benedictine monastery of Le Bec-Hellouin.[31] The content of this part, ordered and annotated by the famous astronomer, helps us to provide the *terminus post quem* of that codex. It contains the *Patefit*, written between 1329 and 1335, and dedicated to Geoffroy Faé, abbot of Le Bec-Hellouin,[32] followed by the additions to the calendar of the same monastery (likely composed by John of Murs); John of Murs's *Sermo de regulis computistarum* (1332); the *Canones eclipsium* (1332); excerpts from the eclipse canons often attributed to John of Saxony; and some materials extracted from the canons of 1327. A short astrological work was copied by the same scribe after the *Additiones kalendarii Beccensis*, and ends abruptly on the same folio. The catchword at the bottom of folio 212^{v} shows that at least one quire is missing. The following work is the *Sermo de regulis*, which is incomplete at the beginning on folio 213ra. This anonymous astrological treatise, entitled in some manuscripts *Sermo de electionibus curationis morborum secundum astronomos* and beginning, 'Quoniam electiones laudabiles sunt salubres' was composed in Paris in 1344, as noted in the explicit of other witnesses containing this work.[33] Thus, *L* was assembled in its final form in the 1340s, and likely a little after 1344 when John of Murs was in Avignon, summoned by Pope Clement VI in order to take part to the reform of the ecclesiastical calendar.[34] The content of this codex is a real testimony of John of Murs's scientific concern at that time: the reform of the calendar, the computation of syzygies and eclipses, as well as astrology.

---

30 Cf. Guy de Chauliac, *La* Grande Chirurgie *de Guy de Chauliac, chirurgien, maistre en médecine de l'université de Montpellier, composée en l'an 1363*, ed. by Alfred Nicaise (Paris: Félix Alcan Éditeur, 1890), p. 624 (tract. 7, doctrine 1, Chapter 6); Guy de Chauliac, *Inventarium sive Chirurgia Magna* ed. by Michael R. McVaugh, Studies in Ancient Medicine, vol. 14 (Leyden: Brill Publishers, 1996).

31 Lawrence Gushee suggested that this manuscript was copied to be presented to the Abbot of Le Bec-Hellouin, in 'New Sources for the Biography of Johannes de Muris', *Journal of the American Musicological Society*, 22 (1969), 3–26.

32 On the syzygy tables of the *Patefit*, see Richard L. Kremer, 'John of Murs', pp. 147–60; José Chabás and Bernard R. Goldstein, 'John of Murs's Tables of 1321', *Journal for the History of Astronomy*, 40 (2009), 297–320.

33 This work is associated with John of Murs's *Patefit* and *Sermo de regulis* in Erfurt, Universitätsbibliothek CA Q360 (s.XIV^{2}), ff. 54–55; Erfurt, Universitätsbibliothek CA Q371 (s.xiv^{2}), ff. 43^{v}–44^{v}; and Metz, BM 285 (s.XV), now lost. Other witnesses of this astrological work are Erfurt, Universitätsbibliothek CA Q386, (s.xiv^{med}), ff. 19^{v}–20^{v}, which also includes John of Saxony's canons and John of Lignères's *Algorismus minutiarum*. Nuremberg, Stadtbibliothek Cent. VI 22, ff. 11ra-12ra (after 1344); Paris, BIU Sorbonne 1037, (s.XV^{in}), ff. 81^{v}–82^{r}, which also includes the eclipse canons attributed to John of Saxony: *Eclipsim solis quantitatem et durationem*. The astrological work ends: *Explicit liber de electionibus medicine editus Parisius anno Domini 1344*.

34 C. Philipp E. Nothaft, 'Science at the Papal Palace: Clement VI and the Calendar Reform Project of 1344/45', *Viator*, 46 (2015), 277–302; Jean-Patrice Boudet, 'Jean des Murs, Astrologer', *Erudition and the Republic of Letters*, 4 (2019), 123–45.

In 1344 and 1345, John of Murs and Firmin de Beauval were in Avignon. Having been the pope's surgeon since 1342, John of Genoa probably met them on this occasion. Admittedly, there is also the possibility that John of Genoa already knew John of Murs from Paris, as was the case of John of Murs and Clement VI (Pierre Roger), who was *provisor* of the Collège de Sorbonne (1326–36). However, the fact that they were both in Avignon in 1344 and 1345 when *L* was completed can explain why John of Murs preferred to insert the *Canones eclipsium* in his manuscript, rather that of his own canons of 1339. Another explanation might be that John of Murs had already ordered *L* by 1339. In any event, the presence of the *Canones eclipsium* in the manuscript commissioned by John of Murs shows that he already knew and valued John of Genoa's eclipse canons.

As a matter of fact, John of Murs was perfectly aware of the *Canones eclipsium* when in 1339 he finished his *Canones tabularum Alfonsii* at the Collège de Sorbonne. Philipp E. Nothaft recently highlighted, in his analysis of the *Canones tabularum Alfonsii*, that John of Murs borrowed some eclipse computation theories from John of Genoa's work.[35] In fact, in the last part of the canons of 1339, dedicated to syzygies and eclipses, the way to derive the digits from the apparent diameters of both luminaries, as well as the parallax correction, are strongly influenced by the *Canones eclipsium*.[36]

This connection is also expressed in the manuscript tradition. The *Canones tabularum Alfonsii* are associated with John of Genoa's works in two English codices. Oxford, Bodleian Library, Digby 97, an early-fifteenth-century manuscript copied in England, contains John of Murs's *Canones tabularum Alfonsii* (ff. 122^r–125^r) followed by John of Genoa's *Canones eclipsium* (ff. 125^r–128^v), and the short canon, *Verum motum solis et lune* (ff. 129^v–130^r)[37] associated with his table (f. 130^v). Moreover, the fifteenth-century Oxford, Hertford College 4 shows the *Canones tabularum Alfonsii* (ff. 140^r–147^r) followed by John of Genoa's canons, *Verum motum solis et lune* (ff. 147^{r-v}) and the table (f. 148^r). In this manuscript, John of Genoa's canons and table are followed by John of Lignères's canons *Quia ad inveniendum loca planetarum* (ff. 148^v–154^r), which was one of the former's sources.

In both manuscripts, John of Murs's canons of 1339 are materially linked to the short canon *Verum motum solis et lune* and the table of solar and lunar hourly velocities, including radii of the Sun, the Moon, and the *variatio umbre*. As we shall see shortly, this table and its rules were created by John of Genoa with the specific aim of computing eclipses. For instance, the velocity set can be used for calculating true syzygies and the radius part of this table was designed partly for computing the eclipse magnitudes. Therefore, the association of those works with John of Murs's canons in the two English manuscripts establishes a coherent corpus related to eclipse computations. They represent a close textual and theoretical connection between John of Murs's canons and John of Genoa's production on eclipses.

Recently, Richard L. Kremer pointed out that John of Murs and Firmin de Beauval used John of Genoa's table of hourly velocities of the Sun and the Moon to establish their *Tabule permanentes*, dedicated to the computation of time from mean to true syzygies.

35 Nothaft, 'Jean des Murs's *Canones tabularum Alfonsii* of 1339', pp. 98–122.

36 Nothaft, 'Jean des Murs's *Canones tabularum Alfonsii* of 1339', pp. 117–22.

37 William Merle's *Regule ad futurum aeris temperiem prenosticandam* (ff. 128^v–29^r) is copied between the *Canones eclipsium* and the canon *Verum motum solis et lune.*

Kremer assigns the composition of the *Tabule permanentes* to the late 1340s.[38] More likely, they were composed in the mid 1340s during Firmin de Beauval and John of Murs's stay in Avignon.[39] In any case, the *Tabule permanentes* provide further evidence that John of Murs was not only aware of the *Canones eclipsium* but also of John of Genoa's table and its associated canons.

If John of Genoa's itinerary is difficult to capture due to the sparse evidence concerning his life, his milieu and background can be read in his works as well as in the influence he had on John of Murs.

2. The *Opus astronomicum*

The fame of John of Genoa as an astronomer can be measured through Simon de Phares' entry in his *Recueil des plus célèbres astrologues*, written between 1494 and 1498. The author's statement is not fully reliable; but the fact that he dedicated a biblio-biographical record to John of Genoa demonstrates the appreciation of his activity as an astronomer.[40] However, as Jean-Patrice Boudet highlighted in his critical edition, Simon de Phares took that information from one of the only manuscript witnesses gathering the whole *Opus astronomicum* of John of Genoa (except the table). In fact, Paris, BnF, lat. 7281 partly copied in the mid-fifteenth century by 'Jo[hannes] B.', who was likely a canon of Cambrai, contains the *Canones eclipsium*, the canon *Verum motum solis et lune*, and the computation of the solar eclipse of March 1337 (*Investigatio eclipsis solis*).[41] This last treatise is followed by William Batecombe's Oxford canons of 1348.[42] Hence Simon de Phares' mistake in attributing the canons of 'Auxonfort' (i.e. Oxford) to John of Genoa.

38 For a critical edition of the tables, see Richard L. Kremer, 'Cracking the *Tabulae permanentes*'; for a critical edition of the canons, see José Chabás and Beatriz Porres, 'John of Murs's *Tabulae permanentes*', *Journal for the History of Astronomy*, 32 (2001) 63–72. It is noteworthy that in Vatican, BAV Pal. lat. 446, f. 218^r and BAV Ottob. lat. 1826, f. 156^r containing John of Genoa's table, we find the canon of the *Contratabule*, dated to 1321, explicitly attributed to John of Murs, and associated with syzygy tables. It begins: 'Canon huius operis est annis christi perfectis deme 1320 et annorum remanentium scias numerum lunationum' and ends (see esp. Vat. Pal. lat. 446): 'Explicit canon super coniunctiones veras solis et lune et oppositiones earundem super meridiam Parisiensis compositas et ordinatas a magistro Johanne de Muris. Deo Laus'.

39 Kremer dates the *Tabule permanentes* to the late 1340s.

40 Jean-Patrice Boudet, *Le Recueil des plus célèbres astrologues de Simon de Phares*, 2 vols (Paris: Honoré Champion, 1997), I, p. 480: 'Johannes de Janua florit en ce temps pour l'experience qu'il avoit de la science de astrologie et quasi eut bruit par tout le monde et fist et composa plusieurs tables et traittés et, entre autres, en composa unes sur Auxonfort en Engleterre, mil IIIcXLVIII, commençant: *Vera loca planetarum*. Cestui predist l'empoisonnement des puis et fontaines que avoient fait les Juifz'.

41 Simon de Phares consulted this manuscript, thanks to Jean Avis, his friend and dean of the faculty of medicine in Paris since 1470. On this manuscript, see Jean-Patrice Boudet, *Lire dans le ciel. La bibliothèque de Simon de Phares, astrologue du xve siècle* (Bruxelles: Centre d'Études des manuscrits, 1994), p. 175–89. The *Canones eclipsium* (ff. 206^r–208^r) were copied directly after John of Lignères' *Canones super tabulas magnas* (ff. 201^v–205^v), beginning *Multiplicis philosophie variis radiis illustrato domino Roberto de Florencia*. The short canon *Verum motum solis et lune* (ff. 208^r) follows the *Canones eclipsium*; the *Investigatio eclipsis solis anno Christi 1337°* (ff. 208^v–210^v) follows.

42 Paris, BnF lat. 7281, ff. 210^v–212^v. On those canons and tables see John D. North, 'The Alfonsine Tables in England', in *Prismata, Naturwissenschaftsgeschichtliche Studien. Festschrift für Willy Hartner*, ed. by Y. Maeyama and W. G. Saltzer (Wiesbaden: Franz Steiner, 1977), pp. 269–301; José Chabás and Bernard R. Goldstein, 'The Moon in the Oxford Tables of 1348', *Journal for the History of Astronomy*, 47 (2016), 159–67.

The presence of the *Opus astronomicum* in the manuscript assembled by 'Jo[hannes] B.' suggests that John of Genoa's scientific production was still appreciated at that time and likely considered as important in the Alfonsine Parisian tradition. The Cambresian compiled this codex with the aim of retracing the history of astronomy from twelfth- and thirteenth-century translations to the adaptations and new works of the fourteenth and fifteenth centuries. In fact, the twenty-six astronomical texts are assembled in chronological order and range from al-Farghani's *Liber de aggregationibus scientie stellarum* and the Toledan Tables to Alfonsine astronomy.[43]

Among the works that can be ascribed to John of Genoa with certainty, we have: 1) the table with columns for the solar and lunar hourly velocities and for the radii of the Sun, the Moon, and the shadow of the Earth; 2) the short canon, *Verum motum solis et lune*, which is a rule for using both parts of the table; 3) the *Canones eclipsium*; and 4) the *Investigatio eclipsis solis ab anno Christi 1337*. I now present these works and establish a chronology of their composition.[44]

3. The table of velocities and radii

The first work of John of Genoa's *Opus astronomicum* is his table with columns for the solar and lunar hourly velocities and for the apparent radii of the Sun, the Moon, and the shadow of the Earth. Tables of lunar and solar velocities are needed in the computation of the times of true syzygy and for the duration of eclipses.[45] In the Alfonsine Latin tradition, one can find tables of solar and lunar velocities in a minute of a day (i.e. $1/60^{d}$) or in one hour, with different intervals in degrees.[46]

The most common velocity tables are found in John of Lignères' Tables of 1322 and derive from al-Battānī's *zīj* via the Toledan Tables.[47] John of Murs also established (perhaps in the 1320s) his own table of velocities, which he later revised. Both are only extant in his 'notebook', now Madrid, Escorial O.II.10, folios 204^{v} and 217^{v}.[48] In this late-thirteenth-century

43 cf. Boudet, *Lire dans le ciel*, pp. 175–89; and Jean-Patrice Boudet, 'A History of Astronomy by Texts in the Fifteenth Century: MS Paris, BnF lat. 7281', in the ALFA workshop *Manuscript Descriptions and Survey of Texts, Tables and other Items for the Study of Alfonsine Astronomy*, 29 September 2018.

44 At least two other works have been ascribed to John of Genoa: the aforementioned *Unguent de gloire* presented by Guy de Chauliac, which does not seem extant; and medical *consilia* contained in a fifteenth-century manuscript, Munich, Bayerische Staatsbibliothek Clm 205, ff. 103^{r}–104^{v}. The *Consilia aliqua* begin: 'Nota istum casum cognata ducis Venetorum fuit gravida' (see TK 0930E). Palémon Glorieux mentioned a treatise called *Astronomica* and ascribed to John of Genoa in Oxford, Bodleian Library, Canon. Misc. 226; however, the manuscript does not contain any astronomical work. See Palémon Glorieux, *La Faculté des Arts et ses Maîtres au 13e siècle* (Paris: Vrin, 1971), pp. 214–15; Olga Weijers, *Le travail intellectuel à la Faculté des arts de Paris: textes et maîtres (ca. 1200-1500)*, 9 vols (Turnhout: Brepols Publishers, 1994-2012), III (2003), pp. 105-6.

45 José Chabás and Bernard R. Goldstein, *A Survey of Astronomical Tables in the late Middle Ages* (Leiden: Brill, 2012), pp. 95–102.

46 The two tables reproducing the minimum and maximum values of solar and lunar velocities in °/mn and °/h have been established according to Chabás and Goldstein, *A Survey of Astronomical Tables*, p. 97.

47 The velocity tables in the 1483 edition of the Parisian Alfonsine tables are quite different from al-Battānī's tables and might be derived from a Hebrew adaption of 1460. See Bernard R. Goldstein 'Solar and Lunar Velocities in the Alfonsine Tables', *Historia mathematica*, 7 (1980), 134–40.

48 On the tables of the manuscript Madrid, Escorial O.II.10, see Matthieu Husson's and Laure Miolo's forthcoming article. I thank José Chabás for making me aware of the table located in f. 217^{r}.

manuscript containing the *Corpus astronomicum* derived from the Toledan tradition, John of Murs wrote from the early 1320s to the 1340s in the margins or in blank folios a large number of *memoranda*, astronomical tables and canons of his own compositions, as well as computations. The tables of solar and lunar velocities are part of these materials.[49]

Other examples of revising and adapting velocity tables can be found by John of Montfort, perhaps in 1332, and the anonymous author of the fifteenth-century manuscript Oxford, Bodleian Library Canon. Misc. 499, folios 41^v–42^r.[50] Little is known about John of Montfort, who was likely a contemporary of John of Genoa. A certain *Johannes de Monte Forti* is mentioned in the *computus* of 1329–30, in which John of Genoa, as well as an individual named *Johannes de Saxonia*, are mentioned. The individual cited in the *computus* was living *in vico Lignorum* (*rue de la Bûcherie*) with five unnamed fellows, as he was likely a master.[51] Although there is no further evidence that he is the same person as the Alfonsine practitioner, we know that the astronomer was responsible for a table for solar and lunar velocities in a minute of a day. According to the manuscript witnesses of this table, he composed it in 1332.[52] The lunar velocity is based on Ptolemy's second lunar model, which is also the case of John of Genoa's table, but John of Montfort's table displays the values at degree intervals and does not show the apparent radii.[53] The proximity of both tables can be illustrated by Paris, BnF, lat. 7283, which displays the works of John of Genoa and John of Montfort.[54] In Krakow, BJ, 613, folio 33^r, which is likely based on lat. 7283, the tables are not displayed, but the short canons related to the respective tables are copied

49 The *corpus astronomicum* was a set of texts usually associated with each other, including the Toledan Tables and their canons. This corpus was likely part of astronomical teaching in the University of Paris from the mid-thirteenth century to the first half of the fourteenth century. See Olaf Pedersen, 'The Corpus Astronomicum and the Traditions of Mediaeval Latin Astronomy', *Studia Copemicana*, 8 (1975), 55–96; on the manuscript Escorial O.II.10, see Guy Beaujouan, 'Observations et calculs astronomiques de Jean de Murs (1321–1344), in *Proceedings of the xivth International Congress of the History of Science*, 4 vols (Tokyo: Science council of Japan, 1975), II, pp. 27–30, repr. as Chapter 7 in Guy Beaujouan, *Par raison des nombres: l'art du calcul et les savoirs scientifiques médiévaux* (Aldershort: Variorum, 1991); Gushee, 'New sources', pp. 2–26; Lawrence Gushee, 'Jehan des Murs and his Milieu', in *Musik und die Geschichte der Philosophie und Naturwissenschaften im Mittelalter*, ed. by Frank Hentschel, (Leiden: Brill, 1998), 339 72.

50 A later note added to those tables attributes them to John of Lignères.

51 For the quotation of John of Genoa, as well as a certain 'John of Saxony', see note 18 above. For John of Montfort, see Courtenay, *Parisian Scholars*, p. 224: *Johannes de Monte Forti cum 5 sociis, 12 s[olidi] solv[antur]*. 'John of Montfort with five fellows, twelve *sous* were paid'.

52 Some manuscript witnesses of the table are as follows: Budapest, OSK MS 62, ff. 45^r–46^r; Erfurt, Uiversitätsbibliothek, CA Q364, ff. 130^r–132^v; Paris, BNF lat. 7283, ff. 43^r–44^r; lat. 13014, ff. 166^v–167^v; lat. 14481, ff. 96^v–97^r. The date of composition of the table is mentioned in four of the witnesses: BnF lat. 14481, f. 96^v: 'Tabula ad sciendum motum solis et lune in uno minuto diei et in motu lune includitur equatio centri et est composita anno domini 1332 incompleto mense Januarii per magistrum Johannem de Monte Forti secundum equationem tabularum Alfoncii'. Two witnesses cite January 1332, the others only the year 1332. January 1332 also corresponds to the month of publication of John of Genoa's *Canones eclipsium*. Since many of the witnesses range from the late fourteenth century to the fifteenth century, the date remains uncertain. The earliest witness, the composite manuscript Erfurt, Universitätsbibliothek CA Q364 cites the year 1321; however, this date seems to refer more to the 'equation of centre for the Moon' in the Parisian Alfonsine tradition than to the date of the table itself (f. 130^r): 'Tabula ad sciendum motum solis et lune in uno minuto diei et in motu lune includitur equatio centri et est composita anno domini .1320. completo mense Januarii per magistrum Johannem de Monte Forti, secundum equationem tabularum Alfonsii'.

53 On John of Montfort's lunar velocity table, see Goldstein, 'Lunar Velocity', p. 11.

54 On this manuscript, see below. John of Montfort's table is situated in ff. 43^r–44^r, including its short canon (f. 43^r); John of Genoa's table and short canons are on ff. 44^v–45^r.

Table 1. Solar and lunar velocities at apogee and perigee in °/mn (degrees per minute of a day i.e. 24hrs/60) and °/h, from Goldstein, 1980; Poulle, 1984, p. 210–11; Goldstein, 1992, p. 11; Goldstein, 1996, p. 196; Chabás and Goldstein, 2002, p. 97. The values for John of Genoa are from my own edition. John of Murs [1] corresponds to a table in Madrid, Escorial O.II.10 f. 204v; John of Murs [2] is from f. 217v of the same codex. Only the values of the lunar velocities vary between the later two tables.

Lunar velocities in °/mn			
Tables	**Intervals**	**Minimum (apogee 0°/1°)**	**Maximum (perigee 180°)**
Editio princeps 1483	3°	0;12,9	0;14,25
John of Montfort	1°	0;11,51,9,11	0;14,44,47,8
John of Genoa	6°	0;11,50,53	0;14,47,33
Lunar velocities in °/h			
al-Battānī	6°	0;30,18	0;36,4
Editio princeps 1483	6°	0;30,21	0;36,25
John of Montfort	1°	0;29,37,52,57	0;36,51,53,50
John of Genoa	6°	0;29,37,13	0;36,58,54
John of Murs [1]	6°	0;30,20	0;36,1
John of Murs [2]	6°	0;30,10	0;36,0
MS. Canon.Misc.499	1°	0;29,37	0;36,53
Solar velocities in °/mn			
Tables	**Intervals**	**Minimum (apogee)**	**Maximum (perigee)**
Editio princeps 1483	3°	0;57	1;2
John of Montfort	1°	0;57,1,12	1;1,30,15
John of Genoa	6°	0;57,0	1;1,28
Solar velocities in °/h			
al-Battānī	6°	0;2,23	0;2,33
Editio princeps 1483	6°	0;2,22,30	0;2,33,40
John of Montfort	1°	0;2,22,33,0	0;2,33,45,37
John of Genoa	6°	0;2,22,30	0;2,33,40
John of Murs [1]	6°	0;2,22	0;2,34
John of Murs [2]	6°	0;2,22	0;2,34
MS. Canon.Misc.499	1°	0;2;23	0;2;34

one after the other.[55] In any event, John of Montfort and John of Genoa belonged to the same milieu and demonstrated a similar interest in creating new astronomical tools, both focusing on true velocities of the Sun and the Moon.

55 Cracow, BJ 613, f. 33ra: the canon for John of Montfort's table begins, 'Cum motu argumenti solis capias motum solis in uno minuto diei, et cum argumento lune capias motum lune in vero minuto diei', and ends, 'sed motus solis verus potest haberi in quocumque die et hora cum argumento solis'. It is directly followed by the beginning of John of Genoa's short canon (see infra).

However, John of Genoa's table is the most widely circulated, according to extant manuscripts. Thirty manuscripts, ranging from the fourteenth to the late-fifteenth century, contain his table, which is quite often displayed in John of Lignères' Tables of 1322.[56]

One of the peculiarities of John of Genoa's table is that it combines columns for the solar and lunar true velocities with columns for the apparent radii of the Sun, Moon, and the Earth's shadow. This structure is quite rare in the Latin tradition, as velocities and radii are often displayed separately.[57] However, this format was used long before in the *zīj* of al-Khwārizmī.[58] In John of Genoa's table, the lunar and solar velocities are arranged at six-degree intervals and are given per hour. Values are determined to three sexagesimal places. The other important innovation of this table is that it is based on Ptolemy's second lunar model.[59] Moreover, Bernard Goldstein and José Chabás highlighted an error in the maximum entry in John of Genoa's lunar velocity table. The value given for 180° is 0;36,58,54$^{\circ/h}$, instead of 0;36,53,20$^{\circ/h}$.[60] It does not seem to be a copyist error, since the manuscript witnesses of this table show the same *lectio*; it is difficult to know whether the variant came directly from John of Genoa or from an early inaccuracy in the manuscript tradition.[61]

Some previous studies have hesitated to ascribe the table to John of Genoa, arguing that the author could have been John of Lignères. Admittedly, in the manuscript tradition, the table is often associated with the Tables of 1322 composed by the latter and the name *Johannes de Janua* is rarely mentioned in the headings of the table.[62] However, some examples extracted from extant codices can strengthen the argument in favour of the ascription to John of Genoa. Two manuscripts combine the table with other works of John of Genoa. Firstly, the aforementioned Oxford, Bodleian Library Digby 97 includes the *Canones eclipsium*, the aforesaid table, and the canon, *Verum motum solis et lune*. Secondly, Douai, BM 715 (mid-fifteenth century) is the only manuscript containing the whole *Opus astronomicum* of John of Genoa. The table is copied after its canon, *Verum motum solis et lune*.[63] The location of the table in both manuscripts appears to be an element supporting this authorship.

---

56 Cf. my edition of John of Genoa's *Opus astronomicum*.

57 It is important to note that John of Murs displays the diameters of the Moon and the shadow of the Earth in the main table of solar and lunar velocities of Escorial O.II.10, f. 204^v, the table of the solar diameter is shown in a small table written on the same folio again displaying the solar true velocity. In the table located on f. 217^v and entitled 'Tabula dyametrorum in eclipsibus', the diameters of the Sun and the Moon are displayed facing, respectively, the columns of their true velocities.

58 Otto Neugebauer, *The Astronomical Tables of al-Khwārizmī: Translation with Commentaries of the Latin Version Edited by H. Suter Supplemented by Corpus Christi College MS 283* (Copenhagen: Ejnar Munksgaard, 1962), pp. 105–6; Gerald J. Toomer, 'A Survey of the Toledan Tables', *Osiris*, 15 (1968), 5–174, esp. 82–84.

59 See Goldstein, 'Lunar Velocity', pp. 11–14.

60 They explained that closer to the maximum value (180°), from 174°, the entries are greater than expected. Chabás and Goldstein, 'Nicholaus de Heybech', p. 22; see also the discussion in Goldstein, 'Lunar Velocity', p. 14.

61 One of the manuscript witnesses, Paris, BnF lat. 7286A, f. 117^r, presents entries closer to Bernard Goldstein's recomputation from 174° (0;36,51,15°/h); cf. Goldstein, 'Lunar Velocity', p. 14 n. 26. However, it is the only extant manuscript showing those values; whereas in Vatican, BAV Pal. lat. 1376, ff. 57^v–60^r, the maximum lunar velocity given at a degree interval in John of Gmunden's table of velocities is 0;36,53,21°/h (cf. f. 60^r). I thank Richard L. Kremer for this information.

62 Cf. José Chabás's and Marie-Madeleine Saby's forthcoming edition of John of Lignères' Tables of 1322.

63 This manuscript of 78 folios was copied in northern France (perhaps in Picardy) and was owned by a physician. The content is mainly dedicated to astrology and medical astrology. Only the rubric opening the *Canones eclipsium* (f. 32^r) and the colophon (f. 35^r) mention John of Genoa: 'Incipiunt canones eclipsium magistri Johannis de Janua [...]',

One finds a direct ascription in an Italian manuscript of the second half of the fourteenth century, Vatican, Ottob. lat. 1826, which displays the table (f. 148^{r}) and the canon, *Verum motum solis et lune* (f. 148^{v}).[64] The rubric above the canon reads, 'Canon tabule sequentis que intitulatur tabula motus diversi solis et lune in una hora et semidyametrorum secundum tabulas Alfonsi'. It is, of course, a direct reference to John of Genoa's table, the incipit of which is similar to this rubric. The canon ends with a mention of the author, which establishes clearly that he made the table as well as the canon: M. J. C [sic] composuit istum canonem et etiam tabulam et hec est tabula quam ipse nominat in principio suorum canonum de eclipsibus. Despite the scribe's confusion in the capital letters of the initials (C instead of G), there is little doubt that the colophon designates John of Genoa, since it also mentions his *Canones eclipsium.*[65] The manuscript also contains short canons which are linked to a table for the entry of the Sun into the signs that provides the dates 1332 and 1334 (ff. 46va-48rb) and mention the meridian of Genoa.[66] However, it is difficult to assert firmly that they are linked to John of Genoa.

Another piece of evidence can be found in a fifteenth-century manuscript copied in France (completed in 1468), containing *inter alia* John of Lignères' canons and tables. Paris, BnF lat. 7282 belonged to Philippe Boulier, master of medicine of the University of Paris, active between 1467 and 1483.[67] John of Genoa's table is located on folio 129^{r}, followed on the verso by the canon, *Verum motum solis et lune.* The explicit reads, 'Nota quod hec tabula est facta super equationes tabularum Alfonsii per magistrum Johannem de Janua'. The *tabula* is of course the one copied on folio 129^{r}. In Paris, BnF lat. 7283, which contains John of Lignères' tables, as well as John of Montfort's table (ff. 43^{r}–44^{r}), one can read a more explicit ascription. This early-fifteenth-century manuscript, copied in Erfurt and quickly moved to Metz (Celestines convent),[68] shows two versions of the table in *vis-à-vis* format, including the canons written in the margins (ff. 44^{v}–45^{r}). The beginning of the canon (f. 44^{v}) specifies, 'Ista tabula non est de tabulis magistri Johannis de Lineriis, sed ipsa est calculata ex tabulis Alphonsii', meaning 'This table is not from master John

---

'Expliciunt canones eclipsis quas magister Johannes de Janua compilavit [...]'. The *Canones eclipsium* are followed by the other works without any ascription to the author: canon *Verum motum solis et lune* (ff. 35^{r-v}); table (f. 36^{r}) and the *Investigatio eclipsis solis* (ff. 36^{v}–44^{r}).

64 The codex also contains John of Lignères' canons, *Quia ad inveniendum loca planetarum* (ff. 41ra-46rb), *Cuiuslibet arcus propositi sinum rectum invenire* (ff. 51ra–61va), and his Tables for 1322 (ff. 113^{v}–140^{v}). See the critical edition of the former by Alena Hadravová and Petr Hadrava in this volume.

65 See *infra* for the references to the table in the *Canones eclipsium.*

66 Vatican, BAV Ottob. lat. 1826, f. 47ra: '1334 imperfecto intravit sol in primum minutum Arietis, die .12. martii, o in horis 8 in minutis et 44 in secundis, non apposita equatione dierum super meridianum Janue secundum veritatem istarum tabularum illustris Regis Alfunsi'; further one reads (f. 47va), 'Canon tabule infrascripte que intitulatur tabula ad sciendum introitum solis in 12 signa. Sciendum quod hec tabula subscripta est facta ad meridianum Janue, et suppono quod distet a meridiano Tolleti, hora una et minutis .18. et est facta ad annum domini 1332 imperfectum'. The table in question is copied in the lower margin of f. 47^{v}.

67 Paris, BnF lat. 7282, f. 62^{r}: 'P. Boulier 1468 14 Junii', he copied ff. 25–131. He also annotated and corrected ff. 1–24. The same Philippe Boulier owned Paris, BnF lat. 6819, containing Dioscorides's *De materia medica* with the commentary of Peter of Padua (i.e. Peter of Abano), f. 70^{v}: 'Explicit P. Boulier 1469 4 Junii .6.' (the final '6' corresponds to the shelfmark of his private library).

68 See the eighteenth-century ex-libris (f. 1^{v}) 'Celestinorum Metis A.80'. On f. 6^{v}, the radices of Paris, Prague, and Erfurt are mentioned; an annotator has added the radix for Metz, 'Radix Metis'. On f. 7^{v}, the radices of Paris, Prague, Erfurt, Vienne, and Cologne are recorded and the same annotator has added the 'Radix Metis'. On f. 60^{v}, the year 1407 *currente* is mentioned.

of Lignères' tables, but it was calculated from the Alfonsine Tables'.[69] Here, it is clear that the table does not belong to John of Lignères' set. At the end of the canon, *Verum motum solis et lune* (f. 45^r), one reads the usual explicit, 'Etiam sciendum quod hec tabula est facta super equationes tabularum Alfonsii per magistrum Johannem de Janua, anno domini 1332 incompleto'. Eventually, Vatican, BAV Reg. lat. 1241, a manuscript copied in 1493 by Guillelmus Copp of Basel[70], presents the table facing the canon *Verum motum solis et lune* on folios 153^v–154^r. At the end there is the explicit, 'Nota quod hec tabula est facta super equationes tabularum Alfonsii per magistrum Johannem de Janua'. A note written by the same scribe follows and says that for computing eclipses, only two sexagesimal (seconds) places are needed, so the third is too precise and useless.[71] Interestingly, the manuscript also contains John of Lignères' canons and tables. Furthermore, it opens with an epistle written by Guillelmus to Conrad Heingarter, who owned two manuscripts including (with other Alfonsine materials) John of Genoa's table.[72]

The authorship of John of Genoa is also attested by the references he made to his table in the *Canones eclipsium*. According to the aforementioned explicit of the manuscript Vatican, BAV Ottob. lat. 1826, the table is quoted from the beginning of the treatise. We read in the first part of the *Canones eclipsium*: 'intra tabulam motus solis in una hora', meaning, 'enter in the table of solar hourly velocity'. Additionally, we see, intra tabulam motus lune in una hora, meaning, 'enter in the table of lunar hourly velocity'.[73] John of Genoa also specifies that the table of lunar velocity was computed by considering what affects the movement of the Moon. This 'anomaly' is described by the 'equation of centre' (*equatio centri*), which is a corrective function making it possible to calculate the true argument of the apogee of the epicycle from the mean argument.[74] One finds almost the same sentence in the canon Verum motum solis et lune.[75]

In two chapters of the *Canones eclipsium* dedicated, respectively, to the magnitude of the solar and lunar eclipses, John of Genoa wrote:

> De quantitate eclipsis [solis]
> Deinde scias dyametrum solis et lune per tabulam quam feci de hoc secundum canonem quam super eam ordinavi, et si non habes tabulam motum solis in una hora [...]

69 The same incipit is in Cracow, BJ 613, f. 33^r.

70 Cf. *Les Manuscrits de la reine de Suède au Vatican. Réédition du catalogue de Montfaucon et cotes actuelles* (Vatican: Biblioteca Apostolica Vaticana, 1964), no. 498.

71 Vatican, BAV Reg. lat. 1241, f. 153^r, 'Nota quod in eclipsibus fit operatio usque ad 2um est satis prescisa operatio quia non indigebit prescisius, primis precisissimis, habes si operatus fueris usque ad tertia'.

72 Cf. Paris, BnF lat. 7295A, f. 137^r, and lat. 7432, f. 255^r.

73 Cf. *Canones eclipsium, L*, f. 214ra.

74 where is the equation of centre (*equatio centri*), the true argument, and the mean argument; cf. Olaf Pedersen, *A Survey of the Almagest, with Annotation and New Commentary by Alexander Jones* (New York: Springer, 2011), pp. 184–94. The equation of centre can be found in lunar equation tables, see Chabás and Goldstein, *A Survey of Astronomical Tables*, pp. 67–73.

75 *Canones eclipsium*, Oxford, Bodleian Library Digby 97, f. 125^v, 'Supposito quod illa tabula sit positum cum motu lune in una hora illud quod convenit ei propter equationem centri in una hora'. A similar sentence can be found in the canon, *Verum motum solis et lune*, f. 130^r: 'Hic sciendum quod in tabula motus lune est computatum illud quod contingit Lune propter equationem centri'. The equation of centre refers to the lunar correction table; see Goldstein, 'Lunar Velocity', p. 4.

> On the quantity [magnitude] of the Sun.
> Then, look for the diameter of the Sun and the Moon with the table that I made for that, and according to the canon that I arranged for that table, and if you do not have the table of solar velocity in one hour [...].[76]
>
> De quantitate eclipsis [lune]
> Deinde quere dyametrum lune et umbre secundum canonem quem ordinavi super tabulam de semidiametris quam feci [...].
>
> On the quantity [magnitude] of the Moon.
> Then, look for the diameter of the Moon and the Earth's shadow, according to the canon that I wrote for the table of radii that I made [...].[77]

Those internal references in the *Canones eclipsium* leave little doubt that John of Genoa is in fact the author of the table of solar and lunar true velocities at syzygy, as well as the table of apparent radii associated with it. They also address the date of composition of the table, as well as that of its short canon. The colophon of the *Canones eclipsium* provides the precise date of their completion: 22 January 1332.[78] Yet, two manuscripts of the table give the date 1332 at the end of the canon, *Verum motum solis et lune*: the aforementioned Paris, BnF lat. 7283 and Vatican, BAV Pal. lat. 1374. However, both codices are late manuscript witnesses, dating to the early fifteenth century. In Vatican, BAV Pal. lat. 1374, which was finished in Prague in 1407, John of Genoa's table (f. 47^r^) is located after John of Lignères' set of tables. At the bottom of the table, two later hands have written excerpts from the canon *Verum motum solis et lune*. The latest — in black ink — has added an unfamiliar explicit, combining the end of the first chapter with the explicit of the canon, followed by the year 1332, which is incomplete: '[...] quod in illa tabula motus lune est computatum illud quod contingit lune propter equationem centri, et est facta super equationes tabularum Alfonsii per magistrum Johannem de Janua anno Christi 1332'. This reading is exactly the same as in Paris, BnF lat. 7283, and we cannot exclude that the annotator of the Vatican manuscript knew the former, or at least, another codex derived from lat. 7283. For the date, one cannot rely on these two manuscripts alone. The year 1332 seems doubtful, as the *Canones eclipsium*, which refer to the table, were finished in January 1332. Thus, the table and its canon were undoubtedly composed before 1332. The eclipse canons often attributed to John of Saxony, dated to 1330, did not use or refer to this table. John of Genoa might have established it between 1330 and 1331, although nothing excludes the fact that the author of the 1330 canons chose to not use it. In that case, the table could have been drawn before 1330. In any event, one can say for certain that the canon and table were completed before the *Canones eclipsium*.

The table of solar and lunar velocities was originally drawn for hourly motions, as stated in the canon, *Verum motum solis et lune*, and the *Canones eclipsium*. The earliest witnesses only include the table of hourly velocities and apparent radii. However, later codices preserve a different version of the table, which combines the hourly velocities and

76 Oxford, Bodleian Library Digby 97, f. 127^r^.
77 Oxford, Bodleian Library Digby 97, f. 127^v^.
78 For the quotation of this colophon, see *infra*.

their equivalence in a minute of a day. One can determine four families of manuscripts, arranging the table in different ways:

1) Manuscripts providing the 'original table', which is the table of luminaries, hourly velocities, and the apparent radii. The earliest manuscripts offer this arrangement: Douai, BM 715 (s. xv^{med}); Paris, BnF lat. 7286C (s. xiv^{ex}); Vatican, BAV Ottob. 1826 (s. xiv^{ex}); Oxford, Bodleian Library Digby 97 (s. xv^{1}). Some later manuscripts provide this oldest *lectio*, such as Oxford, Hertford College 4 (xv^{1}) and Cambridge, UL Ee.III.61 (1482).
2) Later witnesses combining the table of velocities in an hour, the equivalence in a minute of a day, and the table of apparent radii. Examples include Paris, BnF lat. 7282 (1468) and Vatican, BAV Reg. lat. 1241 (s. xv^{ex}).
3) Manuscripts displaying the table of hourly velocities with the equivalence in a minute of a day but no radii, for example, Paris, BnF lat. 7295A (s. xv$^{2/4}$) and Paris, BnF lat. 7432 (1468–88).[79] Both are linked to Conrad Heingarter; Erfurt, Universitätsbibliothek CA folio 37 (1377); Brussels, Bibliothèque Royale 926–40 (s. xv^{med}); Freiburg, Universitätsbibliothek 28 (1428); and Prague, NKCR XIII.C.17 (s. xv^{1}). There is also an early German witness: Vatican, BAV Pal. lat. 446 (s. xv^{1}).
4) Only two witnesses transmit the 'original' version of the table (including radii) plus a table combining the hourly velocities (with the same values as the original table) and their equivalence in a minute of a day (without the radii). Both codices are dated to the early fifteenth century. Paris, BnF lat. 7283 presents both tables facing each other (ff. 44^{v}–45^{r}) and Vatican, BAV Pal. lat. 1374 displays (on f. 47^{r}) the original table, and (on f. 51) the table of hourly velocities and their equivalence in a minute of a day.

Although John of Genoa is the author of the table of hourly velocities, it remains difficult to assert that he is responsible for the table of equivalence. This last table could also be a later adaptation of the astronomer's table. Nothing excludes that, after establishing his hourly table, he later converted it into minutes of a day. However, it is important to stress that the earliest manuscripts transmitting the table, as well as other works composed by the astronomer, offer only the original table. Similarly, in his *Canones eclipsium* and in his computation of the solar eclipse of March 1337, he referred only to the hourly table. Nevertheless, the same complication appears with the short canon, *Verum motum solis et lune*, as we discuss in the following section.

4. The canons to the table

The canon, *Verum motum solis et lune*, is named after its incipit, which in a longer version reads: 'Verum motum solis et lune in una hora et semidiametros solis et lune, et umbre, tempore coniunctionis vel oppositionis per hanc tabulam invenire', meaning, 'In order to find, with this table, the solar and lunar true velocities in an hour, and the radii of the Sun,

79 Paris, BnF lat. 7295A is Conrad Heingarter's student notebook; lat. 7432 is a *deluxe* manuscript, copied at the castle of Belleperche under the supervision of Conrad Heingarter for Jean II, duke of Bourbon. David Juste, *Les Manuscrits astrologiques latins conservés à la Bibliothèque nationale de France à Paris* (Paris: CNRS éditions, 2015), pp. 85, 152–53.

Moon, and the Earth's shadows during the time of the true conjunction or opposition'. This short canon is linked to the early table, of which John of Genoa is the author. Fifteen extant manuscripts present this canon, dating from the fourteenth century to the late fifteenth century. According to the number of manuscripts, the table circulated mainly in a larger set of tables, such as John of Lignères' Tables of 1322, sometimes without its short canon. Fifteen manuscripts contain the table with a version of the canon *Verum motum solis et lune*, whereas three of the canon witnesses do not display the table.[80] In manuscripts providing other works of John of Genoa, the canon follows the *Canones eclipsium* directly, as an appendix would: Paris, BnF lat. 7281 (ff. 208^{r}-^{v}); lat. 7322 (ff. 41ra-^{rb}); Oxford, Bodleian Library Digby 97, folios 129^{v}–130^{r}; Douai, BM 715, folios 35^{r}-^{v}. In the Oxford and Douai manuscripts, the canon is followed by the table, whereas in lat. 7281, the canon is followed by the detailed computation of the solar eclipse of March 1337 (f. 208^{v}). The canon is commonly structured in three chapters:

1) The first provides instructions for using the table of hourly velocities and then the table of apparent radii. It begins, 'Cum argumento solis resoluto in gradus, intra hanc tabulam in lineis numeris' and ends 'et sciendum quod hec tabula est facta super equationes tabularum Alfonsii'.[81]
2) The second treats the geometrical representation of the solar eclipse using the table of apparent radii and is entitled, 'Ad figurandum eclipsim solis'. Instructions for drawing eclipse figures are quite common in eclipse canons. John of Lignères' canons, *Priores astrologi*, contain two chapters devoted to the geometrical figuration of lunar and solar eclipses.[82]
3) The last is dedicated to the representation of the lunar eclipse, 'Ad figurandum eclipsim lune'. Usually shorter than five lines, it simply sends the reader back to the second chapter, specifying at the end: '[...] I do not see a need to repeat what has been said before'.[83]

Therefore, only the first chapter of the canon, *Verum motum solis et lune*, provides rules for using the table. This could explain why most of the manuscripts provide only the first chapter. For example, eight manuscripts, two of which are dated to the second half of the fourteenth century, display the canon without the instructions for drawing the

80 Paris, BnF lat. 7281; lat. 7322 (s. XV^{1}) and Cracow, BJ 613. According to the catchword (f. 41^{v}), at least one quire is missing after the canon in Paris, BnF lat. 7322, so it is cannot be excluded that the table followed it.

81 Oxford, Digby 97, ff. 129^{v}–130^{r}.

82 See Marie-Madeleine Saby, 'Les Canons de Jean de Lignères sur les tables astronomiques de 1321. Édition critique, traduction et étude' (unpublished dissertation, Paris, Ecole Nationale des Chartes, 1987), Ch. 37 'Figuram eclipsis solis depingere', pp. 251–53 and Ch. 39 'Figuram eclipsis lune depingere', pp. 258–60. Those chapters were also printed in the *editio princeps of 1483 (ff. m3r-m4v)* under the titles: 'De figuratione ostendente eclipsim solis' and 'De figuratione ostendente eclipsim lune'. In the Toledan Tables, the canons 'Cb' include such rules: Pedersen, *The Toledan Tables*, pp. 331–499, see Cb199a.

83 The entire chapter reads (Oxford, Bodleian Library Digby 97, f. 130^{r}): *Ad figurandum eclipsis Lune, procede per omnia sicut dixi in figura eclipsis Solis, faciendo de umbra in eclipsi Lune quicquid dixi prius de corpore Solis, nec est in alio differentia, et ideo non video necessitatem repetendi que predicta sunt.* (In order to trace a geometrical figure of the lunar eclipse, proceed according to what I said on the solar eclipse diagram; for the drawing of the shadow of a lunar eclipse, there is no difference with regard to everything I stated previously about the solar disc, and therefore, I do not see a need to repeat what has been said before.)

eclipse figures.[84] Most of them (except Cracow, BJ 613) contain the table.[85] Some give a different incipit from the original *Verum motum solis et lune*, and start directly by the part of the canon beginning: 'Cum argumento solis invenies dyametrum eius'.[86] Moreover, the manuscripts transmitting the full text of the canon, including the two last chapters, are those that include other works written by John of Genoa. Only the English codex, Oxford, Hertford College 4 shows the full text and the table, but does not provide other texts composed by the astronomer. The absence of the *Canones eclipsium* in this manuscript is probably due to the presence of John of Murs's *Canones tabularum Alfonsii* preceding the short canon *Verum motum solis et lune* and the table. It would have been redundant to copy the *Canones eclipsium*, given that John of Murs's canons dedicate a large section to eclipse computation which, moreover, derives from John of Genoa's work. The logical continuation of these canons are of course the short canon and the table; hence their locations in the manuscript. Here again Hertford College 4 highlights the close connection between John of Murs's canons and John of Genoa's works.

The full text of the canons, *Verum motum solis et lune*, seems to belong to the earliest textual family, which circulated this canon at the same time as other compositions of John of Genoa. Douai, BM 715 is likely one of the testimonies of this branch, probably relying on an earlier witness that is no longer extant.

What seems to be an early adaption of John of Genoa's table can be read in Heinrich Selder's *Canones tabularum alphonsinarum* composed in 1365, probably in Paris.[87] It is more an insertion of John of Genoa's table than an adaptation, such as the one Nicholaus de Heybech would elaborate later (*c.* 1400).[88] The manuscript witnesses containing all of the canons written by Heinrich Selder display the table of lunar and solar velocities, accompanied by canons. Both the table and the canon are found in the third part, or *differentia*, of the canons, in the ninth and tenth chapters. The *tertia differentia* provides, among others, rules for finding true syzygies and eclipse parameters.[89] In that framework, John of Genoa's table is situated between or following the ninth and tenth chapters, which begin, respectively, 'Tempus et locum vere coniunctionis et oppositionis solis et lune invenire' and 'Tempus et locum verorum quattuor aspectum solis et lune invenire'.[90] Those chapters had been composed by Selder, who reproduced the table of lunar and

---

84 The late fourteenth-century manuscripts are Vatican, Ottob. lat. 1826 (from Italy) and Paris, BnF lat. 7286C (from the north of France). The others are all fifteenth-century manuscripts: Cambridge, Ee.III.61 (f. 154^{r}), Cracow, BJ 613 (ff. 33^{r-v}), Paris, BnF lat. 7282 (f. 129^{v}) and lat. 7283 (ff. 44^{v}–45^{r}); Vatican, Reg. lat. 1241 (f. 155^{r}); Florence, BNC Conv. Soppr. J.V.4, (s. XIV^{2}) (f. 39^{r}).

85 The manuscript presents the first chapter of the canon on f. 33^{v} and includes a canon related to the adaptation of the table in a minute of a day. A later part copied in a fifteenth-century cursive script has been added; it also includes the canon, *Verum motum solis et lune*, with the original incipit but displays only the first chapter, f. 172^{r}.

86 For example, Cambridge, Ee.III.61, f. 154^{r}; Cracow, BJ 613, f. 33^{v}, Paris, BnF lat. 7286C, f. 56^{v}.

87 On Heinrich Selder and his *Canones tabularum Alphonsinarum*, see C.P.E. Nothaft, '*Vanitas vanitatum et super omnia vanitas:* The Astronomer Heinrich Selder and a Newly Discovered Fourteenth-Century Critique of Astrology', *Erudition and the Republic of Letters*, 1 (2016), 261–304, esp. 261–69.

88 Cf. Chabás and Goldstein, 'Nicholaus de Heybech'.

89 '[…] The third (in 35 chapters) adds to this the techniques necessary for finding the true longitudes and syzygies, eclipse parameters, stellar coordinates, planetary latitudes, as well as a host of other topics'. Nothaft, '*Vanitas vanitatum*', p. 265.

90 For both chapters, see Erfurt, Universitätsbibliothek CA F.37, ff. 73^{r}–74^{r}; Freiburg, Universitätsbibliothek 28, ff. 101^{v}–103^{r}; Prague, NKCR XIII.C.17 (s. XV^{1}).

solar velocities of John of Genoa. The table belongs to the third family displaying the table of hourly velocities with the equivalence in a minute of a day (without including radii). Since Erfurt, CA F37 represents one of the earliest witnesses of this arrangement, one might wonder if the table of solar and lunar velocities in a minute of a day originates from Heinrich Selder's *Canones*. This is the earliest manuscript showing the *Canones tabularum alphonsinarum*; it was completed in Paris on 14 December 1377 by Kristian of Hag, priest and monk of Saint Peter of Salzburg, during the lifetime of the author.[91] However, three later witnesses present this table as being associated with other canons.

Another early-fifteenth-century textual family transmits the first chapter of the canon, *Verum motum solis et lune*, with another canon related to the use of the table of hourly velocities and its equivalence in a minute of a day. Two manuscripts are representative of this tradition: Cracow, BJ 613, folios. 33^{r}-^{v} and Paris, BnF lat. 7283, folio 44^{v}.[92] This canon begins 'Ista tabula non est de tabulis magistri Johannis de Lineriis, sed ipsa est calculata ex tabulis Alfonsii. Et intratur in eam sic quem locum inter solem et lunam tempore vere applicationis est addita 12 eius pars'.[93] The end of each codex is different, as the scribe of Paris, BnF lat. 7283 stops the text after this statement, which shows the same remark as in the canon *Verum motum solis et lune* concerning the role of the lunar correction table in the computation of the lunar hourly velocity table. Consequently, this remark includes a positive statement: 'this table is better than the other and is really important for finding the time of true conjunction'.[94] This statement seems to imply that the author of those additional canons is not John of Genoa. Further evidence in the Cracow manuscript also supports that assumption. In that codex, the text continues 'it is sufficient to find the true positions of the planets, as they are contained in the best tables of their motions, which include their fractions to the third [...]'.[95] This signifies that the author did not compose the tables and thus this canon was likely written after John of Genoa's.

Furthermore, another canon associated with *Verum motum solis et lune* can be found in Paris, BnF lat. 7282, folio 129^{v}. It is also related to the hourly velocities table, with their equivalent in a minute of a day. This neat copy first displays the canon *Verum motum solis et lune*, with the red rubric, 'Canon tabule precedentis quam composuit magister Johannes de Janua'. A line-filler at the end of this canon strictly marks the division with the other text. The other canon also has a red heading, which reads simply, 'Alius canon'. It begins, 'Item alius canon ad inveniendum motum solis et lune in una hora et in uno minuto diei. Intra primo cum argumento equato in tabulam prius' and ends 'dupla horas et adde earum medietatem et habebis'. Again in this case, John of Genoa does not seem to be the author

91 Erfurt, CA F.37, f. 85^{v}: 'Expliciunt canones Henrici Sälder scripti per me, Kristianum de Hag, presbuterum et monachum monasterii Sancti Petri Saltzburge, anno domini 1377, 14a die mensis Decembris, Parisius'.

92 On Cracow, BJ 613, see Grażyna Rosińska, *Scientific Writings and Astronomical Tables in Cracow, XIVth-XVIth Centuries* (Wrocław: Polish Academy of Sciences, 1984), pp. 202–3.

93 Cracow, BJ 613, f. 33^{r}.

94 Paris, BnF lat. 7283, f. 44^{v}: 'Nota hec tabula est calculata ex tabulis Alphonsii et in tabula motus Lune est computatum illud quod contingit lune propter equationem centri quod in aliis tabulis non est. Secundum igitur ipsa est melior inter alias tabulas et deservit maxime ad inveniendum veram coniunctionem solis et lune'. This statement is repeated *verbatim* in Cracow, BJ 613, ff. 33rb-^{va}.

95 Cracow, BJ 613, f. 33va: 'Ad inveniendum verum locum planetarum sufficit, ut capiantur in tabulis meliorum motuum ipsorum fractiones usque ad tertiam inclusive'.

of the canon. The layout chosen by the scribe also tends to underline that both canons are different.

The additional canons were clearly intended to use the aforementioned adaptation of John of Genoa's table of hourly velocities. Anonymous authors apparently wanted to give new instructions for this table of equivalence in a minute of a day. This also suggests that John of Genoa is likely not the originator of the additional columns dedicated to lunar and solar velocities in a minute of a day. This evidence highlights the later need for a table in a minute of a day, as well as a canon ruling it. The presence of John of Genoa's table, in its third family, in Heinrich Selder's *Canones* shows that the former was in circulation from 1365, if not earlier.

5. The *Canones eclipsium*

The table and its canon probably led John of Genoa to compose a more theoretical treatise on eclipse computations. He finished these *canones* on 22 January 1332, as the detailed colophon, found in all the witnesses, attests. The *Canones eclipsium* are contained in seven extant copies, mainly dated to the fifteenth century. Some of these manuscripts have already been quoted for the other works of John of Genoa they feature. I list these extant codices and their sigla:

D Douai, BM 715, ff. 32^{r}–35^{r} (or. France, north; date: s. xv^{med})
F Florence, BML Ashburnham 206, ff. 73^{r}–76^{r} (or. Italy, Padua; date: 1409)
L London, BL Royal MS 12.C.XVII, ff. 214ra-217ra (or. France; date: in or after 1344)
M Melk, Stiftsbibliothek 601, ff. 196ra-197va (or. Germany, Erfurt?; date: s. xv^{med})
O Oxford, Bodleian Library Digby 97, ff. 125^{r}–128^{v} (or. England, date: s. xv^{1})
P1 Paris, BnF lat. 7281, ff. 206^{r}–208^{r} (or. France, Cambrai; date: s. xv^{med})
P2 Paris, BnF lat. 7322, ff. 39va-41va (or. France, north; date: s. xv^{1})

The first transmission of the treatise can be inferred from the earliest manuscript containing this work. I already mentioned that *L*, which was commissioned by John of Murs, is the oldest extant copy of the *Canones eclipsium*. The later history of the manuscript may suggest how this treatise came early to England. As noted above, *L* is a neat copy, including decorated initials and pen-flourished letters, which was perhaps intended for the monastery of Le Bec-Hellouin in Normandy, where John of Murs held an expectative benefice by 1329.[96] If so, it never arrived there or was returned to John of Murs as attested by the annotations and emendations of the volume. John of Murs's hand is found throughout the manuscript. On the upper flyleaf, he drew a circular diagram including the years of indictional cycle from 1344 to 1370. He also inserted a small piece of paper (now folio 212), where he wrote a statement on equal and unequal planetary hours.[97] Some pieces of

96 cf. Gushee, 'New sources', pp. 21–22.
97 See Laure Miolo, 'In Quest of Jean des Murs's Library. An Overview of his Readings and Uses of Manuscripts', *Erudition and the Republic of letters*, 4 (2019), 13–39, esp. 30–31.

evidence suggest that the codex passed to John's relative, Julian of Murs. The latter's peculiar handwriting, including chancery script and English *rotunda* features, is quite recognizable, as well as his notary signature. On a flyleaf, he wrote a record in French related to a house sale in the street of the *Teste Noire* in Paris and annuities to some lords. He conducted these affairs on the behalf of the King of France, to whom he was secretary.[98] A reliable source shows that Julien accompanied King John II of France in his exile to England in 1356, and that he was still there in 1357. It is likely that Julian brought the manuscript to England at that time.[99] The codex had been in Oxford since the fifteenth century according to some additions to the calendar of the *Patefit*.[100] This suggests an initial transmission route of the *Canones eclipsium*.

Another tradition can be considered with *D*, copied in the north of France, likely in Picardy, in the middle of the fifteenth century.[101] This manuscript derives from an early witness offering the entire *Opus astronomicum*, including the table. This manuscript consists of 78 folios copied by only one scribe, who was probably the owner of the codex according to additional notes. For instance, he wrote, at the end of some excerpts from the *Physionomia* ascribed to Hippocrates and Pseudo-Aristotle's *Tractatus de signis et moribus naturalibus hominum*, an unguent recipe followed by several astronomical and astrological considerations in French (including a paragraph on lunar elections, f. 3^{r}), which he likely made.[102] The astrological and medical content of the volume strongly suggests that the commissioner, scribe, and owner was a physician. This volume was probably his 'workbook'. In parallel with the astrological works contained in this volume and the *Opus astronomicum*, one also finds Pseudo-Messahallah's treatise on the astrolabe, William of Saint-Cloud's *Kalendarium regine* and an anonymous treatise on the *Directorium*, an instrument only made for astrological directions.[103] Another French witness is *P2*. This manuscript was copied in the north of France in the early fifteenth century. John of Genoa's canons, the *Canones eclipsium* and *Verum motum solis et lune*, are the only astronomical works. The rest of the codex consists of John of Saxony's commentary on the *Introductorius* of Alcabitius (ff. 1ra–39rb) and of the *Liber novem iudicum* (ff. 42ra–49vb), which has missing parts at the beginning and at the end. This neat copy has at least one quire missing after folio 49. *D*

98 Miolo, 'In Quest', p. 30.

99 On the royal act mentioning Julien of Murs with the king at Westminster in 1357, see Gushee, 'Jehan des Murs', p. 362.

100 On f. 150^{r}, one reads the following additions related to the University of Oxford: 'term', 'terme', i.e. the university terms; 'Fredeswyde', i.e. Saint Frideswide, Patron Saint of Oxford.

101 Manuscript *D* is written on paper. The watermarks, 'P' surmounted by a round cross, can clearly be identified in Charles-Moïse Briquet, *Les filigranes. Dictionnaire historique des marques du papier dès leur apparition vers 1282 jusqu'en 1600*, 4 vols (Geneva: A. Jullien, 1907), III, n°8593. This mark is found in a manuscript in the Douaisis dated early 1460s. It should also be noted that *P1* shows similar watermarks.

102 *D*, ff. 2^{v}–3^{r}. This manuscript also contains the astrological medical treatise composed by William the Englishman in 1220, *De urina non visa*, cf. Laurence Moulinier-Brogi, *Guillaume l'Anglais, le frondeur de l'uroscopie médiévale (XIIIe siècle)* (Geneva: Droz, 2011).

103 *D*, Ps.-Messahallah, *De utilitatibus astrolabii* (ff. 9^{r}–14^{r}); *William of Saint-Cloud, Kalendarium regine, (ff. 16r-28r)* and the anonymous text on the *Directorium*, *Utilitates instrumenti quod dicitur directorium* (ff. 28^{v}–29^{v}) beginning, 'Cum per istud instrumentum volueris dirigere significatores', the last treatise is followed on f. 30^{r} by a note in French providing the latitude of 51°. Two treatises on a similar instrument devoted to astrological practices, called *Directorium*, are respectively attributed by Simon de Phares to John of Lignères and Peter of Saxony. I am grateful to Jean-Patrice Boudet for this information.

and *P2*, are quite representative of the association of astrological materials with John of Genoa's works.

Among the French manuscripts we also have *P1*, which was probably copied in Cambrai.[104] This codex was intended by its owner — Jo[hannes] B. (now abridged, 'J.B') — as a historical anthology of Latin astronomy. It is also the only witness beside *D* to provide the whole *Opus astronomicum*, except the table. *P1* clearly depends on *D*, and was copied quasi contemporarily, as can be testified by the similar paper employed. Contrarily to his habit, J.B did not alter the content of the *Canones eclipsium* by shortening it or adding materials, and he did not annotate it either. He even kept the rubrics of the chapters. The only intervention he made was to insert his monogram in the heading of the canons (f. 206^{r}). J.B copied John of Genoa's treatises because he considered him part of the Alfonsine Parisian circle. Every text chosen by him is intended to offer the most important or representative texts produced during a particular time, from Toledan astronomy to Nicole Oresme, in order to provide an overview of the innovation and adaptation made over this extensive period of time.[105] Not surprisingly, he copied all of the texts produced by John of Genoa, as J.B was acting as a real philologist, collecting and assembling texts.

As previously mentioned, the first transmission was likely from France to England. This is suggested by *O*, which is an early-fifteenth-century witness copied in England. This codex written by one scribe — in an English *rotunda textualis* script — consists of astrological and astronomical treatises. The astronomical part includes a fragment of the Parisian Alfonsine Tables (ff. 5^{r}–13^{v}), William Reed's tables and canons (fols 14^{r}–32^{v} and 64^{v}–71^{r}), William Batecombe's latitudes tables (fols 33^{v}–39^{v}), a long text based on John of Saxony's canons to these tables (ff. 240^{v}–90^{v}), John of Murs's canons in the shortened version (ff. 122^{r}–25^{r}) and John of Genoa's canons (ff. 120^{v}–30^{v}) and table.[106] This manuscript belonged to a court astrologer in the fifteenth century, as suggested by differently dated additions. In the margin of the Alfonsine Tables, the astrologer drew two *figure celi* devoted to the birth of John Sutton's son, John Junior, in 1452.[107] John Sutton, 1st Baron Dudley, (d. 1487) was a courtier and a diplomat under the reigns of Henry VI and Edward IV. His son (d. 1502) was a justice of the peace in Sussex.[108] Although the identity of the astrologer remains unknown, one can assume that he practiced at the royal court. More information about this former owner is provided by pen trials added to the last flyleaf of the manuscript. Among those notes, the practitioner wrote this sentence about the nativity of his wife,

104 See *supra*.

105 Jean-Patrice Boudet has described J.B as a great scribe, high-level astronomer and historian of astronomy. He attributed, for example, the *Expositio* to John of Murs, the *Tractatus super defectibus tabularum Alfonsii* to Geoffrey of Meaux, etc. See Jean-Patrice Boudet, 'A history of astronomy by texts'; Boudet, *Lire dans le ciel*, Appendix II.

106 The codex also contains John of Saxony's commentary on the *Introductorius* of Alcabitius (ff. 165^{r}–240^{r}). On William Reed's table, see José Chabás *Computational Astronomy in the Middle Age: Sets of Astronomical Tables in Latin* (Madrid: Consejo Superior de Investigaciones Científicas, 2019), pp. 207–13.

107 *O*, f. 32^{r}: 'Ista figura est nativitas Junioris J. S. [...]. In the centre of the first *figure celi*, 'Figura nativitatis J. Sutton' Junioris anno christi 1452 imperfecto [...]'; in the second horoscope, 'hic sunt partes quas pater habuit in nativitatis filii sui in figura precedente, anno christi 1452 imperfecto. J. S. Junioris'.

108 On John Sutton and his son John, see Hugh Collins, 'Sutton, John [John Dudley], first Baron Dudley (1400–1487), courtier and diplomat', in the Oxford Dictionary of National Biography, (Oxford: Oxford University Press; 2008), accessed: 20 September 2019 <https://www.oxforddnbcom.ezp.lib.cam.ac.uk/view/10.1093/ref:odnb/9780198614128.001.0001/odnb-9780198614128-e-8155>

'Nativitas Agnetis uxoris mee 20 die Aprilis ante nonam [...]. Anno domini 1432'.[109] *O* can be considered as an astrological practitioner's book.

The last two manuscript witnesses attest the presence of the *Canones eclipsium* in Italy and Germany. *F* was written in 1409 by the polymath Prodoscimo de' Beldomandi (d. 1428), who studied at the Universities of Bologna and Padua. As a professor of arts and medicine, he taught in Padua from 1422–28. Between 1404 and 1425, Prodoscimo produced treatises on the four disciplines of the *quadrivium*. He is well-known for his *Algorismus de integris* (1410) and his music treatises.[110] Among his nine astronomical treatises, he compiled a set of Alfonsine Tables with canons in 1424.[111] Manuscript *F* is autographical and was written during his student years. Its content is representative of the Paduan curriculum at that time in the faculties of arts and medicine, which is surely the reason why he copied the statutes of the Paduan College of Arts and Medicine of 1330 (revised in 1409). The statutes are copied just before John of Genoa's *Canones eclipsium*, on folios 72vb-73ra.[112] They were later crossed out by Prodoscimo. It should be noted that John of Saxony's canons of 1327 and his commentary of the *Introductorius* of Alcabitius are part of this manuscript. According to the statutes, the latter was prescribed in Padua at that time. *F* can certainly be considered as Prodoscimo's student textbook. He copied the codex, continued to annotate it, and crossed out passages: in other words, he kept revising it. The treatment of the *Canones eclipsium* is quite representative of this kind of intervention in the text. Prosdocimo did not only copy the text; he wrote a commentary on it within the text and in the margins. His copy suggests that he also had access to a copy of the canon, *Verum motum solis et lune*, as he appended the two chapters devoted to the geometrical figuration of lunar and solar eclipses in the text of the *Canones eclipsium*.[113] At the same time, he annotated them as he did for the rest of the *Canones eclipsium*. In a way, he created a new version of the *Canones* by adding parts of the canon, *Verum motum solis et lune*, and by commenting on the whole treatise. This textbook shows that the *Canones eclipsium* were read in a university context, perhaps due to their succinctness. However, it was clearly intended for readers with a certain level of astronomical knowledge, because John of Genoa did not develop parts he considered rudimentary.

The last witness is *M*, a vast composite manuscript of 300 folios that gathers treatises on astronomical instruments, such as Richard of Wallingford's *Tractatus Albionis* or Prophatius Judeus's new quadrant; astronomical tables and canons; and astrological treatises, such as Pseudo-Ptolemy's *Centiloquium* or the judgement on the great conjunction of 1345 written by John of Murs and Firmin de Beauval. *M* was copied by several scribes and

---

109 *O*, f. 292^{r}. He also wrote the name 'Edwardus' several times, but this could correspond to King Edward IV.

110 Jan Herlinger, 'Prosdocimus de Beldemandis', in *Grove Music Online* (Oxford: Oxford University Press, 2001), Accessed: 20 September 2019.
<https://www- oxfordmusiconline- com.ezp.lib.cam.ac.uk/grovemusic/view/10.1093/gmo/9781561592630.001.0001/omo-9781561592630-e-0000022430>.

111 cf. José Chabás, 'From Toledo to Venice: The Alfonsine Tables of Prosdocimo de' Beldomandi of Padua (1424)', *Journal for the History of Astronomy*, 38 (2007), 269–81; Chabás, *Computational Astronomy*, Chapter 16.

112 These statutes are studied in Nancy Siraisi, *Arts and Sciences at Padua: The Studium of Padua before 1350* (Toronto: Pontifical Institute of Medieval Studies, 1973), pp. 15–32; see also: Pearl Kibre, *Scholarly Privileges in the Middle Ages: The Rights, the Pprivileges, and Immunities of Scholars at Bologna, Padua, Paris and Oxford* (Cambridge: Medieval Academy of America, 1962), p. 246.

113 See *F*, ff. 76ra-rb.

likely completed *c.* 1461, as the date is mentioned several times.[114] Other parts of the manuscript may have been written earlier and bound with the rest of the codex at a later date.[115] This codex is likely linked to Erfurt, for one finds on folio 192^v a set of three tables dedicated to the solar and lunar mean conjunctions for the Erfurt meridian.[116] Similarly, the Parisian Alfonsine Tables on folios 219^r–252^r are based on the Erfurt meridian. Here, John of Genoa's *Canones eclipsium* is located between a short treatise on sines and excerpts from the *Almagest*. The copy of the text is quite neat in two columns, despite the scribe's additions — in the margins — of forgotten parts of the canons. Other Alfonsine material of the volume consists of John of Lignères' canons on the prime mover, the *Priores astrologi* canons and John of Saxony's canons of 1327, and the Alfonsine star catalogue.[117] Nothing is known about the commissioner of this manuscript. The content suggests that he held a deep interest in astronomy and was perhaps an astronomer himself. By 1517, the manuscript was in the library of the monastery of Melk.[118]

The textual milieus in which the *Canones eclipsium* are copied seem quite diverse. However, they are often associated with other works written in the Parisian Alfonsine circle, except in the physician's volume *D*. The former owners of those witnesses shared a common practice of astronomy, whether it was for astronomical computation, for their studies, or more often for astrological purposes.

I now turn to the structure of the *Canones eclipsium*. This is a short work divided into five chapters. All the witnesses display the rubrics related to those parts, which are the following: 1) *De vera coniunctione*; 2) *De diversitate aspectus*; 3) *De quantitate eclipsis solis*; 4) *De eclipsi lune*; 5) *De quantitate eclipsis lune*.

The first three canons are related to solar eclipses. Thus, the first part is devoted to the procedure for finding the time of solar and lunar true conjunction. Following this, a larger canon is entirely dedicated to the computation of the lunar parallax, including a detailed explanation of the parallax correction method. Furthermore, the canon entitled *De quantitate eclipsis solis* is devoted to the calculation of the magnitude of the solar eclipse, including the duration. The last two chapters are shorter than the others. They give instructions for finding the time of true opposition and then for calculating the quantity of the lunar eclipse. In these closing canons, John of Genoa did not repeat the process to find the time of true opposition, as it is similar to the time of true conjunction. He only refers to the first chapter of the *Canones eclipsium*, which he calls 'canon de vera coniunctione'.[119] The

---

114 The astronomical table on f. 190^v starts in 1461; the precessional increment is 1461 in ff. 68^r, 204^v, 253^r. See David Juste, 'MS Melk, Stiftsbibliothek, 601 (olim 51)' (update: 22.11.2018), *Ptolemaeus Arabus et Latinus. Manuscripts*, URL: http://ptolemaeus.badw.de/ms/73.

115 On f. 218^v, the year 1393 is mentioned, 'anno domini 1393 finitus et completus per manus Wewue (?)'; on f. 251^r, and at the end of the Alfonsine Tables, there is the same note mentioning the year 1438, 'et sic est finis tabularum primi mobilis anno domini 1438 incompleto. 17 die mensis Maii'.

116 *M*, f. 192^v: 'Tabula mediarum coniunctionis solis et lune super meridianum Erfordiensis'.

117 John of Lignères' canons on the first mover (ff. 1ra-9ra); Id, canons *Priores astrologi* (ff. 86^v–105^v); John of Saxony, canons of 1327 (ff. 255^r–267^v); the Alfonsine star catalogue (ff. 54^r–68^r).

118 *Catalogus codicum manu scriptorum, qui in bibliotheca monasterii Mellicensis O.S.B. servantur* (Vienna: Alfred Hoelder, 1889), pp. 95–101. An owner note on f. 1^r mentions the year 1506, which is associated with the name Hannsenn Geyrs.

119 *O*, f. 127^v: '[...] addatur vero loco lune et provenit argumentum latitudinis 2° equatum tempore medie eclipsis cui tempori addatur equatio dierum quam quere cum gradu solis, ut dictum fuit in canone de vera coniunctione et provenit tempus vere oppositionis sive medie eclipsis diebus equatis'.

instructions for computing the quantity of the lunar eclipse are more developed than they are in the canon for finding the true opposition. The didactic character of this work also appears in John of Genoa's attempt to provide two different ways of computing a given problem. When he mentioned a specific table, such as his own table, he recognized that not everyone owned it and thus proposed another alternative.[120]

The *Canones eclipsium* end with four paragraphs that justify the author's choices in this treatise, including his sources. I now present one of four passages:

> It should be known, concerning what has been mentioned above, that in Chapter 42, Albategni presents another method for calculating the true conjunction, subtracting from the longitude the sixth and the eighth, and this method is repeated again in the *Almagesti minor* in Book 5, in the chapter on the computation of the true conjunction. However, the method explained here is more precise, and so I did not repeat that of Albategni.[121]

Here, John of Genoa refers to al-Battānī 's Chapter 42 (for finding the time of true syzygy) and to the *Almagesti minor*, book VI, Chapter 3 'Tempus et locum vere applicationis solis et lune verum preoccupare'.[122] It is noteworthy that John of Genoa is aware that the author of the *Almagesti minor* simply repeats the method explained by al-Battānī.[123] The passage in question in both works is dedicated to the method of computing the true distance between the Sun and the Moon at the time of mean conjunction.[124] However, in the *Canones eclipsium*, John of Genoa's method seems close to the one described by John of Saxony in 1327. Therefore, John of Genoa appears to have favoured this process for computing the time of the true conjunction, which is 'more precise'.[125]

The other three passages refer to the methods used for the parallax correction, material that John of Genoa seems to have extracted from the *Almagesti minor*, the purpose of which is to find the maximum obscuration of the eclipse and to compute the exact duration of the eclipse.[126] On this last point, he instructs the reader to rely on al-Battānī and John of

---

120 *O*, f. 127r: 'Deinde scias dyametrum solis et lune per tabulam quam feci [...] et si non habes tabulam motum Solis in una hora [...]'.

121 *O*, f. 128r: 'Circa predicta sciendum quod Albategni capitulo .420 ponit alium modum equandi veram coniunctionem subtrahendo a longitudinis .6am. et .8vam., et idem modus repetitur in minori Almagesti in libro .50. capitulo de equatione vere coniunctionis. Tamen, modus hic positus est precisior et ideo non repetivi modum Albategni'. On al-Battānī's method and the factor 1/6 +1/8 = 7/24, which is repeated in the *Almagesti minor*, see n. 38 in Richard L. Kremer, 'Cracking the *Tabulae permanente*'.

122 cf. al-Battānī, *al-Battānī sive Albatenii Opus atronomicum*, ed. by Carlo Alfonso Nallino, 3 vols (Frankfurt: 1899), I, pp. 92–96, for the commentary: pp. 273–74; Henry Zepeda, *The First Latin Treatise on Ptolemy's Astronomy: The Almagesti minor (c. 1200)* (Turnhout: Brepols, 2018), pp. 446–55. See also José Chabás and Bernard. R. Goldstein, 'Computational Astronomy: Five Centuries of Finding True Syzygy', *Journal for the History of Astronomy*, 28 (1997), 93–105.

123 Cf. Zepeda, *The First Latin Treatise* p. 568.

124 The *primus locus* is the result of the addition of the lunar argument at the time of the mean conjunction to the solar longitude or of a subtraction of this same argument less the lunar longitude; cf. *Canones eclipsium*, Canon 1.

125 On John of Saxony's method for finding the time of true syzygy, see above n. 2.

126 *O*, f. 128r: '*Secundo, sciendum circa diversitatem aspectus in longitudine quod istum modum extraxi ex quibusdam diffinitionibus positis in principio libri .5. minoris Almagesti, licet in querendo a eandem diversitatem aspectus non viderim nec ibi, nec in Albategni, nec alibi. Tamen, si bene consideres, videbis hunc modum valde brevi et precisum et facilem.*' This refers to the *Almagesti minor*, book VI, see: Zepeda, *The First Latin Treatise*; Nothaft, 'Jean des Murs's *Canones tabularum Alfonsii*', p. 120.

Lignères, whose canons are 'more long-winded than difficult'.[127] This last reference to al-Battānī was probably a direct borrowing from John of Lignères, who also relied explicitly on the Arabic astronomer.

John of Genoa's care for his sources seems more an attempt to compose a reliable and accurate treatise on eclipses than to legitimize his work. Similarly, the colophon of his *Canones eclipsium* is well known, because it is one of the rare examples of such a commitment in detailing sources. It reads as follows:

> End of the eclipse canons that master John of Genoa compiled, excerpting part from the *canones communes*, part from Albategni, part from the *Almagesti minor*, part from John of Sicily's commentary on the Toledan Tables, especially for eclipse digits, half-duration (*minuta casus*) and half-duration of the totality (*minuta more*). Year 1332 incomplete, 22 January.[128]

The *canones communes*, al-Battānī, the *Almagesti minor*, and John of Lignères are explicitly mentioned within the text; there is no direct reference to John of Sicily's commentary on the canons of the Toledan Tables, except in the colophon. The 'common canons' is a vague and generic formula, which could refer to several texts. However, it seems that John of Genoa is referring to the canons of the Toledan Tables here.[129] The fact that John of Lignères is always explicitly named in the text suggests that the *canones communes* do not refer to him. There are four occurrences related to those canons. For instance, at the opening of the *Canones eclipsium*, John of Genoa writes, 'To find a solar eclipse, first look for the mean conjunction when you have found the possibility of the eclipse according to what is taught in the 'common canons. For that time, equate the Sun and the Moon very precisely according to the most truthful tables, as it is taught in the common canons'.[130] When the author states simply, 'equate the Sun and the Moon', he therefore assumes that his readers know the *canones communes* and how to equate the true positions of both luminaries, and so he largely abbreviates. Similarly, he does not describe the method for calculating those true positions since he is referring to the common canons. Although for these specific parts, John of Genoa could have relied on other Alfonsine sources, such as John of Lignères' *Priores astrologi*, John of Saxony's canons of 1327 (canon 22) or the eclipse canons commonly attributed to John of Saxony and dated to 1330, in the end he seemed

---

127 *O*, ff. 128^{r}-^{v}: 'Quarto, notandum quod non est precise tantum tempus a principio eclipsis Solis ad medium, sicut a medio ad finem propter diversitatem aspectus que non augmentatur, vel diminuitur uniformiter, vel vero manet eadesemper in istis temporibus, modum autem rectificandi, vide in Albategni et in canonibus magistri Johannis de Lineriis, quia magis est prolixum quam difficile, et ideo hoc ad presens supersedeo.' He certainly relied on Canon 35 from John of Lignères' canons *Priores astrologi*. See Saby, *Les Canons de Jean de Lignères*, pp. 231–49.

128 *O*, f. 128^{v}: 'Expliciunt Canones eclipsis quas magister Johannes de Janua compilavit, extrahendo eos partim a Canonibus communibus, partim ab Albategni, partim a minori Almagesti, partim a magistro Johanne de Sicilia in scripto suo super Tabulas Toletanas, et specialiter quantum ad puncta eclipsis, minuta casus ac etiam minuta more. Anno .1332. incompleto.22. *die Januarii.*'

129 More specifically, the canons 'Cb', cf. Pedersen, *The Toledan Tables*, pp. 331–499.

130 *O*, f. 125^{r}: 'Ad sciendum eclipsim Solis primo quere coniunctionem mediam ad illud tempus ad quod possibilitatem eclipsis invenisti secundum quod docetur in canonibus communibus. In quo tempore equa Solem et Lunam precisissime secundum tabulas veriores, secundum quod docetur in canonibus communibus.'

to come back to the earlier Toledan Canons.[131] All the references to the common canons are related to processes that Alfonsine Parisian astronomers, such as John of Lignères or John of Saxony, borrowed from the Toledan Canons, such as: 1) finding the time of a mean conjunction at the time of a possibility of eclipse;[132] 2) computing the true positions of the Moon and the Sun;[133] 3) drawing a figure of the solar eclipse; and 4) drawing a figure of a lunar eclipse.[134] Furthermore, John of Genoa also refers to the *Tabule communes*, which are also likely to be the Toledan Tables. When referring to the table of lunar velocities in one hour, he first mentions his own table and explicitly states that it was designed taking into account the 'equation of centre'. He also explains the process of computation with a table from the 'common tables', which is not designed as his own table, but is based on Ptolemy's simple lunar model.[135] Furthermore, other evidence supporting the identification of the 'common canons and tables' with the Toledan material is the colophon in itself. Indeed, the other sources mentioned in this paragraph belonged to a pre-Alfonsine tradition.

In what way John of Genoa relies on John of Sicily's *Scriptum super canones Azarchelis de tabulis Toletanis* (1290) is difficult to say, as both John of Genoa and John of Sicily have common sources. In fact, John of Sicily relied on the *Almagesti minor* as well as al-Battānī's *Opus*.[136] However, one passage in the *Canones eclipsium* is closer to John of Sicily than the *Almagesti minor* or al-Battānī, suggesting that John of Genoa excerpted this part from the commentary of 1290. It is related to the computation of the half-duration of the eclipse:[137]

John of Sicily's Scriptum super canones Azarchelis. Pedersen, 1986, p. 238	John of Genoa *Canones eclipsium* *O*, f. 127^{v}
Quod si minuta dimidii more per se volueris distincte cognoscere, **solam diametrum umbre quadra et ex eius quadrato minue lune latitudinem** in se ductam, et **residui radicem sume, et habebis minuta dimidii more; que subtrahe a toto aggregato ex minutis casus et ex minutis dimidii more prius invento**, et **supererunt minuta casus.**	Et si eclipsis habet moram **tunc ex quadrato dyametri umbre subtrahe quadratum latitudinis lune et residui sinus radicem quadratam, et residui sume radicem quadratam et provenit minuta dimidie more, que subtrahe a toto aggregato ex minutis casus et dimidie more et remanent minuta casus.**

131 For John of Lignères' *Priores astrologi*, cf. Saby, *Les canons de Jean de Lignères*, I, esp. pp. 222–26; Emmanuel Poulle, *Les Tables Alphonsines*, pp. 80–87; for the two eclipse canons attributed to John of Saxony see *Tabule astronomice*, 1483 (cf. *supra*, n.3) and *Traducción castellana anónima de los Cánones de Juan de Sajonia: las tablas de los movimientos de los cuerpos çelestiales del Iluxtrisimo Rey don Alono de Castilla seguidas de su Additio*, transcribed and translated by José Martinez Gázquez (Murcia: Universidad de Murcia, 1989), pp. 96–129.

132 Pedersen, *The Toledan Tables*, Cb167–69; furthermore, according to Pedersen, in the *Priores astrologi* John of Lignères borrowed materials from Canon 33 of Cb167–69.

133 Pedersen, *The Toledan Tables*, Cb171a-b.

134 Pedersen, *The Toledan Tables*, Cb199a-c and Cb207–8b; these parts from the Toledan Canons are a direct borrowing from al-Battānī. It should be noted that in his canon, *Verum motum solis et lune*, in the parts dedicated to the diagrams of the solar and lunar eclipses, John of Genoa refers to al-Battānī and not to the *canones communes*.

135 *O*, f. 125^{v}: '[…] Et cum producto intra tabulam motus lune in una hora. Supposito quod in illa tabula sit positum cum motu lune in una hora illud quod convenit ei propter equationem centri in un hora. Si autem non sit ita, secundum quod sunt tabule communes […]'.

136 Pedersen, 'Johannes de Sicilia', 2–268.

137 Pedersen had already highlighted the proximity of these passages. Cf. Pedersen, 'Johannes de Sicilia', p. 18.

Although John of Genoa is indebted to the *Almagesti minor* for some part of the *Canones eclipsium*, al-Battānī and John of Lignères are the only individuals that he mentioned in his other works. Indeed, al-Battānī is mentioned clearly at the end of the second chapter of the canon, *Verum motum solis et lune*. John of Genoa refers to the geometrical diagrams displayed in al-Battānī's treatise in those terms: 'Albategni offers many of them [the geometrical figures] on the shadow of the zenith and on others [...]'.[138]

John of Lignères is the only Parisian master mentioned by John of Genoa. It is probably for this reason that Pierre Duhem assumed that John of Genoa was John of Lignères' disciple.[139] That cannot be excluded, given the association of their respective works in extant manuscripts and all the references one finds to John of Lignères in the *Canones eclipsium* as well as in the *Investigatio eclipsis solis*. In his last work composed in 1337, John of Genoa refers to the master in the opening paragraph related to the motion of the argument of the lunar latitude in one lunation (*motum argumenti latitudinis lune in una lunatione*); that is, he explains, 'how much the argument of the latitude is displaced from one mean conjunction to the other'. He pursues, 'I displayed this motion in my canon on the mean conjunction and the eclipse possibility [i.e. the *Canones eclipsium*], and it is also in the table of mean conjunction and opposition of master John of Lignères'.[140] The method to which John of Genoa alludes can be found in John of Lignères' canons, *Priores astrologi*, Canon 40, which is partly related to this issue.[141] John of Genoa was surely indebted to John of Lignères' canons, as well as to the other works he mentioned in his sources. However, it confirms that the astronomer evolved in this dynamic Parisian milieu. There is also some sparse external evidence that he was linked to John of Murs and perhaps to John of Montfort, whose tables present similarities to those of John of Genoa.[142] It should be noted that John of Montfort's table can be found in Paris, BnF lat. 7283, folios 43^r–44^r, just before John of Genoa's table. The date provided — January 1332 — (ff. 43^r, 44^r) can perhaps be linked to John of Genoa's *Canones eclipsium*, although nothing excludes confusion on the scribe's part. Furthermore, Goldstein had already stated the similar mention of the equation of centre in both tables.[143] At the very least, all these pieces of evidence show that John of Genoa was well established in this Parisian circle.

The *Canones eclipsium* were intended to provide an easy-to-use text for astronomers. John of Genoa assembled and condensed the knowledge available at this time for calculating eclipses more precisely. However, his innovative table was surely the first mover, which incited him to write this work. In 1337, looking at past works, he explained why he wrote

138 *O*, f. 130^r: 'Multa alia ponit Albategni et de cenith umbre et de aliis, sed nunc dimitto, vera ista sufficiunt ad ostendendum ad oculum illud de quo contentamur.'

139 Duhem, *Le Système du Monde*, IV, pp. 74–75.

140 Cambridge, UL Ee.III.61, f. 75^r: 'Tunc oportet videre motum argumenti latitudinis Lune in una lunatione, scilicet quantum argumentum latitudinis movetur ab una coniunctione media usque ad aliam. Quem motum posui in canone meo de media coniunctione et de possibilitate eclipsim invenienda, etiam positus in tabula magistri Johannis de Lineriis de mediis coniunctionibus et oppositionibus.'

141 Saby, *Les Canons de Jean de Lignères*, I, pp. 260–64.

142 On John of Montfort's table of solar and lunar velocities, see Goldstein, 'Lunar Velocity', pp. 3–17. See also José Chabás and Bernard R. Goldstein, *The Alfonsine Tables of Toledo* (Dordrecht: Springer, 2003), pp. 289–90.

143 Goldstein, 'Lunar Velocity', p. 4.

the *Canones eclipsium*: 'When I was younger, I wrote eclipse canons in order to lighten as well as to shorten the *opus* [*i.e.* eclipse canons in general]'.[144]

6. The *Investigatio eclipsis solis anno Christi 1337*

His last known work is the *Investigatio eclipsis solis* — the title derives from its incipit — which consists of the calculation of the full solar eclipse of the 3 March 1337.[145] This solar eclipse drew the attention of other masters, such as John of Murs, Levi ben Gerson, and Geoffrey of Meaux.[146] It is the most detailed extant computation of a solar eclipse in Latin for that time. Every step of the computation is detailed by John of Genoa. Three manuscripts have survived:

Ca	Cambridge, UL Ee.III.61, ff. 75ʳ–81ʳ (or. England; date: 1482)
D	Douai, BM 715, ff. 36ᵛ–44ʳ (or. France, north; date: s. xv^{med})
P1	Paris, BnF lat. 7281, ff. 208ᵛ–210ʳ (or. France, Cambrai; date: s. xv^{med})

Only, *P1* specifies that the author is John of Genoa[147]. In *D*, the *Investigatio* directly follows the other works of the astronomer so one can assume that the commissioner of the manuscript was aware of his authorship. *P1* offers an abridged version of the *Investigatio*, which is likely the result of J.B's intervention in the text. He kept the whole calculation but abbreviated the sentences. Similarly, he seemed more concerned by the results than by the detailed method, so he kept the main steps of the computation. *D* displays the text in a specific format which is probably closer to the archetype; each calculation step is explained in a short paragraph, then in front of it, the result is expressed in Arabic numbers. This codex can be considered as the oldest witness of this treatise. It is likely that *Ca* was copied from a member of the same textual family or, if not, from *D*.

The manuscript *Ca* was fully assembled by Lewis of Caerleon (d. *c.* 1495) and mainly copied by him. This native of Wales was a physician trained in Cambridge, who served the Lancastrian court, including Lady Margaret Beaufort, before becoming a royal physician of King Henry VII after his incarceration in the Tower of London (between 1484-85), at

144 Cf. The *Investigatio eclipsis solis*, Cambridge, UL Ee.III.61, f. 78ʳ: '[...] *Et ego cum essem junior fieri unum canonem de eclipsibus in quo credidi alleviare simul et abreviare opus* [...].'

145 I include the critical edition of this text in the corpus of the *Opus astronomicum* of John of Genoa.

146 On John of Murs's statement on this eclipse, which can be read in the manuscript Escorial O.II.10: Beaujouan, 'Observations et calculs'; Matthieu Husson, 'Exploring the Temporality of Complex Computational Practice: Two Eclipse Notes by John of Murs in the ms. Escorial O.II.10', *Centaurus*, 58 (2016), 46–65. On Levi ben Gerson's observation at Orange see Bernard R. Goldstein, 'Medieval Observations of Solar and Lunar Eclipses', *Archives internationales d'histoire des sciences*, 29 (1979), 101–56. Geoffrey of Meaux wrote a judgement on the eclipse, which is more astrological than astronomical. He also wrote a judgement on the comet of 1337. Both treatises show the connection medieval scholars made between eclipses and comets. Lynn Thorndike, *Latin Treatises on Comets between 1238 and 1368 A.D.* (Chicago: University of Chicago Press, 1950), pp. 215–25.

147 *P1*, f. 210ᵛ: '*Explicit doctrina ad inveniendum eclipsim solis anno domini 1337, 2a die Martii data a magistro Johanne de Saxonia [de Saxonia crossed out] Janua.*'

the end of the Wars of the Roses.[148] Besides being a physician, Lewis of Caerleon was also a great astronomer and a bibliophile, or rather a philologist. Indeed, his manuscripts or the books he annotated show another side to the individual. He seems more interested in performing calculations, annotations, or in collecting former astronomical texts than in medical literature. Lewis of Caerleon's medical practice in the Lancastrian circle, and then in the royal court, provided him with the financial support necessary to exercise his interest in astronomy.

Ca is a large codex of 192 (partly paper, partly parchment) folios. Originally this volume was composed of three different parts, mainly copied by the physician who assembled his notebook between 1481 and 1484. Indeed, the manuscript includes several dated computations, such as for the solar eclipse of 28 May 1481 (ff. 12^v–15^v), the solar eclipse of 17 May 1482 (f. 1^r), the solar eccentricity for the same year (ff. 189^v–190^r) and the conjunction of Mars and Jupiter in 1484 (f. 188^v). The content of this codex is primarily devoted to astronomy (tables, canons, instrumentation treatises), except for the presence of an anonymous treatise on the nativity of Henry VI, beginning 'Cum rerum motu' (1441).[149] This is the only work in the manuscript that Lewis did not copy, although he did annotate it.

With this manuscript, Lewis of Caerleon assembled a compendium largely devoted to eclipse computation on which he certainly depended for writing his own works about eclipses, some of which are in *Ca*. His works can be divided into three parts: 1) the canons, or more theoretical works on the composition of tables; 2) his tables; and 3) his eclipse computations.

Lewis collected many works written by fourteenth-century and fifteenth-century astronomers. He was probably a great admirer of Richard of Wallingford. Among the fourteenth-century astronomers from Merton College Oxford, Lewis of Caerleon copied the works of Simon Bredon and John Killingworth's *Algorismus*. Lewis of Caerleon was also interested in John Somer's *Kalendarium*. He extracted tables from this work, which he revised and corrected. The parallax table copied below the table of hourly velocities of the Sun, the Moon, and the radii of both luminaries and the shadow of the Earth designed by John of Genoa is said to have been extracted from John Somer: 'de copia manus proprie Ffratris Sommer'.[150] Lewis of Caerleon was not only interested in English astronomers' works. Hence, he also copied some of John of Lignères' works, including the *Algorismus minutiarum*, eclipse canons extracted from his canon of 1322, *Priores astrologi*, and a part of his canons on the prime mover, *Cuiuslibet arcus*. The manuscript ends with some propositions excerpted from John of Saxony's numerical applications of John of Lignères' prime mover canons: *Exempla super tabulas primi mobilis et canones Johannes de Lineriis*. It is interesting to note that the excerpts from John of Lignères' prime mover, as well as John of Saxony's propositions are all related to the first propositions dedicated to the arc sine. John of Lignères probably inspired Lewis of Caerleon, who wrote a short text on

148 See Pearl Kibre, 'Lewis of Caerleon, Doctor of Medicine, Astronomer and Mathematician (d. 1494?)', *Isis*, 43/2 (1952), 100–08. Keith Snedegar, 'Caerleon, Lewis (d. in or after 1495), physician and astronomer' in the *Oxford Dictionary of National Biography* (Oxford: Oxford University Press, 2004), Retrieved 26 Sep. 2019, from https://www.oxforddnb.com.ezp.lib.cam.ac.uk/view/10.1093/ref:odnb/9780198614128.001.0001/odnb9780198614128-e-4324; Hilary M. Carey, 'Henry VII's Book of Astrology and the Tudor Renaissance', *Renaissance Quarterly*, 65 (2012) 661–710.

149 Cf. Hilary M. Carey, *Courting Disaster: Astrology at the English Court and University in the Later Middle Ages* (London: Palgrave Macmillan, 1992), pp. 138-53 and 255-56.

150 *Ca*, f. 154^r.

the same topic. John of Genoa's *Investigatio* is copied after the astrological text, *Trutina Hermetis,* and is directly followed by the eclipse chapters from John of Lignères' canon *Priores astrologi* (ff. 82^r–86^r), which is annotated by Lewis of Caerleon.

It should be noted that the *Investigatio* in *Ca* displays the original text, including some additional corrections by Lewis. His intervention is more explicit in the part dedicated to the parallax computation. At the end of the treatise, Lewis added a short paragraph explaining why he corrected the *Investigatio.* He writes, 'Note and indubitably understand that, I, Lewis Caerleon, doctor of medicine, I proved by my own computation all the things written above, and I did that because I found mistakes in his [John of Genoa's] division, when he divided the excess of the third lunar parallax in latitude to the second, because I believed that from this subtle mistake, several other and more important mistakes followed [...]'.[151] This example shows a precise reading and awareness of the texts he collected. He used these sources for building and reinforcing his knowledge on eclipses. Lewis likely copied the *Investigatio* for the purpose of practicing calculation as well as for having a model to follow for his own eclipse calculations.

Similarly, it should be noted that in *P1* a sentence was written by J.B at the end of John of Saxony's *Exempla super tabulas primi mobilis et canones Johannis de Lineriis* (*c.* 1335), among other notes, on folio 232^r: 'Good example of eclipses by John of Genoa, in the year of our Lord 1337, beginning: *Ad investigandum eclipsim solis oportet primo procedere,* that is contained in this book', followed by the corresponding folio (f. 208) added by a later annotator.[152] It suggests that this text was used as an example for practitioners.

The structure of the text and the calculation are indeed very clear. It is quite a didactic computation, which provides all the key steps for completing an eclipse computation. The *Investigatio* is divided into two parts. The first is devoted to the calculation of the time of true conjunction. After three iterations, John of Genoa succeeded in finding the time of the true conjunction.[153] The second part is dedicated to the calculation of the parallax, the magnitudes, and the duration of the eclipse. It is specifically within the computation of the parallax that John of Genoa had inserted a warning to the reader concerning the error of the 'canons' in their instructions for calculating the parallax after three iterations. The canons mentioned by John of Genoa are not his own, but he remained careful and did not quote the author. He specified that he did not blame his 'masters', but rather himself, as he did not quite understand the content of those canons.[154] The canons in question are likely those of John of Lignères (*Priores astrologi*), who describes this method in canon 35.[155]

151 *Ca*, f. 81^v: 'Nota et indubitanter scias quod ego Lodowycus Caerlyon in medicinis doctor singula prescripta calculo proprio probavi, et hoc feci quia inveni errores in divisione sua quando divisit excessum 3e diversitatis aspectus lune in latitudine super secundam, quia credidi ex illo modico errore plures maiores errores secutoros.'

152 *P1* f. 232^r: 'Bonum exemplum de eclipsibus a Io[hanne] de Ianua anno domini 1337, incipit: *Ad investigandum eclipsim solis oportet primo querere* habetur in isto libro'.

153 This iterative process is studied in John of Saxony's canons: cf. Poulle, *Les Tables Alphonsines,* pp. 208–19; Chabás and Goldstein, 'Nicholaus de Heybech'; pp. 13–16; Kremer, 'Thoughts on John of Saxony', pp. 263–77.

154 *Ca*, f. 78^r: 'Advertat hic quilibet volens operari secundum canones, quia canones convenientes in hac parte operis errant [...] Et ego cum essem junior fieri unum canonem de eclipsibus in quo credidi alleviare simul et abreviare opus et posui in eo eundem errores sequentes vias aliorum. Sed ex nunc confiteor me male intellexisse dicta meorum magistrorum, non quod ipsi erraverunt absit. Sed ego male intellexi ipsorum verba. Revoco ergo errorem meum et dico quod error est operari secundum quod sonant verba canonum nisi intelligantur sicut exposui.'

155 Saby, *Les Canons de Jean de Lignères,* pp. 231–49.

Thus, among the masters to whom John of Genoa refers, we find the author of the canons *Priores astrologi*. His debt to and admiration for John of Lignères' work could explain why John of Genoa did not name him in his warning. It is also interesting to note that the astronomer questioned the theory of those canons when he applied their instructions in the computational practice. The *Investigatio* was probably the last work of John of Genoa, who likely left for the curia in Avignon to practice medicine and surgery. As far as we know, his treatise is the conclusion of an astronomical production entirely dedicated to eclipses.

John of Genoa's work had a certain posterity. One of his contemporaries, John of Murs, relied on it for his canons of 1339 and the *Tabule permanentes*. Similarly, a short time later, Heinrich Selder employed John of Genoa's table, as would Nicholaus de Heybech *c.* 1400, who can also be considered as a witness to the reception of John of Genoa's table.[156] Furthermore, at least two bibliophiles were interested in gathering the whole works of John of Genoa: the unidentified J.B from Cambrai in the mid-fifteenth century and the anonymous French physician who commissioned the manuscript *D*. Eventually, more than a century later, in England, the aforementioned astronomer and physician, Lewis of Caerleon, constituted one of the examples of the reception of the table and the *Investigatio*. He walked in the footsteps of John of Genoa. During the troubled times of the War of the Roses and then as a royal physician, Lewis devoted his work to eclipses, establishing new eclipse and parallax tables as well as canons for these specific purposes. John of Genoa's table and computation were likely inspiring to him as he gathered this material in his notebook.

The originality of John of Genoa's astronomical works mainly lies in his table of velocities and radii based on Ptolemy's second lunar model. All his texts have a link to this table. He composed it before 1332 to renew the old table of lunar and solar velocities used by the Parisian circle, which were based on Ptolemy's simple lunar model (except for John of Montfort's velocities table). With his new table, John of Genoa renewed (in a way) the computation of syzygies and eclipses. The importance of this table in the Alfonsine corpus is attested by the number of extant witnesses containing it.

Manuscript sources

Brussels, Bibliothèque Royale, 926–40
Budapest, Országos Széchényi Könyvtár, 62
Cambridge, University Library, Ee.III.61
Cracow, Biblioteka Jagiellońska, 613
Douai, Bibliothèque municipale, 715
Erfurt, Universitätsbibliothek, CA 2° 37
Erfurt, Universitätsbibliothek, CA 4° 360
Erfurt, Universitätsbibliothek, CA 4° 364
Erfurt, Universitätsbibliothek, CA 4° 371

156 Chabás and Goldstein, 'Nicholaus de Heybech'.

Erfurt, Universitätsbibliothek, CA 4° 386
Florence, Biblioteca Medicea Laurenziana, Ashburnham 206
Florence, Biblioteca Nazionale Centrale, Conv. Soppr. J.V.4
Freiburg, Universitätsbibliothek, 28
London, British Library Royal, 12.C.XVII
Melk, Stiftsbibliothek, 601
Metz, Bibliothèque municipale, 285
Munich, Bayerische Staatsbibliothek, Clm 205
Nuremberg, Stadtbibliothek, Cent. VI 22
Oxford, Bodleian Library, Digby 97
Oxford, Bodleian Library, Can. Misc. 226
Oxford, Hertford College, 4
Paris, Bbiliothèque interuniversitaire de la Sorbonne, Reg. 2. 1
Paris, Bbiliothèque interuniversitaire de la Sorbonne, 1037
Paris, Bibliothèque nationale de France, lat. 6819
Paris, Bibliothèque nationale de France, lat. 7281
Paris, Bibliothèque nationale de France, lat. 7322
Paris, Bibliothèque nationale de France, lat. 7282
Paris, Bibliothèque nationale de France, lat. 7283
Paris, Bibliothèque nationale de France, lat. 7295A
Paris, Bibliothèque nationale de France, lat. 7432
Paris, Bibliothèque nationale de France, lat. 7286A
Paris, Bibliothèque nationale de France, lat. 7286C
Paris, Bibliothèque nationale de France, lat. 13014
Paris, Bibliothèque nationale de France, lat. 14481
Prague, Národní knihovna, XIII.C.17
Vatican, Biblioteca Apostolica Vaticana, Ottob. lat. 1826
Vatican, Biblioteca Apostolica Vaticana, Pal. lat. 446
Vatican, Biblioteca Apostolica Vaticana, Pal. lat. 1374
Vatican, Biblioteca Apostolica Vaticana, Pal. lat. 1376
Vatican, Biblioteca Apostolica Vaticana, Reg. lat. 1241

Bibliography

al-Battānī, *al-Battānī sive Albatenii Opus atronomicum*, ed. by Carlo Alfonso Nallino, 3 vols (Milan: Ulrico Hoepli, 1899–1907).

Beaujouan, Guy, 'Observations et calculs astronomiques de Jean de Murs (1321–1344), in *Proceedings of the XIVth International Congress of the History of Science*, 4 vols (Tokyo: Science Council of Japan, 1975), II, pp. 27–30, repr. as Chapter 7 in Guy Beaujouan, *Par raison des nombres: l'art du calcul et les savoirs scientifiques médiévaux* (Aldershort: Variorum, 1991).

Boudet, Jean-Patrice, *Lire dans le ciel. La bibliothèque de Simon de Phares, Astrologue du XV*[e] *siècle* (Brussels: Centre d'Études des manuscrits, 1994).

———, *Le Recueil des plus Célèbres Astrologues de Simon de Phares*, 2 vols (Paris: Honoré Champion, 1997-1999).

———, 'A History of Astronomy by Texts in the Fifteenth Century: MS Paris, BnF lat. 7281', talk presented at the ALFA Workshop 'Manuscript Descriptions and Survey of Texts, Tables and Other Items for the Study of Alfonsine Astronomy', 29 September 2018.

———, 'Jean des Murs, Astrologer', *Erudition and the Republic of Letters*, 4 (2019), 123–45.

Briquet, Charles-Moïse, *Les filigranes. Dictionnaire historique des marques du papier dès leur apparition vers 1282 jusqu'en 1600*, 4 vols (Geneva: A. Jullien, 1839–1918).

Carey, Hilary M., *Courting Disaster: Astrology at the English Court and Universities in the Late Middle Ages*, (London: MacMillan, 1992).

———, 'Henry VII's Book of Astrology and the Tudor Renaissance', *Renaissance Quaterly*, 65 (2012), 661–710.

Catalogus codicum manu scriptorum, qui in bibliotheca monasterii Mellicensis O.S.B. servantur (Vienna: Alfred Hoelder, 1889).

Chabás, José, 'From Toledo to Venice: The Alfonsine Tables of Prosdocimo de' Beldomandi of Padua (1424)', *Journal for the History of Astronomy*, 38 (2007), 269–81.

———, *Computational Astronomy in the Middle Ages. Sets of Astronomical Tables in Latin* (Madrid: Consejo Superior de Investigaciones Científicas, 2019).

———, and Bernard R. Goldstein, 'Nicholaus de Heybech and His Table for Finding True Syzygy', *Historia mathematica*, 19 (1992), 265–89.

———, and———, 'Computational Astronomy: Five Centuries of Finding True Syzygy', *Journal for the History of Astronomy*, 28 (1997), 93–105.

———, and ———, *The Alfonsine Tables of Toledo* (Dordrecht: Kluwer, 2003).

———, and ———, 'John of Murs's Tables of 1321', *Journal for the History of Astronomy*, 40 (2009), 297–320.

———, and ———, *A Survey of Astronomical Tables in the Late Middle Ages* (Leiden: Brill, 2012).

———, and ———, 'The Moon in the Oxford Tables of 1348', *Journal for the History of Astronomy*, 47 (2016), 159–67.

———, and Beatriz Porres, 'John of Murs's *Tabulae permanentes*', *Journal for the History of Astronomy*, 32 (2001), 63–72.

Chartularium Universitatis Parisiensis, ed. by Henri Denifle and Émile Châtelain, 4 vols (Paris: Delalain, 1889–1897).

Collins, Hugh, 'Sutton, John [John Dudley], first Baron Dudley (1400–1487), courtier and diplomat', in the on-line *Oxford Dictionary of National Biography*, Oxford: Oxford University Press, 2008; accessed 20 September 2019.

Courtenay, William J., *Parisian Scholars in the Early Fourteenth Century. A Social Portrait* (Cambridge: Cambridge University Press, 1999).

———, ed., Rotuli Parisienses: *Supplications to the Pope from the University of Paris, I: 1316–1349* (Leiden: Brill, 2002).

Duhem, Pierre, *Le Système du monde. Histoire des doctrines cosmologiques de Platon à Copernic*, 10 vols (Paris: A. Hermann et fils, 1916–1963).

Glorieux, Palémon, *La Faculté des Arts et ses Maîtres au 13e siècle* (Paris: Vrin, 1971).

Goldstein, Bernard R., 'Medieval Observations of Solar and Lunar Eclipses', *Archives internationales d'histoire des sciences*, 29 (1979), 101–56.

———, 'Solar and Lunar Velocities in the Alfonsine Tables', *Historia mathematica*, 7 (1980), 134–40.

———, 'Lunar Velocity in the Ptolemaic Tradition', in *The Investigation of Difficult Things. Essays on Newton and the History of the Exact Science in Honour of D. T. Whiteside*, ed. by P. M. Harman and Alan E. Shapiro (Cambridge: Cambridge University Press, 1992), pp. 3–18.

———, 'Lunar Velocity in the Middle Ages: A Comparative Study', in *From Baghdad to Barcelona: Studies in the Islamic Exact Sciences in Honour of Prof. Juan Vernet*, 2 vols, ed. by J. Casulleras and J. Samsó (Barcelona: Instituto 'Millás Vallicrosa, 1996), I, pp. 181–94.

Gorochov, Nathalie, *Le Collège de Navarre, de sa fondation (1305) au début du xve siècle (1418). Histoire de l'institution, de sa vie intellectuelle et de son recrutement* (Paris: Honoré Champion, 1997).

Guillemain, Bernard, *La Cour pontificale d'Avignon (1309–1376). Étude d'une société* (Paris: De Boccard, 1962).

Gushee, Lawrence, 'New Sources for the Biography of Johannes de Muris', *Journal of the American Musicological Society*, 22 (1969), 3–26.

———, 'Jehan des Murs and his Milieu', in *Musik und die Geschichte der Philosophie und Naturwissenschaften im Mittelalter*, ed. by Frank Hentschel (Leiden: Brill, 1998), pp. 339–72.

Guy de Chauliac, *La Grande chirurgie … composée en l'an 1363*, ed. by Alfred Nicaise (Paris: Félix Alcan Éditeur, 1890).

———, *Inventarium sive Chirurgia Magna* ed. by Michael R. McVaugh, Studies in Ancient Medicine, vol. 14 (Leiden: Brill Publishers, 1996).

Herlinger, Jan, 'Prosdocimus de Beldemandis', in *Grove Music Online*, Oxford: Oxford University Press, 2001; accessed 20 September 2019.

Husson, Matthieu, 'Exploring the Temporality of Complex Computational Practice: Two Eclipse Notes by John of Murs in the ms. Escorial O.II.10', *Centaurus*, 58 (2016), 46–65.

Jean XXII (1316–1334). Lettres Communes. Analysées d'après les registres dits d'Avignon et du Vatican, VIII, ed. by Guillaume Mollat (Paris: E. De Boccard, 1924).

Juste, David, *Les manuscrits astrologiques latins conservés à la Bibliothèque nationale de France à Paris* (Paris: CNRS éditions, 2015).

Juste, David, 'MS Melk, Stiftsbibliothek, 601 (olim 51)' (updated: 22.11.2018), *Ptolemaeus Arabus et Latinus. Manuscripts*, URL: http://ptolemaeus.badw.de/ms/73.

Kibre, Pearl, 'Lewis of Caerleon, Doctor of Medicine, Astronomer and Mathematician (d. 1494?)', *Isis*, 43 (1952), 100–08.

———, *Scholarly Privileges in the Middle Ages: The Rights, the Privileges, and Immunities of Scholars at Bologna, Padua, Paris and Oxford* (Cambridge: Medieval Academy of America, 1962).

Kremer, Richard L., 'Thoughts on John of Saxony's Method for Finding Times of True Syzygy', *Historia mathematica*, 30 (2003), 263–77.

———, 'John of Murs, Wenzel Faber and the Computation of True Syzygy in the Fourteenth and Fifteenth Centuries', in *Mathematics Celestial and Terrestrial: Festschrift für Menso Folkerts zum 65. Geburtstag*, ed. by Joseph W. Dauben and others (Stuttgart: Wissenschaftliche Verlagsgesellschaft, 2008), pp. 147–60.

———, 'Cracking the *Tabulae permanentes* of John of Murs and Firmin of Beauval with Exploratory Data Analysis', in *Editing and Analysing Astronomical Tables: Towards a Digital Information System for the History of Astral Sciences*, ed. Matthieu Husson, Clemency Montelle and Benno van Dalen (Turnhout: Brepols, 2022), pp. 363-422.

Les manuscrits de la reine de Suède au Vatican. Réédition du catalogue de Montfaucon et cotes actuelles, Vatican: Biblioteca Apostolica Vaticana, 1964.

Miolo, Laure, 'In Quest of Jean des Murs's library. An overview of his readings and uses of manuscripts', *Erudition and the Republic of letters*, 4, 2019, 13–39.

Miolo, Laure, *John of Genoa, Opus astronomicum. Edition, Translation and Commentary*, forthcoming.

Moulinier-Brogi, Laurence, *Guillaume l'Anglais, le frondeur de l'uroscopie médiévale (XIIIe siècle)* (Geneva: Droz, 2011).

Neugebauer, Otto, *The Astronomical Tables of al-Khwārizmī. Translation with Commentaries of the Latin version edited by H. Suter Supplemented by Corpus Christi College MS 283* (Copenhagen: Ejnar Munksgaard, 1962).

North, John D.,'The Alfonsine Tables in England', in *Prismata, Naturwissenschaftsgeschichtliche Studien: Festschrift für Willy Hartner*, ed. by Y. Maeyama and W. G. Saltzer (Wiesbaden: Franz Steiner, 1977), pp. 269–301.

Nothaft, C. Philipp E., 'Science at the Papal Palace: Clement VI and the Calendar Reform Project of 1344/45', *Viator*, 46 (2015), 277–302.

———, '*Vanitas vanitatum et super omnia vanitas*: The Astronomer Heinrich Selder and A Newly Discovered Fourteenth-Century Critique of Astrology', *Erudition and the Republic of Letters*, 1 (2016), 261–304.

———, 'Jean des Murs's *Canones tabularum Alfonsii* of 1339', *Erudition and the Republic of Letters*, 4 (2019), 98–122.

Pansier, Pierre, 'Les Médecins des Papes d'Avignon (1308–1403)', *Janus*, 14 (1909), 405–34.

Pedersen, Fritz S., *The Toledan Tables. A Review of the Manuscripts and the Textual Versions with an Edition* (Copenhagen: C.A. Reitzels Forlag, 2002).

———, 'Johannes de Sicilia: Scriptum super canones Azarchelis', *Cahiers de l'Institut du Moyen-Âge grec et latin*, 51–52 (1986), 2–268.

———, 'Anonymous Parisian Astronomer of 1290', *Cahiers de l'Institut du Moyen-Âge grec et latin*, 72 (2001), 169–269; 73 (2002), 61–166.

Pedersen, Olaf, 'The Corpus Astronomicum and the Traditions of Mediaeval Latin Astronomy', *Studia Copemicana*, 8 (1975), 55–96.

———, *A Survey of the Almagest*, ed. Alexander Jones (New York: Springer, 2011).

Poulle, Emmanuel, *Les Tables Alphonsines avec les canons de Jean de Saxe* (Paris: Éditions du CNRS, 1984).

Rosińska, Grażyna, *Scientific Writings and Astronomical Tables in Cracow, XIVth-XVIth centuries* (Wrocław: Polish Academy of Sciences, 1984).

Saby, Marie-Madeleine, 'Les Canons de Jean de Lignères sur les tables astronomiques de 1321. Édition critique, traduction et étude' (unpublished dissertation, Paris, Ecole Nationale des Chartes, 1987).

Schäfer, K. H., *Die Ausgaben der apostolischen Kammer unter Benedikt XII, Klemens VI, und Innocenz VI* (Paderborn: F. Schöningh, 1914).

Siraisi, Nancy, *Arts and Sciences at Padua: The Studium of Padua before 1350* (Toronto: Pontifical Institute of Medieval Studies, 1973).

Snedegar, Keith, 'Caerleon, Lewis (d. in or after 1495), physician and astronomer' in the on-line *Oxford Dictionary of National Biography*, Oxford: Oxford University Press, 2004; accessed 26 September 2019.

Thorndike, Lynn, *Latin Treatises on Comets between 1238 and 1368 A.D.* (Chicago: University of Chicago Press, 1950).

Teeuwen, Mariken, *The Vocabulary of Intellectual Life in the Middle Ages* (Turnhout: Brepols, 2003).

Traducción castellana anónima de los Cánones de Juan de Sajonia: las tablas de los movimientos de los cuerpos çelestiales del Iluxtrisimo Rey don Alono de Castilla seguidas de su Additio, transcribed and translated by José Martinez Gázquez, Murcia: Universidad de Murcia, 1989.

Toomer, Gerald J., 'A Survey of the Toledan Tables', *Osiris*, 15 (1968), 5–174.

Verger, Jacques, 'Teachers', in *A History of the University in Europe, I: Universities in the Middle Ages*, ed. by Walter Rüegg (Cambridge: Cambridge University Press, 1992), pp. 144–69.

Weijers, Olga, *Le travail intellectuel à la Faculté des arts de Paris: textes et maîtres (ca. 1200-1500)*, 9 vols (Turnhout: Brepols Publishers, 1994-2012).

Wickersheimer, Ernest, *Dictionnaire biographique des médecins en France au Moyen Âge* (Geneva: Droz, 1979).

Zepeda, Henry, *The First Latin Treatise on Ptolemy's Astronomy: The Almagesti minor (c. 1200)* (Turnhout: Brepols, 2018).

Appendix: Concordance of manuscripts copying John of Genoa's work(s)

Manuscript witnesses	Table[157] (before 1332)	Canon *VMSL* (before 1332)	Canon Anon. 1 (15th c)	Canon Anon. 2 (15th c)	Heinrich Selder's *Canones* (1365)	Canones eclipsium (1332)	Investigatio eclipsis solis (1337)
B Brussels, Bibliothèque Royale, 926–40	X				X		
Ca Cambridge, UL, Ee.III.61	X	X					X
Cr Cracow, BJ 613		X	X				
D Douai, BM, 715	X	X				X	X
E Erfurt, Universitätsbibliothek, BA, Fol. CA 2° 37	X				X		
F Florence, BML, Ashburnham 206		X				X	
Fr Freiburg, Universitätsbibliothek, 28	X				X		
L London, BL, Royal MS 12. Royal 12.C.XVII						X	
M Melk, Stiftsbibliothek, 601						X	
O Oxford, Bodleian Library Digby 97	X	X				X	
O2 Oxford, Hertford College 4	X	X					
P1 Paris, BnF, lat. 7281		X				X	X
P2 Paris, BnF, lat. 7322,		X				X	
P3 Paris, BnF, lat. 7282	X	X		X			
P4 Paris, BnF, lat. 7283	X	X	X				
P5 Paris, BnF, lat. 7295A	X						
P6 Paris, BnF, lat. 7432	X						
P7 Paris, BnF, lat. 7286C	X	X					
Pr Prague, NKCR, XIII.C.17	X				X		
V1 Vatican, BAV, Ottob. lat. 1826	X	X					
V2 Vatican, BAV, Pal. lat. 446	X	X					
V3 Vatican, BAV, Pal. lat. 1374	X	X					
V4 Vatican, BAV, Reg. lat. 1241	X	X					

157 For the different families of the table, see *supra*.

GLEN VAN BRUMMELEN

All In: Fifteenth-Century Manuscripts Devoted to Giovanni Bianchini's Astronomy

Introduction

One of the many positive outcomes of this volume is the emergence of a potentially helpful classification of fourteenth- and early-fifteenth-century astronomical manuscripts into three major categories.[1] The first category includes student manuscripts, which were collections of texts used to train scholars in the discipline. The works included were often dissonant, containing different approaches to the same topic; for instance, a set of canons (instructions) for a set of tables might be matched with a different set of tables. Texts from the second category, toolbox manuscripts, were collated by astronomers or astrologers for their own personal use. Often, they focused on a particular discipline or task, with the texts providing the practitioner with various options and methods. In the third category, presentation manuscripts were documents assembled with great care and featured ornamentations, of which the primary purpose was to ensure an impressive appearance. They were rarely solicited for day-to-day use, and may rather have been thought of as performance pieces for patrons.

At first glance, toolbox manuscripts appear to belong to the category from which new knowledge would have been gained. The attitude conveyed is peculiar for the time; their goal tended to be to produce results for clients, rather than insight for a scholarly community. Manuscripts were a means to an end, such as the casting of a horoscope or the

* This work was supported by the ERC project ALFA: Shaping a European scientific scene, Alfonsine astronomy, CoG 723085, PI Matthieu Husson.
Often an author states in the acknowledgements that the paper could not have been completed without the support of certain people. This statement is especially true in this case. In the process of learning new research skills for this project, I had the great fortune to be surrounded by colleagues who selflessly gave their time and expertise: in particular, José Chabás, Marilyn Lavin, Laure Miolo, Darcy Otto, and Alexandre Tur. My greatest thanks are owed to Matthieu Husson, who cajoled me into moving in this direction, and then provided the resources, background, and advice that inevitably became indispensable. Matthieu, for everything, I thank you. Thanks also to the Fund for Historical Studies at the Institute for Advanced Study (Princeton), which provided funding to make this project possible.

1 This categorization works according to the purposes of the manuscripts. Others are possible; for instance, one might group them according to the types of works that are gathered together, by discipline, or by author.

Glen Van Brummelen • Trinity Western University, Vancouver, Canada

Alfonsine Astronomy: The Written Record, ed. by Richard L. Kremer, Matthieu Husson, José Chabás, Turnhout, 2022 (*Alfonsine Astronomy: Studies and Sources*, 1), pp. 381-400
© BREPOLS PUBLISHERS 10.1484/M.ALFA.5.124933

prediction of an eclipse. Just as a typical carpenter might own a collection of tools obtained from many different manufacturers, a typical manuscript contains works by a number of different authors. The point was to bring together sources that a reader could use, rather than promote a given author's point of view. This makes the manuscript tradition of the texts of one author in particular, Giovanni Bianchini (mid-fifteenth-century Ferrara) almost unique for its time. While there are a couple of dozen manuscripts containing his work alongside the works of other authors, and another couple of dozen that contain a single Bianchini text, eight particular manuscripts contain multiple works by Bianchini on different topics, and nothing or very little else. Our aim in this paper is to explore these eight manuscripts (not his entire corpus), find commonalities among them, and form hypotheses regarding how Bianchini came to be among the earliest astronomers to receive this special treatment.

1. A typical toolbox manuscript of the time

The following is a table of contents of a typical toolbox manuscript, Vatican, BAV Reg. lat. 1904, dating from sometime between the mid-fifteenth and -sixteenth centuries:[2]

- Giovanni Bianchini, *Flores Almagesti* treatises 3 (plane trigonometry), 4 (proportions), 5 (spherical trigonometry and astronomy), 1 (arithmetic), and 2 (algebra); treatises 6–10 are not present (ff. 1^{r}–56^{r})
- Thebit Bencora, *De hiis que indigent expositione antequam legatur Almagesti*, C.1.1 (ff. 57^{r}–61^{r})
- Anonymous, *Kardaga est portio circuli constans ex 15 gradibus*… (ff. 61^{r}–62^{v})
- Thebit Bencora, *De figura sectore* (ff. 62^{v}–70^{r})
- Jordanus of Nemore, *Elementa super demonstrationem ponderum* (ff. 70^{v}–73^{r})
- Jordanus of Nemore, *Liber de canonio* (ff. 73^{v}–76^{v})
- Thebit (attributed), *Queritur in longitudine* (ff. 76^{v}–77^{r})
- Pseudo-Euclid, *De ponderoso et levi* (ff. 77^{r}–78^{r})
- Pseudo-Thebit Bencora, *De motu octave spere* (ff. 79^{r}–83^{v})
- Anonymous, *Data elevationis cum instrumento ab orizonte* (ff. 84^{v})
- Anonymous, *Nota quod habita notitia arcus dici et altitudinis Solis* (ff. 85^{r})
- Euclid, *Elements* (excerpts from Gerard of Cremona's translation) (ff. 86^{r}–90^{r})
- John of Lignères, *Canones primi mobilis* (ff. 91^{r}–102^{v})

A user could rely on this collection of texts when working in spherical astronomy. Arithmetical and algebraic underpinnings, the calculation of sine tables, spherical trigonometry, and examples of both theoretical and practical spherical astronomy are included. Virtually any elementary problem in the discipline can be answered by appealing to one or another of the texts in this manuscript. Many of the texts are extracts. In the largest text (the *Flores Almagesti*), the material following spherical astronomy and comprising half the book is discarded, and the chapters are not in order. Clearly, this was a reference source for a discipline that was considered by the compiler to stand above the particular approaches of the individual authors.

2 See David Juste, MS Vatican, Biblioteca Apostolica Vaticana, Reg. lat. 1904 (updated: 12.01.2018). *Ptolemaeus Arabus et Latinus. Manuscripts*, URL = http://ptolemaeus.badw.de/ms/471.

Authors were of course accustomed to such cut-and-paste practices, and often anticipated them. For instance, in the canons to his 1322 tables, John of Lignères uses incipits and explicits in various sections throughout the treatise to specify where the cuttings might be made most effectively.[3] The 1339 canons of John of Murs continues this practice, with three explicits and a colophon.[4]

2. Who was Giovanni Bianchini?

The first and largest of the treatises in the above manuscript, the *Flores Almagesti*, was authored by Giovanni Bianchini (*c.* 1410–69) of the Duchy of Ferrara, today part of northern Italy. A Venetian merchant, he came into the administrative service of the d'Este family in 1427. He served under three members of the family: Niccolò (1427–41), Leonello (1441–50), and Borso (1450-*c.* 1470).[5] He was active in astronomy from around 1440 to 1460, first under Leonello (known primarily as a patron of the arts and sciences), and then under Borso, for whom he seems to have performed some form of military service.[6] Today, he is known mostly for a series of letters he exchanged with Regiomontanus between 1463 and 1464.

Bianchini authored five major texts:

- the *Flores Almagesti*, a treatise in honour of Ptolemy's *Almagest* that expounds the mathematical theory of astronomy (seven manuscripts; see the appendix);[7]
- the *Tabulae astronomiae*, a set of tables on the motions of the planets and instructions for their use (thirty-eight manuscripts);[8]
- two treatises, both entitled *Tabulae primi mobilis* on spherical astronomy, which we shall call *A* and *B*, and the associated *Tabulae magistrales* (five manuscripts of *A*, ten of *B*, and eight of the *Tabulae magistrales*);[9]
- the *Tabulae eclypsium* (ten manuscripts), on predictions of eclipses.[10]

3 See Marie-Madeleine Saby, 'Les Canons de Jean de Lignères sur les tables astronomiques de 1321' (unpublished thesis, Paris, École Nationale des Chartes, 1987).

4 C. Philipp E. Nothaft, 'Jean des Murs's *Canones tabularum Alfonsii* of 1339', *Erudition and the Republic of Letters*, 4 (2019), pp. 98–122 (100-101).

5 See José Chabás and Bernard R. Goldstein, *The Astronomical Tables of Giovanni Bianchini* (Leiden/Boston: Brill, 2009), p. 13.

6 From the explicit by Arnaud de Bruxelles in Paris BnF lat. 10267, f. 106^v, and the explicit to the *Tabulae astronomiae* in Paris BnF lat. 7271, f. 25^v.

7 Incipit *Arithmetica dico quod determinatur per numeros…*

8 Incipit *Consideranti mihi, dive Leonelle, et principatum tuum…* This text was the subject of Chabás and Goldstein, 2009.

9 In Glen Van Brummelen, 'The End of an Error: Bianchini, Regiomontanus, and the Tabulation of Stellar Coordinates', *Archive for History of Exact Sciences*, 72 (2018), 547–63, I described Bianchini's repair of a mathematical error in the conversion of stellar coordinates in the *Tabulae primi mobilis*. Within the canons, Bianchini states that he had composed an earlier work containing the error. Since the publication of that article, I have discovered the original work in five manuscripts, only one of which is complete. Both the canons and the tables overlap in roughly one quarter of their contents. See Glen Van Brummelen, 'Before the End of an Error: Giovanni Bianchini's Original Flawed Treatise on the Conversion of Stellar Coordinates', *Archive for History of Exact Sciences*, 75 (2021), 109–124. Both treatises share the incipit *Non veni solvere legem sed illuminare hiis qui in tenebris sedent…* On the *Tabulae magistrales*, see José Chabás, 'An Analysis of the *Tabulae magistrales* by Giovanni Bianchini', *Archive for History of Exact Sciences*, 70 (2016), 543–52.

10 Incipit *In libro Florum Almagesti per me Ioannem Blanchinum demonstratum est…* On the manuscripts of all of Bianchini's works, see José Chabás, *Computational Astronomy in the Middle Ages: Sets of Astronomical Tables in Latin* (Madrid: Consejo Superior de Investigaciones Cientificas, 2019). On the manuscripts of the *Flores Almagesti*, see

The earliest of these works, and the one that received the greatest attention, may have been the *Tabulae astronomiae*, which are based on the Alfonsine Tables, but employ innovative structures that ease the user's computational burden. This early hint at his creativity began to flower with the *Tabulae primi mobilis A*, which introduces the exclusive use of decimal notation for entries in astronomical tables. The *Flores Almagesti*, a regrettably understudied magnus opus of astronomical theory, formed the backbone of the rest of Bianchini's work and contained a number of seeds that would revolutionize Ptolemaic mathematical astronomy. It follows the path of the *Almagest*, up to but not including the motions of the planets. Based on the mathematics of the *Flores Almagesti*, the *Tabulae primi mobilis B* both repairs the most serious errors in the *Tabulae primi mobilis A* and contains a version of decimal fractional notation that would evolve into today's numeral system.[11] It is closely associated with the *Tabulae magistrales*, the earliest European collection of auxiliary astronomical tables. These tables were also employed in the *Tabulae eclypsium*, Bianchini's last major work, which also deserves further study. His letters exchanged with Regiomontanus, some of the earliest scientific correspondence in Europe, were written near the end of his life. Bianchini's contributions seem to have been written to attract Regiomontanus's attention to his achievements, notably in trigonometry and spherical astronomy.[12]

These texts appear in the primary literature in various combinations. Roughly twenty-five manuscripts contain one of these five works, most frequently the *Tabulae astronomiae*, or fragments of it, along with a number of texts by other authors. Around twenty manuscripts consist of a single work, usually the *Flores Almagesti* or the *Tabulae astronomiae*. The other eight contain multiple Bianchini works, and nothing or very little else. Here is the table of contents of one of the most complete manuscripts, Paris, BnF, 7271:[13]

- Bianchini, a short text on a surveying instrument (fragment, ff. 1^{r}–1^{v})[14]
- Bianchini, *Tabulae astronomiae* (ff. 8^{r}–146^{v})
- Bianchini, *Tabulae primi mobilis B* (ff. 147^{r}–161^{r}, 181^{r}–216^{v}, 222^{r}–237^{r}, 242^{r}–245^{v})
- Bianchini, *Tabulae eclypsium* (ff. 169^{r}–180^{r}, 237^{v}–240^{v})

---

the appendix.

11 On the correction of the errors in *Tabulae primi mobilis A*, see Van Brummelen, 'The End of an Error'; on the invention of decimal fractional numeration, see Grażyna Rosińska, 'Decimal Positional Fractions: Their Use for Surveying Purposes, Ferrara 1442', *Kwartalnik historii nauki i techniki*, 40 (4) (1995), 17–32, and Glen Van Brummelen, 'Distinctions of Magnitude: Numbers and Quantities in the Fifteenth Century and the Invention of Decimal Positional Arithmetic' (forthcoming).

12 The correspondence was edited in Maximilian Curtze, 'Der Briefwechsel Regiomontans mit Giovanni Bianchini, Jacob von Speier und Christian Roder', *Abhandlungen zur Geschichte der mathematischen Wissenschaften*, 12 (1902), 185–336; see also Armin Gerl, *Trigonometrisch-astronomisches Rechnen kurz vor Copernicus: Der Briefwechsel Regiomontanus-Bianchini* (Stuttgart: Steiner, 1988).

13 A couple of isolated tables (motions of the Sun, Moon, and planets in collected years) are scattered throughout the manuscript.

14 *Compositio instrumenti* is a short work on the construction of an instrument to measure stellar altitudes that was edited in Paolo Garuti, 'Giovanni Bianchini, *Compositio instrumenti* (Cod. lat. 145 = α.T.6.19) della Biblioteca Estense di Modena', *Istituto Lombardo (Rend. Lett.)*, 125 (1991), 95–127. This text appears to describe the same instrument.

This table of contents stands in sharp contrast to Vatican Reg. lat. 1904; rather than a unity of topics and diverse range of authors, we have a single author and a diversity of topics. The three different groups of tables are mixed together somewhat in this manuscript, which was common for table sets at this time. See the appendix for a table of the eight manuscripts and the major treatises that they contain.

The eight multiple-treatise Bianchini manuscripts fall into a continuum between the categories of presentation and toolbox: none appear to be student manuscripts. In this work, we discuss two manuscripts at each end of the spectrum in detail. Following this, we summarize the other six.

A Bianchini presentation manuscript: Florence, BML Plut. 29.33[15]

This ornately decorated manuscript contains the following works:
- *Tabulae astronomiae* canons (ff. 1^r–14^r)
- *Tabulae astronomiae* tables (ff. 15^r–121^r)
- *Tabulae primi mobilis A* canons (ff. 122^r–134^v)
- *Tabulae primi mobilis A* tables (ff. 138^r–149^v)

This is the only manuscript of the eight to include the earlier *Tabulae primi mobilis A*. The *Tabulae astronomiae* were likely Bianchini's earliest major work, so this manuscript might have been composed and presented while Bianchini was still an active astronomer.

The manuscript seems to have been assembled all at once; there are no indications that some of the pages ever existed separately. The same semi-gothic Italian hand is used throughout for both the canons and the tables. There are several well-drawn geometric diagrams near the beginning of the *Tabulae primi mobilis A*. The text wraps around one of them (f. 123^r), and space is held within the text for two of the others (f. 123^v), so we can deduce that they were intended to be part of the text from the beginning.

A number of signs indicate the care with which the manuscript was composed. The document is made entirely of a thin parchment, which is a statement of luxury.[16] The book opens with a beautiful miniature of a figure seated at a desk holding a quadrant, surrounded by various instruments (f. 1^r, Figure 1). The halo, the robe, and the bare feet indicate that the figure represents a divine entity. The writing on this page has faded significantly, suggesting that it was on display for a time. The initial letters of chapters in the canons are elaborately ornamented, alternating between red and blue ink (the latter being expensive to obtain).[17] The initial letter of the *Tabulae primi mobilis A* (f. 122^r) is richly decorated with red, blue, green, and yellow ink. The modern scribe who wrote the folio numbers was clearly respectful of what he saw: they are written unobtrusively in the bottom right corners of the pages.

15 For the catalogue entry of this manuscript, see Angelo Maria Bandi, *Catalogus codicum latinorum Bibliothecae Laurentianae* (Florence: Praesidibus adnventibus, 1775), II, cols 50–52.

16 The leaves were arranged so that two facing pages would exhibit the same side of the parchment: either on the hair side or the flesh side.

17 Initial letters are missing from ff. 13^v and 14^r at the end of the planetary canons, but they reappear in the canons of the *Tabulae primi mobilis A*.

Figure 1. The beginning of the canons of the *Tabulae astronomiae* in Florence, BML Plut. 29.33, folio 1r. Reproduced by permission of Italian Ministry of Cultural Heritage and Activities; any further reproduction by any means is prohibited.

The manuscript is in very good condition, which suggests that its intended purpose was public display rather than astronomical work. This is supported by the absence of marginal notes, other than an insertion where some text had been accidentally omitted (f. 12^{v}). The edges of the paper are sharp and unworn; since the binding is very old (perhaps original), this further supports our perception that the book was not used as a reference source. The pairing of the two treatises seems to be based on the promotion of the author rather than scientific reasons. While the two topics do have some small degree of interaction, there is no direct link between the two, such as shared tables or methods. Furthermore, there are no references to one treatise from another.

The overall impression is that of a manuscript that had a public role to play, rather than a scientific one. It would have been expensive to produce. It might have stood as a formal statement of approval of the astronomical and astrological work that Bianchini had presented to date, or perhaps as a symbol of the result of Leonello d'Este's patronage.

A Bianchini toolbox manuscript: Cracow, BJ MS 556[18]

This toolbox has a much more complex structure than our previous manuscript, with four hands acting on it at various times. The table of contents is more complicated than that of Florence, BML Plut. 29.33:

- Bianchini, *Additiones canonum primi mobilis* (ff. 5r–8r)
- Bianchini, *Tabulae primi mobilis B* canons (ff. 9r–23v)
- Bianchini, *Tabulae eclypsium* canons, Ch. 39, and an astrological chapter (ff. 23v–24v)
- Bianchini, *Tabulae eclypsium* canons (ff. 25r–34v)
- Bianchini, Two new chapters related to the astrology in the *Tabulae primi mobilis B* (ff. 34v–36v)
- Bianchini, Several planetary tables (ff. 37r–39r, 40r)
- Bianchini, *Tabulae primi mobilis B* table (f. 40v)
- Bianchini, *Tabulae eclypsium* tables (ff. 40v–42r)
- Bianchini, *Tabulae primi mobilis B* tables (ff. 47r–105v)
- Ptolemy's star table in Alfonsine calculation, precessed 17;08 (ff. 107r–117r)
- Two astrolabe star tables, dated 1464 and 1500, respectively (ff. 117r–118r)
- Alfraganus, *Liber de aggregationibus scientiae stellarum et principiis celestium motuum* (fragment) (ff. 119r–119v)
- Messahala, *Epistola de rebus eclipsium et coniunctionibus* and another text (fragment) (ff. 120r–122v)
- Guillelmus Anglicus, *Tractatus astrolabii universalis* (ff. 122v–123v)
- Blank (ff. 1r–4r, 8v, 39v, 42v–46v, 106r–106v, 118v, 124r–126r, 127r–128v)

The *Additiones* to the *Tabulae primi mobilis*, found only in this manuscript and one other (Vat. lat. 2228), are attributed explicitly to Bianchini. The last three chapters of twenty occur after the phrase, '*Finis canonum tabularum primi mobilis Johannis Blanchini*', so they may not have been considered to be part of the *Additiones* itself. The extra chapters in folios 34v–36v are similarly attributed to Bianchini. The end of the canons to the *Tabulae eclypsium* (ff. 23v–24v) appears, oddly, before the rest of the work and in a darker coloured ink than the rest of the text. Marginal notes where the main eclipse text ends and the final fragment begins alert the reader to the existence of the extra section.

The first of the four hands in the document is that of Gregorius de Cracovia, astrologer of Pope Paul II, dated June 25, 1468 in Rome on folio 117r. He is responsible for the text of the canons, and for some of the tables (ff. 37r–56v and from f. 105v to the end). The handwriting is cursive, and it verges on casual in a number of places (see Fig. 2). Although the writing exhibits a less careful appearance, it is usually more scientifically accurate than other manuscripts, using a correct word when others are incorrect. On the other hand, the bulk of the tables were written with more precision by another scribe, Franciscus Quatuor

18 On this manuscript, see Maria Kowalczyck *et al.*, *Catalogus codicum manuscriptorum medii aevi latinorum qui in Bibliotheca Jagellonica Cracoviae asservantur* (Wrocław: Institutum Ossolinianum Officina Editoria Academiae Scientiarum Polonae, 1984), III, pp. 376–82 and Grażyna Rosińska, *Scientific Writings and Astronomical Tables in Cracow: A Census of Manuscript Sources (XIVth-XVIth Centuries)* (Wrocław: Polish Academy of Sciences Press, 1984), entries 28, 108, 113, 424, 606, 607, 875, 973, 1150, 1269, 1486, 2121, 2136, 2173, and 2341.

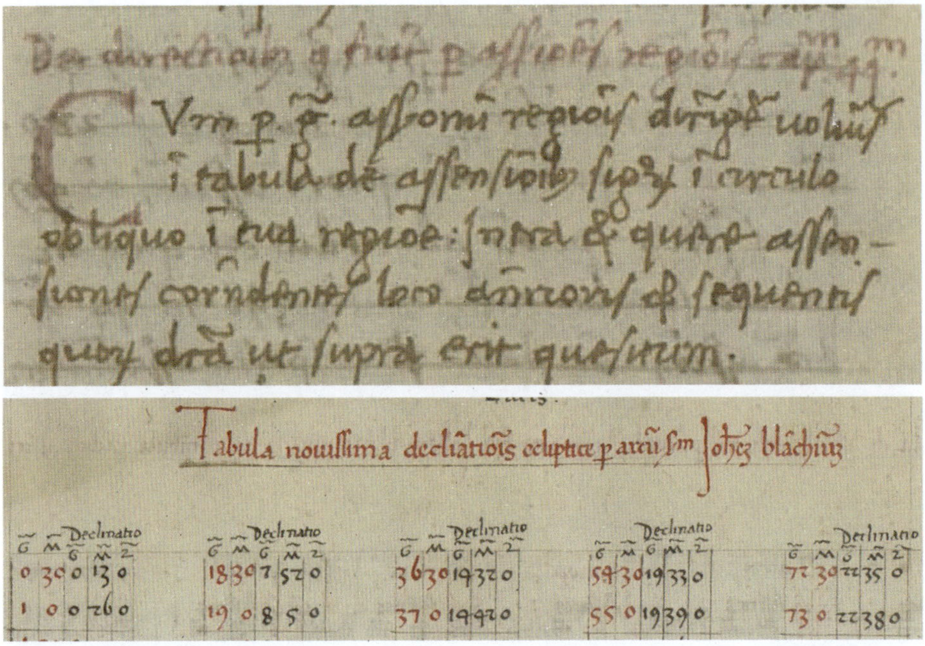

Figure 2. Gregorius de Cracovia and Franciscus Quatuor e Castris's hands in Cracow, BJ MS 556, folios 22[v] and 59[r]. Source jbc.bj.uj.edu.pl//dlibra/.

e Castris (indicated at the bottom of f. 104[v]). This reverses a common pattern, in which the canons are written more carefully than the tables.

At various points throughout the manuscript, we find additional notes written in the margins in yet another hand, associated with the date 1500 (f. 118[r]). The marginal notes, usually written in a darker ink than the rest of the manuscript, tend to cluster around passages related to astrology. This unnamed commentator seems to be responsible for a number of diagrams in the margins, which are unique to this manuscript and provide significant insights into the field of astronomy under discussion. Sometimes the diagrams are drawn on top of the text, demonstrating that they were added later (see, for example, Fig. 3). Evidence that the commentator engaged with the astronomical content of the manuscript is present in a number of places, including two entries in the Alfonsine star table at the end of the manuscript (f. 109[r]), where he writes 'falsa' with drawings of two eyes, indicating either that his own observations of the stars failed to conform with the table, or that these entries do not match with another star table that he had on hand.

Finally, the fly leaves at the front and back of the manuscript (ff. Ir, IIv) reveal yet another, very different presence. The leaves are parchment; the rest of the manuscript is paper. This unknown scribe has also filled folio 126[v] with text describing various authors' views on the length of the year and the solar apogee, amid some blank pages. These writings contain astrological and astronomical text not directly related to the content of the book itself. The hand, much more careful than the text within the book, is Italian semi-gothic with frequent flourishes. The book remains within an old binding, assuring us that these additions must have been made early in the history of the manuscript.

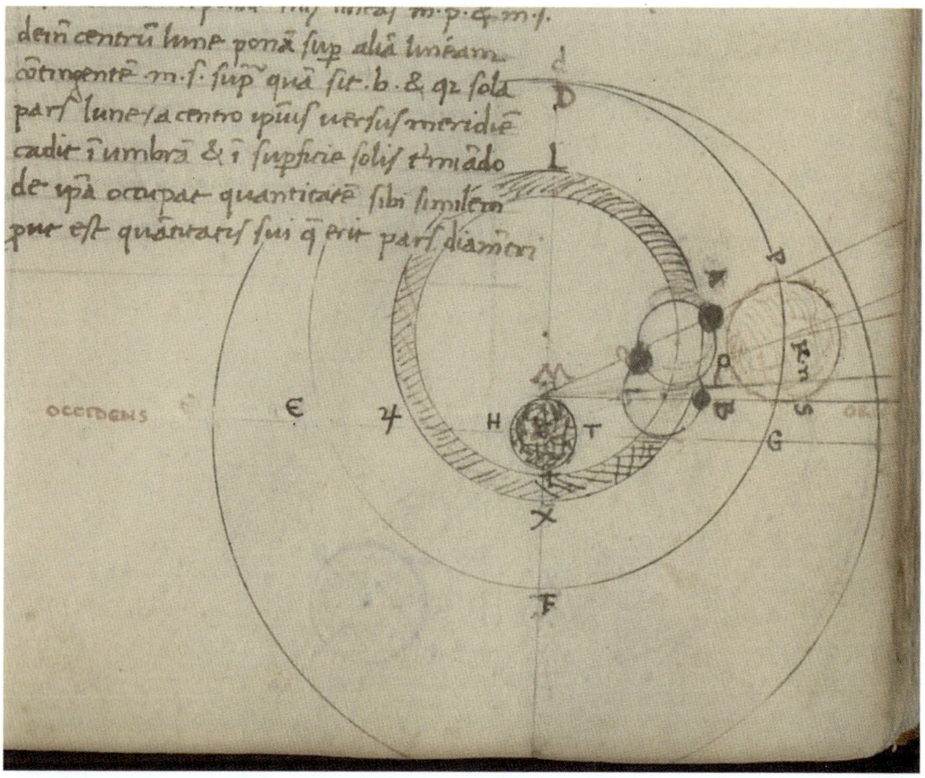

Figure 3. A diagram drawn by an unknown annotator around the text in *Tabulae eclypsium*, Cracow, BJ MS 556, folio 26^{r}. Source jbc.bj.uj.edu.pl//dlibra/.

The structure of the quires, especially in the early part of the manuscript containing the canons to the *Tabulae primi mobilis B* and the *Tabulae eclypsium*, provides extra insight into the assembly of the manuscript. Folio 5^{r} (the beginning of quire I) is darker than 4^{v}, suggesting that it was exposed to the elements before eventually being bound. Folio 36^{v} (the end of Quire III) is also dark, so the first three quires containing the canons may have existed as a booklet for a while, separate from the rest of the treatise. This would have been useful; in this form, the quires containing the canons could have been placed side by side with the tables for study. Quires I (ff. 5^{r}–14^{v}) and II (ff. 15^{r}–24^{v}) were intended to contain the *Tabulae primi mobilis B* canons. Quire III (ff. 25^{r}–36^{v}) was intended for the eclipse canons, but Gregorius stopped at 34^{v}. The rest of Quire III contains about four pages of extra material related to the end of the canons of the *Tabulae primi mobilis B* (the chapter number in this section indicates that it might have been an additional chapter), while the rest of Quire II contains the last chapter of the eclipse canons and some other material. It seems that Gregorius filled up the blank pages at the end of Quires II and III as new material became available, and he wrote where he could find space. The tables themselves begin at the beginning of Quire IV (f. 37^{r}), with scribe Franciscus Quatuor e Castris taking over from Gregorius.

Our overall impression is of a living, working, and growing document. The separation of the canons from the tables would have been extremely helpful to the user, especially in the case of Bianchini's unique tables, which would have required some effort to learn. The additions at the end of the second and third quires, as well as the more formal additions at the beginning of the first quire, suggest a possible connection to Bianchini himself as he generated more material — or at least that the author of the new material was extremely well versed in Bianchini's approach, since the added material fits in seamlessly as part of Bianchini's oeuvre. The third hand, by the author of the scribal notes and the diagrams, reveals that the manuscript continued to serve as a source of scholarly effort, at least to the end of the fifteenth century. Finally, the pairing of the *Tabulae primi mobilis B* and the *Tabulae eclypsium* might be related to the fact that both canons make use of the *Tabulae magistrales*, a set of auxiliary tables also included in the document.

The other six joint Bianchini manuscripts

Having described the two manuscripts that stand at opposite ends of the spectrum, we now outline the other six in order from presentation style to toolbox style. Space precludes detailed accounts of all six; instead, we highlight the most salient features.

Paris, BnF lat. 7270: this carefully copied fifteenth-century manuscript (although the tables have been rendered with less care, as was common) is reported by Quentin-Bauchart to have been part of the collection of Henri II in the mid-sixteenth century.[19] The manuscript begins with an ornate first letter opening the *Tabulae Astronomiae* (ff. 1^{r}–142^{r}). The canons to the *Tabulae primi mobilis B* (ff. 147^{r}–167^{r}) and the *Tabulae eclypsium* (ff. 167^{r}–181^{r}) follow, with the tables related to these latter two treatises concluding the codex (ff. 183^{r}–236^{r}). Curiously, the *Tabulae primi mobilis B* canons are preceded by eight pages of a false start to the same treatise (ff. 143^{r}–146^{r}). This attempt breaks off at the end of a quire, complete with a catchword that would have continued the treatise on the next page; but instead, the *Tabulae primi mobilis B* starts over. Perhaps someone was dissatisfied with the first attempt, and a new quire was obtained to start over. The completed version has larger text and margins: and unlike the fragment, the initial letter has been completed. Hardly any marginal notes appear.

Vatican, BAV Vat. lat. 2228: the explicit (f. 120^{r}) gives the scribe's name as Johannes Carpensis and states that the manuscript was completed on 4 December, 1470 in Ferrara. An embellished title page in red, green, blue, and yellow ink announces the presence of the canons to the eclipse tables (ff. 1^{r}–16^{r}; see Figure 4), which begin, lavishly illuminated, on the next page. The *Flores Almagesti* follows next (ff. 16^{r}–51^{v}, ff. 78^{r}–120^{r}), with the *Tabulae primi mobilis B* canons embedded within it at an appropriate place (ff. 52^{r}–78^{r}), just as the *Flores Almagesti* has completed its spherical astronomy.[20] Although the canons describing

---

19 Henri Quentin-Bauchart, *La Bibliothèque de Fontainebleau et les livres des derniers valois à la Bibliothèque nationale (1515–89)* (Paris: Ém. Paul, Huard et Guillemin, 1891) p. 93. See also Lynn Thorndike, 'Giovanni Bianchini in Paris Manuscripts', *Scripta Mathematica*, 16 (1950), 5–12, 169–80 (pp. 171–80).

20 In fact, an enthusiastic scribe seems to have commenced drawing a diagram for the following chapter of the *Flores Almagesti* on f. 52^{r} before realizing that the *Tabulae primi mobilis B* was to come next.

Figure 4. The opening of BAV, Vat. lat. 2228, folio IIIv. Reproduced with permission of the Biblioteca Apostolica Vaticana.

the tables' uses are complete, no tables are to be found.[21] Although many of the diagrams are in the margins, a couple are embedded in the text, which wraps around them. The quire breaks do not match with the beginnings of treatises or chapters, so the document was likely assembled all at once. The text occasionally varies from other manuscripts, sometimes containing passages of up to half a page that are not found in the other manuscripts.

21 It was not unusual to find canons circulating separately from the tables, although perhaps rarer for Bianchini since his tables were unique. It is possible that this manuscript was paired with another one containing the tables, and which is now lost.

While the opulent title page might imply a purpose of presentation, several signs indicate that the manuscript was also taken seriously as a scholarly work. The insertion of the *Tabulae primi mobilis B* in the right place, the well-executed diagrams, and a series of marginal notes (especially the calculations in the section of the *Flores Almagesti* that describes Bianchini's arithmetic and algebra) reveal that their composer made a significant effort to come to terms with Bianchini's system of astronomical calculation, which uses decimal numbers (including fractions) that would have been utterly novel at the time.[22]

Bologna, MS B 1601: this fifteenth-century manuscript is a curious mixture of presentation and toolbox styles. The first 30 folios contain the canons of the *Tabulae primi mobilis B* and the *Tabulae eclypsium*. The two sets of canons are carefully written in two different hands, the first being closer to semi-gothic, and the second more humanist. The initial letters of paragraphs alternate in colour between red and blue. These are signs of presentation. However, the canons end on the penultimate page of a quire, and the last page is blank. This raises the possibility that the canons and tables (written in yet another hand) were initially separate, which would have been convenient for practical use. There are no diagrams and not many marginal notes, but those that do exist (e.g. f. 27ʳ, which also contains a large stain) are in a different colour ink and indicate some interaction with the content of the text.[23]

Paris, BnF lat. 10265 and lat. 10267: according to the toolbox practice of composing canons and tables in separate documents that we have seen, it is possible that these two manuscripts were intended to be together. The document BnF lat. 10267 consists entirely of canons. These include in particular the *Tabulae astronomiae*, the *Tabulae primi mobilis B*, and the *Tabulae eclypsium*.[24] BnF lat. 10265 contains the tables related to these three treatises. The two manuscripts have a long and mostly shared history. BnF 10267 has an explicit by Arnaud of Bruxelles dated April 8, 1468 in Naples; BnF lat. 10265 has no recorded date. The paper for both documents derives from the south of Italy.[25] These volumes, along with a number of others with similar shelfmarks, are from a collection copied by Arnaud de Bruxelles. In the first half of the eighteenth century, they were part of the collection of Bernard Collot, principal of the Collège de Fortet, where our pair of manuscripts bore the shelfmarks '48' and '47', respectively. Collot donated the collection to what would eventually become the Bibliothèque nationale de France in 1751.[26]

The two manuscripts also reveal significant differences. BnF lat. 10267 opens with a flourish (f. 1ʳ; see Fig. 5) and maintains a precise writing style throughout the document with minimal abbreviations. As we have seen before, the initial letters alternate in colour between red and blue. Space was set aside in a couple of places for diagrams (ff. 84ʳ, 85ʳ),

22 See Lynn Thorndike, 'Giovanni Bianchini in Italian Manuscripts', *Scripta Mathematica*, 19 (1953), 5–17 (pp. 5–6); and David Juste, MS Vatican, Biblioteca Apostolica Vaticana, Vat. lat. 2228 (update: 10 August 2017). *Ptolemaeus Arabus et Latinus. Manuscripts*, URL = http://ptolemaeus.badw.de/ms/476.

23 See Thorndike, 'Giovanni Bianchini in Italian Manuscripts', pp. 13–17; and Fausto Mancini, *Inventari dei manoscritti delle biblioteche d'Italia*, vol. 79: *Bologna* (Florence: Olschki, 1954), p. 58.

24 Another treatise with astrological content, *Canones super tabulas directionum*, opens the manuscript. Its separate foliation (ff. I–VIII) suggests that it might have been added after the original composition, but the watermark of the paper of these folios is similar to that of ff. 45–108. See Emmanuel Poulle, *La Bibliothèque scientifique d'un imprimeur humaniste au xve siècle* (Geneva: Librairie Droz, 1963), p. 74.

25 See Poulle, *La Bibliothèque scientifique*, p. 10.

26 Poulle, *La Bibliothèque scientifique* is a survey of the Arnaud de Bruxelles collection. On the Collot shelfmarks, see p. 29; on BnF lat. 10265, see pp. 59–60; on BnF lat. 10267, see pp. 73–74.

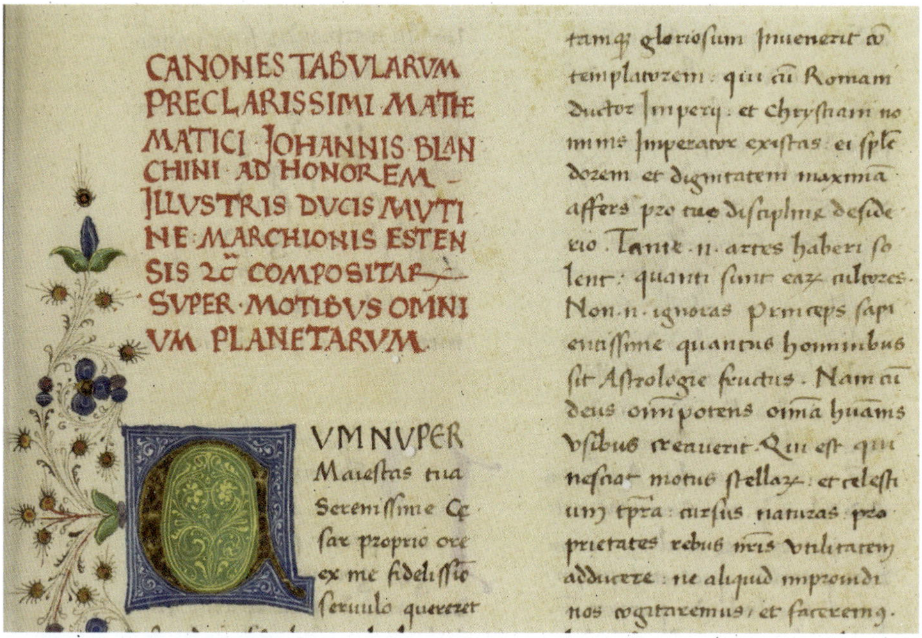

Figure 5. Paris, BnF lat. 10267, f. 1r (Source gallica.bnf.fr).

but they were never completed. Several simple diagrams appear in the lower margins (ff. 95^{v}, 96^{r}). The signs here point to a meticulously constructed, clean document.

The manuscript BnF lat. 10265, on the other hand, gives the appearance of a depository. It is around three times the size of BnF lat. 10267 (about 300 folios versus 100), and it contains blank folios in various places, including ten at the beginning. Unlike BnF lat. 10267, the eclipse tables come before the *Tabulae primi mobilis B*. From folio 242 onwards, a number of tables are incomplete or empty. Tables by other authors are featured, including those of Campanus and Regiomontanus, predominantly near the end. There are a couple of extensive marginal notes (ff. 147^{r}, 235^{r}). One is a lengthy excursion, with a diagram regarding sines from John of Murs on a page displaying Bianchini's decimal sine table, immediately preceding a sexagesimal table by another author (see Fig. 6). The compiler of this manuscript showed a deep interest in Bianchini, wanting to understand how he worked and comparing his approaches with those of others.

Paris, BnF lat. 7271: located next to Paris, Bn Flat. 7270 on the shelves of the Bibliothèque nationale de France and containing the same major treatises, this manuscript could not be more different. The explicit to the planetary canons (f. 25^{v}) dates it to 1458 in Ferrara; in the late fifteenth century, it was part of the collection of Lanzalao da Pisinis, physician to the Duke of Calabria.[27] In addition to the *Tabulae astronomiae*, it also contains the *Tabulae primi*

27 In this manuscript, see Tammaro de Marinis, *La Biblioteca Napoletana dei re d'Aragona*, 4 vols (Milan: Hoepli, 1947), II, p. 29; Thorndike, Giovanni Bianchini in Paris Manuscripts, 5–12, 169–80 (pp. 6–7, 9, 171–72); and Marie-Pierre Laffitte, 'La bibliothèque "professionnelle" d'un médecin napolitan du xve siècle: Les manuscrits de Lanzalao de Pisinis conservés à la Bibliothèque nationale de France', in *La Rigueur et la passion: Mélanges en l'honneur de Pascale*

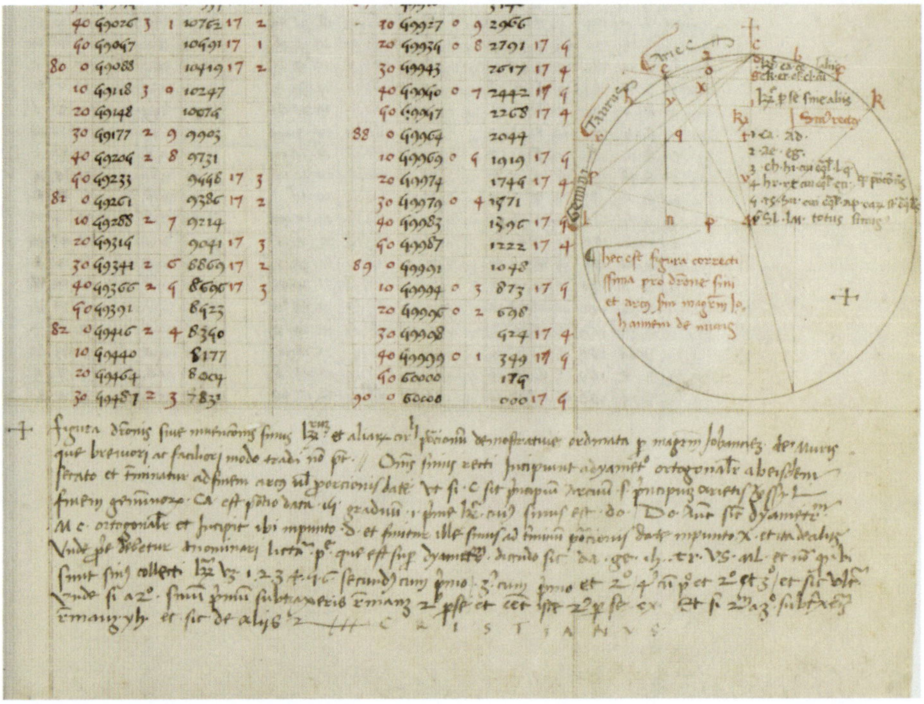

Figure 6. An extensive marginal note beside Bianchini's sine table, Paris, BnF lat. 10265, folio 235r. Source gallica.bnf.fr.

mobilis B and the *Tabulae eclypsium*. The manuscript actually begins with an incomplete attempt at a treatise on the construction of an altitude measurement instrument.[28] Only the first two pages are filled in, followed by twelve blank pages (one ruled for a table). This pattern continues throughout; between two and sixteen blank pages can be found between most sections.

Other signs of the working nature of this manuscript are plentiful. For instance, we see marginal notes in different colours throughout. A couple of pages are crossed out with a diagonal line (ff. 172^{v}, 177^{v}). Incomplete or empty table grids are frequent; for instance, a large sexagesimal multiplication table with increments of 0;10 for one of the arguments is started and apparently abandoned between the *Tabulae astronomiae* and the canons to the *Tabulae primi mobilis B* (ff. 143^{r}–146^{v}). In the canons to the *Tabulae eclypsium*, notes have been made for diagrams to be drawn, but the diagrams themselves are not forthcoming. A collection of rough notes, many of them crossed out, concludes the volume (f. 250^{r}). Another set of folio numbers suggests that the canons to the *Tabulae astronomiae* and the pages for the incomplete treatise on Bianchini's surveying instrument may have been added to the beginning of the manuscript after its original assembly.

Bourgain ed. by Cédric Giraud and Dominique Poirel (Turnhout, Belgium: Brepols, 2016), pp. 765–80.

28 As noted earlier, Garuti, 'Giovanni Bianchini *Compositio instrumenti* (Cod. lat. 145 = α.T.6.19) della Biblioteca Estense di Modena' contains an edition of a complete treatise on this instrument.

3. Taking an 'all in' approach to Bianchini: A hypothesis

Several conclusions may be drawn from our examination of the Bianchini manuscripts. Firstly, the range of dates is extraordinarily compressed: the earliest is from 1458, and all of them are dated before the end of the fifteenth century. The last specified date of composition is 1470. In fact, it is possible that Bianchini was alive during the production of all the dated manuscripts, two of which were actually assembled in Ferrara. One wonders if he kept an eye on at least some of the scribes as they worked. The compactness of the dates might be explained by considering Bianchini's successor, Regiomontanus, who became more widely known and soon overshadowed him. One of Bianchini's most important accomplishments, the conversion of stellar coordinates in the *Tabulae primi mobilis B*, was copied in Regiomontanus's *Tabulae directionum*; the latter was very popular and had a life in print of more than a century. There would have been no need for astronomers to consult a manuscript when a printed copy was readily available.

The range of places is similarly compressed. All the manuscripts with recorded locations were composed within Italy. During the sixteenth century, we find them being distributed more widely around Europe. This suggests that the only astronomers who devoted themselves to mastering Bianchini's entire astronomical system up to several decades after his death were Italian. However, his influence would later spread elsewhere.

The coherence of this group of eight manuscripts is thus in part a result of the reaction to Bianchini that occurred within Italy during a period ending several decades after his death. This activity consists of two major features. Firstly, as a man, Bianchini was honoured and celebrated for his achievements through the gathering and formal presentation of his work, a type of recognition for which there was precedent.[29] Secondly, as a scientist, Bianchini was studied by astronomers, in some cases quite deeply. On the other hand, students may have been exposed to individual works, but were not assigned the task of learning his entire system.

Why did Bianchini, in particular, receive such exceptional treatment from his readers? A simple explanation might be that these works are so tightly bound together in time and place that they just did not have the time to drift apart as other authors' works had done. However, more critical is the nature of Bianchini's project itself; each of his major works is an aspect of a single integrated intellectual approach to astronomy and astrology. The *Flores Almagesti*, not so much a commentary on the *Almagest* as an incomplete attempt to write another one, formed the mathematical underpinnings of almost all of his astronomy. The canons of the *Tabulae primi mobilis* and the *Tabulae eclypsium* refer to it frequently for mathematical support.[30] Bianchini's various innovations, often announced by the

29 See, for instance, the John of Gmunden manuscripts Österreichische Nationalbibliothek Cod. 5268, British Library Add mss 24070, and Add Ms 24071, referenced in Beatriz Porres de Mateo, 'Les Tables astronomiques de Jean de Gmunden: Édition et étude comparative', unpublished PhD thesis, Paris, École Pratique des Hautes Études, 2003, pp. 118–23, 87–94 (personal communication, Matthieu Husson). On 'originalia' in other disciplines, see, for instance, Donatella Nebbiai, 'L'*originale* et les *originalia* dans les bibliothèques médiévales', in *Auctor et auctoritas: Invention et conformisme dans l'écriture médiévale*, ed. by Michel Zimmerman (Paris: École Nationale des Chartes, 2001), pp. 487–503.

30 The *Tabulae astronomiae* do not, because they were likely written before the *Flores Almagesti*. In any case, Bianchini does not seem to have survived long enough to cover the planets in the latter work.

appearance of his own name in a title or in an explanation, usually require the reader to move between his works to understand them. Finally, his use of a novel decimal system of arithmetic in all his works except the *Tabulae astronomiae* was a unifying factor that brought readers to a crossroads: either invest some energy to understand and work with the new arithmetic, or turn away and use traditional calculation. Those who made the initial investment would find working with Bianchini's other texts much easier. Some chose not to, and moved on to other authors.

In the end, the Bianchini manuscripts with multiple treatises represent two groups of people: those who wished to commemorate the man himself, and the astronomers and astrologers who made the commitment to go 'all in' on Bianchini's unified approach. In both cases, they gave their respect to an unusually creative and restless mind that has, in recent centuries, all too often been forgotten.

Manuscript sources

Bologna, Biblioteca comunale dell'Archginnasio, B 1601
Bologna, Biblioteca Universitaria, 198 (293)
Cracow, Biblioteka Jagiellońska, 556
Cracow, Biblioteka Jagiellońska, 558
Cracow, Biblioteka Jagiellońska, 601
Florence, Biblioteca Medicea Laurenziana, Plut. 29.33
London, British Library, Add Ms 24070
London, British Library, Add Ms 24071
Paris, Bibliothèque nationale de France, lat. 7270
Paris, Bibliothèque nationale de France, lat. 7271
Paris, Bibliothèque nationale de France, lat. 10253
Paris, Bibliothèque nationale de France, lat. 10265
Paris, Bibliothèque nationale de France, lat. 10267
Perugia, Biblioteca Comunale Augusta, 1004 (M. 27)
Vatican City, Biblioteca Apostolica Vaticana, Reg. lat. 1904
Vatican City, Biblioteca Apostolica Vaticana, Vat. lat. 2228
Vienna, Österreichische Nationalbibliothek, Cod. 5268

Bibliography

Bandi, Angelo Maria, *Catalogus codicum latinorum Bibliothecae Laurentianae* (Florence: Praesidibus Adnventibus, 1775).

Chabás, José, 'An analysis of the *Tabulae magistrales* by Giovanni Bianchini', *Archive for History of Exact Sciences*, 70 (2016), 543–52.

———, *Computational Astronomy in the Middle Ages: Sets of Astronomical Tables in Latin* (Madrid: Consejo Superior de Investigaciones Cientificas, 2019).

———, and Bernard R. Goldstein, *The Astronomical Tables of Giovanni Bianchini* (Leiden/Boston: Brill, 2009).

Curtze, Maximilian, 'Der Briefwechsel Regiomontans mit Giovanni Bianchini, Jacob von Speier und Christian Roder', *Abhandlungen zur Geschichte der mathematischen Wissenschaften*, 12 (1902), 185–336.

de Marinis, Tammaro, *La Biblioteca napoletana dei re d'Aragona*, 4 vols (Milan: Hoepli, 1947).

Garuti, Paolo, 'Giovanni Bianchini *Compositio instrumenti* (Cod. lat. 145 = α.T.6.19) della Biblioteca Estense di Modena', *Istituto Lombardo (Rend. Lett.)*, 125 (1991), 95–127.

Gerl, Armin, *Trigonometrisch-astronomisches Rechnen kurz vor Copernicus: Der Briefwechsel Regiomontanus-Bianchini* (Stuttgart: Steiner, 1988).

Juste, David, 'MS Vatican, Biblioteca Apostolica Vaticana, Vat. lat. 2228 (updated: 10.08.2017)', *Ptolemaeus Arabus et Latinus. Manuscripts*, URL = http://ptolemaeus.badw.de/ms/476.

———, 'MS Vatican, Biblioteca Apostolica Vaticana, Reg. lat. 1904 (updated: 12.01.2018)', *Ptolemaeus Arabus et Latinus. Manuscripts*, URL = http://ptolemaeus.badw.de/ms/471.

———, 'Giovanni Bianchini, *Flores Almagesti* (updated: 19.06.2019)', *Ptolemaeus Arabus et Latinus. Works*, URL = http://ptolemaeus.badw.de/work/79.

Kowalczyck, Maria *et al.*, *Catalogus codicum manuscriptorum medii aevi latinorum qui in Bibliotheca Jagellonica Cracoviae asservantur* (Wrocław: Institutum Ossolinianum Officina Editoria Academiae Scientiarum Polonae, 1984).

Laffitte, Marie-Pierre, 'La bibliothèque «professionnelle» d'un médecin napolitan du XVe siècle: Les manuscrits de Lanzalao de Pisinis conservés à la Bibliothèque nationale de France', in *La Rigueur et la passion: Mélanges en l'honneur de Pascale Bourgain* ed. by Cédric Giraud and Dominique Poirel (Turnhout: Brepols, 2016), pp. 765–80.

Mancini, Fausto, *Inventari dei manoscritti delle biblioteche d'Italia*, vol. 79: *Bologna* (Florence: Olschki, 1954).

Nebbiai, Donatella, 'L'originale et les originalia dans les bibliothèques médiévales', in *Auctor et auctoritas: Invention et conformisme dans l'écriture médiévale*, ed. by Michel Zimmerman (Paris: École Nationale des Chartes, 2001), pp. 487–503.

Nothaft, C. Philipp, 'Jean des Murs's *Canones Tabularum Alfonsii* of 1339', *Erudition and the Republic of Letters*, 4 (2019), 98-122.

Poulle, Emmanuel, *La Bibliothèque scientifique d'un imprimeur humaniste au XVe siècle* (Geneva: Librairie Droz, 1963).

Porres de Mateo, Beatriz, 'Les Tables astronomiques de Jean de Gmunden: Édition et étude comparative', unpublished PhD thesis, Paris, École Pratique des Hautes Études, 2003.

Quentin-Bauchart, Henri, *La Bibliothèque de Fontainebleau et les livres des derniers Valois à la Bibliothèque nationale (1515–1589)* (Paris: Ém. Paul, Huard et Guillemin, 1891).

Rosińska, Grażyna, *Scientific Writings and Astronomical Tables in Cracow: A Census of Manuscript Sources (XIVth-XVIth Centuries)* (Wrocław: Polish Academy of Sciences Press, 1984).

———, 'Decimal Positional Fractions: Their Use for Surveying Purposes, Ferrara 1442', *Kwartalnik historii nauki i techniki*, 40 (4) (1995), 17–32.

Saby, Marie-Madeleine, 'Les Canons de Jean de Lignères sur les tables astronomiques de 1321', (unpublished thesis, Paris, École Nationale des Chartes, 1987).

Thorndike, Lynn, 'Giovanni Bianchini in Paris Manuscripts', *Scripta Mathematica*, 16 (1950), 5–12, 169–80.

———, 'Giovanni Bianchini in Italian Manuscripts', *Scripta Mathematica*, 19 (1953), 5–17.

Van Brummelen, Glen, 'The End of an Error: Bianchini, Regiomontanus, and the Tabulation of Stellar Coordinates', *Archive for History of Exact Sciences*, 72 (2018), 547–63.

———, 'Before the End of an Error: Giovanni Bianchini's Original Flawed Treatise on the Conversion of Stellar Coordinates', *Archive for History of Exact Sciences*, 75 (2021), 109-124.

———, Glen, 'Distinctions of Magnitude: Numbers and Quantities in the 15th Century and the Invention of Decimal Positional Arithmetic', forthcoming.

Appendix:

Bianchini's treatises in the eight manuscripts

	Flores Almagesti	*Tabulae astronomiae*	*Tabulae primi mobilis A*	*Tabulae primi mobilis B*	*Tabulae eclypsium*
Bologna B 1601				■	■
Cracow, BJ MS 556				■	■
Florence, Plut. 29.33		■	■		
Paris, BnF lat. 7270		■		■	■
Paris, BnF lat. 7271		■		■	■
Paris, BnF lat. 10265		■		■	■
Paris, BnF lat. 10267		■		■	■
Vatican, Vat. lat. 2228	■			■	■

Note: Paris, BnF lat. 10265 consists entirely of tables, BnF lat. 10267 and Vatican, Vat. lat. 2228 entirely of text.

Manuscripts of the *Flores Almagesti*:[31]

- Bologna, BU 198 (293), ff. 3^{r}–125^{v}.
- Cracow, BJ MS 558, ff. 1^{r}–116^{r}.
- Cracow, BJ MS 601, ff. 62^{v}–68^{v} (Chapter 2, and a small fragment of the canons to *Tabulae primi mobilis A*).
- Paris, BnF lat. 10253, ff. 6^{r}–138^{v}.
- Perugia Biblioteca Comunale Augusta, 1004 (M. 27), ff. 1^{r}–77^{r}.
- Vatican, BAV, Reg. lat. 1904, ff. 1^{r}–56^{r} (first five chapters).
- Vatican, BAV, Vat. lat. 2228, ff. 16^{r}–51^{v}, 78^{r}–120^{r}.

31 See also: David Juste, 'Giovanni Bianchini, *Flores Almagesti* (updated: 19.06.2019)', *Ptolemaeus Arabus et Latinus. Works*, URL http://ptolemaeus.badw.de/work/79.

Postface

engineers to collaboratively define or re-define those categories. We can refer to this as the 'internal' effect caused by modelling. This process, as much as the constitution of the data set, and ultimately the constitution of common tools and study, led to a highly collaborative effort and workflow and produced opportunities for an increased openness of the field. This contributes to the shaping and extending of research communities and it embodies erudition as the property of a research collective rather than simply an individual researcher. We call these collective effects. Several instances of both types of effects, internal and collective, are described below.

In an initial modelling step, it was necessary for us to define description types matching our research needs. Three different digital projects have resulted from this first analysis, corresponding to three different levels of granularity; they include i) the perimeter and the depth of our corpus description or the survey, ii) the digitial information system for the history of astral sciences (DISHAS), and iii) the manuscript description programme. The 'survey' digital project aims to reference manuscripts and early printed editions witnessing Alfonsine works. It currently records more than 900 manuscripts containing copies of around 400 different works from nearly 100 different authors. This referencing effort is already a very significant research result in itself, but we have defined two further levels of description. In order to study the diffusion of astronomical practices, we launched DISHAS in collaboration with partner projects that share our needs. This second digital project goes deeper into the textual layers of our corpus. It makes it possible to edit and analyse, in detail, the content of astronomical tables, and to describe the tabular contents of a manuscript in a more sophisticated manner. Finally, it is also important for us to analyse witness materiality in order to pinpoint the actors, milieus, and practices attached to individual manuscripts. To this end, we designed a detailed descriptive tool enabling us to analyse the complex relations between the material and intellectual dimension of the manuscripts in our corpus. For each of these three digital projects, the method of producing data from documents is different. We outline each one individually.

Survey of the corpus of Alfonsine texts

1.1 Delimiting objects and corpora

The idea of designating a general corpus of Alfonsine texts is both a primordial step of the ALFA project and an opportunity to provide a new and important resource for the community. This survey aims to collect both the Document (also referred to as Primary Source in the following paragraphs) and the Works of the Alfonsine corpus, following the classic categorisation norms of Anglo-American textual criticism.[8] In this article, we do

Willard McCarty, 'Modeling: A Study in Words and Meanings', in *Companion to Digital Humanities*, ed. by Susan Schreibman, Ray Siemens, and John Unsworth (Oxford: Blackwell Publishing Professional, 2004), Sects. 2 and 4 <http://www.digitalhumanities.org/companion/view?docId = blackwell/9781405103213/9781405103213.xml&chunk.id = ss1-3-7> [accessed: 05.05.2020].

8 Although the aim of this article is not to discuss these concepts, a general discussion about 'Works', 'Documents', and digital editions can be found in Bárbara Bordalejo, 'Work and Document', *Ecdotica: Rivista Distudi Testuali*, 10 (2013), pp. 7–93.

not discuss the boundaries and identification method of the survey, but rather the process to turn the identified objects and conceptual results into data.

Using library catalogues to search for an 'Alfonsine astronomy document' did not prove sufficient or conclusive; finding this level of precision is rare in cataloguing practices and identifying Alfonsine production is an exigent task.[9] The identification and classification of mathematical astronomy texts and tables is further complicated by the nature of the sources that transmit the Alfonsine tradition, which are essentially multiple-text manuscripts.[10] In most cases, and at least in the Latin traditions, these manuscripts demonstrate little if any concern for the integrity of works and table sets. Therefore, not only is the corpus formed essentially of multiple texts manuscripts, but inside a given manuscript, texts and table sets are often not copied one after the other, but various parts are mixed according to different ordering conventions. Moreover, the identification and classification of mathematical astronomy texts requires expert knowledge, the ability to analyse aspects deeply embedded in the content of documents, and, in some cases, the use of statistical tools.

Currently, the survey is still a work in progress; joining competences in Alfonsine astronomy and codicology, experts have identified roughly 1000 manuscripts and early printed editions in more than one hundred European libraries.[11] They include 398 different Works of the Alfonsine tradition - primarily understood to be a unique conceptual creation of historical actors and described in detail later in this article.

This resource enriches the perception we have of Alfonsine astronomy far beyond that given by the early printed editions or the manuscripts produced in Alfonso's Castilian workshop and helps to fill a void in terms of corpus and work identification. Beyond identification, the survey is also an important working tool used to elaborate further studies of individual texts, manuscripts, or authors, as well as to quantitatively identify the most important works, milieus, and periods of the Alfonsine astronomical tradition. The tool also identifies less prevalent works of the corpus and their specificities and it analyses the circulation of different kinds of astronomical content.

1.2 From handwritten notes to a digital survey

The survey began as a list of notes on Alfonsine works and where to find them (in which manuscript and in which library). These personal research archives had been assembled over time by individual ALFA team members and were available in various digital and paper formats. Soon, however, it became clear that for project management and quality

9 Rameau, the French national subject authority thesaurus, http://rameau.bnf.fr/index_en.htm [accessed: 05.05.2020], contains a fairly rich list of authorities containing the word 'astronomie', but nothing specifically designating 'astronomie Alphonsine'. This exemplifies, for the Alfonsine corpus and for the French national Library, a more universal situation concerning the general cataloguing of mathematical astronomy texts in European patrimonial libraries.

10 The term 'multiple-text manuscripts' was recently shaped in codicological scholarship, see Alessandro Bausi, Michael Friedrich, and Marilena Maniaci, *The Emergence of Multiple-Text Manuscripts* (Berlin, Boston: De Gruyter, 2019) <https://doi.org/10.1515/9783110645989> [accessed: 05.05.2020].

11 In the introduction, we describe the scientific criteria on which we relied to establish these authority lists and different case studies in the book exemplify how we approached manuscript complexity in general. The survey was produced by consulting the actual documents in heritage libraries whenever possible and by relying on digital surrogates when originals were not accessible.

reasons, a digital tool would be needed to assist in the collection of references and to host the survey; furthermore, the datafication of the collected information would lead to new opportunities in terms of both data analysis and diffusion. Data analysis was the main priority, because having a digital format would open doors to software, methods of statistical analysis, and data-visualization tools and media. A secondary yet nonetheless significant aspect was the diffusion of a web-based dynamic list of records associated with faceted research methods as well as the survey itself as an open and reusable digital object.

The datafication of the survey refers to a process of transformation and development, of which we specifically discuss the modelling aspect in this article. For the survey, we chose not to follow 'curation-driven models' designed by a cultural heritage institution, but instead offered a 'research-driven' modelling of the object.[12] This model is thus unique and the features were specifically selected for the purpose of the ALFA survey. As a digital representation of the handwritten survey, it distinguishes two main objects — Works and Primary Sources — defined by a short list of features (described hereafter), which makes it possible to cover a large number of documents. In this regard, it is worth noting that the ALFA survey does not account for all the 'hints' that helped to qualify the Alfonsine documents; these often lie in a deeper level of textuality than we wish to explore in this context. The survey may thus occasionally erase a certain level of specificity (kept in our research archives) in favour of consistency and adequacy.

1.3 A conceptual and logical model of the survey objects

The survey database is divided into two main objects of study and is centred on the relation between them: manuscripts (and early printed editions) — referred to as Primary Sources — and Works. The first step towards selection and modelling consists of defining those objects.

In modelling the relationship between the manuscript artefact and its intellectual content, we were inspired by curatorial practices, especially by the Functional Requirements for Bibliographic Records (FRBR) recently integrated in the IFLA-LRM model for bibliographic records.[13] The first Primary Source is quite straightforward in that it refers to the material document, the artefact itself, or, following the FRBR definition, the 'Item'. Traditionally, a Primary Source is identified by a library shelfmark. On the other hand, the task of defining a Work can be a little more subtle. Our definition of a Work was partially inspired by the FRBR definition:

---

12 The distinction between the two types of models based on their audience and use was theorized during the Women Writers Project Workshop at Brown University and the Centre for Digital Editions at the University of Würzburg: 'Curation-driven modellers also make assumptions about what features of the digital objects are of interest for most users and in most use cases, while research-driven modellers typically concentrate more (though not exclusively) on the needs of their own project' (Flanders and Jannidis 2015, p. 4-5).

13 The FRBR WEMI describes a quadripartite definition of a bibliographic record: the Work, the Expression, the Manifestation, and the Item. Manuscripts are a specific kind of bibliographic resource in which the Manifestation can be summarized in one unique Item. In this article, we follow the definitions provided in the 2009 correction of the 1998 Final Report, originally published as *Functional Requirements for Bibliographic Records: Final Report*, ed. by the IFLA Study Group on the Functional Requirements for Bibliographic Records and International Federation of Library Associations and Institutions (Munich: K.G. Saur, 1998), p. 17 (online access to 2009 corrected edition: <https://www.ifla.org/files/assets/cataloguing/frbr/frbr_2008.pdf>).

> A *work* is an abstract entity; there is no single material object one can point to as the *work*. […] When we speak of Homer's Iliad as a work, our point of reference is not a particular recitation or text of the work, but the intellectual creation that lies behind all the various expressions of the work." But, as this definition states out, "Because the notion of a work is abstract, it is difficult to define precise boundaries for the entity. The concept of what constitutes a work and where the line of demarcation lies between one work and another may in fact be viewed differently from one culture to another.[14]

This uncertainty is reflected in our research process; rather than defining precisely where to draw the line between two Works, we chose to follow the historic and bibliographic tradition, meaning that we identify Works by their *incipit* and table sets by the recognition of a specific core group of tables. We refine the identification of Works via an iterative process as our understanding of them develops. In the survey, different 'adaptations', up to the modification of astronomical information for a specific place, are considered as part of the same Work. Works are clearly typed according to their textual mode; we distinguish among tabular, instrumental, and different kinds of textual works. These distinctions are important because they allow us to analyse the often very distinct manuscript traditions of table sets and their related canons. They can also help us to identify different kinds of manuscripts according to the type of content they primarily convey. A manuscript witnessing only tables is different from a manuscript featuring only texts, for instance.[15]

According to the aforementioned research ambitions, the features were selected to give insights into the intellectual content, the milieu of production, the actors involved, and the material aspects of the artefact. The following questions are raised: What work does this manuscript contain? Which author's work was most widely circulated? How can we describe the geographic and temporal circulation of the main works of Alfonsine astronomy?

We chose a limited list of characteristics of interest for the description of the Primary Source, and at first we were confronted by the lack of uniform and unique identifiers for manuscripts. Although shelfmarks may be inaccurate, subject to evolution, and hard to normalize — especially considering the accepted variant inside a given library — we kept them as identifiers for Primary Sources in the survey.[16] They carry out two functions: identifying a manuscript and giving access to information on the physical document. Their normalization is not an easy process and we rely on curatorial tools offered by the Equipex Biblissima to manage the ambiguity and the varying accepted forms. For a

---

14 *Functional Requirements*, p. 17.

15 Paris, BnF latin 7281 is an example of a manuscript containing mainly text; Lisbon, Ajuda MS 52-XII-35 is an example of a manuscript containing mainly tables.

16 The International Standard Manuscript Identifier (ISMI) is a work in progress that intends to deliver unique identifiers for manuscripts themselves; it is true that the 'usual way of referring to manuscripts: shelfmark (at least in the occidental world …), is not adapted to online references and links. It is too complicated to work on a harmonisation of shelfmarks systems in all online resources; we need direct and easy links.' Matthieu Cassin, *ISMI: International Standard Manuscript Identifier. Project of Unique and Stable Identifiers for Manuscripts* (2018), https://www.manuscript-cultures.uni-hamburg.de/files/mss_cataloguing_2018/Cassin_abstr.pdf [accessed: 01.05.2020], p. 1.

physical description, the extent and dimensions of the manuscript are provided, as they are quite revealing of the potential use of the document and give a quick mental image of the object. The manuscript, as an object, is also situated between historical bounds that give a minimum (*Terminus Post Quem*) and maximum (*Terminus Ante Quem*) time of creation.[17] As stated above, this description of a manuscript might seem quite simplistic. Considering the survey in its capacity to to designate corpora, we decided to provide, when available, a hyperlink to the curating institution's online catalogue record, in order to provide more specific and detailed information. Furthermore, in the context of IIIF 360°, a partnership with the French 'Equipex Biblissima' project offered new opportunities to give direct access to the scanned reproductions of our corpus through a Mirador IIIF viewer.[18]

The Works in the survey are also broadly defined by a very restricted number of features. When identifying Works, we cannot rely on a classic authority list as this survey is the first attempt of its scope in this field. Apart from the most famous Works (such as Ptolemy's Almagest, for instance), the identification of astronomical treatises is still a matter of expertise and one of the scientific targets of the survey. Where possible, we collect the identifier of the works from institutional references.[19]

Works are traditionally identified by a title that either comes from a medieval tradition or has been forged by the historiography (for instance, John of Lignères's *Tabule magne* or the *Alfonsine Tables of Toledo*). We also collect an incipit (or the most common version, as the *incipit* may differ from one witness to another).[20] Works are attributed to an author with manageable levels of certitude. Each work is localized in space and time - with a minimum and maximum date of conception. As stated above, Works are clearly categorized; we distinguish tabular, diagrammatic, and different textual types.

The singular asset of this modelling lies in the link between the works and the manuscripts and in the research opportunities it grants to the researchers. One Primary Source may contain many Works, and each of them may be witnessed in many Sources. This results in a cross-list of Primary Sources and Works, which enables finer quantitative queries. This cardinality assists the discovery of a Work's precise impact. For example, to what extent has it been copied and over what timescale? The Works are also precisely located in the Source: this is an indication of the completeness, or, on the contrary, the fragmentary state of a given witness. It also helps us to distinguish mixed manuscripts from single-work or single-author manuscripts. The 'from folio' and 'to folio' metadata are divided into two different values that allow a graphic visualization of the partition of the manuscript; our precision goes up to the foliation (recto or verso), but not to the column, as they are not consistently used in the manuscript template of the corpus. Here, we knowingly chose to

17 For multiple-part manuscripts or rebound manuscripts, we only considered the manuscript parts containing Alfonsine productions.

18 IIIF360 is a service providing expertise about IIIF protocols provided by a consortium by Biblissima, Campus Condorcet, and Huma-Num https://projet.biblissima.fr/fr/ressources/iiif-360 [accessed: 01.05.2020].

19 The most famous works of historic astronomy are well identified in the major curatorial institutions. See, for instance, the Authority File of the Library of Congress for the Work, 'Almagest', http://id.loc.gov/authorities/names/n. 83229412.html [accessed: 20.20.2020].

20 For the incipit, we refer the reader to Lynn Thorndike and Pearl Kibre, *A Catalogue of Incipits of Mediaeval Scientific Writings in Latin*, rev. and augmented ed. (Cambridge: The Mediaeval Academy of America, 1963).

erase some of the precision provided by the researchers and cataloguers, as this variety of description would not be manageable by the machine.

Finally, this relation between a Work and a Primary Source is attributed to the scholar in charge of consulting the original document; we also record the medium of access to the manuscript (e.g. physical, digital). These more 'administrative' metadata guarantee the quality of the information and its source.

1.4 The physical model: Selecting a database technology - a restrictive choice and major chain link for research results

To move from the conceptual model to the physical implementation of the database, we must choose the most appropriate technology according to the model, its intended use, the resources of the project, and the skills and practices of its members. Sometimes, the availability of well-designed products may interfere with the original conceptual model, as it can be worth adapting them to a specific purpose rather than building from scratch a database structure associated with visualization interfaces. Both Zotero, the widespread citation management tool, and spreadsheet software were once considered as turnkey solutions for the survey database, but they had to be discarded. Zotero is highly specialized in the description of book Manifestations - according to FRBR framework - and did not fit our model very well in terms of Work modelling.[21] Spreadsheets, despite being easy to implement, proved hard to handle in terms of project management. The sharing, versioning and merging of spreadsheets filled by different participants tended to challenge the quality and completeness of the data; the frequent repetition of values and the lack of precise input control would also be prone to error. Finally, the use of spreadsheets for diffusion and extraction of the information is limited in comparison with web-based systems.

We turned to a custom SQL*-based relational database inspired by the architecture of Excel spreadsheets, with two main tables for Work and Primary Source associated with metadata tables.[22] This choice of technology allows a direct web integration, which guarantees, to a certain extent, a better diffusion and more tools for computer-based analytics and visualization of the corpus than would an Excel database.

The choice of such technologies might appear costly, especially at the beginning of a project; construction requires multiple skill sets in terms of data architecture, server-side treatment, and web development. However, they prove the best suited to the given research needs. In a way, the choice of the database technology could be seen as constraining the modelling choice to a certain extent, as well as being less epistemological and more practical. The digital aspects of the data tend to make us forget its materiality, but the implementation choices are central in terms of project management, available processing and analysis technology, and, following such processing, the end results of the research.

---

21 Zotero is a free, open-source software for citation management developed by the Corporation for Digital Scholarship <https://www.zotero.org/> [accessed: 04.05.2020].

22 An introduction to the relational model used in the context of digital humanities can be found in Stephen Ramsay, 'Databases', in *Companion to Digital Humanities*, ed. by Susan Schreibman, Ray Siemens, and John Unsworth (Oxford: Blackwell Publishing Professional, 2004) <http://www.digitalhumanities.org/companion/> .

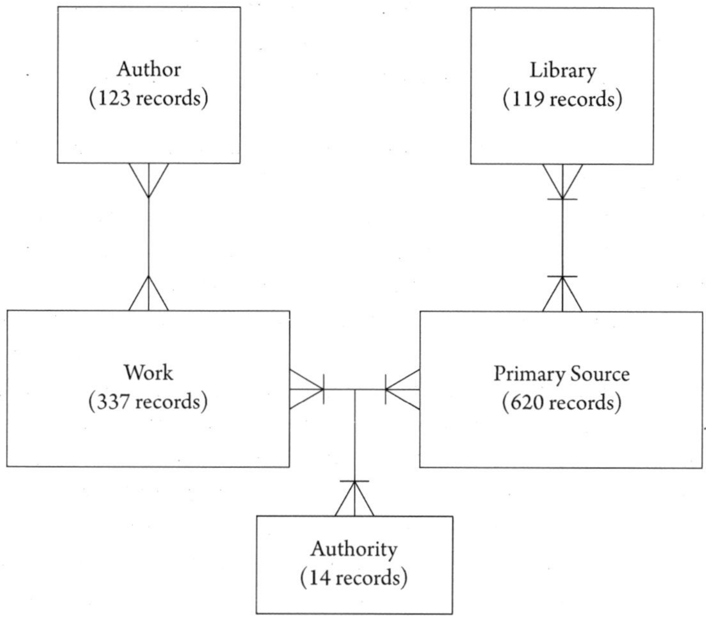

Figure 1. Simplified data model of the ALFA survey.

Finally, one may ask about the sustainability of such an implementation. A database and a website require a great deal of maintenance, both in terms of software and hardware. After developping an initial hand-made and self-hosted proof of concept, we turned to the Heurist project, an all-inclusive, free, open-source, and research-driven solution.[23] In the framework of the partnership between Heurist and Huma-Num, the French CNRS-supported infrastructure for humanities research data, the survey is now hosted by a third party public organization, which will guarantee the maintenance of the database, as well as the accessibility and availability of the data.[24]

DISHAS: Building an information system for sources of historical astronomy

2.1 What is DISHAS?

The survey allows us to identify Alfonsine works and the manuscripts witnessing them. However, for the purpose of the project, ALFA needs to go deeper in the analysis. Alfonsine astronomy is inscribed in a long and far-ranging transmission history that connects to a much broader corpus. Its table formats, some of its models, and parameters have a shared history with sources found in Arabic, Hebrew, and in some cases even Sanskrit corpora. ALFA therefore established partnerships with other projects and colleagues studying

23 http://heuristnetwork.org/ [accessed: 05.05.2020].
24 https://www.huma-num.fr/about-us [accessed: 28.04.2020].

mathematical astronomy corpora in different periods and on a Eurasian scale in order to build an open data repository where research data from our different projects can be retrieved and analysed with common mathematical and editorial tools.[25] This is the general purpose of DISHAS (Digital Information System for the History of Astral Sciences). This project, which deeply implements methods and practices from digital humanities, was built to use mathematical analysis to study astronomical witnesses from the post-Ptolemaic period from as broad a geographical area as possible.

Fundamentally, DISHAS is based on three layers. Firstly, the database of astronomical objects contains a digital representation of astronomical witnesses faceted to answer the research questions stated above (their main features are described later). Secondly, DISHAS offers tools to search for, visualize, and manipulate those data. The project started with integrated tools that help to critically edit and mathematically analyse the data, keeping in mind that the structure was built so that other tools and software could be easily interfaced with the database, thanks to a powerful search API*. The third layer features the results produced by those tools concerning the data; according to the main developed tools, they consist of native digital critical editions, the identification of table models and parameters, and the specification of table computation scenarios freely accessible on the platform. Hence, DISHAS can be understood as three different web-based structures: (1) an online database that stores research data in the field of the history of astronomy targeted to answer a specific set of questions; (2) a repository of open source tools and algorithms that assist the analytical sharing processes within the scholarly community interested in astronomical tables; and (3) an open-access platform for scientific production dealing with the aforementioned data and tools.

2.2 From historical documents to data

In the different fields of inquiry related to DISHAS, mathematical astronomy documents can contain three kinds of interrelated content: texts in prose or verse, diagrams, and numerical tables. Among these, the DISHAS project started with astronomical tables, in particular, with tables that translate a mathematical function and are used in the context of computations. There are several factors that led to this choice. Firstly, tables are a fundamental object of analysis; they often convey not only very important clues on transmission histories concerning tabular values but also parameters and models on which they rely, as well as computation practices that are attached to them. Tables are the core of the Alfonsine corpus and appear in all the other traditions that are considered by DISHAS. Furthermore, tables from all traditions are objects that share a certain quality, which makes them easier to distinguish, align, and compare than texts or diagrams.[26]

---

25 ALFA's main scientific partnerships in building DISHAS are the project PAL https://ptolemaeus.badw.de/start [accessed: 05.05.2020] and the project HAMSI <http://www.hamsi.org.nz/> [accessed: 05.05.2020]. DISHAS tools focus on tables from the eighth to the eighteenth centuries. Other tools addressing tables from earlier corpora have been developed for private use, such as Mathieu Ossendrijver's tools for cuneiform sources or Christopher Cullen's for Han sources.

26 Their intellectual unit is clearly defined by a template; they can be rather easy to abstract as they translate deeply structured and meaningful mathematical objects, associating values with a single or double argument. They contain quantities that are often clearly expressed and are close enough to our own tabulated objects to offer a quick

Finally, dealing digitally with these types of objects offers new opportunities for scientific editing and publication. The difficulty of pursuing paper-based editions is well-known in our field. DISHAS offers the framework for new attempts at creating editions based on digital readings.[27]

Our analysis of how a table should be modelled and its main features relies on a basic distinction. One may think of a table as a grid of rows and columns. One may also think of a numerical table as an abstract mathematical object instantiating the relation of two quantities in the form of a discrete function. These two conceptions lead to very different modellings. Under the first, visual aspects of tables as they appear in the documents are central, while the actual numerical content of the table is secondary. The reverse is true for the second option.

We have favoured the second, more mathematically oriented view of astronomical tables. Here, we present the principal different reasons for this choice. Firstly, DISHAS is designed to study the transmission history of astronomical tables. It consists of comparing table contents across different traditions of templating tables via written media. Models focused on the mathematical aspect of astronomical tables produce objects where these layout differences are not considered and they offer the opportunity to look beyond these differences. Another aim of DISHAS is to propose different kinds of mathematical exploration and analysis tools, including a statistical method of extracting astronomical parameters from table content, for which only the numbers are taken into account. Finally, we require DISHAS to implement tables as the computation tools that they once provided for historical actors; this also necessitates easily retrievable table content.

This definition led to a construction where the 'table as data' refers to the numerical values themselves, ordered as a JSON* object containing one or two arguments and their associated values. For calculation purposes, this digital table contains both the quantities as read in the text and their floating number translations.

A single table as data is linked to a series of contextual metadata. A detailed explanation of the modelling choices of each of the entities is beyond the scope of this article; we offer here, in a nutshell, a description of their historical, astronomical, and editorial features.[28]

Firstly, historical metadata qualify the physical witness (manuscript, author…) and the intellectual tradition (Works). Contrary to the survey model, where the central units are both the Work and the Primary Source, DISHAS goes deeper in the granularity of the manuscript. The central unit (the table) is described historically as a Table Witness, understood as a given table instance written in a particular manuscript. This Table Witness stands as a link between the Work and Primary Source, which makes DISHAS much more precise than the survey in terms of manuscript composition and the comprehensiveness analysis of the work. The Table Witness also contains time

understanding and digital analysis.

27 See, for instance, José Chabás's and Marie-Madeleine Saby's contribution to this book. The largest recent effort in this direction is Fritz S. Pedersen, *The Toledan Tables: A Review of the Manuscripts and the Textual Versions* (Copenhagen: C.A. Reitzels Forlag, 2002).

28 A graphical and interactive simplified representation of the database is accessible on the DISHAS website at: https://dishas.obspm.fr/glossary.

and place information rather than Primary Source information, making it possible to finely describe each part of a composite manuscript. This level of granularity allows the expression of historical expertise regarding the source. It is fundamental to critical editions resting on the comparison of manuscript witnesses. Finally, it anchors the study of transmission in manuscript evidence.

Secondly, astronomical and mathematical metadata, read from the table or calculated from its values, help to build an understanding that surpasses time and space criteria to classify and define scientific models and categories of analysis. The 'parameter sets' attached to the table are a common list of astronomical values that are shared, notwithstanding the historical context. This sharing helps in describing the geographical and historical dispersion of astronomical parameters based on their source tables.

Finally, we store author and editorial information. It has indeed been clear since the beginning of the project that the reading and imputing of a table is always a matter of choices by a specific reader, this process being itself a creation of the mind. This statement might be true for all datafication processes, but even more so in the case of DISHAS. Technically, we allow a single table from a given manuscript to be inputted into DISHAS more than once, according to various authoring choices. The table as data is never considered to be the strict expression of a given historical document but rather a reading by a specific author for a given purpose. Thanks to this modelling, we acknowledge the fact that the reading of a table requires a high level of expertise. Some readings are difficult and might be oriented by the knowledge of the reader; but DISHAS also allows a single reader to produce multiple versions, depending on the level of correction they choose to bring for the later use of the data. For instance, a close-to-the-text version is needed to build a critical edition of the document; however a reading where the most obvious scribal errors are corrected is more useful for statistical tests.[29] Defining these editions implied the creation of a norm for transcription and the common practices that we discuss in the next section.

2.3 The datafication process and its side effects

Encoding historical tables from the manuscript to the DISHAS database required a modelling effort that led to the clarification of certain scholarly practices, vocabulary, and processes. We present a few representative instances here.

Resting on the assumption that table reading is a skilled enterprise and can be done in different ways for different purposes, we had to distinguish three levels of 'edition' to collectively harmonize our practices so that the digital product becomes more obvious to the user. A first, close-to-the-text reading or transcription renders the numbers as they are read in the table, with no mathematical corrections, witnessing the historical document and its specificity. The second level of table reading in DISHAS allows the user to compare sources and (automatically) create critical editions. The third level of input is a mathematically corrected version of the table. In this case, the large aberration

29 This 'close-to-the-text' reading is not, however, a diplomatic transcription, in particular because many layout features of the witness (for instance the colour of numbers) will not be conserved in DISHAS, except in the form of 'free text' comments on the table.

outliers attributed to scribal errors are polished, and the table is recomputed according to a recomputation scenario defined by the modern scholars as a result of their analysis. This clarification, with respect to our table reading practices, led to a profound transformation of the very notion of critical edition concerning tables. DISHAS now embodies native digital critical editions as flexible tools used to explore and analyse the manuscript tradition of a table according to different purposes.

DISHAS modelling also transforms fundamental notions, such as those of the Author, Work, or Primary Source. In the context of DISHAS these are metadata of a given Table Witness, different from the hierarchical 'Work-Expression-Item-Parts' vision of FRBR. In other words, because a table witness is a fundamental unit of the modelisation, Works are now defined bottom up from the sets of table witnesses attached to them in different manuscripts rather than the other way around. This allows us to meticulously study the reception of a Work. This also provides a much more refined tool to describe the tabular content of a given manuscript and proposes a concrete answer to the very specific kind of manuscript transmission witnessed for mathematical astronomy, at least in the Latin traditions. Similarly, the notion of an Author related to a table becomes metadata of a given Table Witness in a given Primary Source. This makes it possible to problematize the relation of astronomers to their tabular production in a sophisticated manner (use of the 'same' table in different contexts several times).

Building the DISHAS database also raises the question of the nature of the data itself in opposition to algorithms. Understood as discrete functions, tables, for instance, can be considered either as a process (the model of the table is a function) or as a data set. In DISHAS, the codes of the historical models are generated by the researchers using a graphic interface and they are managed the same way as the data. The models interact with the same objects stored in the database and are qualified by the same contextual, historical, mathematical, and editorial metadata that describe the tables. In this context, the method of calculation is both a way to reconstruct the table and a method of defining it; as such, models are fully regarded as performative data in our system of information. This management of the model conforms with the DISHAS ambition to provide access to both the data and the process of analysing them, resulting in a shared repository of practices and research objects. This entanglement of data and processes led us also to clarify our practices in this respect. For instance, we have defined two kinds of astronomical parameters that make it possible to classify tables in a broader astronomical tradition. We have defined explicit parameters as particular table values, like 'maximum' or 'minimum', which are usually characteristic of specific historical traditions in astronomy. In this case, numerical information from the table content itself becomes key metadata within DISHAS. Another kind of astronomical parameter defined in DISHAS is the implicit parameter. These are usually extracted from all the tabulated values with statistical methods and relate the table to the (geometrical) model on which it relies. In this case, the numerical metadata are generated via an internal DISHAS process. Implicit parameters can, in turn, be used to recompute the table and in some cases control the values of the explicit parameters.

2.4 Tailor-made tools and sustainability

One major difficulty of a digital research project is where to put the cursor between all-inclusive, well-sustained tools and tailor-made developments. On one hand, we want the projects to last, to be sustainable, and to remain accessible in the long term. On the other hand, we want them to be innovative, to go further, and to explore new paths and new methodologies. This question applies to each layer of the digital project: the choice of metadata scheme, the format, and the vocabulary; the programming frameworks and software; the data-hosting solution, and the website. Unfortunately, sustainability, accessibility, and research adequacy often require distinct and incompatible solutions.

With DISHAS, we sought to find the right balance for each of these concerns. We have already discussed the data aspect where we opted for a custom modelling that would include, as far as possible, external references for future alignment with more general databases and export options to open file formats such as CSV. The back-end framework, namely Symfony*, was chosen as one of the most famous and implemented php solutions, which guarantees mid-term sustainability of the website. However, we essentially believe that the website and interfaces are 'momenta' in the life cycle of the data. What we want to guarantee is not necessarily long-term access to the website interface, but rather the safe management of the research data, and options for their long-lasting archiving. This is why our data management plan relies on Huma-Num, a third-party institution with robust expertise in data sustainability.

3. Manuscript descriptions and partial editions

The identification of the corpus of Alfonsine texts and the analysis of some of its most significant components are two important aims of ALFA. They are both also steps towards a better understanding of the history of Alfonsine astronomy, which is focused on the recovery of historical actors' practices. The modelling enforced in the survey offers opportunities to study the distribution of works among manuscript witnesses, but neither its level of granularity nor its set of studied features would allow the precise study of the practices of the actors in relation to the manuscript. By going deeper in granularity, DISHAS offers a finer understanding, oriented towards the comparison of table content and the mathematical practices of the author. DISHAS indeed relies on a distinction between the content of a table and its manuscript form, which is fundamental for its purpose but not very efficient when trying to recover the practices of those involved from the manuscript evidence. For instance, the way manuscripts display several tables in one and the same grid can be significant in terms of the meaning of this group of tables or their use in the computation of a given astronomical quantity. The layout of these grids, and thus the way they can be read is also very important and has consequences, for instance, on the algebraic sign of the quantities read in the tables. Layout can also be studied in terms of 'editorial practices'; their dissemination can reveal a stream of scribal practices. In other words, when considering precisely how the meaning of a table is formed, a clear-cut distinction between form and content is not achievable. More generally, the intellectual organization of astronomical manuscripts is a fundamental clue when it comes to analysing practices.

The nature of the collection of texts grouped in a manuscript, the order in which they are organized, the way they are cut in different parts, and the history of composition of the manuscript often point to different kinds of uses, users, and milieus.

On this more general level, it is similarly inefficient to cut a clear line between the intellectual content of a manuscript and its material form. To give but one significant argument, the content of the manuscript may have been defined by a seventeenth-century librarian who was rebinding a collection of old documents. In such a case, it is obviously difficult to deduce anything about fourteenth- and fifteenth-century participants via analysis of the manuscript as a collection of texts. Regarding the table, it appears that an analysis aiming to recover historical actors' practices needs to consider the manuscript in a holistic way. The relationships between different dimensions of the document (material, graphic, intellectual) need to be studied in order to disclose, to the greatest extent possible, the source's history and the ways in which it could have been meaningful for the various actors who used it within different contexts of astronomical practices. Offering a descriptive manuscript tool that makes it possible to tackle this issue is the general aim of ALFA's third digital project.

In principle, each manuscript is unique, an individual with its own history. Each codex offers specific clues to the historian about different contexts in which mathematical astronomy was practiced, about the motivations and values that a range of historical actors associated with these practices, and about the connection between these different contexts. Each can teach us something about how mathematical astronomy was a part of medieval intellectual cultures and societies. The very nature of the in-depth and intimate analysis necessary to retrieve these clues from the document makes it unrealistic to cover a corpus of nearly 600 manuscripts at this level of detail. Thus, while developing this detailed descriptive tool, we also decided that it was to be applied only to a small sample of the manuscripts of the Alfonsine corpus. Manuscripts in this sample are selected either because they represent a very common type in the corpus (presentation manuscript, university manual, student notebook, astrological toolbox ...), or because they are truly exceptional and thus deserve a detailed description. The sample is then to be used heuristically as a palette allowing the user to quickly understand, when looking at a new manuscript, how it is positioned and what kind of clues it can offer to the historian of Alfonsine astronomy. The sample list is not closed and new manuscripts can always be described.

Modelling manuscripts is a compelling task, as they are rather complex objects; they often result from successive acquisition and their composition is often the result of a long and eventful history. Moreover, as they are unique objects, the depth of the description that should be pursued and the specific features of the documents to be identified will differ in each case. This raises a specific question: how can we achieve a data model to represent objects that are valued precisely for their uniqueness?[30] ALFA can rely on a long and lively tradition of the scholarly analysis of manuscripts in order to tailor a descriptive method suited to its scientific needs. The recent research on manuscript description proposes an approach in which different layers of analysis are considered.[31] They can be broadly

30 Flanders and Jannidis, p. 26.

31 Patrick Andrist, Paul Canart, and Marilena Maniaci, *La syntaxe du codex : essai de codicologie structurale* (Turnhout: Brepols, 2013).

grouped in material, graphical and intellectual layers. Material aspects of the manuscript are fundamental: what are the geometrical dimensions of the object? What is the physical nature of the writing support (for us, paper or parchment in most cases)? What other kinds of material composes the object (for us, often wood, leather, different kind of metal pieces, threads mainly in the binding)? What inks and pigments are used, and how are all of these elements arranged together to constitute one manuscript (for us, this often means the study of quire composition and binding, but other manuscript cultures might have very different configurations than the codex for their manuscripts)? Graphical aspects of the document are also very important: how are the pages laid out? What kind of ruling, if any, was prepared? What decorations and illustrations are used? What types of script are present? How many hands or official institutional owners left traces in the documents? Finally, the intellectual aspects of the documents, as we have described briefly above, should not be overlooked. Although all these elements of description can be useful, it is not, however, by accumulating information about the document that an understanding of its history will emerge. Rather, recent research on manuscripts reveals the relations between these different elements of description and the identification of locus in the manuscript where several layers concomitantly show discontinuity.

This approach to manuscripts allows us to produce descriptions disclosing the different steps in the progressive formation of the codex. For each of these steps, these analyses make it possible to understand the nature of the modifications that transformed the state of the document. For the manuscripts of the Alfonsine corpus, some of these modifications imply astronomical practices (for instance, a marginal set of notes to a canon displaying a computed example or a justification to some procedures) or show how the document could be used in the context of astronomy (for instance, adding a set of radices for a new city). This descriptive methodology is also perfectly suited to the needs of the project, but it needs to be implemented.

Finally: why do we want this manuscript description to be digitized? If we consider manuscript description as a free and literary exercise, a print edition (or online version, as a static-pdf) would translate it in a clear yet unstructured way, without the need for a computer or database medium. However, our manuscript descriptions go down to partial editions of the text, and like any scholarly edition, the content is not to be read in a linear way. Digital editions offer new opportunities to the scholar editor, with specific ways for the reader-user to target, browse, and access the para-textual content.[32] Furthermore, turning the text into a database offers the ability to explore the document, extract its features, and compare it with other documents modelled the same way. With the help of dedicated programs, such a digital encoding of the selected manuscripts would easily produce a cartography of their places of production, a comparative diagram of the hand variety on each manuscript, a study on the length of paragraphs, and more.

32 This change of behaviour, from 'reader' to 'user', is often described as a result of the digitisation of scholarly editions. See, among others, Tim McLoughlin, 'Bridging the Gap', in *Jahrbuch für Computerphilologie 10*, ed. by Georg Braungart, Peter Gendolla, and Fotis Jannidis (Paderborn: Mentis-Verlag, 2010), pp. 37–54. The question of the evolution from print edition to digital scholarly edition, both in terms of use and methodology, is often debated within the framework of the digital humanities; see, among others, Elena Pierazzo, *Digital Scholarly Editing: Theories, Models and Methods* (Farnham, Surrey: Routledge, 2015).

3.1 Modelling manuscripts

Modelling manuscripts in this way implies a great flexibility both in terms of granularity (no control or restrictions on the hierarchical structure of the material and intellectual units) and in terms of the variety of the analysis features. In the framework of our project, we wish to describe the manuscript in a fine-grained way, such that some parts of the text will be edited, as witnesses of the author's practice, both in terms of astronomical content or material practices. However, in a hierarchy of content, the 'level' of the edition might vary depending on the structure of the text (chapter, paragraph, column, etc. versus plain student notes). The same goes for the material units comprising the manuscript, which can presumably be resumed as a single quire, while other composite codices can be described as a complex material architecture. These requirements deeply influence the physical structure of the data and point towards a document-oriented database associated with a rich and well-designed metadata scheme rather than a relational structure that was chosen for both the survey and DISHAS.

Text-oriented databases based on XML technology are well suited for managing these kinds of text-driven, hierarchical resources that mix both textual content and metadata lists of features, as XML documents are fundamentally text-document themselves. The hierarchical structure offers the opportunity to convey the logical architecture of any text, both in terms of external or internal architecture. The tag structure enables the encoder to stress particular aspects and add semantic information to the text, which can be easily extracted and manipulated by the computer as a datum piece.[33]

Building clever and extensive models of these complex objects from scratch would be far from manageable within the framework of ALFA, all the more so because these manuscript descriptions should be as inclusive as possible. They are thought of as collaborative works, and should accept analyses from different fields of expertise in all the pertinent auxiliary sciences.

Fortunately, neither the study of manuscripts nor the elaboration of digital description and edition are specific to the ALFA project and we have been able to benefit from multiple model standards for manuscript description and edition designed for research communities. We turned to the *Text Encoding Initiative* (TEI) XML specification, which presented multiple assets for the project.[34] Its specialization in textual documents, including medieval manuscripts, was especially relevant. A specific module was developed — the msDesc element and its children, 'contain[ing] a description of a single identifiable manuscript or other text-bearing object'.[35] This standard has been collaboratively produced and maintained by researchers and organizations since 1988. With approximately 500 elements for describing

33 'Humanities scholars represent their documents in XML for two reasons: (i) XML is a formal model designed to represent an ordered hierarchy, and (ii) to the extent that human documents are logically ordered and hierarchical, they can be formalized and represented easily as XML documents. Computers can operate quickly and efficiently on trees (ordered hierarchies), much more quickly and efficiently than they can on non-hierarchical text. This means that if we can model the documents we need to study as trees, we can manage and manipulate large amounts of data efficiently'. David J. Birnbaum, 'What Is XML and Why Should Humanists Care? An Even Gentler Introduction to XML' (2015) <http://dh.obdurodon.org/what-is-xml.xhtml> [accessed: 18.09.2019].

34 https://tei-c.org/ [accessed: 05.05.2020].

35 https://www.tei-c.org/release/doc/tei-p. 5-doc/en/html/ref-msDesc.html [accessed: 05.05.2020].

material, graphical, and textual aspects of the document, TEI is still growing and being improved and promoted by one of the major digital humanities communities. As an XML specification, the TEI project can also benefit from other XML-oriented metadata schemes and models such as MEI (for music encoding) or MathML for mathematical notations.

Choosing a standard guarantees a strong foundation for the project, but does not mean that we can avoid refining the model in order to better fit the studied document and the research targets; a specification based on TEI can restrict the field of application and/or specify a delimited understanding or use for a given element. This is one of the strengths of TEI and what makes it the primary model for text description in the humanities, a so-called institution associated with a social community of practice.[36]

> TEI is a common model for representing the observed reality (i.e. texts and documents in the humanities domain), but it also leaves the interpreter free to define his or her own model of the text(s)/document(s) by choosing the features to be in focus in the computational representation.[37]

The ALFA specification is based on a set of selected TEI elements related to the studied documents (manuscript and early printed editions from the fourteenth into the sixteenth centuries) and the scope of our research. This restriction allows a better definition of the project targets, helps the collective data generation and management, and facilitates the tagging process.

The ALFA model also incorporates a specific TEI-compliant implementation of codicological appreciation derived from *La Syntaxe du Codex*.[38] Inspired by other attempts designed by related research projects, the model loosely adapts the *Syntax* specifications, mainly preserving the idea of an informative relationship between the material and the intellectual unit. The TEI model implies a clear-cut distinction between material and intellectual content descriptions. The manuscript description is led through major categories: content (intellectual), physical description, history, and so on. These sections do not always comply with our own scope of analysis (e.g. would diagrams fit better in intellectual content, or images in physical descriptions?). Moreover: how would this strict distinction satisfy our will to trace a history of the conception, production, and use of the manuscript?

To allow a better circulation of the information and to build a consistent rhetoric based on the internal and external analysis of the document, relational structures are brought forward in the data, allowing elements from different parts of the description to be associated through identifiers. We exemplify this method in the following caption from the detailed description of the manuscript BnF lat. 7281.

In this example, we see how a text (also called a witness) contained in the manuscript (here identified as MS_BnF_Latin_7281_item2 is described as both an intellectual and material unit. On the intellectual side, we associate it with an author (auth_0002) and a

---

36 Flanders and Jannidis, p. 28.

37 Francesca Tomasi, 'Modelling in the Digital Humanities: Conceptual Data Models and Knowledge Organization in the Cultural Heritage Domain', *Historical Social Research/Historische Sozialforschung*. Supplement, 31 (2018), 170–77 (p. 172).

38 Andrist, Canart, and Maniaci.

```
<msItem xml:id="MS_BnF_Latin_7281_item2" n="002" resp="#copyist_02">
   <locus from="17r" to="28v" target="#unit_prod_2"/>
   <textLang mainLang="la"/>
   <title type="uniform" xml:lang="la">Lectiones tabularum Toletanarum secundum
      Arzachelis, Canones Ca</title>
   <title type="original" xml:lang="la"><locus from="17r"/>Incipiunt Lectiones
      tabulraum Toletanarum secundum Arrachiel scilicet Hyspanus secundum rationem
      Caldeorum.</title>
   <incipit><locus from="17r"/>In domine Domini. Scito quod annus lunaris sit ex
      354 diebus</incipit>
   <explicit><locus from="28v"/>erit dies revolutionis priusquam dies radices ex
      hoc mense duobus diebus.</explicit>
   <author cert="high" ref="#auth_0002">Azarchel</author>
</msItem>
```

Figure 2. ALFA's TEI modelling of a Paris manuscript.

'uniform' title. This representation embodies the link between this text and a more abstract Work, *Lectiones tabularum Toletanarum secundum Arzachelis, Canones Ca.*

At the level of this witness, we gathered information about the way in which the text is actually written in this manuscript, with an alternative title and a specific incipit and explicit at specific folios of the document. Finally, on a material level, this text is linked to a specific unit of production of the document (unit_prod_2). In this manuscript, the textual and production units exactly overlap and correspond to different hands, which give us clues about the way the manuscript was created: by successive addition.

These links suggest an undetermined relation between the material description, the intellectual content metadata, and the partially edited text. Each part of the text contains multiple identifying references that link them to a specific part of the manuscript, one or many authors, a writer's hand, etc. A partial edition of the text or of marginalia can then be linked to such a textual part of the manuscript. Tagging these relations in this way, we expose the data as a body of evidence, and the reader is free to interpret the correspondence.

3.2 Implementation of the model: Project management

Compared to relational databases where the structure of the object and its features are strictly determined in advance, the XML-TEI model offers more flexibility. This advantage does, however, lead to a major drawback in terms of project management; to preserve this flexibility, the data input has to be done manually in the XML database. In practical terms, this implies that the annotation encoding must be done by the researchers themselves. If more and more young scholars are trained in the digitization of their literary productions — either through LaTeX* or in an XML*-based format — these procedures still require close collaboration with encoding technicians. A proper modelling will however facilitate this process, as a data engineer would provide the tools and definition to enforce the modelling choices during the input.[39]

39 Specific XML encoding software is of great help in this process; based on a detailed scheme declaration, this software is able to autocomplete the tag's name and suggest the correct child element depending on the hierarchy of the tree. It also checks the validity of the document and registers the definition of each tag.

Conclusion

In this chapter we explain the various digital projects developed by ALFA in order to provide new methods of constituting, analysing, and exposing the Alfonsine corpus. We focus on our modelling choices and what they would provoke in terms of epistemology: how this change, from the document to the data, alters the researcher's view of their object; what kind of tools are made accessible once the document is transformed into digital data, and finally, the management of the project.

As we began to consider digital tools, it became clear that the definition of the object of research and the granularity of the description would differ according to the questions asked to the object and their quantity. Three models were built, from the most distant to the closest to the object. The survey seeking a general constitution of the Alfonsine corpus, would offer general information about the localization of the corpus and its intellectual content. The DISHAS project would look more closely at tables and, while gaining more granularity, would broaden the scope of the corpus, allowing comparison to tables from other traditions on a large historical and geographical scale. The manuscript description project considers manuscripts as global objects. Its modelling of the source offers a more detailed and entangled analysis of the link between the external aspect and the internal content of the source among a selection of manuscripts. Formalizing these three ways of describing the source of the Alfonsine corpus led to an important methodology change for the researchers on the team, with the most easily relatable effect being the standardization of a common vocabulary and the definition of objects and concepts. The need to collaborate with engineers in a multidisciplinary setting has forced the development of an explicit and objective discourse rather than an intimate and implicit experience of the source. Finally, the dialogue with the machine prompted a need for these definitions to be translated in a strict and restrictive way, limiting ambiguity as much as possible.[40] On the surface, this explicitness may appear to limit interpretative freedom, the flexibility of concepts and, ultimately, their evolution and the emergence of new categories of analysis. However, we believe that this clarification effort is not only necessary to 'feed the monster' but also to establish a base on which collaborative work can be built, opening a space where new concepts and approaches can be created.

In terms of tools and expected results, this modelling effort is all the more likely to be effective if the models are adapted to the research questions. Statistics on the distribution and volume of works for the survey, the automated critical edition of tables for DISHAS, or the dynamic visualization of production and circulation units that constitute a manuscript, are all tools and research objectives that will allow the research group to better pose and answer its questions.

In conclusion, it is clear that the issue of the datafication of research objects implies a major change in terms of collaboration. Collaborations between researchers occur no

40 This new and more objective way of looking at the document is often described as a first clear effect of the datafication of the object by humanities researchers: 'how does the process of data modelling change, reveal, or consolidate one's relation to one's research materials or methods? There are various ways in which this might happen: by making assumptions more explicit; by changing our understanding of language or what aspects of language are perceptible to us; and by changing our understanding of 'text' (from product to process)', Flanders and Jannidis, p. 28.

longer at the level of a common work but at the level of the data. The results produced with the help of these tools depend on the quality of each person's participation. This requires a continuous collaboration between researchers for the creation of data sets, an agreement on objects and objectives as well as on new working methods and co-working tools, and, finally, partnerships between researchers, engineers, and technicians. By developing such numerical modelling, we of course will open up the history of astronomy to other methods and new participants.

Appendix: Technical vocabulary

Application programming interface (API): in the context of a web application, an API offers the tools needed to interact with the raw data - usually output in JSON or XML - without using a graphical interface. It can be used by distant software to search, analyse, or display the data. Typically, the weather forecast on a smartphone is not calculated by the device; the smartphone gets the data from a weather forecast API and is in charge of displaying it on the screen. In the context of DISHAS, our search API is a resource for a distant python library used by the DISHAS Interactive Parameter Squeezer (DIPS), which is designed to manage the statistical analysis of the table content.

Comma-separeted values (CSV): this format is a standard for tabular values, both numerical and textual. It is read by the most classic spreadsheet programs, statistical analysis tools, and libraries.

JavaScript Object Notation (JSON): nowadays, it is used in particularly in web-oriented developments in concurrence with XML to describe objects as a string. Both XML and JSON allow a hierarchical representation of objects.

LaTeX (pronociation: \la.tɛk\) is both a free language and a composer process that helps in the designing of documents. As opposed to text processors such as Microsoft Word, LaTeX distinguishes the text and its semantics from the actual graphical output. It is used quite extensively for scholarly editing in the fields of mathematics and physics.

Structured Query Language (SQL): This language is used in the context of relational databases. SQL statements are used for the query, definition, control, and manipulation of the data.

Symfony: an open-source PHP framework for building websites originally developped by Sensiolabs. Symfony is based on the 'model-view-controller' paradigm. It is currently used by Drupal, one of the leading content management systems.

Extensible Markup Language (XML) is a markup language used to provide semantic information to textual data. XML does not define vocabulary or syntax but rather a structure for hierarchic data.

Bibliography

Andrist, Patrick, Paul Canart, and Marilena Maniaci, *La syntaxe du codex: essai de codicologie structurale* (Turnhout: Brepols, 2013).

Bausi, Alessandro, Michael Friedrich, and Marilena Maniaci, *The Emergence of Multiple-Text Manuscripts* (Berlin, Boston: De Gruyter, 2019) <https://doi.org/10.1515/9783110645989> [accessed: 05:05.2020].

Birnbaum, David J., 'What Is XML and Why Should Humanists Care? An Even Gentler Introduction to XML' (2015) <http://dh.obdurodon.org/what-is-xml.xhtml> [accessed: 18.09.2019].

Bordalejo, Bárbara, 'Work and Document', *Ecdotica : Rivista Distudi Testuali*, 10 (2013), 7–93.

Cassin, Matthieu, *ISMI: International Standard Manuscript Identifier. Project of Unique and Stable Identifiers for Manuscripts*, 2018, p. 1 <https://www.manuscript-cultures.uni-hamburg.de/files/mss_cataloguing_2018/Cassin_abstr.pdf> [accessed: 01.05.2020].

Flanders, Julia, and Fotis Jannidis, 'Knowledge Organization and Data Modeling in the Humanities' (2015), pp. 1-38 <http://www.wwp.northeastern.edu/outreach/conference/kodm2012/ flanders_jannidis_datamodeling.pdf>[accessed: 05.05.2020].

Functional Requirements for Bibliographic Records: Final Report, ed. by IFLA Study Group on the Functional Requirements for Bibliographic Records and International Federation of Library Associations and Institutions (Munich: K.G. Saur, 1998).

Kassell, Lauren, 'Paper Technologies, Digital Technologies: Working with Early Modern Medical Records', in *Edinburgh Companion to the Critical Medical Humanities*, ed. by Anne Whitehead and Angela Woods (Edinburgh: Edinburgh University Press, 2016), pp. 120–35 <https://www.universitypressscholarship.com/view/10.3366/edinburgh/9781474400046.001.0001/upso-9781474400046-chapter-6> [accessed: 05.05.2020].

Kräutli, Florian, and Matteo Valleriani, 'CorpusTracer: A CIDOC Database for Tracing Knowledge Networks', *Digital Scholarship in the Humanities*, 33.2 (2018), pp. 336–46 <https://doi.org/10.1093/llc/fqx047> [accessed: 01.05.2020].

McCarty, Willard, 'Modeling: A Study in Words and Meanings', in *Companion to Digital Humanities*, ed. by Susan Schreibman, Ray Siemens, and John Unsworth (Oxford: Blackwell Publishing Professional, 2004) <http://www.digitalhumanities.org/companion/view?docId = blackwell/9781405103213/9781405103213.xml&chunk.id = ss1-3-7> [accessed: 02.05.2020].

McLoughlin, Tim, 'Bridging the Gap', in *Jahrbuch für Computerphilologie 10*, ed. by Georg Braungart, Peter Gendolla, and Fotis Jannidis (Paderborn: Mentis-Verlag, 2010), pp. 37–54 .

Minsky, Marvin L., 'Matter, Minds, and Models', in *Semantic Information Processing*, ed. by Marvin L. Minsky (Cambridge: MIT Press, 1968), pp. 425–31.

Pedersen, Fritz S., *The Toledan Tables: A Review of the Manuscripts and the Textual Versions* (Copenhagen: C.A. Reitzels Forlag, 2002).

Pierazzo, Elena, *Digital Scholarly Editing: Theories, Models and Methods* (Farnham, Surrey: Routledge, 2015).

Poulle, Emmanuel, and Owen Gingerich, 'Les positions des planètes au Moyen Age : application du calcul électronique aux tables alphonsines', *Comptes rendus des séances de l'Académie des Inscriptions et Belles-Lettres*, 111 (1967), pp. 531–48 <https://doi.org/10.3406/crai.1967.12173>.

Ramsay, Stephen, 'Databases', in *Companion to Digital Humanities*, ed. by Susan Schreibman, Ray Siemens, and John Unsworth, Blackwell Companions to Literature and Culture (Oxford: Blackwell Publishing Professional, 2004) <http://www.digitalhumanities.org/companion/> [accessed: 02.05.2020].

Rentier, Bernard, *Open Science: The Challenge of Transparency* (Brussels: Académie royale de Belgique, 2019).

Schreibman, Susan, Ray Siemens, and John Unsworth, *Companion to Digital Humanities* (Oxford: Blackwell Publishing Professional, 2004) <http://www.digitalhumanities.org/companion/> [accessed: 02.05.2020].

Thorndike, Lynn, and Pearl Kibre, *A Catalogue of Incipits of Mediaeval Scientific Writings in Latin*, rev. and augmented ed. (Cambridge: The Mediaeval Academy of America, 1963).

Tomasi, Francesca, 'Modelling in the Digital Humanities: Conceptual Data Models and Knowledge Organization in the Cultural Heritage Domain', *Historical Social Research / Historische Sozialforschung, Supplement*, 31 (2018), pp. 170–79.